Lecture Notes in Physics

Volume 884

Founding Editors

W. Beiglböck
J. Ehlers
K. Hepp
H. Weidenmüller

Editorial Board

B.-G. Englert, Singapore, Singapore
P. Hänggi, Augsburg, Germany
W. Hillebrandt, Garching, Germany
M. Hjorth-Jensen, Oslo, Norway
R.A.L. Jones, Sheffield, UK
M. Lewenstein, Barcelona, Spain
H. von Löhneysen, Karlsruhe, Germany
M.S. Longair, Cambridge, UK
J.-F. Pinton, Lyon, France
J.-M. Raimond, Paris, France
A. Rubio, Donostia, San Sebastian, Spain
M. Salmhofer, Heidelberg, Germany
S. Theisen, Potsdam, Germany
D. Vollhardt, Augsburg, Germany
J.D. Wells, Geneva, Switzerland

For further volumes:
www.springer.com/series/5304

The Lecture Notes in Physics

The series Lecture Notes in Physics (LNP), founded in 1969, reports new developments in physics research and teaching—quickly and informally, but with a high quality and the explicit aim to summarize and communicate current knowledge in an accessible way. Books published in this series are conceived as bridging material between advanced graduate textbooks and the forefront of research and to serve three purposes:

- to be a compact and modern up-to-date source of reference on a well-defined topic
- to serve as an accessible introduction to the field to postgraduate students and nonspecialist researchers from related areas
- to be a source of advanced teaching material for specialized seminars, courses and schools

Both monographs and multi-author volumes will be considered for publication. Edited volumes should, however, consist of a very limited number of contributions only. Proceedings will not be considered for LNP.

Volumes published in LNP are disseminated both in print and in electronic formats, the electronic archive being available at springerlink.com. The series content is indexed, abstracted and referenced by many abstracting and information services, bibliographic networks, subscription agencies, library networks, and consortia.

Proposals should be sent to a member of the Editorial Board, or directly to the managing editor at Springer:

Christian Caron
Springer Heidelberg
Physics Editorial Department I
Tiergartenstrasse 17
69121 Heidelberg/Germany
christian.caron@springer.com

V.K.B. Kota

Embedded Random Matrix Ensembles in Quantum Physics

Springer

V.K.B. Kota
Physical Research Laboratory
Ahmedabad, India

ISSN 0075-8450 ISSN 1616-6361 (electronic)
Lecture Notes in Physics
ISBN 978-3-319-04566-5 ISBN 978-3-319-04567-2 (eBook)
DOI 10.1007/978-3-319-04567-2
Springer Cham Heidelberg New York Dordrecht London

Library of Congress Control Number: 2014934167

Printed on acid-free paper

Springer is part of Springer Science+Business Media (www.springer.com)

To my parents

Preface

Random matrix theory was introduced into physics by E.P. Wigner in 1955, and consolidated with deeper and wider ranging investigations in the last three decades, it has become an integral part of quantum physics. As aptly stated by H.A. Weidenmüller in a recent commentary: "although used with increasing frequency in many branches of physics, random matrix ensembles sometimes are too unspecific to account for important features of the physical system at hand. One refinement which retains the basic stochastic approach but allows for such features consists in the use of embedded ensembles." This new class of random matrix ensembles, the embedded random matrix ensembles, were introduced in the context of the nuclear shell model in early 1970. As stated by J.B. French: "the GOE, now almost universally regarded as a model for a corresponding chaotic system, is an ensemble of multi-body, not two-body interactions. This difference shows up both in one-point (density of states) and two-point (fluctuations and smoothed transition strengths) functions generated by the nuclear shell model. For a better a priori model we can choose an ensemble of k-body interactions ($k = 2$ is an interesting case) by generating a GOE in k-particle space and using it in the space of m-particles. For most purposes the resulting embedded GOE (or EGOE) is very difficult to deal with, but by good luck, we can use it to study the questions we have posed and the answers are different from, and much more enlightening than, those which would come from GOE."

Research over the last two decades in particular has produced a large body of new results for embedded ensembles and it is clear that these random matrix ensembles are indispensable in the study of finite many-particle quantum systems such as atoms, nuclei, quantum dots, small metallic grains, lattice spin models for quantum computers, and so on. In this book, starting with an easy-to-read introduction to general random matrix theory, all the necessary concepts for embedded random matrix ensembles are developed from scratch and the reader is then carried to the frontiers of present-day research. The first chapter gives a general introduction and the next two chapters deal with some general aspects of classical random matrix ensembles. Eight chapters in the remaining part of the book give results for a variety of embedded ensembles, mainly classified according to the Lie symmetries of the

Hamiltonian of a finite many-body quantum system, while four chapters are devoted to applications. The last chapter provides a summary and future prospects.

The starting point for this book was a series of lectures given by the author at Andhra University, Visakhapatnam (India) in 2002. Efforts have been made to give enough detail in every chapter to ensure that an advanced graduate student can follow the mathematics and understand the results of 'computer experiments' for embedded ensembles. On the other hand, the book gives an exhaustive review of the field so that a research student can use the material to start working on new questions in the subject of embedded ensembles itself and in their application to many-body quantum physics.

Over the last three decades I have had the pleasure of collaborating with many people, and discussed the topics of this book with many others. First of all, I would like to specially thank the late J.B. French for a long and profitable collaboration on statistical nuclear spectroscopy. Embedded random matrix ensembles have grown out of this subject and the present work is complementary to the book *Statistical Spectroscopy and Spectral Distributions in Nuclei* by R.U. Haq and myself, published in 2010 by World Scientific. The influence of J.B. French on my way of thinking about random matrix theory in physics is surely visible in several parts of the present book.

I was fortunate in having A. Pandey, J.C. Parikh, V. Potbhare, and S. Tomsovic as collaborators in my early years of research on random matrix theory. Regarding the topics discussed in several chapters of this book, I have collaborated intensively with R. Sahu, N.D. Chavda, and my former Ph.D. student Manan Vyas. Without that collaboration, this book would not have been possible. I have also benefited from collaboration and discussions with many colleagues, friends, and students, and in particular with Dilip Angom, B. Chakrabarti, J.M.G. Gomez, R.U. Haq, K. Kar, D. Majumdar, the late J. Retamosa, S. Sumedha, and Y.M. Zhao. I am especially indebted to H.A. Weidenmüller and the late O. Bohigas for discussions and encouragement. I am thankful to N.D. Chavda and Manan Vyas for preparing some of the figures and thank Manan Vyas once again for typing some parts of the book. Thanks are also due to the directors of the Physical Research Laboratory (Ahmedabad, India) for facilities and support. There are many others who have directly or indirectly contributed to my work on embedded ensembles and I sincerely thank them. Copyright permission for using some of the figures, from the American Physical Society, the American Institute of Physics, Elsevier Science, the Institute of Physics, Springer-Verlag, and World Scientific is gratefully acknowledged. I am also thankful to all the authors who have given permission to use figures from their publications. Special thanks are due to the editors at Springer-Verlag for their efforts in bringing out this book. And lastly and most importantly, I am indebted to my wife Vijaya for her unfailing support since 1980.

Physical Research Laboratory, Ahmedabad, India
November 2013

V.K.B. Kota

Contents

Chapter 1
Introduction

Wishart [1] introduced random matrices in 1928 in the context of multi-variate statistics. However, it was Wigner [2–4] who introduced random matrix ensembles into physics in 1955, in his quest to derive information about the level and strength of fluctuations in compound nucleus resonances. As stated by Wigner [5, p. 203]:

> The assumption is that the Hamiltonian which governs the behavior of a complicated system is a random symmetric matrix with no particular properties except for its symmetric nature.

And French adds [6]:

> With one short step beyond this, specifically replacing 'complicated' by 'non-integrable', this paper would have led to the foundations of quantum chaos. Perhaps it should be so regarded even as it stands.

Dyson [7–10] provided the tripartite classification of random matrix ensembles giving the classical random matrix ensembles, i.e., the Gaussian orthogonal (GOE), unitary (GUE) and symplectic (GSE) ensembles. The classical random matrix ensembles were developed and applied in nuclear (and to a lesser extent atomic) physics over the period 1955–1972 by Dyson, Mehta, Porter, and others. Porter's book [11] gives an excellent introduction to random matrix theory and also contains a collection of papers on this subject, published up until 1965, including all the original papers of Wigner and Dyson. Later, Mehta in his book [12], first published in 1967 and with a third edition in 2004, described the mathematical foundations of the classical ensembles, and this has since become a standard reference in work relating to random matrices. The Albany conference in 1972 [13] changed the course of research activity in applications of this field to quantum physics. From 1972 to 1983, developments in the subject were due in particular to French, Bohigas, Pandey, Wong, and others [14].

Random matrix theory (RMT) has become a common theme in quantum physics, with the recognition that it is relevant to quantum systems whose classical analogues are chaotic. The Bohigas–Giannoni–Schmit (BGS) conjecture [15, 16] put forward in 1984 is the cornerstone for this. This asserts that *the spectra of time-reversal-invariant systems whose classical analogs are K systems show the same fluctuation properties as predicted by GOE*. Furthermore, as stated by BGS, *if this conjecture*

V.K.B. Kota, *Embedded Random Matrix Ensembles in Quantum Physics*,
Lecture Notes in Physics 884, DOI 10.1007/978-3-319-04567-2_1,

happens to be true, the 'universality of the laws of level fluctuations' in quantal spectra, already found in nuclei and to a lesser extent in atoms, will then have been established. As a consequence, they should also be found in other quantal systems, such as molecules, hadrons, etc. Recently, Heusler et al. [17–19] gave a proof of the BGS conjecture using semi-classical methods. On the other hand, Berry [20, 21] showed that integrable (or regular) systems follow Poisson. Combining BGS with the work of Berry one can conclude as summarized by Altshuler in the abstract of the colloquium he gave in memory of J.B. French at the university of Rochester in 2004:

> Classical dynamical systems can be separated into two classes—integrable and chaotic. For quantum systems this distinction manifests itself, e.g., in spectral statistics. Roughly speaking, integrability leads to a Poisson distribution for the energies, while chaos implies Wigner–Dyson statistics of levels, which are characteristic for the ensemble of random matrices. The onset of chaotic behavior for a rather broad class of systems can be understood as a delocalization in the space of quantum numbers that characterize the original integrable system.

Haake's book [22] is the best available reference for quantum chaos and its relation to RMT. Similarly, laboratory tests of RMT for wave chaos using microwave billiards are discussed in [23], while Efetov [24] introduced a supersymmetry approach for RMT. Besides books, there are some good review articles on RMT in physics by Brody et al. [14], Guhr et al. [25], Mirlin [26], and Weidenmüller et al. [27, 28], but see also the articles in a special issue of the Journal of Physics A [29]. In order to study symmetry-breaking effects on level and strength fluctuations, order–chaos transitions, etc., one must consider interpolating/deformed random matrix ensembles, i.e., ensembles with more information. The earliest examples are banded random matrix ensembles, the Rosenzweig–Porter model, 2×2 GOE due to Dyson, and so on; see [11]. A large class of random matrix ensembles is now being studied and applied to all branches of physics.

With the revival of RMT research in physics from 1984, large scale research on random matrices also began in probability theory. Developments on the mathematical and statistics side are due not only to the Wigner–Dyson classical random matrix ensembles, but more importantly to Wishart's original paper [1] and the work by Pastur on covariance matrices [30, 31]. The result of all this research led to applications of RMT to many diverse fields such as quantum information science, econophysics, multivariate statistics, information theory, wireless communication, neural networks, biological networks, number theory, and so on. This has also led to many mathematical books on the subject over the last 5 years. These are due to Anderson, Bai, Dieft, Pastur, Sarnak, and others [32–41]. In addition, there are books giving details of RMT applications to physics in particular and to science in general. See, for example, [42–44].

While the above developments were under way, a new class of random matrix ensembles, called embedded random matrix ensembles, began to receive special attention in quantum physics [45–48]. Isolated finite many-particle quantum systems such as nuclei, atoms, quantum dots, small metallic grains, spin models for quantum computer cores, BEC, etc., share one common property—their constituents interact via interactions of low body rank (see Chap. 4 for definitions) and they are mostly two-body in nature. Besides this, the particles move in a mean-field, giving a one-body term in the Hamiltonian operator. Representation of the many-particle Hamiltonian by classical ensembles implies many-body interactions. In fact, *the GOE, now almost universally regarded as a model for a corresponding chaotic system, is an ensemble of multi-body, not two-body interactions. This difference shows up both in one-point (density of states) and two-point (fluctuations and smoothed transition strengths) functions generated by the nuclear shell model.* Two-body interactions imply that many of the many-particle Hamiltonian matrix elements should be zero (see Fig. 1.1 for an example). Therefore it is more realistic to consider random interactions and then generate many-particle Hamiltonian matrices using the geometry of the many-particle Hilbert space. With say k-body interactions (for an m particle system $k < m$), random interactions imply that we represent the Hamiltonian matrix in k-particle spaces by a classical random matrix ensemble (or a deformed version of such). As a classical ensemble is embedded in many-particle spaces, these are generically called *embedded ensembles* (EE) or random interaction matrix models (RIMM). When the embedding matrix is one of the classical Gaussian ensembles, they are called *embedded Gaussian ensembles* (EGE). Thus, with GOE embedding we have EGOE and similarly EGUE and EGSE. In addition, with k-body interactions, we have EE(k), EGOE(k), EGUE(k), and EGSE(k).

With two-body interactions, EEs are often called two-body ensembles (TBRE). In 1970, TBREs with angular momentum J symmetry were introduced by French and Wong [50–52] and Bohigas and Flores [53, 54] following the observation that nuclear-shell-model Hamiltonians give a Gaussian eigenvalue density, in contrast to the semi-circle density generated by classical ensembles. As French states [55]:

> For a better a priori model we can choose an ensemble of k-body interactions ($k = 2$ is an interesting case) by generating a GOE in k-particle space and using it in the space of m-particles. For most purposes the resulting embedded GOE (or EGOE) is very difficult to deal with, but by good luck, we can use it to study the questions we have posed and the answers are different from, and much more enlightening than, those which would come from GOE.

The EGOE(k), discussed in detail by Mon and French in 1975 [56], were explored in a limited manner and exclusively in nuclear physics, up until the early 1990s [57]. However, with the progress in mesoscopic physics and quantum chaos, research work on two-body random matrix ensembles started growing very quickly from 1996 onwards, with a flurry of papers from the research groups of Alhassid, Flambaum, Izrailev, Kota, Shepelyansky, and Zelevinsky [58–67]. As stated by Altshuler, Bohigas, and Weidenmüller in a workshop on the chaotic dynamics of many-body systems held at ECT*, Trento, in February 1997:

> The study of quantum manifestations of classical chaos has known important developments, particularly for systems with few degrees of freedom. Now we understand much better how the universal and system-specific properties of 'simple chaotic systems' are connected with the underlying classical dynamics. The time has come to extend, from this perspective, our understanding to objects with many degrees of freedom, such as interacting many-body systems. Problems of nuclear, atomic, and molecular theory as well as the theory of mesoscopic systems will be discussed at the workshop.

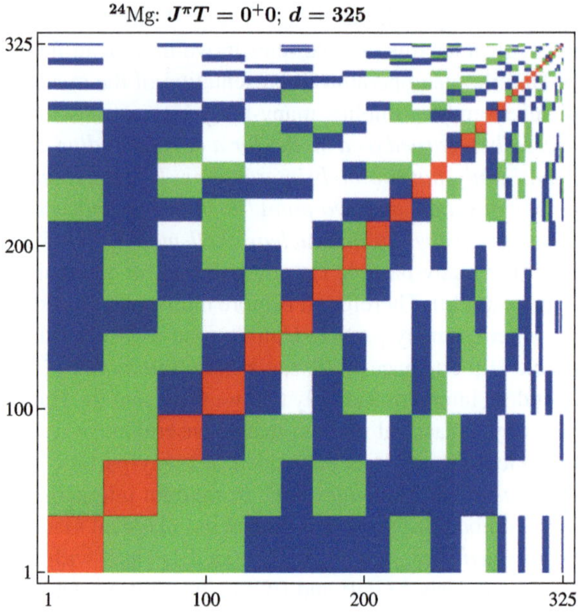

Fig. 1.1 Block matrix structure of the H matrix of the ^{24}Mg nucleus, with $J = 0, T = 0$, displaying two-body selection rules. Here ^{24}Mg is described by 4 protons and 4 neutrons in the shell model $^{1}d_{5/2}$, $^{2}s_{1/2}$ and $^{1}d_{3/2}$ orbits with H preserving angular momentum J and isospin T. The total number of blocks is 33, each labeled by the spherical configurations (m_1, m_2, m_3). The diagonal blocks are shown in *red*, and within these blocks there will be no change in the occupancy of the nucleons in the three sd orbits. *Green* corresponds to the region (in the matrix) connected by the two-body interaction that involves change of occupancy of one nucleon. Similarly, *blue* corresponds to change of occupancy of two nucleons. Finally, white corresponds to the region forbidden by the two-body selection rules. This figure is taken from [49] with permission from Springer

Thus, as will be made clear in Chaps. 4–15, many-body quantum chaos is modeled by embedded random matrix ensembles, whence there has been an explosion of research activity analyzing a wide variety of EEs over the past 15 years. Three reviews focusing exclusively on EEs are currently available [45–47] and there are several other review articles in which a good part is devoted to EE [14, 27, 48, 66]. Besides applications in nuclear, atomic, and mesoscopic physics, it has been recognized more recently that EEs are important in quantum information science (QIS) [68] and in understanding the thermodynamics of isolated finite quantum systems [69–71]. The embedded ensembles are analyzed analytically using the binary correlation approximation, perturbation theory, the Wigner–Racah algebra of the Lie algebra defining the embedding, and trace propagation methods for spectral variances. They are also analyzed numerically on a much larger scale using the Monte-Carlo method.

French recognized that random matrix theory based on embedded ensembles gives a complete statistical theory for quantum systems—it gives both the spectral distributions of various physical quantities and their fluctuations, the latter coinciding with those generated by classical random matrix ensembles. It is worth recalling French's own words [72]:

> The striking features of spectral distribution methods are their wide applicability and the connections which they display, and make use of, between statistical behavior, unitary and other symmetries and their associated geometries, and information content and propagation. As for the future, it might be good for us not to think of the methods we are discussing as forming a separate domain with its own special tricks, devices, methods, and assumptions. It is probably wise instead to think of the whole subject as forming a sub-domain of statistical mechanics in which special attention is paid to the nature of the model space.

Although there are now many books on random matrices, as cited above, none of them have a discussion on EEs, and this includes the most recent book entitled *The Oxford Handbook of Random Matrix Theory* [44]. Therefore, there is clearly a need for a book on EEs, and the purpose of the present book is to fill this gap.

The aim here is thus to give an easy-to-understand introduction to EEs so that young researchers can take up this subject, develop it much further, and apply it to a whole range of problems in quantum physics. As the book has to be self-contained, we will give a user-friendly introduction to some of the results of classical ensembles in two chapters. We use this to introduce the so-called binary correlation approximation (BCA), along with many other concepts, definitions, and notations that are used in the later chapters. At present, the BCA is the main physically understandable mathematical method available for analyzing EEs.

So let us now give a short preview. Chapter 2 classifies the classical GOE, GUE, and GSE ensembles, and to get started with their properties, nearest neighbor spacing distributions (NNSD) for the 2×2 matrix version of these ensembles are derived. In addition, one- and two-point functions (in the eigenvalues) are presented for general $N \times N$ GOEs and GUEs, these being derived using the so-called binary correlation approximation. Some aspects of data analysis are discussed for measures of fluctuations given by the ensembles, together with the wavefunction structure generated by the ensembles. The discussion of GSEs is kept to a minimum, as EGSEs are not yet addressed in the literature.

Chapter 3 deals with various interpolating and deformed classical ensembles, emphasizing their applications in physics. The results in Chaps. 2 and 3 provide the necessary background in random matrix theory that is essential in order to follow the results and discussion on embedded ensembles presented in the remainder of this book.

Chapters 4–15 are devoted to EGEs with Chaps. 4–8 describing exclusively fermion systems and Chaps. 9 and 10 boson systems. In Chap. 4, EGOE(2) and more general EGOE(k) for *spinless fermion systems* are defined and a method for their construction is described. The one-point function (eigenvalue density) and some aspects of the two-point function for the eigenvalues generated by EGOE(k) are discussed using the binary correlation approximation. The asymptotic form of transition strength densities, which are also two-point functions, generated by transition matrix elements, is also discussed.

Chapter 5 introduces EGOE($1+2$) for spinless fermion systems, i.e., for EGOEs generated in many-particle spaces by random two-body interactions in the presence of a mean field, discussing in particular the transition (or chaos) markers generated by this ensemble. In the limiting situations with interactions much stronger than the mean field, EGOE(1 + 2) reduces to EGOE(2).

Chapter 6 deals with EGOE(1 + 2)-**s** for fermions with *spin* degrees of freedom, discussing some general properties of this ensemble. Chapter 7 is devoted to applications of EGOE(1 + 2) and EGOE(1 + 2)-**s**, discussed in particular (i) simple applications to mesoscopic systems and (ii) the EGOE basis for statistical spectroscopy in nuclei and atoms. Chapter 8 describes EGOEs with parity symmetry. The corresponding ensemble is called EGOE(1 + 2)-π. This ensemble is important in the study of parity ratios in nuclear level densities.

Chapter 9 is devoted to embedded ensembles for spinless boson systems and Chap. 10 to two-species boson systems and bosons carrying a spin-one degree of freedom.

In Chap. 11, we consider GUE versions of embedded ensembles for both fermion and boson systems. Using the Wigner–Racah algebra of the Lie algebras defining the embedded ensembles, a general formulation for the lower order moments of the one- and two-point functions for the ensembles with $U(\Omega) \otimes SU(r)$ embedding and random two-body Hamiltonians with $SU(r)$ symmetry has been developed, and this formulation is presented with examples for fermion systems with $r = 1$, 2, and 4, and likewise for boson systems with $r = 1$, 2, and 3. Results for EGUE(k) and BEGUE(k) with $k < m$ are also presented.

Chapter 12 presents numerical results, for embedded ensembles, for self-correlations and more importantly for cross-correlations, which are absent in the classical ensemble (GOE/GUE/GSE) description of many-particle systems.

Going beyond the embedded ensembles considered in Chaps. 4–12, various other extended embedded ensembles, explored analytically only to a very limited extent in the literature, are briefly considered in Chap. 13, while Chap. 14 discusses the new paradigm of regular structures generated by random interactions, which is a quite different application of embedded ensembles. Chapter 15 focuses on the application of EEs to time dynamics and entropy production, as well as the question of thermalization in isolated finite interacting quantum systems. Finally, Chap. 16 discusses the outlook for the future. There are eight appendices and the survey of the literature for this book goes up to 31 March 2013.

The main emphasis in this book is on analytical results derived for EEs, while the discussion of numerical results is kept to a minimum. Although most physics examples are taken from nuclear physics, some examples from atomic and mesoscopic physics are also discussed. This book complements our earlier book on *Spectral Distributions in Nuclei and Statistical Spectroscopy*, where the focus was on spectral distribution theory, without going into the details of the random matrix basis for the various spectral distributions [73].

References

1. J. Wishart, The generalized product moment distribution in samples from a normal multivariate population. Biometrika **20A**, 32–52 (1928)

2. E.P. Wigner, Characteristic vectors of bordered matrices with infinite dimensions. Ann. Math. **62**, 548–564 (1955)
3. E.P. Wigner, Characteristic vectors of bordered matrices with infinite dimensions II. Ann. Math. **65**, 203–207 (1957)
4. E.P. Wigner, Statistical properties of real symmetric matrices with many dimensions, in *Can. Math. Congr. Proc.* (University of Toronto Press, Toronto, 1957), pp. 174–184
5. E.P. Wigner, The probability of the existence of a self reproducing unit, reproduced from the logic of personal knowledge: essays in honor of Michael Polanyi, Chap. 19 (Routledge and Kegan Paul, London, 1961), in *Symmetries and Reflections: Scientific Essays of E.P. Wigner*, ed. by W.J. Moore, M. Scriven (Indiana University Press, Bloomington, 1967), Reprinted by (Ox Bow Press, Woodbridge, 1979), pp. 200–208
6. J.B. French, Elementary aspects of nuclear quantum chaos, in *A Gift of Prophecy—Essays in the Celebration of the Life of R.E. Marshak*, ed. by E.C.G. Sudarshan (World Scientific, Singapore, 1994), pp. 156–167
7. F.J. Dyson, Statistical theory of energy levels of complex systems I. J. Math. Phys. **3**, 140–156 (1962)
8. F.J. Dyson, Statistical theory of energy levels of complex systems II. J. Math. Phys. **3**, 157–165 (1962)
9. F.J. Dyson, Statistical theory of energy levels of complex systems III. J. Math. Phys. **3**, 166–175 (1962)
10. F.J. Dyson, The threefold way: algebraic structures of symmetry groups and ensembles in quantum mechanics. J. Math. Phys. **3**, 1199–1215 (1963)
11. C.E. Porter, *Statistical Theories of Spectra: Fluctuations* (Academic Press, New York, 1965)
12. M.L. Mehta, *Random Matrices*, 3rd edn. (Elsevier, Amsterdam, 2004)
13. J.B. Garg (ed.), *Statistical Properties of Nuclei, Proceedings of the International Conference on Statistical Properties of Nuclei, Held at Albany (N.Y.), August 23–27, 1971* (Plenum, New York, 1972)
14. T.A. Brody, J. Flores, J.B. French, P.A. Mello, A. Pandey, S.S.M. Wong, Random matrix physics: spectrum and strength fluctuations. Rev. Mod. Phys. **53**, 385–479 (1981)
15. O. Bohigas, M.J. Giannoni, C. Schmit, Characterization of chaotic quantum spectra and universality of level fluctuation laws. Phys. Rev. Lett. **52**, 1–4 (1984)
16. O. Bohigas, Random matrix theories and chaotic dynamics, in *Chaos and Quantum Physics*, ed. by M.J. Giannoni, A. Voros, J. Zinn-Justin (Elsevier, Amsterdam, 1991), pp. 87–199
17. S. Heusler, S. Müller, A. Altland, P. Braun, F. Haake, Periodic-orbit theory of level correlations. Phys. Rev. Lett. **98**, 044103 (2007)
18. S. Müller, S. Heusler, A. Altland, P. Braun, F. Haake, Periodic-orbit theory of universal level correlations in quantum chaos. New J. Phys. **11**, 103025 (2009)
19. P. Braun, Beyond the Heisenberg time: semiclassical treatment of spectral correlations in chaotic systems with spin 1/2. J. Phys. A **45**, 045102 (2012)
20. M.V. Berry, M. Tabor, Level clustering in the regular spectrum. Proc. R. Soc. Lond. A **356**, 375–394 (1977)
21. M.V. Berry, Semiclassical theory of spectral rigidity. Proc. R. Soc. Lond. A **400**, 229–251 (1985)
22. F. Haake, *Quantum Signatures of Chaos*, 3rd edn. (Springer, Heidelberg, 2010)
23. H.-J. Stöckmann, *Quantum Chaos: An Introduction* (Cambridge University Press, New York, 1999)
24. K.B. Efetov, *Supersymmetry in Disorder and Chaos* (Cambridge University Press, New York, 1997)
25. T. Guhr, A. Müller-Groeling, H.A. Weidenmüller, Random-matrix theories in quantum physics: common concepts. Phys. Rep. **299**, 189–425 (1998)
26. A.D. Mirlin, Statistics of energy levels and eigenfunctions in disordered systems. Phys. Rep. **326**, 259–382 (2000)
27. H.A. Weidenmüller, G.E. Mitchell, Random matrices and chaos in nuclear physics: nuclear structure. Rev. Mod. Phys. **81**, 539–589 (2009)

28. G.E. Mitchell, A. Richter, H.A. Weidenmüller, Random matrices and chaos in nuclear physics: nuclear reactions. Rev. Mod. Phys. **82**, 2845–2901 (2010)
29. P.J. Forrester, N.C. Snaith, J.J.M. Verbaarschot (eds.), Special issue: random matrix theory. J. Phys. A **36**, R1–R10 and 2859–3646 (2003)
30. V.A. Marčenko, L.A. Pastur, Distribution of eigenvalues for some sets of random matrices. Math. USSR Sb. **1**, 457–483 (1967)
31. L.A. Pastur, On the spectrum of random matrices. Theor. Math. Phys. **10**, 67–74 (1972)
32. N.M. Katz, P. Sarnak, *Random Matrices, Frobenius Eigenvalues and Monodromy*, Colloquium Publication, vol. 45 (Am. Math. Soc., Providence, 1999)
33. P. Deift, *Orthogonal Polynomials and Random Matrices: A Riemann-Hilbert Approach*, Courant Lecture Notes, vol. 3 (Am. Math. Soc. and Courant Institute of Mathematical Sciences at New York, Providence, 2000)
34. A.M. Tulino, S. Verdú, *Random Matrix Theory and Wireless Communication* (Now Publishers, Hanover, 2004)
35. J. Baik, T. Kriecherbauer, L. Li, K.D.T.-R. McLaughlin, C. Tomei (eds.), *Integrable Systems and Random Matrices*, Contemporary Mathematics, vol. 458 (Am. Math. Soc., Providence, 2008)
36. G. Blower (ed.), *Random Matrices: High Dimensional Phenomena*, London Mathematical Society Lecture Series, vol. 367 (Cambridge University Press, Cambridge, 2009)
37. P. Deift, D. Gioev, *Random Matrix Theory: Invariant Ensembles and Universality* (Am. Math. Soc. and Courant Institute of Mathematical Sciences at New York, Providence, 2009)
38. G.W. Anderson, A. Guionnet, O. Zeitouni, *An Introduction to Random Matrices* (Cambridge University Press, New York, 2010)
39. Z. Bai, J.W. Silverstein, *Spectral Analysis of Large Dimensional Random Matrices*, 2nd edn. (Springer, New York, 2010)
40. L. Pastur, M. Shcherbina, *Eigenvalue Distribution of Large Random Matrices*, Mathematical Surveys and Monographs, vol. 171 (Am. Math. Soc., Providence, 2011)
41. R. Couillet, M. Debbah, *Random Matrix Methods for Wireless Communications* (Cambridge University Press, New York, 2012)
42. M. Wright, R. Weiver, *New Directions in Linear Acoustics and Vibrations: Quantum Chaos, Random Matrix Theory and Complexity* (Cambridge University Press, New York, 2011)
43. P.J. Forrester, *Log-Gases and Random Matrices* (Princeton University Press, Princeton, 2010)
44. G. Akemann, J. Baik, P. Di Francesco (eds.), *The Oxford Handbook of Random Matrix Theory* (Oxford University Press, Oxford, 2011)
45. V.K.B. Kota, Embedded random matrix ensembles for complexity and chaos in finite interacting particle systems. Phys. Rep. **347**, 223–288 (2001)
46. L. Benet, H.A. Weidenmüller, Review of the k-body embedded ensembles of Gaussian random matrices. J. Phys. A **36**, 3569–3594 (2003)
47. T. Papenbrock, H.A. Weidenmüller, Random matrices and chaos in nuclear spectra. Rev. Mod. Phys. **79**, 997–1013 (2007)
48. J.M.G. Gómez, K. Kar, V.K.B. Kota, R.A. Molina, A. Relaño, J. Retamosa, Many-body quantum chaos: recent developments and applications to nuclei. Phys. Rep. **499**, 103–226 (2011)
49. M. Vyas, V.K.B. Kota, Random matrix structure of nuclear shell model Hamiltonian matrices and comparison with an atomic example. Eur. Phys. J. A **45**, 111–120 (2010)
50. J.B. French, S.S.M. Wong, Validity of random matrix theories for many-particle systems. Phys. Lett. B **33**, 449–452 (1970)
51. J.B. French, S.S.M. Wong, Some random-matrix level and spacing distributions for fixed-particle-rank interactions. Phys. Lett. B **35**, 5–7 (1971)
52. J.B. French, Analysis of distant-neighbour spacing distributions for k-body interaction ensembles. Rev. Mex. Fis. **22**, 221–229 (1973)
53. O. Bohigas, J. Flores, Two-body random Hamiltonian and level density. Phys. Lett. B **34**, 261–263 (1971)
54. O. Bohigas, J. Flores, Spacing and individual eigenvalue distributions of two-body random Hamiltonians. Phys. Lett. B **35**, 383–387 (1971)

55. J.B. French, Special topics in spectral distributions, in *Moment Methods in Many Fermion Systems*, ed. by B.J. Dalton, S.M. Grimes, J.P. Vary, S.A. Williams (Plenum, New York, 1980), pp. 91–108
56. K.K. Mon, J.B. French, Statistical properties of many-particle spectra. Ann. Phys. (N.Y.) **95**, 90–111 (1975)
57. J.B. French, V.K.B. Kota, A. Pandey, S. Tomsovic, Statistical properties of many-particle spectra VI. Fluctuation bounds on N-N T-noninvariance. Ann. Phys. (N.Y.) **181**, 235–260 (1988)
58. V.V. Flambaum, A.A. Gribakina, G.F. Gribakin, M.G. Kozlov, Structure of compound states in the chaotic spectrum of the Ce atom: localization properties, matrix elements, and enhancement of weak perturbations. Phys. Rev. A **50**, 267–296 (1994)
59. V.V. Flambaum, G.F. Gribakin, F.M. Izrailev, Correlations within eigenvectors and transition amplitudes in the two-body random interaction model. Phys. Rev. E **53**, 5729–5741 (1996)
60. V.V. Flambaum, F.M. Izrailev, Statistical theory of finite Fermi systems based on the structure of chaotic eigenstates. Phys. Rev. E **56**, 5144–5159 (1997)
61. M. Horoi, V. Zelevinsky, B.A. Brown, Chaos vs thermalization in the nuclear shell model. Phys. Rev. Lett. **74**, 5194–5197 (1995)
62. V. Zelevinsky, B.A. Brown, N. Frazier, M. Horoi, The nuclear shell model as a testing ground for many-body quantum chaos. Phys. Rep. **276**, 85–176 (1996)
63. Ph. Jacquod, D.L. Shepelyansky, Emergence of quantum chaos in finite interacting Fermi systems. Phys. Rev. Lett. **79**, 1837–1840 (1997)
64. B. Georgeot, D.L. Shepelyansky, Breit-Wigner width and inverse participation ratio in finite interacting Fermi systems. Phys. Rev. Lett. **79**, 4365–4368 (1997)
65. V.K.B. Kota, R. Sahu, Information entropy and number of principal components in shell model transition strength distributions. Phys. Lett. B **429**, 1–6 (1998)
66. Y. Alhassid, Statistical theory of quantum dots. Rev. Mod. Phys. **72**, 895–968 (2000)
67. V.K.B. Kota, R. Sahu, Structure of wavefunctions in $(1+2)$-body random matrix ensembles. Phys. Rev. E **64**, 016219 (2001)
68. W.G. Brown, L.F. Santos, D.J. Starling, L. Viola, Quantum chaos, delocalization and entanglement in disordered Heisenberg models. Phys. Rev. E **77**, 021106 (2008)
69. M. Rigol, V. Dunjko, M. Olshanii, Thermalization and its mechanism for generic isolated quantum systems. Nature (London) **452**, 854–858 (2008)
70. L.F. Santos, M. Rigol, Onset of quantum chaos in one dimensional bosonic and fermionic systems and its relation to thermalization. Phys. Rev. E **81**, 036206 (2010)
71. L.F. Santos, M. Rigol, Localization and the effects of symmetries in the thermalization properties of one-dimensional quantum systems. Phys. Rev. E **82**, 031130 (2010)
72. J.B. French, Elementary principles of spectral distributions, in *Moment Methods in Many Fermion Systems*, ed. by B.J. Dalton, S.M. Grimes, J.P. Vary, S.A. Williams (Plenum, New York, 1980), pp. 1–16
73. V.K.B. Kota, R.U. Haq, *Spectral Distributions in Nuclei and Statistical Spectroscopy* (World Scientific, Singapore, 2010)

Chapter 2
Classical Random Matrix Ensembles

2.1 Hamiltonian Structure and Dyson's Classification of GOE, GUE and GSE Random Matrix Ensembles

The discussion in this section is largely from Porter's book [1]. Original contributions here are due to Wigner and Dyson and all their papers were reprinted in [1]. More recent discussion on Dyson's classification of classical random matrix ensembles is given by Haake [2]. The classification is based on the properties of time reversal operator in quantum mechanics [3]. Appendix A gives a brief discussion on time reversal and the results given there are used in this section. In finite Hilbert spaces, the Hamiltonian of a quantum system can be represented by a $N \times N$ matrix. Now we will consider the properties of this matrix, with regard to time-reversal T and angular momentum J symmetries.

Firstly, imagine there is no time reversal invariance. Then we know nothing about the H matrix except that it should be complex Hermitian. And all such H's should be complex Hermitian in any representation differing from any other by a unitary transformation.

Now, consider H to be T invariant. Then we have, from the results in Appendix A, $T^2 = \pm 1$. Let us say T is good and $T^2 = 1$. Then, it is possible to construct an orthogonal basis ψ_K such that $T\psi_K = \psi_K$ is satisfied. Let us start with a normalized state Φ_1 and construct,

$$\Psi_1 = a\Phi_1 + Ta\Phi_1 \tag{2.1}$$

where a is an arbitrary complex number. This gives trivially

$$T\Psi_1 = Ta\Phi_1 + T^2 a\Phi_1 \overset{T^2=1}{\rightarrow} \Psi_1. \tag{2.2}$$

Now consider Φ_2 such that $\langle \Psi_1 \Phi_2 \rangle = 0$ and construct

$$\Psi_2 = a'\Phi_2 + Ta'\Phi_2. \tag{2.3}$$

V.K.B. Kota, *Embedded Random Matrix Ensembles in Quantum Physics*, Lecture Notes in Physics 884, DOI 10.1007/978-3-319-04567-2_2,

Then,

$$\begin{aligned}\langle\Psi_1|\Psi_2\rangle &= a'\langle\Psi_1|\Phi_2\rangle + \langle\Psi_1\big|Ta'\Phi_2\rangle = \big(a'\big)^*\langle\Psi_1|T\Phi_2\rangle \\ &= \big(a'\big)^*\big\langle T\Psi_1\big|T^2\Phi_2\big\rangle^* = \big(a'\big)^*\langle\Psi_1|\Phi_2\rangle^* = 0.\end{aligned} \tag{2.4}$$

Here we used Eq. (A.12) and $T^2 = 1$. Continuing this way an entire set of orthogonal Ψ_i can be produced satisfying

$$T\Psi_i = \Psi_i \tag{2.5}$$

and they can be normalized. In this basis all H's that are T invariant, i.e. $THT^{-1} = H$ will be real symmetric,

$$H_{kl} = \langle\Psi_k \mid H \mid \Psi_l\rangle = \langle T\Psi_k \mid TH\Psi_l\rangle^* = \big\langle T\Psi_k \,\big|\, THT^{-1}T\Psi_l\big\rangle^* = H^*_{kl}. \tag{2.6}$$

Therefore for systems with $T^2 = 1$, all H's that are T invariant can be made, independent of J, real symmetric in a single basis.

Let us consider the situation with T is good, J is good and $T^2 = -1$. Now, say $\overline{T} = \exp(-i\pi J_y)T$ and then using $T = \exp(i\pi S_y)K$ as given by Eq. (A.23), we have

$$\begin{aligned}(\overline{T})^2 &= \exp(-i\pi J_y)\exp(i\pi S_y)K\exp(-i\pi J_y)\exp(i\pi S_y)K \\ &= \exp(-i\pi L_y)\exp(-i\pi L_y)K^2 \\ &= \exp(-2i\pi L_y) = 1,\end{aligned} \tag{2.7}$$

as L_y is an integer in the (L^2, L_y) diagonal basis. Now we have $(\overline{T})^2 = 1$ and therefore we can proceed to construct a Γ_k basis with $\overline{T}\Gamma_i = \Gamma_i$ just as in the situation with $T^2 = 1$. In the Γ_i basis, a T invariant H will be real symmetric,

$$\begin{aligned}H_{k\ell} &= \langle\Gamma_k|H|\Gamma_\ell\rangle = \langle T\Gamma_k \mid TH\Gamma_\ell\rangle^* \\ &= \big\langle e^{-i\pi J_y}T\Gamma_k \,\big|\, e^{-i\pi J_y}TH\Gamma_\ell\big\rangle^* = \big\langle \overline{T}\Gamma_k \,\big|\, e^{-i\pi J_y}THT^{-1}T\Gamma_\ell\big\rangle^* \\ &\overset{THT^{-1}=H}{\longrightarrow} \big\langle\Gamma_k \,\big|\, e^{-i\pi J_y}He^{i\pi J_y}e^{-i\pi J_y}T\Gamma_\ell\big\rangle^* \\ &\overset{J_iHJ_i^{-1}=H}{\longrightarrow} \langle\Gamma_k \mid H \mid \overline{T}\Gamma_\ell\rangle^* = \langle\Gamma_k \mid H \mid \Gamma_\ell\rangle^* \\ &= H^*_{k\ell}.\end{aligned} \tag{2.8}$$

Therefore if H is invariant under both J and T, the H matrix can be made symmetric. In fact all such H's will be simultaneously real symmetric in the Γ_k basis and they remain so by an orthogonal transformation.

The final situation is where T is good, J is not good but $T^2 = -1$. In the situation we still have Kramer's degeneracy and given a $|\psi\rangle$, the $|\psi\rangle$ and $|T\psi\rangle$ are orthogonal. With a basis of $2N$ states, $(|i\rangle, T|i\rangle)$ with $i = 1, 2, \ldots, N$, consider

$$|\psi\rangle = \sum_m \big[C_{m+}|m\rangle + C_{m-}|Tm\rangle\big]. \tag{2.9}$$

Then,

$$T|\psi\rangle = \sum_m \left[-C_{m-}^* |m\rangle + C_{m+}^* |Tm\rangle \right] \tag{2.10}$$

and here we have used $T^2 = -1$ with $TC_0 = C_0^*$ for any number C_0. As $T = UK$, Eq. (2.10) then gives $U = \left[\begin{smallmatrix} 0 & -1 \\ 1 & 0 \end{smallmatrix}\right]$ for each ($|m\rangle$, $|Tm\rangle$) pair. Also $T^2 = -1 \Rightarrow UKUK = -1$ and then $UU^* = -1$. Also $UU^\dagger = 1$ and therefore $U = -\widetilde{U}$. Any U such that $U = -\widetilde{U}$ can be brought to the form

$$U = \begin{bmatrix} 0 & -I \\ I & 0 \end{bmatrix} \tag{2.11}$$

by a similarity transformation. We can also chose

$$U = \begin{bmatrix} 0 & -1 & & & \\ 1 & 0 & & & \\ & & 0 & -1 & \\ & & 1 & 0 & \\ & & & & \ddots \end{bmatrix}. \tag{2.12}$$

Now we consider a unitary matrix S that commutes with $T = UK$. Then

$$\begin{aligned} SUK &= UKS \\ SU &= UKSK^{-1} \\ &= US^* \\ \Rightarrow \quad U &= SU(S^*)^{-1} = SU(S^{-1})^* = SUS^{\dagger *} \\ &= SU\widetilde{S}. \end{aligned} \tag{2.13}$$

Therefore with

$$Z = \begin{bmatrix} 0 & -I \\ I & 0 \end{bmatrix}, \tag{2.14}$$

S must be

$$SZ\widetilde{S} = Z. \tag{2.15}$$

The S that are unitary and satisfying Eq. (2.15) are called symplectic matrices. We will now construct H matrices that are invariant under S, i.e. symplectic transformations. To-wards this end consider

$$\mathcal{T}_1 = -i\begin{bmatrix} 0 & I \\ I & 0 \end{bmatrix}$$
$$\mathcal{T}_2 = Z = \begin{bmatrix} 0 & -I \\ I & 0 \end{bmatrix} \tag{2.16}$$
$$\mathcal{T}_3 = -i\begin{bmatrix} I & 0 \\ 0 & -I \end{bmatrix}$$

and H in the form, with $I = \left[\begin{smallmatrix} I & 0 \\ 0 & I \end{smallmatrix}\right]$,

$$\begin{aligned} H &= H_0 I + \sum_{k=1}^{3} H_k \mathcal{T}_k \\ &= \begin{bmatrix} H_0 - iH_3 & -iH_1 - H_2 \\ -iH_1 + H_2 & H_0 + iH_3 \end{bmatrix}. \end{aligned} \tag{2.17}$$

Then

$$H = H^{\dagger} \quad \Rightarrow \quad H_0^{\dagger} = H_0, \qquad H_k^{\dagger} = -H_k. \tag{2.18}$$

Now $THT^{-1} = H$ and $T = UK = \left[\begin{smallmatrix} 0 & -K \\ K & 0 \end{smallmatrix}\right]$ will give,

$$\begin{aligned} H &= \begin{bmatrix} H_0^* - iH_3^* & -iH_1^* - H_2^* \\ -iH_1^* + H_2^* & H_0^* + iH_3^* \end{bmatrix} \\ &\Rightarrow \quad H_i = H_i^*, \quad i = 0, 1, 2, 3. \end{aligned} \tag{2.19}$$

Comparing Eqs. (2.18) and (2.19) we have,

$$H_0 = H_0^* = \widetilde{H}_0, \qquad H_k = H_k^* = -\widetilde{H}_k. \tag{2.20}$$

Now we will prove that, if H is T invariant, then SHS^{-1} is also T invariant,

$$\begin{aligned} T\left[SHS^{-1}\right]T^{-1} &= TSHS^{-1}T^{-1} \\ &= STHT^{-1}S^{-1} \\ &= SHS^{-1}. \end{aligned} \tag{2.21}$$

Therefore the quaternion structure of H given by Eq. (2.17), valid for T invariant H with $T^2 = -1$ and J may not be good, will be preserved by symplectic transformations, that is by S that are unitary and satisfying the condition $SZ\widetilde{S} = Z$. Together with Eq. (2.20), the H's are quaternion real (QR) matrices.

The results proved above will give the Hamiltonian form and the corresponding group structure under (J, T) invariance as follows:

Table 2.1 Classification of classical random matrix ensembles

Ensemble	Transformation matrices	Hamiltonian structure
GOE	Real orthogonal matrices O $O\widetilde{O} = I$	$H = H^* = \widetilde{H}$
GUE	Unitary matrices U $UU^\dagger = I$	$H = H_0 + iH_1$ $H_0 = H_0^* = \widetilde{H}_0$ $H_1 = H_1^* = -\widetilde{H}_1$
GSE	Symplectic matrices S $SZ\widetilde{S} = Z$, $SS^\dagger = I$	$H = H_0 I + \sum_{k=1}^{3} H_k \mathcal{T}_k$ $H_0 = H_0^* = \widetilde{H}_0$ $H_k = H_k^* = -\widetilde{H}_k$, $k = 1, 2, 3$

1. For T not good and J is good or not good, the Hamiltonian is complex Hermitian and the canonical group of transformations is $U(N)$, the unitary group in N dimensions (N is the dimension of the H matrix).
2. For T is good and J is good, the Hamiltonian is real symmetric and the canonical group of transformations is $O(N)$, the orthogonal group in N dimensions.
3. For T is good and J is not good but J is a integer, the Hamiltonian is real symmetric and the canonical group of transformations is again $O(N)$.
4. For T is good and J is not good but J is a half-odd integer, the Hamiltonian is quaternion real and the canonical group of transformations is $Sp(2N)$, the symplectic group in $2N$ dimensions (note that here we are using the H matrix dimension as $2N$ as it must be even).

Note that in (1)–(4) above, in a single basis all H's can be simultaneously made real symmetric, QR or complex Hermitian as appropriate. In the absence of any other information except invariance with respect to J and T are known, one can represent the H matrix of a given quantum system by an ensemble of $N \times N$ matrices with structure as given by (1)–(4) above. The matrix elements are then chosen to be independent random variables. Note that for the $U(N)$, $O(N)$ and $Sp(2N)$ systems mentioned above, the number of independent variables (note that for a complex number there are two variables—one for the real part and other for the complex part) will be N^2, $N(N+1)/2$ and $N(N+1)$ respectively. In the classical random matrix ensembles, called GUE, GOE and GSE respectively, the matrix elements are chosen to be independent Gaussian variables with zero center and variance unity (except that the diagonal matrix elements—they are real—have variance 2). Then these ensembles will be invariant under $U(N)$, $O(N)$ and $Sp(2N)$ transformations respectively and accordingly they are called Gaussian orthogonal (GOE), unitary (GUE) and symplectic (GSE) ensembles. Table 2.1 gives for these three ensembles, the corresponding transformation matrices and the mathematical structure of the Hamiltonians. In order to make a beginning in deriving the properties of GOE, GUE and GSE, we will start with the simplest 2×2 matrix version of these ensembles. Hereafter, zero centered Gaussian variables with variance v^2 will be denoted by $G(0, v^2)$.

2.1.1 2 × 2 GOE

For a 2×2 GOE, the Hamiltonian matrix is

$$H = \begin{bmatrix} X_1 + X_2 & X_3 \\ X_3 & X_1 - X_2 \end{bmatrix} \tag{2.22}$$

and the joint distribution for the independent variables X_1, X_2 and X_3 is

$$p(X_1, X_2, X_3)\, dX_1\, dX_2\, dX_3 = P_1(X_1)P_2(X_2)P_3(X_3)\, dX_1\, dX_2\, dX_3. \tag{2.23}$$

The X_i in Eq. (2.22) are $G(0, v^2)$. Then, $(X_1 + X_2)$ is $G(0, 2v^2)$ and $(X_1 - X_2)$ is $G(0, 2v^2)$. Let the eigenvalues of the H matrix be λ_1 and λ_2. Using the properties of the sum and product of eigenvalues, we have $\lambda_1 + \lambda_2 = 2X_1$ and $\lambda_1\lambda_2 = X_1^2 - X_2^2 - X_3^2$. This gives

$$S^2 = (\lambda_1 - \lambda_2)^2 = 4X_2^2 + 4X_3^2. \tag{2.24}$$

Now $x_2 = 2X_2$ is $G(0, 4v^2)$, $x_3 = 2X_3$ is $G(0, 4v^2)$ and they are independent. Therefore

$$P(x_2, x_3)\, dx_2\, dx_3 = \frac{1}{2\pi(4v^2)} \exp -\frac{(x_2^2 + x_3^2)}{8v^2} dx_2 dx_3. \tag{2.25}$$

Transforming the variables x_2, x_3 to S, ϕ where $x_2 = S\cos\phi$, $x_3 = S\sin\phi$ we have

$$P(S)dS = \frac{e^{-S^2/8v^2} S dS}{8\pi v^2} \int_0^{2\pi} d\phi. \tag{2.26}$$

Then the NNSD is,

$$P(S)dS = \frac{S}{4v^2} \exp -\frac{S^2}{8v^2} dS; \quad 0 \le S \le \infty. \tag{2.27}$$

Note that, with $\overline{D}$ denoting mean (or average) spacing,

$$\int_0^\infty P(S)dS = 1; \qquad \overline{D} = \int_0^\infty SP(S)dS = \sqrt{2\pi}\, v. \tag{2.28}$$

In terms of the normalized spacing $\hat{S} = S/\overline{D}$,

$$P(\hat{S})d\hat{S} = \frac{\pi\hat{S}}{2} \exp\left(-\frac{\pi\hat{S}^2}{4}\right) d\hat{S}; \quad \overline{\hat{S}} = 1. \tag{2.29}$$

Thus, GOE displays linear level repulsion with $P(S) \sim S$ as $S \to 0$. This is indeed the von Neumann-Wigner level repulsion discussed in 1929 [4] signifying that quantum levels with same quantum numbers will not come close. The variance of $P(S)$ is

$$\sigma^2(0) = \overline{\hat{S}^2} - 1 = \frac{\overline{S^2}}{\overline{D}^2} - 1$$
$$= \frac{8v^2}{2\pi v^2} - 1 = \frac{4}{\pi} - 1 \simeq 0.272. \tag{2.30}$$

Here, used are Eqs. (2.24) and (2.28).

2.1.2 2 × 2 GUE

Here the Hamiltonian matrix is,

$$H = \begin{bmatrix} X_1 + X_2 & X_3 + iX_4 \\ X_3 - iX_4 & X_1 - X_2 \end{bmatrix} \tag{2.31}$$

with X_i being $G(0, v^2)$ and independent. Solving for the eigenvalues of H, we get $\lambda_1 + \lambda_2 = 2X_1$ and $\lambda_1\lambda_2 = X_1^2 - X_2^2 - X_3^2 - X_4^2$. Therefore $S^2 = (\lambda_1 - \lambda_2)^2 = 4(X_2^2 + X_3^2 + X_4^2)$. With $x_2 = 2X_2$, $x_3 = 2X_3$ and $x_4 = 2X_4$, we have x_2, x_3 and x_4 to be independent $G(0, 4v^2)$ variables. The joint probability distribution function for these is

$$P(x_2, x_3, x_4)dx_2dx_3dx_4 = \frac{1}{(2\pi)^{3/2}(2v)^3} \exp\left(-\frac{x_2^2 + x_3^2 + x_4^2}{8v^2}\right)dx_2dx_3dx_4. \tag{2.32}$$

Transforming to spherical co-ordinates i.e. $x_2 = S\sin\theta\sin\phi$, $x_3 = S\sin\theta\cos\phi$ and $x_4 = S\cos\theta$ with $dx_2dx_3dx_4 = S^2dS\sin\theta d\theta d\phi$,

$$P(S)dS = \frac{S^2}{(2\pi)^{3/2}(2v)^3}dS \int_0^\pi \sin\theta d\theta \int_0^{2\pi} d\phi$$
$$= \frac{S^2}{4\sqrt{2\pi}v^3} \exp\left(-\frac{S^2}{8v^2}\right)dS. \tag{2.33}$$

Note that $\int_0^\infty P(S)dS = 1$ and $\overline{D} = \int_0^\infty SP(S)dS = 8v/\sqrt{2\pi}$. With $\hat{S} = S/\overline{D}$,

$$P(\hat{S})d\hat{S} = \frac{32\hat{S}^2}{\pi^2} \exp\left(-\frac{4\hat{S}^2}{\pi}\right)d\hat{S}; \quad \overline{\hat{S}} = 1. \tag{2.34}$$

Thus, GUE gives quadratic level repulsion with $P(S) \sim S^2$ for S small. The variance of NNSD for GUE is,

$$\sigma^2(0) = \overline{\hat{S}^2} - 1 = \frac{\overline{S^2}}{\overline{D}^2} - 1 = \frac{3\pi}{8} - 1 \simeq 0.178. \tag{2.35}$$

2.1.3 2 × 2 GSE

Here H is quaternion real defined by Eqs. (2.17) and (2.20). These equations and the choice

$$H_0 = \begin{pmatrix} a & b \\ b & c \end{pmatrix}, \qquad H_1 = \begin{pmatrix} 0 & -x \\ x & 0 \end{pmatrix}, \qquad H_2 = \begin{pmatrix} 0 & y \\ -y & 0 \end{pmatrix},$$
$$H_3 = \begin{pmatrix} 0 & -z \\ z & 0 \end{pmatrix} \tag{2.36}$$

will give

$$H = \begin{bmatrix} a & 0 & b+iz & ix-y \\ 0 & a & ix+y & b-iz \\ b-iz & y-ix & c & 0 \\ -y-ix & b+iz & 0 & c \end{bmatrix}. \tag{2.37}$$

Eigenvalue equation for this matrix is,

$$(a-\lambda) \begin{vmatrix} a-\lambda & ix+y & b-iz \\ y-ix & c-\lambda & 0 \\ b+iz & 0 & c-\lambda \end{vmatrix} + (b+iz) \begin{vmatrix} 0 & a-\lambda & b-iz \\ b-iz & y-ix & 0 \\ -y-ix & b+iz & c-\lambda \end{vmatrix}$$
$$+ (y-ix) \begin{vmatrix} 0 & a-\lambda & ix+y \\ b-iz & y-ix & c-\lambda \\ -y-ix & b+iz & 0 \end{vmatrix} = 0. \tag{2.38}$$

Simplifying, we obtain $\{(a-\lambda)(c-\lambda) - (b^2+z^2+y^2+x^2)\}^2 = 0$ which implies that λ's are doubly degenerate and they are given by,

$$\lambda = \frac{(a+c) \pm [(a-c)^2 + 4(b^2+z^2+y^2+x^2)]^{1/2}}{2}. \tag{2.39}$$

This gives $S = |\lambda_1 - \lambda_2| = [(a-c)^2 + 4(b^2+z^2+y^2+x^2)]^{1/2}$. Let us define $X_1 = a+c$, $X_2 = a-c$, $X_3 = 2b$, $X_4 = 2x$, $X_5 = 2y$ and $X_6 = 2z$. The X_i's are independent Gaussian variables $G(0, 4v^2)$. Note that a and c are $G(0, 2v^2)$ and b, x, y, z are $G(0, v^2)$. Thus $S^2 = X_2^2 + X_3^2 + X_4^2 + X_5^2 + X_6^2$. Transforming to spherical polar co-ordinates in 5 dimensions with hyper-radius S gives $X_2 = S\cos\theta_1\cos\theta_2\cos\theta_3\cos\theta_4$, $X_3 = S\cos\theta_1\cos\theta_2\cos\theta_3\sin\theta_4$, $X_4 = S\cos\theta_1\cos\theta_2\sin\theta_3$, $X_5 = S\cos\theta_1\sin\theta_2$ and $X_6 = S\sin\theta_1$ (volume element being $dv = S^4 dS \cos^3\theta_1 \cos^2\theta_2 \cos\theta_3 \, d\theta_1\, d\theta_2\, d\theta_3\, d\theta_4$). Then,

$$P(S)dS = \frac{S^4 dS}{(2\pi)^{5/2}(4v^2)^{5/2}} \exp\left(-\frac{S^2}{8v^2}\right) \int_{-\pi/2}^{+\pi/2} \cos^3\theta_1 d\theta_1 \int_{-\pi/2}^{+\pi/2} \cos^2\theta_2 d\theta_2$$
$$\times \int_{-\pi/2}^{+\pi/2} \cos\theta_3 d\theta_3 \int_0^{2\pi} d\theta_4. \tag{2.40}$$

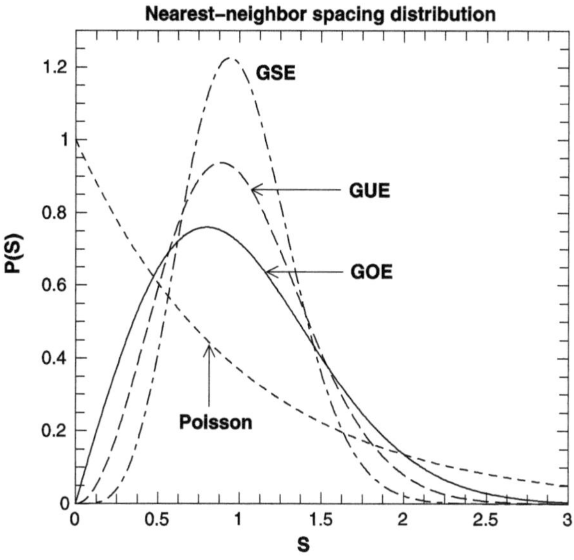

Fig. 2.1 Nearest-neighbor spacing distributions for GOE, GUE and GSE ensembles compared with Poisson

Simplifying, we have

$$P(S)dS = \frac{S^4}{48\sqrt{2\pi}\,v^5} \exp\left(-\frac{S^2}{8v^2}\right) dS. \tag{2.41}$$

Note that $\int_0^\infty P(S)dS = 1$ and $\overline{D} = \int_0^\infty S P(S)dS = 32v/3\sqrt{2\pi}$. With $\hat{S} = S/\overline{D}$, the NNSD is

$$P(\hat{S})d\hat{S} = \frac{2^{18}}{3^6\pi^3} \hat{S}^4 \exp\left(-\frac{64\hat{S}^2}{9\pi}\right) d\hat{S}; \quad \overline{\hat{S}} = 1. \tag{2.42}$$

Thus, GSE generates quartic level repulsion with $P(S) \sim S^4$ for S small. The variance of the NNSD is

$$\sigma^2(0) = \overline{\hat{S}^2} - 1 = \frac{\overline{S^2}}{\overline{D}^2} - 1 = \frac{45\pi}{128} - 1 \simeq 0.105. \tag{2.43}$$

Figure 2.1 shows NNSD for GOE, GUE and GSE as given by Eqs. (2.29), (2.34) and (2.42) respectively. More importantly, these random matrix forms for NNSD are seen in many different quantum systems such as nuclei, atoms, molecules etc. In addition, the RMT results for NNSD are also seen in microwave cavities, aluminum blocks, vibrating plates, atmospheric data, stock market data and so on. Figure 2.2 shows some examples. It is important to stress that the simple 2×2 matrix results are indeed very close, as proved by Mehta (see [5]), to the NNSD for any $N \times N$ matrix ($N \to \infty$). Thus, level repulsion given by random matrices is well established in real systems.

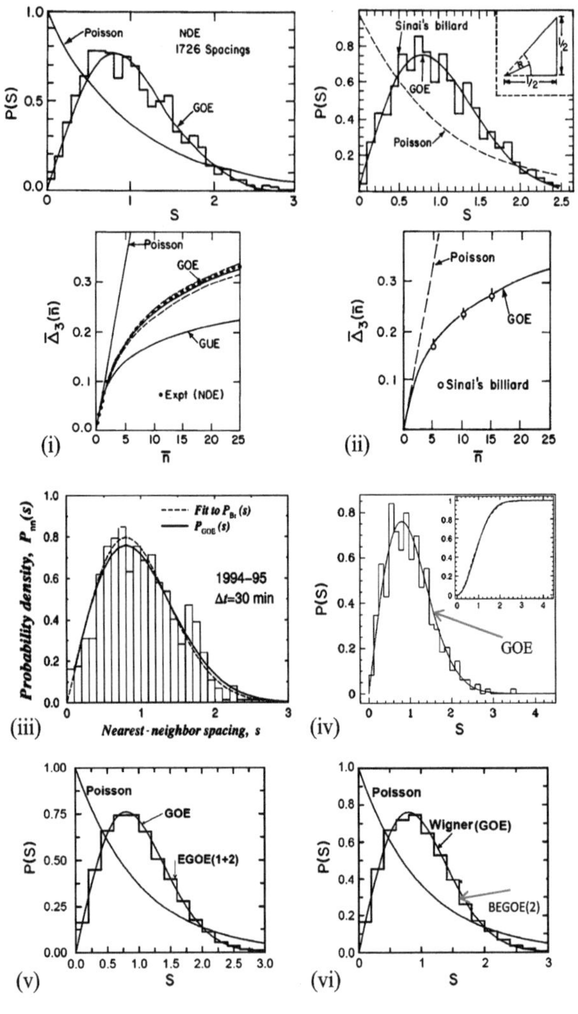

Fig. 2.2 Figure showing NNSD for different systems and their comparison with RMT. (**i**) Nuclear data ensemble [6]; (**ii**) chaotic Sinai billiard [7]; (**iii**) example from Econophysics [8]; (**iv**) example from atmospheric science [9]; (**v**) EGOE(1 + 2) ensemble for fermion systems [10]; (**vi**) BEGOE(2) ensemble for boson systems [11]. In (**i**) and (**ii**) results for Δ_3 statistic are also shown and these are from [12] and [7] respectively. In (**iii**), the NNSD is for the eigenvalues of the cross correlation matrix for 30-min returns of 1000 US stocks for the 2-yr period 1994–1995. Here a fit to the Brody distribution [Eq. (3.47)] is also shown. Similarly in (**iv**) the NNSD is for the eigenvalues of the correlation matrix for monthly mean sea-level pressure for the Atlantic domain from 1948 to 1999. Shown in the insect is the cumulative distribution for monthly and daily averaged correlation matrix. Finally, the embedded ensembles EGOE(1 + 2) in (**v**) and BEGOE(2) in (**vi**) are discussed in detail in Chaps. 5 and 9 respectively. Figures (**i**)–(**iv**) (except the NNSD figure for nuclear data ensemble and this is taken from [7] with permission from Springer) are taken from the above references with permission from American Physical Society and figures (**v**) and (**vi**) with permission from Elsevier

2.2 One and Two Point Functions: $N \times N$ Matrices

For more insight into the Gaussian ensembles and for the analysis of data, we will consider one and two point functions in the eigenvalues. Although we consider only GOE in this section, many of the results extend to GUE and GSE [13]. Also Appendix B gives some properties of univariate and bivariate distributions in terms of their moments and cumulants and these are used throughout this book. A general reference here is Kendall's book [14].

2.2.1 One Point Function: Semi-circle Density

Say energies, in a spectra, are denoted by x (equivalently, the eigenvalues of the corresponding H matrix). Then, number of states in a given interval Δx around the energy x defines $\rho(x)dx$, where $\rho(x)$ is normalized to unity. In fact the number of states in Δx interval around x is $I(x)\Delta x = d\rho(x)dx$ where d is total number of states. Carrying out ensemble averaging (for GOE, GUE and GSE), we have $\overline{\rho(E)}$ defined accordingly. Given $\overline{\rho(E)}$, the average spacing $\overline{D(E)} = [d\overline{\rho(E)}]^{-1}$. If we start with the joint probability distribution for the matrix elements of Gaussian ensembles and convert this into joint distribution for eigenvalues $P_N(E_1, E_2, \ldots, E_N)$, one sees [1]

$$P_N(E_1, E_2, \ldots, E_N) \propto \prod_{i<j} |E_i - E_j|^\beta \exp\left\{-\alpha \sum_i E_i^2\right\}. \tag{2.44}$$

Here $\beta = 1, 2$ and 4 for GOE, GUE and GSE respectively. Then $\overline{\rho(E)}$ is the integral of $\rho(E_1, E_2, \ldots, E_N)$ over all E's except one E. For completeness, let us point out that $\rho(x) = \langle \delta(H - x) \rangle = d^{-1} \sum_{i=1}^d \delta(E_i - x)$. One can construct $\overline{\rho(E)}$ via its moments $\overline{M_p} = \overline{\langle H^p \rangle} = d^{-1}\overline{\text{Trace}(H^p)}$.

Using the binary correlation approximation (BCA), used first by Wigner [15], it is possible to derive for $\overline{\langle H^p \rangle}$ a recursion relation. In the present book most results, both for classical and embedded ensembles, are derived using BCA. In fact, for the embedded ensembles BCA is the only tractable method available at present for deriving formulas for higher moments (even though this also has limitations as discussed in later chapters).

In BCA, only terms that contain squares (but not any other power) of a given matrix element are considered and in the $N \to \infty$ limit, only these terms will survive. As the matrix elements are zero centered, all $\overline{M_p}$ for p odd will be zero. Firstly, for $p = 2$ we have

$$\langle H^2 \rangle = d^{-1} \sum_{i,j} H_{ij} H_{ji} = d^{-1} \sum_{i,j} (H_{ij})^2. \tag{2.45}$$

Let us say that the variance of the matrix elements for the GOE matrices is v^2. Then the ensemble averaged second moment (note that all the moments are central moments as the centroid is zero) is,

$$\overline{M_2} = \overline{\langle H^2 \rangle} = v^2 d = \langle \underbrace{H H} \rangle . \tag{2.46}$$

In Eq. (2.46) we have introduced a notation for correlated matrix elements. Also note that we have ignored the fact that the diagonal elements have variance $2v^2$ as this will give a correction of $1/d$ order and this will vanish as $d \to \infty$. The first complicated moment is M_4 or $\langle H^4 \rangle$. Explicitly,

$$\langle H^4 \rangle = d^{-1} \sum_{i,j,k,l} H_{ij} H_{jk} H_{kl} H_{lk} . \tag{2.47}$$

Then, ensemble average gives three terms in the above sum (it contains product of four matrix elements): (i) first two H matrix are correlated and similarly the last two matrix elements; (ii) the first and third and similarly the second and the fourth matrix element are correlated; (iii) the first and fourth and similarly the second and the third matrix element are correlated. Symbolically they can be written as $\langle \underbrace{H H}\,\underbrace{H H} \rangle$, $\langle \underbrace{H \underbrace{H H} H}_{1,3;\,2,4} \rangle$ and $\langle \underbrace{H \underbrace{H H} H}_{1,4;\,2,3} \rangle$. Their values are

$$\begin{aligned}
\overline{\langle \underbrace{H H}\,\underbrace{H H} \rangle} &= d^{-1} \sum_{ijl} \overline{H_{ij} H_{ji} H_{il} H_{li}} = v^4 d^2 = \left(\overline{\langle H^2 \rangle} \right)^2 \\
\overline{\langle \underbrace{H \underbrace{H H} H}_{1,3;\,2,4} \rangle} &= d^{-1} \sum_{ij} \overline{H_{ij} H_{ji} H_{ij} H_{ji}} = v^4 d = d^{-1} \left(\overline{\langle H^2 \rangle} \right)^2 \\
\overline{\langle \underbrace{H \underbrace{H H} H}_{1,4;\,2,3} \rangle} &= d^{-1} \sum_{ijk} \overline{H_{ij} H_{jk} H_{kj} H_{ji}} = v^4 d^2 = \left(\overline{\langle H^2 \rangle} \right)^2 .
\end{aligned} \tag{2.48}$$

Thus the second term that involves cross correlation with odd number of H's inside [in the second term in Eq. (2.48), there is one H in between] will vanish as $d \to \infty$. Then, we have

$$\overline{M_4} = \overline{\langle H^4 \rangle} = v^4 d^2 = 2 \overline{\langle \underbrace{H H}\,\underbrace{H H} \rangle} = 2 \left(\overline{\langle H^2 \rangle} \right)^2 . \tag{2.49}$$

Note that in Eqs. (2.45)–(2.49), the correlated H's are joined by the symbol '⊔'. Continuing the above procedure we have [16, 17], valid for all three ensembles with the normalization $\overline{\langle H^2 \rangle} = 1$,

$$\begin{aligned}
\overline{M_p} = \overline{\langle H^p \rangle} &= \sum_{\tau=0}^{p-2} \overline{\langle \underbrace{H H^{\tau} H} H^{p-\tau-2} \rangle} \\
&= \sum_{\tau=0}^{p-2} \overline{\langle H^2 \rangle \langle H^{\tau} \rangle \langle H^{p-\tau-2} \rangle} \\
&= \sum_{\tau=0}^{p-2} \overline{M_{\tau} M_{p-\tau-2}} .
\end{aligned} \tag{2.50}$$

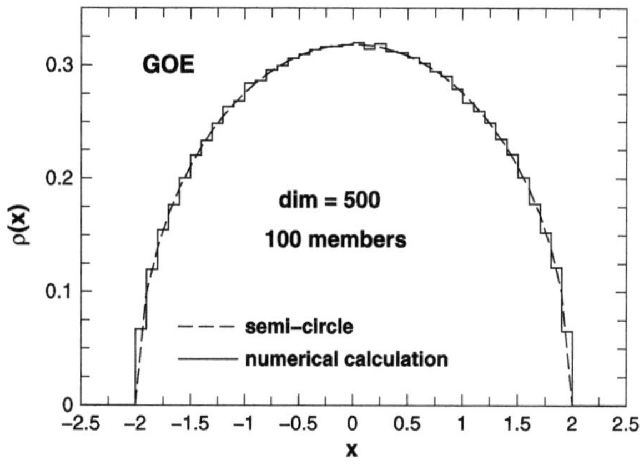

Fig. 2.3 Semi-circle density of eigenvalues for a 500 dimensional GOE ensemble. The x-axis corresponds to standardized (zero center and unit variance) eigenvalues

The solution is $M_{2\nu+1} = 0$ and $M_{2\nu} = (\nu+1)^{-1}\binom{2\nu}{\nu}$. They are the Catalan numbers and it can be verified easily that they are the moments of a semi-circle. Thus $\overline{\rho(E)}$ is a semi-circle,

$$\overline{\rho(x)} = \frac{1}{2\pi}\left(4 - x^2\right)^{1/2} = \frac{1}{\pi}\sin\psi(x). \tag{2.51}$$

Here $\psi(x)$ is the angle between the x axis and the radius vector; $x = -2\cos\psi(x)$, $0 \le \psi \le \pi$. Note that $\overline{\rho(x)}$ vanishes for $|x| > 2$ and $M_2 = 1$. Figure 2.3 shows an example for the semi-circle.

Given a $\rho(x)$, one can define the distribution function $F(x)$,

$$F(x) = \int_{-\infty}^{x} \rho(y)dy. \tag{2.52}$$

With $\rho(x)$ being discrete, as the spectrum is discrete, $F(x)$ is a staircase. Note that $F(x)$ counts the number of levels up to the energy x and therefore increase by one unit as we cross each energy level (if there are no degeneracies).

Another important property is that the exact density $\rho(x)$ can be expanded in terms of $\overline{\rho(x)}$ by using the polynomials defined with respect to $\overline{\rho(x)}$,

$$\rho(x) = \overline{\rho(x)}\left\{1 + \sum_{\zeta \ge 1} S_\zeta P_\zeta(x)\right\}. \tag{2.53}$$

If we know the moments M_r of $\overline{\rho(x)}$, the polynomials $P_\zeta(x)$ defined by $\overline{\rho(x)}$ can be constructed [18] such that $\int_{-\infty}^{+\infty} \overline{\rho(x)}\, P_\zeta(x)\, P_{\zeta'}(x)dx = \delta_{\zeta\zeta'}$. Using this one can study level motion in Gaussian ensembles as described ahead.

2.2.2 Two Point Function $S^\rho(x, y)$

The two point function is defined by

$$S^\rho(x,y) = \overline{\rho(x)\rho(y)} - \overline{\rho(x)}\,\overline{\rho(y)}. \tag{2.54}$$

If there are no fluctuations $S^\rho(x, y) = 0$. Thus S^ρ measures fluctuations. The moments of S^ρ are

$$\begin{aligned} M_{pq} &= \int\int x^p y^q S^\rho(x,y)dxdy \\ &= \overline{\langle H^p\rangle\langle H^q\rangle} - \overline{\langle H^p\rangle}\,\overline{\langle H^q\rangle}. \end{aligned} \tag{2.55}$$

To derive $S^\rho(x, y)$ (also M_{pq}) we consider, as in [13, 17] the polynomials defined by the semi-circle, i.e. Chebyshev polynomials of second kind $U_n(x)$. They are defined in terms of the sum,

$$U_n(x) = \sum_{m=0}^{[n/2]} (-1)^m \binom{n-m}{m} (2x)^{n-2m}. \tag{2.56}$$

They satisfy the orthonormality condition $\int_{-1}^{+1} \omega(x)U_n(x)U_m(x)dx = \pi/2\,\delta_{mn}$ with $\omega(x) = \sqrt{1-x^2}$. Substituting $y = 2x$ in the orthonormality condition and using $V_n(y) = U_n(y/2)$, we obtain,

$$\begin{aligned} &\int_{-2}^{+2} dy\overline{\rho(y)}V_n(y)V_m(y) = \delta_{nm}; \\ &\overline{\rho(y)} = \frac{1}{\pi}\sqrt{1-\frac{y^2}{4}} = \frac{\sin\psi(x)}{\pi}. \end{aligned} \tag{2.57}$$

Note also that $U_n(x) = \frac{1}{a_n\omega(x)}\frac{\partial^n}{\partial x^n}[\omega(x)[g(x)]^n]$; $a_n = (-1)^n 2^{n+1}\frac{\Gamma(n+3/2)}{(n+1)\sqrt{\pi}}$, $g(x) = 1-x^2$. Similarly, $V_\zeta(x) = (-1)^\zeta[\sin\psi(x)]^{-1}\sin(\zeta+1)\psi(x)$.

Returning to M_{pq} it is seen that in $\overline{\langle H^p\rangle\langle H^q\rangle}$ evaluation (again we use BCA) it should be recognized that the correlations come when say ζ number of H's in H^p correlate with ζ number of H's in H^q. When $\zeta = 0$ we get $\{\overline{\langle H^p\rangle}\}\{\overline{\langle H^q\rangle}\}$. Therefore

$$M_{pq} = \overline{\langle H^p\rangle\langle H^q\rangle} - \overline{\langle H^p\rangle}\,\overline{\langle H^q\rangle} = \sum_{\zeta=1}^{<(p,q)} \mu_\zeta^p\mu_\zeta^q\overline{\langle H^\zeta\rangle\langle H^\zeta\rangle} \tag{2.58}$$

and μ_ζ^p are obtained by a counting argument. French, Mello and Pandey [17] showed that (see also [19]),

$$\mu_\zeta^p = \binom{p}{(p-\zeta)/2} = -\zeta^{-1}\int x^p\frac{d}{dx}\{\overline{\rho(x)}V_{\zeta-1}(x)\}dx. \tag{2.59}$$

Now let us evaluate $\overline{\langle H^{\zeta}\rangle\langle H^{\zeta}\rangle}$. Firstly $\langle H^{\zeta}\rangle = d^{-1}\sum_{i,j,k,\ldots=1}^{d} H_{ij}H_{jk}\ldots H_{ni}$. Then the number of indices are ζ and the number of terms in the sum are d^{ζ}. In $\overline{\langle H^{\zeta}\rangle\langle H^{\zeta}\rangle}$ there will be d^{ζ} terms of the type $\langle H_{ij}H_{ji}\rangle\langle H_{jk}H_{kj}\rangle\ldots$. Choosing $v^2 d = 1$, we have $\langle H_{ij}^2\rangle = v_{ij}^2 = v^2 = 1/d$. Therefore $\overline{\langle H^{\zeta}\rangle\langle H^{\zeta}\rangle} = \frac{1}{d^2}d^{\zeta}/d^{\zeta} = 1/d^2$. However for every H_{ij} there are $\langle H_{ij}H_{ij}\rangle$ and $\langle H_{ij}H_{ji}\rangle$ for GOE. Both give v^2 and therefore $\overline{\langle H^{\zeta}\rangle\langle H^{\zeta}\rangle} = 2/d^2$ for GOE. In case of GUE $\overline{H_{ij}H_{ij}} = \overline{(a+ib)(a+ib)} = \overline{a^2} - \overline{b^2} + 2i\overline{a}\overline{b} = 0$ and $\overline{H_{ij}H_{ji}} = \overline{(a+ib)(a-ib)} = \overline{a^2} + \overline{b^2} = 1$ ($|H_{ij}|^2 = v_{ij}^2 = v^2$ and $v^2 d = 1$). Thus $\overline{\langle H^{\zeta}\rangle\langle H^{\zeta}\rangle} = 2/\beta d^2$. In addition there is cyclic invariance and therefore there is an additional ζ factor [for example, $H_{12}H_{21} \oplus H_{21}H_{12}$, $H_{12}H_{23}H_{31} \oplus H_{23}H_{31}H_{12} \oplus H_{31}H_{12}H_{23}$]. Then the final result is,

$$\overline{\langle H^{\zeta}\rangle\langle H^{\zeta}\rangle} = \frac{2\zeta}{\beta d^2} \tag{2.60}$$

with $\beta = 1$ for GOE and $\beta = 2$ for GUE. Putting this result in the expression for M_{pq} we see that

$$\begin{aligned}\overline{M_{pq}} &= \sum_{\zeta \geq 1} \frac{2\zeta}{\beta d^2 \zeta^2} \iint x^p y^q \frac{\partial}{\partial x}\frac{\partial}{\partial y}\overline{\rho(x)}\,\overline{\rho(y)}V_{\zeta-1}(x)V_{\zeta-1}(y)dxdy \\ &= \frac{2}{\beta d^2}\sum_{\zeta \geq 1} \zeta^{-1} \iint x^p y^q \frac{\partial}{\partial x}\frac{\partial}{\partial y}\overline{\rho(x)}\,\overline{\rho(y)}V_{\zeta-1}(x)V_{\zeta-1}(y)dxdy.\end{aligned} \tag{2.61}$$

Then by inversion we get,

$$\begin{aligned} S^{\rho}(x,y) &\overset{\beta=1;\text{GOE}}{=} \frac{2}{d^2}\frac{\partial}{\partial x}\frac{\partial}{\partial y}\left\{\overline{\rho(x)}\,\overline{\rho(y)}\sum_{\zeta\geq 1}\zeta^{-1}V_{\zeta-1}(x)V_{\zeta-1}(y)\right\} \\ \Rightarrow \quad S^F(x,y) &= \frac{2}{d^2}\overline{\rho(x)}\,\overline{\rho(y)}\left\{\sum_{\zeta\geq 1}\zeta^{-1}V_{\zeta-1}(x)V_{\zeta-1}(y)\right\} \\ &= \frac{2}{\pi^2 d^2}\sum_{\zeta\geq 1}\zeta^{-1}\sin\zeta\psi_1(x)\sin\zeta\psi_2(y). \end{aligned} \tag{2.62}$$

Note that $S^F(x,y) = \int_{-\infty}^{x}\int_{-\infty}^{y} S^{\rho}(x',y')dy'dx' = \overline{F(x)F(y)} - \overline{F(x)}\,\overline{F(y)}$. The sum can be simplified by introducing a cut-off $e^{-\alpha\zeta}$ and extending the sum to $\zeta = 1$ to ∞,

$$\begin{aligned} &\sum_{\zeta=1}^{d}\zeta^{-1}\sin\zeta\psi_1\sin\zeta\psi_2 \\ &\quad = \sum_{\zeta=1}^{\infty}\zeta^{-1}\exp(-\alpha\zeta)\sin\zeta\psi_1\sin\zeta\psi_2 \end{aligned}$$

$$
\begin{aligned}
&= \sum_{\zeta=1}^{\infty}\left[\int_{\alpha}^{\infty} e^{-z\zeta} dz\right] \sin\zeta\psi_1 \sin\zeta\psi_2 \\
&= \text{Re}\left\{\frac{1}{2}\sum_{\zeta=1}^{\infty}\left[\int_{\alpha}^{\infty} e^{-z\zeta} dz\right]\left[e^{-i\zeta(\psi_1-\psi_2)} - e^{-i\zeta(\psi_1+\psi_2)}\right]\right\} \\
&= \text{Re}\left\{\frac{1}{2}\sum_{\zeta=1}^{\infty}\int_{\alpha}^{\infty}\left\{e^{-\zeta(z+i(\psi_1-\psi_2))} - e^{-\zeta(z+i(\psi_1+\psi_2))}\right\}dz\right\} \\
&= \text{Re}\frac{1}{2}\left[\ln\left[1-e^{-(z+i(\psi_1-\psi_2))}\right]_{\alpha}^{\infty} - \ln\left[1-e^{-(z+i(\psi_1+\psi_2))}\right]_{\alpha}^{\infty}\right] \\
&= \frac{1}{4}\ln\frac{1+e^{-2\alpha}-2e^{-\alpha}\cos(\psi_1+\psi_2)}{1+e^{-2\alpha}-2e^{-\alpha}\cos(\psi_1-\psi_2)},
\end{aligned}
\tag{2.63}
$$

where we have used the property that $\text{Re}[\ln(1+z)] = \frac{1}{2}\ln[(1+z)(1+z^*)]$. Finally have,

$$
S^F(x,y) = \frac{1}{2\pi^2 d^2}\ln\frac{1+e^{-2\alpha}-2e^{-\alpha}\cos(\psi_1+\psi_2)}{1+e^{-2\alpha}-2e^{-\alpha}\cos(\psi_1-\psi_2)}. \tag{2.64}
$$

Let us consider the structure of $S^F(x,y)$ when $x \sim y$. Firstly the number of levels r in the energy interval $\Delta x = dx = x - y$ is $d\,\rho(x)dx$. But $1 - x^2/4 = \sin^2\psi$ gives $x = 2\cos\psi$. Therefore, $dx = 2\sin\psi d\psi$ and

$$
r = d\rho(x)dx = \left[\frac{d}{\pi}\sin\psi\right] 2\sin\psi d\psi. \tag{2.65}
$$

Then, $S^F(x,y)$ is, with $x \sim y \Rightarrow \sin\psi$, $\psi_1 + \psi_2 \sim 2\psi$ and $\psi_1 - \psi_2 \sim \delta\psi$,

$$
\begin{aligned}
S^F(x,y) &\overset{\alpha\sim 1/d}{\rightarrow} \frac{1}{2\pi^2 d^2}\ln\frac{2-2\cos 2\psi}{2-2\cos\delta\psi} \\
&= \frac{1}{\pi^2 d^2}\ln\frac{\sin\psi}{\sin\delta\psi/2} \simeq \frac{1}{\pi^2 d^2}\ln\frac{2\sin\psi}{\delta\psi} \\
&= \frac{1}{\pi^2 d^2}\ln\frac{4d\sin^3\psi}{\pi r}.
\end{aligned}
\tag{2.66}
$$

In the last step, Eq. (2.65) is used. Therefore the behavior of $S^F(x,y)$ is that it behaves as $\ln r$. Now let us consider the self-correlation term $S^F(x,x)$ which determines the level motion $\delta x^2/\overline{D}^2 = d^2 S^F(x,x)$ in GOE,

$$\begin{aligned}
S^F(x,x) &= \frac{1}{2\pi^2 d^2} \ln \frac{1+e^{-2\alpha} - 2e^{-\alpha}\cos 2\psi}{1+e^{-2\alpha} - 2e^{-\alpha}} \\
&\stackrel{\alpha\sim 1/d}{\Rightarrow} \frac{1}{2\pi^2 d^2} \ln \frac{2(1-\cos 2\psi)}{(1-e^{-\alpha})^2} e^{-\alpha} = 1-\alpha \\
&= \frac{1}{2\pi^2 d^2} \ln \frac{4\sin^2\psi}{\alpha^2} \\
&= \frac{1}{2\pi^2 d^2} \ln (2d\sin\psi)^2 \\
\Rightarrow \quad S^F(x,x) &= \frac{1}{\pi^2 d^2} \ln 2d\sin\psi = \frac{1}{\pi^2 d^2} \ln 2\pi d\overline{\rho(x)}.
\end{aligned} \tag{2.67}$$

Before going further, it is useful to point out that the moment method with BCA used for deriving the asymptotic form of one and two point functions extends to many other random matrix ensembles. A recent discussion on the power of the moment method in random matrix theory is given in [20].

2.2.3 Fluctuation Measures: Number Variance $\Sigma^2(r)$ and Dyson-Mehta Δ_3 Statistic

$P(S)$, the nearest neighbor spacing distribution and its variance $\sigma^2(0)$ are measures of fluctuations. Using $S^F(x,y)$ we can define a new measure called 'number variance' $\Sigma^2(r)$. Say in a energy interval x to y there are r levels, then $r = d[F(y) - F(x)]$. The statistic $\Sigma^2(r)$ is ensemble averaged variance of r and

$$\begin{aligned}
\Sigma^2(r) &= \overline{r^2} - (\overline{r})^2 \\
&= d^2\left[\overline{\left(F(y)-F(x)\right)^2} - \left(\overline{F(y)-F(x)}\right)^2\right] \\
&= d^2\left[\overline{F^2(y)} - \overline{F(y)}^2 + \overline{F^2(x)} - \overline{F(x)}^2 - 2\left(\overline{F(x)F(y)} - \overline{F(x)}\,\overline{F(y)}\right)\right] \\
&= d^2\left[S^F(x,x) + S^F(y,y) - 2S^F(x,y)\right].
\end{aligned} \tag{2.68}$$

Thus $\Sigma^2(r)$ is an exact "two-point measure". Using the asymptotic expressions for $S^F(x,x)$ and $S^F(x,y)$, we obtain

$$\begin{aligned}
\Sigma^2(r) &= d^2\left[\frac{2}{\pi^2 d^2}\ln 2d\sin\psi - \frac{2}{\pi^2 d^2}\ln\frac{4d\sin^3\psi}{\pi r}\right] \\
&= \frac{2}{\pi^2}\ln\frac{\pi r}{2\sin^2\psi}.
\end{aligned} \tag{2.69}$$

Thus $\Sigma^2(r)$ behaves as $\ln r$ and hence the GOE spectrum is rigid. The exact results valid at the center of the semi-circle are:

$$\begin{aligned}\Sigma^2_{\mathrm{GOE}}(r) &= \frac{2}{\pi^2}\left[\ln 2\pi r + 1 + \gamma - \frac{\pi^2}{8}\right] + O\left(\frac{1}{r}\right)\\ \Sigma^2_{\mathrm{GUE}}(r) &= \frac{1}{\pi^2}[\ln 2\pi r + 1 + \gamma]\end{aligned} \tag{2.70}$$

where γ is Euler's constant. Other important measure is Dyson-Mehta Δ_3 statistic and importantly, it is related to $\Sigma^2(r)$ and hence it is also a two-point measure.

Dyson and Mehta Δ_3 statistic [21] is defined as the mean square deviation of $F(E)$, of the unfolded spectrum, from the best fit straight line and $\Delta_3(\overline{n})$ corresponds to the same but over a spectrum of length $\overline{n}\overline{D}$. The ensemble averaged $\overline{\Delta_3}(\overline{n})$ is then defined similarly,

$$\overline{\Delta_3}(\overline{n}) = \overline{\Delta_3}(2L) = \min\left[\frac{1}{2L}\int_{x-L}^{x+L}\left[dF(y) - Ay - B\right]^2 dy\right]_{(A,B)}. \tag{2.71}$$

Here $L = \frac{\overline{n}}{2}$. The $\overline{\Delta_3}(\overline{n})$ statistic is an exact two-point measure. In fact it can be written as an integral involving $\Sigma^2(r)$. This is proved in Appendix C. Using the GOE (similarly GUE) expression for $\Sigma^2(r)$ and applying Eq. (C.8), we obtain the following expression for $\overline{\Delta_3}(\overline{n})$ for GOE and GUE,

$$\begin{aligned}\left[\overline{\Delta_3}(\overline{n})\right]_{\mathrm{GOE}} &= \frac{1}{\pi^2}\left[\ln(2\pi\overline{n}) + \gamma - \frac{5}{4} - \frac{\pi^2}{8}\right] + O\left(\frac{1}{\pi^2\overline{n}}\right),\\ \left[\overline{\Delta_3}(\overline{n})\right]_{\mathrm{GUE}} &= \frac{1}{2\pi^2}\left[\ln(2\pi\overline{n}) + \gamma - \frac{5}{4}\right] + O\left(\frac{1}{\pi^2\overline{n}}\right).\end{aligned} \tag{2.72}$$

For a novel application of $\overline{\Delta_3}(\overline{n})$ statistic, see [22].

2.3 Structure of Wavefunctions and Transition Strengths

2.3.1 Porter-Thomas Distribution

Given a transition operator T (this should not be confused with the time reversal operator 'T' used before or the isospin label 'T'), transition strength connecting two eigenstates is defined by $|\langle E_f \mid T \mid E_i\rangle|^2$. In nuclei T's of interest are electromagnetic (magnetic dipole, electric quadrupole for example), one particle addition or removal, Gamow-Teller operator and so on. Similarly, in atoms and molecules dipole operator is very important. It is also important to recognize that the widths of resonances also measure transition strengths. Leaving detailed discussion on transition strength distributions to Ref. [13], here we will give only some basic results. One can think of $T|E\rangle$ to be a compound state and represent it by a basis state $|i\rangle$. Therefore, statistical properties of transition strengths will be same as those of the expansion coefficients C_i^E of the eigenstates in terms of the basis states $|i\rangle$,

$$|E\rangle = \sum_{i=1}^{d} C_i^E |i\rangle. \tag{2.73}$$

Now an important question is: what is the distribution of $|x_i|^2 = |C_i^E|^2$. First, let us consider the joint distribution $P(x_1, x_2, x_3, \dots, x_d)$ of x_1, x_2, $x_3, \dots, x_d$ for GOE. Because the GOE is an orthogonally invariant ensemble, the eigenvectors uniformly cover the d-dimensional unit sphere. Then the normalization condition $\sum_i |x_i|^2 = 1$ gives,

$$P(x_1, x_2, x_3, \dots, x_d) = \frac{\Gamma(d/2)}{(\pi)^{d/2}} \delta\left(\sum_{i=1}^{d} x_i^2 - 1\right). \tag{2.74}$$

Now, integrating over all but one variable (say x_1 and denote it by x) will give

$$\rho(x)dx = \frac{1}{\sqrt{\pi}} \frac{\Gamma(\frac{d}{2})}{\Gamma(\frac{d-1}{2})} \left(1 - x^2\right)^{\frac{d-3}{2}} dx. \tag{2.75}$$

Then, in the $d \to \infty$ limit we obtain,

$$\rho(x)dx = \sqrt{\frac{d}{2\pi}} \exp - \frac{dx^2}{2} dx, \quad -\infty \le x \le \infty. \tag{2.76}$$

Thus, asymptotically x will be zero centered Gaussian variables with variance $1/d$. As $\overline{|C_i^E|^2}$ should not depend on the index i and $\sum_i |C_i^E|^2 = 1$ will give us the significant result that for GOE $\overline{|C_i^E|^2} = 1/d$. Therefore, the distribution of the renormalized strengths $z = |C_i^E|^2 / \overline{|C_i^E|^2}$ is, putting $dx^2 = z$ in Eq. (2.76),

$$\rho(z)dz = \frac{1}{\sqrt{2\pi}} z^{-\frac{1}{2}} \exp\left(-\frac{z}{2}\right) dz, \quad 0 \le z \le \infty \tag{2.77}$$

and this is nothing but χ_1^2 distribution. Equation (2.77), for GOE, is called Porter-Thomas (P-T) law for strengths [23]. Thus locally renormalized strengths, $z = |C_j^i|^2 / \overline{|C_j^i|^2}$, follow P-T law; note that $\int_0^\infty z\rho(z)dz = 1$. The GOE P-T law was well tested in many examples as shown in Fig. 2.4. Similar to GOE, for GUE the P-T law is χ_2^2 as C_j^i will have real (say A) and complex (say B) parts with each of them being $G(0, v^2 = 1/d)$ and independent; then $|C_j^i|^2 = A^2 + B^2$.

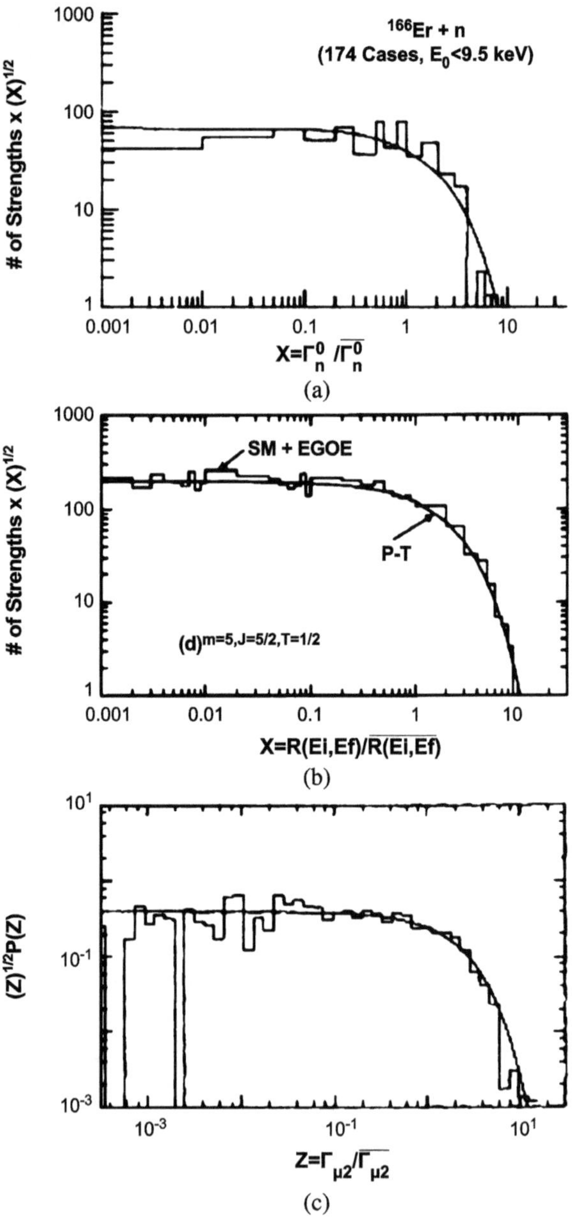

Fig. 2.4 Porter-Thomas distribution for strengths compared with: (**a**) neutron resonance widths in ^{167}Er [13]; (**b**) nuclear shell model transition strengths generated by a special two-body transition operator [10]; (**c**) widths from a microwave resonator [24]. Figures (**a**) and (**c**) are reproduced with permission from American Physical Society and (**b**) with permission from Elsevier

2.3.2 NPC, S^{info} and Strength Functions

With eigenfunctions expanded in terms of some basis states (they form a complete set) $|k\rangle$, let us define the following: (i) NPC (denoted as ξ_2)—number of principal components; (ii) S^{info}—information entropy or ℓ_H—localization length. They are, with $|E\rangle = \sum_{k=1}^{d} C_k^E |k\rangle$,

$$\xi_2(E) = \left[\frac{1}{d\,\rho(E)} \sum_{E'} \sum_{k=1}^{d} |C_k^{E'}|^4 \delta(E - E') \right]^{-1},$$

$$\ell_H(E) = \exp\left[\left(S^{info}\right)_E\right] / (0.48d), \tag{2.78}$$

$$S^{info}(E) = -\frac{1}{d\rho(E)} \sum_{E'} \sum_{k=1}^{d} |C_k^{E'}|^2 \ln |C_k^{E'}|^2 \delta(E - E').$$

In Eq. (2.78), degeneracies of the eigenvalues E are taken into account. It is important to stress that S^{info} is the first and NPC [or the inverse participation ratio (IPR)] the second Rényi entropy introduced in chaos literature [25, 26]. For this reason NPC is denoted by $\xi_2(E)$. Similarly, information entropy is also called Shannon entropy [27]. As we shall see ahead, S^{info} and $\ln \xi_2(E)$ carry the same information and their difference, called structural entropy, is also some times used (with S^{info} or ξ_2) as a chaos measure; see [26] and references therein. For GOE, $\overline{|C_k^E|^2} = \frac{1}{d}$ and C_k^E are Gaussian variables $G(0, \frac{1}{d})$. Therefore $\overline{|C_k^E|^4} = 3(\overline{|C_k^E|^2})^2 = \frac{3}{d^2}$ and then

$$\overline{\xi_2(E)} = \left(\sum_{k=1}^{d} \frac{3}{d^2} \right)^{-1} = \frac{d}{3} \quad \text{for GOE.} \tag{2.79}$$

Thus $\overline{\xi_2(E)}$ is independent of E for GOE. Similarly $\overline{S^{info}(E)} = -d\overline{|C_k^E|^2 \ln |C_k^E|^2}$ where C_k^E are $G(0, \frac{1}{d})$. Then,

$$\begin{aligned}
\overline{S^{info}(E)} &= -\frac{d}{\sqrt{2\pi}\sigma} \int_{-\infty}^{\infty} x^2 \ln x^2 e^{-\frac{x^2}{2\sigma^2}} dx; \quad \sigma^2 = \frac{1}{d} \\
&= -\frac{8d\sigma^2}{\sqrt{\pi}} \int_0^{\infty} y^2 [\ln y + \ln \sqrt{2}\sigma] e^{-y^2} dy; \quad x = \sqrt{2}\sigma y \\
&= -\frac{8}{\sqrt{\pi}} \left[\int_0^{\infty} y^2 \ln y e^{-y^2} dy + \ln \sqrt{\frac{2}{d}} \int_0^{\infty} y^2 e^{-y^2} dy \right] \\
&= -\frac{8}{\sqrt{\pi}} \int_0^{\infty} y^2 \ln y e^{-y^2} dy - \ln \frac{2}{d}
\end{aligned}$$

$$\begin{aligned}
\Rightarrow \quad \exp\left(S^{info}\right) = \frac{d}{2} \exp -\zeta, \qquad \exp -\zeta &= \exp\left(-\frac{8}{\sqrt{\pi}} \int_0^{\infty} x^2 \ln x e^{-x^2} dx \right) \\
&= 4 \exp \gamma - 2 \simeq 0.964 \\
\Rightarrow \quad \exp S^{info}(E) = 0.48d, \qquad \ell_H(E) &= 1.
\end{aligned} \tag{2.80}$$

Thus for GOE, $S^{info} = \ln(0.48d)$ independent of E and the localization length is unity by definition. The NPC and S^{info} formulas are well verified in numerical examples.

Besides NPC and S^{info}, localization properties of wavefunctions can be inferred from strength functions. Given C_k^E, strength functions are defined by,

$$F_k(E) = \sum_{E'} |C_k^{E'}|^2 \delta(E - E') = \overline{|\mathscr{C}_k^E|^2}\, d\rho(E), \tag{2.81}$$

where $\overline{|\mathscr{C}_k^E|^2}$ denotes the average of $|C_k^E|^2$ over the eigenstates with the same energy E. A commonly used form for strength functions is the Breit-Wigner (BW) form and its derivation is given in Appendix D. Many other aspects of transition strengths and strength fluctuations are discussed in [13, 28–30]; see also Appendix E.

2.4 Data Analysis

2.4.1 Unfolding and Sample Size Errors

In the analysis of data one has to pay attention to the following facts: (1) data is available for a given system (say a nucleus); (2) the sample size (number of levels) is usually small (∼100); (3) over the sample, the density may vary. Point (3) is true for shell model data or any other model data. To take care of (1) and (2) it is possible to invoke ergodicity and stationarity properties of the Gaussian ensembles (GE-GOE, GUE or GSE) [13, 31]. Due to ergodicity, we have a permit to compare ensemble averaged results from GE to spectral averaged results from a given spectrum. Similarly due to stationarity, the measures (statistics) of fluctuations will be independent of which part of the spectrum one is looking at. To the extent that the spectra are "complex", we can use "stationarity" to combine the values of the measures (with appropriate weights) to increase the sample size. Let us first consider point #(3). When $\rho(x)$ is varying, it is necessary to remove the "secular variation" before the data is analyzed. This is called "unfolding". For this we have to map the given energies E_i to new energies ε_i such that the ε_i spectrum has constant density. Say $\varepsilon_i = g(E_i)$ such that ε_i has unit mean spacing on the average in the interval $E_i \pm \frac{\Delta E}{2}$. Then, with ΔN levels in the interval $\Delta\varepsilon$,

$$\begin{aligned} \frac{\Delta\varepsilon}{\Delta N} &= 1 = \frac{1}{\Delta N}\left[g\left(E + \frac{\Delta E}{2}\right) - g\left(E - \frac{\Delta E}{2}\right)\right] \\ &= \frac{\Delta E}{\Delta N} g'(E) = \frac{1}{N\overline{\rho(E)}} g'(E), \\ \Rightarrow\quad & g'(E) = N\overline{\rho(E)}. \end{aligned} \tag{2.82}$$

Now integrating Eq. (2.82) on both sides will give the map,

$$g(E) = N\overline{F(E)} \quad \Rightarrow \quad \varepsilon_i = N\overline{F(E_i)}. \tag{2.83}$$

The ε_i's given by Eq. (2.83) will have $\overline{D} = 1$. Significance of the physically (theoretically) defined $\overline{\rho(E)}$ in unfolding has been discussed by Brody et al. [13] and more recently by Jackson et al. [32]. Normally one tries to fit $F(E)$ to a smooth curve, by a least square procedure, to obtain $\overline{F(E)}$, but often this is not proper as fluctuations depend on $\overline{F(E)}$ used in the unfolding procedure.

Coming to the sample size errors, let us consider a measure w calculated over say p levels. Say the theoretical value of w (for infinite sample size) is $\overline{w}$ and its variance $\overline{\text{var}(w)}$. Then the figure of merit $f = \frac{\sqrt{\text{var}(w)_p}}{\langle w \rangle_p} \times 100$. Now $\overline{w} \to \overline{w}\{1 \pm \frac{f}{100}\}$. For $\Sigma^2(\overline{n})$, the Poisson estimate gives $f \sim \sqrt{\frac{2\overline{n}}{p}} \times 100$, thus $f \to 0$ for $p \to \infty$. In practice also used are overlapping intervals. Then it is seen via Monte-Carlo calculations that f reduces by 0.65. Applying the same size error to the GOE analytical results, it is seen that theory and experiment agree almost exactly in the case of Nuclear Data Ensemble (NDE) constructed by combining neutron resonance data from many nuclei [6, 12, 33]. Finally, in practice, for $\overline{\Delta_3(\overline{n})}$ and $\overline{D}$ calculations, it is more useful to use the simple formulas given by French, Pandey, Bohigas and Giannoni [13, 34]. Given a sequence of (ordered) energies $(E_1, E_2, \ldots, E_n)$, the mean spacing $\overline{D}$ is [13],

$$\overline{D} = \frac{12}{n(n^2-1)} \sum_{i=1}^{n} \left(i - \frac{n+1}{2} \right) E_i . \tag{2.84}$$

Similarly consider $\overline{\Delta_3(L)}$ over an energy interval α to $\alpha + L$. Defining $\widetilde{E}_i$ to be $\widetilde{E}_i = E_i - (\alpha + \frac{L}{2})$, we have [34],

$$\begin{aligned} \overline{\Delta_3(\alpha; L)} = {} & \frac{n^2}{16} - \frac{1}{L^2} \left[\sum_{i=1}^{n} \widetilde{E}_i \right]^2 + \frac{3n}{2L^2} \left[\sum_{i=1}^{n} \widetilde{E}_i^2 \right] \\ & - \frac{3}{L^4} \left[\sum_{i=1}^{n} \widetilde{E}_i^2 \right]^2 + \frac{1}{L} \left[\sum_{i=1}^{n} (n - 2i + 1) \widetilde{E}_i \right]. \end{aligned} \tag{2.85}$$

2.4.2 Poisson Spectra

When comparing GOE (or GUE, GSE) results with data, it is important to consider the Poisson case, i.e. uncorrelated spectra so that the effects due to GOE correlations will be clear. We can generate a Poisson spectrum as follows. First generate a set $\{s\}$ such that s is a random variable following $\exp -x$ probability distribution. Then choose $x_1 = 0$ and $x_{n+1} = x_n + s_n$; $n = 1, 2, 3, \ldots$ and draw s_n from $\{s\}$. Now the nearest neighbor spacing $S = s_n$. Therefore $P(S)dS$ for the sequence $(x_1, x_2, x_3, \ldots)$ is

$$P(S)dS = \exp -S dS. \tag{2.86}$$

An important question is what is $\Sigma^2(\overline{n})$ and $\overline{\Delta_3}(\overline{n})$ for a Poisson spectrum. Before proceeding further let us mention that random superposition of several independent spectra leads to Poisson as proved by Porter and Rosenzweig [35]. In order to derive the results for $\Sigma^2(\overline{n})$ and $\Delta_3(\overline{n})$ for a Poisson spectrum, it is useful to consider the $R_2(E_1, E_2)$ and $Y_2(r)$ correlation functions for an unfolded spectrum. $R_2(E_1, E_2)$ is the integral of $P(E_1, E_2, \ldots, E_n)$ over $E_3, E_4, \ldots, E_n$ and Y_2 is simply related to R_2,

$$\begin{aligned} R_2(x_1, x_2) &= N(N-1) \int dx_3 dx_4 \ldots dx_N P_N(x_1, x_2, \ldots, x_N), \\ Y_2(x_1, x_2) &= -R_2(x_1, x_2) + R_1(x_1) R_1(x_2). \end{aligned} \tag{2.87}$$

Note that $R_1(x_1) = 1 = Y_1(x_1)$ and the x's are unfolded energies, i.e. mean spacing $\overline{D} = 1$. The important point is that for a Poisson $Y_2(r) = 0$ and therefore, as $\Sigma^2(L) = L - \int_0^L (L-r) Y_2(r) dr$,

$$\Sigma^2(L) = L, \qquad \overline{\Delta_3}(L) = \frac{L}{15} \tag{2.88}$$

for a Poisson. It is also useful to mention that for pseudo-integrable systems (which possess singularities and are integrable in the absence of these singularities) [36–39] follow semi-Poisson statistics [40, 41],

$$P(S)dS = 4S \exp -2S dS. \tag{2.89}$$

This form is also seen recently in the low-energy part of the spectra generated by two-body interactions [42].

2.4.3 Analysis of Nuclear Data for GOE and Poisson

Bohigas, Haq and Pandey [6, 12] analyzed slow neutron resonance, proton resonance and (n, γ) reaction data (for level and width fluctuations) and established that GOE describes, within sample size errors, almost exactly the experimental data. They constructed nuclear data ensemble (NDE) with 1762 resonance energies corresponding to 36 sequences of 32 different nuclei and they contain: (i) slow-neutron resonance $1/2^+$ levels from ^{64,66,68}Zn, ^{114}Cd, ^{152,154}Sm, ^{154,156,158,160}Gd, ^{160,162,164}Dy, ^{166,168,170}Er, ^{172,174,176}Yb, ^{182,184,186}W, ^{186,190}Os, ^{232}Th and ^{238}U targets; (ii) proton resonance $1/2^+$ levels ($1/2^-$ also for Ca) from ^{44}Ca, ^{48}Ti and ^{56}Fe targets; (iii) (n, γ) reaction data on ^{177,179}Hf and ^{235}U giving two sequences of J^π levels. Similarly considered also are 1182 widths corresponding to 21 sequences of the above neutron resonance data. Comparisons are made with the GOE Wigner's law $P(S)dS = (\pi S/2) \exp(-\pi S^2/4) dS$ for nearest neighbor spacing distribution (NNSD) and Porter Thomas (P-T) χ^2_1 law $P(x)dx = (1/\sqrt{2\pi x}) \exp(-x/2) dx$ for widths; S is in units of average mean spacing ($\overline{D}$) and x is width (rate of transition from a initial state to a channel) in units of average width. Similarly, the GOE number variance $\Sigma^2(L)$ and $\Delta_3(L)$ are also seen to agree (for $L \leq 20$) extremely well with NDE. In the analysis sample size corrections are made as discussed earlier. Unlike the resonance energies, the resonance widths analysis appears to have some uncertainties (see [6] and Appendix E).

Garrett et al. [43] analyzed NNSD for high-spin levels near the yrast line in rare-earth nuclei. Considered are 3130 experimental level spacings from deformed even-even and odd-A nuclei with Z = 62–75 and A = 155–185. As expected $P(S)$ is seen to follow regular Poisson form. Following this study, Enders et al. [44] analyzed NNSD for scissors mode levels in deformed nuclei and found Poisson behavior as scissors mode is a well defined collective mode. Used here are 152 levels from 13 heavy deformed nuclei [^{146,148,150}Nd, ^{152,154}Sm, ^{156,158}Gd, ^{164}Dy, ^{166,168}Er, ^{174}Yb, ^{178,180}Hf] in the energy range $2.5 < E_x < 4$ MeV with the constraint that there must be at least 8 levels in the given energy interval in a given nucleus. Thus low-lying levels of well deformed nuclei and scissor states, being regular, follow Poisson as expected. Finally, Enders et al. [45] analyzed also the electric pigmy dipole resonances located around 5–7 MeV in four $N = 82$ isotones. They made an ensemble of 184 1^- states and an analysis, though difficult due to many missing levels, has been carried out. The authors conclude that there is GOE behavior. Thus, level fluctuations bring out the expected difference between the scissor mode and pigmy dipole resonance.

References

1. C.E. Porter, *Statistical Theories of Spectra: Fluctuations* (Academic Press, New York, 1965)
2. F. Haake, *Quantum Signatures of Chaos*, 3rd edn. (Springer, Heidelberg, 2010)
3. R.G. Sachs, *The Physics of Time Reversal* (University of Chicago Press, Chicago, 1987)
4. J. von Neumann, E.P. Wigner, On the bahavior of eigenvalues in adiabatic processes. Phys. Z. **30**, 467–470 (1929)
5. M.L. Mehta, *Random Matrices*, 3rd edn. (Elsevier, Amsterdam, 2004)
6. O. Bohigas, R.U. Haq, A. Pandey, Fluctuation properties of nuclear energy levels and widths: comparison of theory and experiment, in *Nuclear Data for Science and Technology*, ed. by K.H. Böckhoff (Reidel, Dordrecht, 1983), pp. 809–813
7. O. Bohigas, M.J. Giannoni, C. Schmit, Characterization of chaotic quantum spectra and universality of level fluctuation laws. Phys. Rev. Lett. **52**, 1–4 (1984)
8. V. Plerou, P. Gopikrishnan, B. Rosenow, L.A.N. Amaral, T. Guhr, H.E. Stanley, Random matrix approach to cross correlations in financial data. Phys. Rev. E **65**, 066126 (2006)
9. M.S. Santhanam, P.K. Patrta, Statistics of atmospheric correlations. Phys. Rev. E **64**, 016102 (2001)
10. V.K.B. Kota, Embedded random matrix ensembles for complexity and chaos in finite interacting particle systems. Phys. Rep. **347**, 223–288 (2001)
11. K. Patel, M.S. Desai, V. Potbhare, V.K.B. Kota, Average-fluctuations separation in energy levels in dense interacting boson systems. Phys. Lett. A **275**, 329–337 (2000)
12. R.U. Haq, A. Pandey, O. Bohigas, Fluctuation properties of nuclear energy levels: do theory and experiment agree? Phys. Rev. Lett. **48**, 1086–1089 (1982)
13. T.A. Brody, J. Flores, J.B. French, P.A. Mello, A. Pandey, S.S.M. Wong, Random matrix physics: spectrum and strength fluctuations. Rev. Mod. Phys. **53**, 385–479 (1981)

14. A. Stuart, J.K. Ord, *Kendall's Advanced Theory of Statistics: Distribution Theory* (Oxford University Press, New York, 1987)
15. E.P. Wigner, Characteristic vectors of bordered matrices with infinite dimensions. Ann. Math. **62**, 548–564 (1955)
16. E.P. Wigner, Statistical properties of real symmetric matrices with many dimensions, in *Can. Math. Congr. Proc.* (University of Toronto Press, Toronto, 1957), pp. 174–184
17. J.B. French, P.A. Mello, A. Pandey, Statistical properties of many-particle spectra II. Two-point correlations and fluctuations. Ann. Phys. (N.Y.) **113**, 277–293 (1978)
18. G. Szegő, *Orthogonal Polynomials*, Colloquium Publications, vol. 23 (Am. Math. Soc., Providence, 2003)
19. K.K. Mon, J.B. French, Statistical properties of many-particle spectra. Ann. Phys. (N.Y.) **95**, 90–111 (1975)
20. A. Bose, R.S. Hazra, K. Saha, Patterned random matrices and method of moments, in *Proceedings of the International Congress of Mathematics 2010*, vol. IV, ed. by R. Bhatia (World Scientific, Singapore, 2010), pp. 2203–2231
21. F.J. Dyson, M.L. Mehta, Statistical theory of energy levels of complex systems IV. J. Math. Phys. **4**, 701–712 (1963)
22. R.G. Nazmitdinov, E.I. Shahaliev, M.K. Suleymanov, S. Tomsovic, Analysis of nucleus-nucleus collisions at high energies and random matrix theory. Phys. Rev. C **79**, 054905 (2009)
23. C.E. Porter, R.G. Thomas, Fluctuations of nuclear reaction widths. Phys. Rev. **104**, 483–491 (1956)
24. H. Alt, H.-D. Gräf, H.L. Harney, R. Hofferbert, H. Lengeler, A. Richter, R. Schardt, H.A. Weidenmüller, Gaussian orthogonal ensemble statistics in a microwave stadium billiard with chaotic dynamics: Porter-Thomas distribution and algebraic decay of time correlations. Phys. Rev. Lett. **74**, 62–65 (1995)
25. A. Rényi, On measures of information and entropy, in *Proceedings of the 4th Berkeley Symposium on Mathematics, Statistics and Probabilities*, vol. 1 (1961), pp. 547–561
26. I. Varga, J. Pipek, Rényi entropies characterizing the shape and the extension of the phase space representation of quantum wave functions in disordered systems. Phys. Rev. E **68**, 026202 (2003)
27. C.E. Shannon, W. Weaver, *The Mathematical Theory of Communication* (University of Illinois Press, Champaign, 1949)
28. N. Ullah, On a generalized distribution of the Poles of the unitary collision matrix. J. Math. Phys. **10**, 2099–2104 (1969)
29. P. Shukla, Eigenfunction statistics of complex systems: a common mathematical formulation. Phys. Rev. E **75**, 051113 (2007)
30. B. Dietz, T. Guhr, H.L. Harney, A. Richter, Strength distributions and symmetry breaking in coupled microwave billiards. Phys. Rev. Lett. **96**, 254101 (2006)
31. A. Pandey, Statistical properties of many-particle spectra III. Ergodic behavior in random matrix ensembles. Ann. Phys. (N.Y.) **119**, 170–191 (1979)
32. A.D. Jackson, C. Mejia-Monasterio, T. Rupp, M. Saltzer, T. Wilke, Spectral ergodicity and normal modes in ensembles of sparse matrices. Nucl. Phys. A **687**, 405–434 (2001)
33. O. Bohigas, R.U. Haq, A. Pandey, Higher-order correlations in spectra of complex systems. Phys. Rev. Lett. **54**, 1645–1648 (1982)
34. O. Bohigas, M.J. Giannoni, Level density fluctuations and random matrix theory. Ann. Phys. (N.Y.) **89**, 393–422 (1975)
35. N. Rosenzweig, C.E. Porter, Repulsion of energy levels in complex atomic spectra. Phys. Rev. **120**, 1698–1714 (1960)
36. D. Biswas, S.R. Jain, Quantum description of a pseudointegrable system: the $\pi/3$-rhombus billiard. Phys. Rev. A **42**, 3170–3185 (1990)
37. G. Date, S.R. Jain, M.V.N. Murthy, Rectangular billiard in the presence of a flux line. Phys. Rev. E **51**, 198–203 (1995)

38. P.R. Richens, M.V. Berry, Pseudointegrable systems in classical and quantum mechanics. Physica D **2**, 495–512 (1998)
39. A.M. Garcïa-Garcïa, J. Wang, Semi-Poisson statistics in quantum chaos. Phys. Rev. E **73**, 036210 (2006)
40. E.B. Bogomolny, U. Gerland, C. Schmit, Models of intermediate spectral statistics. Phys. Rev. E **59**, R1315–1318 (1999)
41. E.B. Bogomolny, C. Schmit, Spectral statistics of a quantum interval-exchange map. Phys. Rev. Lett. **93**, 254102 (2004)
42. J. Flores, M. Horoi, M. Mueller, T.H. Seligman, Spectral statistics of the two-body random ensemble revisited. Phys. Rev. E **63**, 026204 (2000)
43. J.D. Garrett, J.Q. Robinson, A.J. Foglia, H.-Q. Jin, Nuclear level repulsion and order vs. chaos. Phys. Lett. B **392**, 24–29 (1997)
44. J. Enders, T. Guhr, N. Huxel, P. von Neumann-Cosel, C. Rangacharyulu, A. Richter, Level spacing distribution of scissors mode states in heavy deformed nuclei. Phys. Lett. B **486**, 273–278 (2000)
45. J. Enders, T. Guhr, A. Heine, P. von Neuman-Cosel, V.Yu. Ponomarev, A. Richter, J. Wambach, Spectral statistics and the fine structure of the electric pygmy dipole resonance in $N = 82$ nuclei. Nucl. Phys. A **41**, 3–28 (2004)

Chapter 3
Interpolating and Other Extended Classical Ensembles

Changes in the nature of level fluctuations in the situations such as (i) a symmetry is gradually broken, (ii) two good symmetry subspaces are gradually admixed, (iii) ordered (integrable) spectra gradually become chaotic and so on are studied by using interpolating and/or partitioned random matrix ensembles [1–7]. A simple yet useful approach for deriving the NNSD's for interpolating ensembles is to extend, as pointed out in [8–12], the simple Wigner's 2×2 matrix formalism. The appropriate 2×2 random matrix ensemble for Poisson to GOE and GUE and GOE to GUE transitions is [8, 12],

$$H = \begin{bmatrix} \alpha(X_1 + X_2) + pv\lambda & \alpha X_3 + i\alpha' X_4 \\ \alpha X_3 - i\alpha' X_4 & \alpha(X_1 - X_2) - pv\lambda \end{bmatrix}. \tag{3.1}$$

In Eq. (3.1) X_1, X_2, X_3 and X_4 are $G(0, v^2)$ variables and the usefulness of p and λ will later become clear. The H matrix in Eq. (3.1) for $\lambda = 0$, $\alpha' = 0$ is GOE, $\lambda = 0$, $\alpha' = \alpha$ is GUE, and $X_i = 0$ and λ a Poisson gives a Poisson spectrum. Thus the matrix in Eq. (3.1) interpolates Poisson, GOE and GUE (in fact also a uniform spectrum). Given λ_1 and λ_2, the two eigenvalues of the H matrix, we have

$$(\lambda_1 - \lambda_2)^2 = S^2 = 4\big[(\alpha X_2 + pv\lambda)^2 + \big(\alpha^2 X_3^2 + \alpha'^2 X_4^2\big)\big]. \tag{3.2}$$

Let us define

$$\begin{aligned} x_2 &= 2\alpha X_2 + 2pv\lambda \to G\big(2pv\lambda, (2\alpha v)^2\big), \\ x_3 &= 2\alpha X_3 \to G\big(0, (2\alpha v)^2\big), \\ x_4 &= 2\alpha' X_4 \to G\big(0, \big(2\alpha' v\big)^2\big). \end{aligned} \tag{3.3}$$

Therefore,

$$P(x_2, x_3, x_4)\, dx_2\, dx_3\, dx_4 = \frac{dx_2\, dx_3\, dx_4}{(2\pi)^{3/2}(2\alpha v)^2(2\alpha' v)} \times \exp -\left(\frac{(x_2 - 2pv\lambda)^2 + x_3^2}{2(2\alpha v)^2} + \frac{x_4^2}{2(2\alpha' v)^2} \right). \tag{3.4}$$

V.K.B. Kota, *Embedded Random Matrix Ensembles in Quantum Physics*, Lecture Notes in Physics 884, DOI 10.1007/978-3-319-04567-2_3,

Changing the variables (x_2, x_3, x_4) to (S, θ, ϕ) such that $x_2 = S\sin\theta\cos\phi$, $x_3 = S\sin\theta\sin\phi$, $x_4 = S\cos\theta$, we get

$$P(S)\,dS = \frac{S^2 dS}{(2\pi)^{3/2}(2\alpha v)^2(2\alpha' v)}\exp\left(-\frac{p^2\lambda^2}{2\alpha^2}\right)\int_0^{2\pi}\exp\left\{\frac{p\lambda}{2v\alpha^2}S\sin\theta\cos\phi\right\}d\phi$$
$$\times\int_0^{\pi}\exp -\frac{1}{2}\left[\frac{S^2\sin^2\theta}{4\alpha^2 v^2}+\frac{S^2\cos^2\theta}{4\alpha' v^2}\right]\sin\theta\, d\theta. \tag{3.5}$$

The integral over ϕ is $2\pi I_0(\frac{p\lambda}{2v\alpha^2}S\sin\theta)$ where I_0 is Bessel function. With $S\cos\theta = z$, the final result is,

$$P(S:\lambda)\,dS = \frac{S\,dS}{4v^3\alpha^2\alpha'\sqrt{2\pi}}\exp\left(-\frac{p^2\lambda^2}{2\alpha^2}-\frac{S^2}{8v^2\alpha^2}\right)$$
$$\times\int_0^S dz\, I_0\left(\frac{p\lambda}{2v\alpha^2}\sqrt{S^2-z^2}\right)\exp\left[\frac{(\alpha')^2-\alpha^2}{8v^2\alpha^2(\alpha')^2}z^2\right]. \tag{3.6}$$

With $\lambda = 0$, $\alpha = 1$ and $\alpha' \to \alpha$ we have GOE to GUE transition. Similarly, assuming a distribution $f(\lambda)d\lambda$ for λ (with λ independent of X_i, $i = 1, 2, 3, 4$), Eq. (3.6) defines for example the Poisson to GOE and GUE interpolations. Combining Eq. (3.6) with

$$P(S)dS = \left[\int_{-\infty}^{+\infty} P(S:\lambda)f(\lambda)d\lambda\right]dS$$
$$\text{for Poisson } f(\lambda)d\lambda = e^{-\lambda}d\lambda \quad \text{for } 0 \le \lambda \le \infty \quad \text{and} \quad 0 \quad \text{for } \lambda < 0 \tag{3.7}$$

gives the spacing distributions for Poisson to GOE and GUE. With $f(\lambda) = 1$ for $0 \le \lambda \le 1$ and 0 for $\lambda < 0$ and also for $\lambda > 1$ will give uniform to GOE and GUE transitions; Ref. [13] gives a numerical study of uniform to GOE and GUE transitions. It is also possible to consider $f(\lambda) = \delta(\lambda - \lambda_c)$. Note that we have always $\int_{-\infty}^{\infty} f(\lambda)\,d\lambda = 1$. Before going further, it is important to mention that an extension of the matrix in Eq. (3.1) including GSE with 4×4 matrices was given in [14].

Before going further, it is important to point out that the results in Refs. [15–18] are used in the simplifications of various integrals we need ahead. A list of some useful integrals are,

$$\begin{aligned}
I_n(a,c) &= \int_0^\infty x^n\left[\exp -ax^2\right]\Phi(cx)\,dx, \\
I_1 &= \frac{c}{2a(a+c^2)^{1/2}}, \\
I_2 &= \frac{1}{2\sqrt{\pi}}\left[\frac{1}{a^{3/2}}\tan^{-1}\frac{c}{a^{1/2}}+\frac{c}{a(a+c^2)}\right], \\
I_3 &= \frac{1}{2a^2}\frac{c}{\sqrt{a+c^2}}+\frac{c}{4a(a+c^2)^{3/2}}.
\end{aligned} \tag{3.8}$$

In Eq. (3.8), $\Phi(x) = \frac{2}{\sqrt{\pi}} \int_0^x \exp -t^2 dt$ is the Error function. An integral with the Bessel function I_0 is,

$$\int_0^\infty \left[\exp -a^2t^2\right] t^{\mu-1} I_0(bt)\, dt = \frac{\Gamma(\mu/2)}{2a^\mu} \, {}_1F_1\left(\mu/2, 1, b^2/4a^2\right). \tag{3.9}$$

Finally,

$$\int_0^\infty [\exp -at] \Phi(bt) dt = \frac{1}{a} \exp\left(\frac{a^2}{4b^2}\right)\left[1 - \Phi\left(\frac{a}{2b}\right)\right]. \tag{3.10}$$

The hyper-geometric function ${}_1F_1$ in Eq. (3.9) is also denoted as $M(\mu/2,\ 1,\ b^2/4a^2)$.

3.1 GOE-GUE Transition

3.1.1 2 × 2 *Matrix Results*

Substituting $\lambda = 0$, $\alpha = 1$ and $\alpha' \to \alpha$ in Eq. (3.6), spacing distribution interpolating GOE to GUE is obtained [1, 8],

$$P_{\text{GOE-GUE}}(S)\, dS = dS \frac{S}{4v^2(1-\alpha^2)^{1/2}} \exp - \frac{S^2}{8v^2}\, \Phi\left[\sqrt{\frac{1-\alpha^2}{8\alpha^2 v^2}}\, S\right]. \tag{3.11}$$

Using Eq. (3.8) it is seen that $P(S)$ is normalized to unity and the average spacing $D_\alpha = \int_0^\infty SP(S) dS$ is,

$$D_\alpha = \frac{1}{\sqrt{\pi(1-\alpha^2)}}\left[\sqrt{8v^2} \tan^{-1} \frac{\sqrt{1-\alpha^2}}{\alpha} + \sqrt{8\alpha^2 v^2 \left(1-\alpha^2\right)}\right]. \tag{3.12}$$

Note that D_0 is the average spacing between the unperturbed levels,

$$D_0 = \sqrt{2\pi}\, v. \tag{3.13}$$

To proceed further it is useful to introduce the transition parameter Λ which is the r.m.s. admixing GUE matrix element $\alpha^2 v^2$ divided by D_0^2,

$$\Lambda = \frac{\alpha^2 v^2}{D_0^2} = \frac{\alpha^2}{2\pi}. \tag{3.14}$$

Note that $\Lambda = 0$ gives GOE and $\Lambda = 1/2\pi$ gives GUE. The importance of the Λ parameter is that it will allow us to extend the 2×2 matrix results to $N \times N$ matrices; see [1, 8, 12] and the results ahead. Although this was pointed out first in [8], it was rediscovered in [9, 10]. It is easy to see that $P(x)dx$ with $x = S/D_0$ will depend only on Λ. For example, $\overline{x} = \overline{S}/D_0$ is

$$\begin{aligned}\overline{x} &= \frac{2}{\pi\sqrt{1-2\pi\Lambda}}\left[\tan^{-1}\sqrt{\frac{1-2\pi\Lambda}{2\pi\Lambda}} + \sqrt{2\pi\Lambda(1-2\pi\Lambda)}\right] \\ &\overset{\Lambda\ll 1}{\longrightarrow} (1+\pi\Lambda) + O\left(\Lambda^{3/2}\right).\end{aligned} \tag{3.15}$$

Now we can write down the expression for the variance $\sigma^2(0:\Lambda)$ of the NNSD. Note that from Eq. (3.2) we have easily $\overline{S^2} = 8v^2 + 4\alpha^2 v^2$ and then,

$$\begin{aligned}\sigma^2(0:\Lambda) &= \frac{\overline{S^2}}{(\overline{S})^2} - 1 \\ &= \frac{\overline{S^2}}{(D_0)^2\overline{x}^2} - 1 = \frac{4(1+\pi\Lambda)}{\pi\overline{x}^2} - 1 \\ &\overset{\Lambda\ll 1}{\longrightarrow} \left(\frac{4}{\pi} - 1\right) - 4\Lambda \\ &= \sigma^2(0:0) - 4\Lambda.\end{aligned} \tag{3.16}$$

Equation (3.16) extends to any $N \times N$ matrix and for most purposes this small Λ result is adequate for data analysis.

A different parametrization that gives GOE for $\Lambda = 0$ and GUE for $\Lambda = \infty$ is to put in Eqs. (3.1), (3.6), $\lambda = 0$, $\alpha \to \alpha + \sqrt{1-\alpha^2}$, $\alpha' \to \alpha$ and finally divide all the matrix elements by $\sqrt{1-\alpha^2}$. This gives GOE + $[\alpha/\sqrt{1-\alpha^2}]$ GUE ensemble. Then

$$\Lambda = \frac{1}{D_0^2}\left[\frac{\alpha}{\sqrt{1-\alpha^2}}\right]^2 v^2, \quad D_0^2 = \sqrt{2\pi}\,v. \tag{3.17}$$

Now, with $\hat{S} = S/D_0$, the NNSD is

$$P(\hat{S})\,d\hat{S} = d\hat{S}\,\frac{\pi}{2}\sqrt{1+2\pi\Lambda}\,\hat{S}\exp -\frac{\pi}{4}\hat{S}^2\,\Phi(\hat{S}/\sqrt{8\Lambda}). \tag{3.18}$$

Equation (3.18) gives correctly the GOE and GUE NNSD for $\Lambda = 0$ and $\Lambda = \infty$; note that $\Phi(ax) \to 2(ax)/\sqrt{\pi}$ as $a \to 0$. The variance of the NNSD, with Λ defined by Eq. (3.17), is

$$\begin{aligned}&\sigma^2_{\text{GOE-GUE}}(0:\Lambda) \\ &= \frac{\pi(1+3\pi\Lambda)}{[(1+2\pi\Lambda)\tan^{-1}\{(2\pi\Lambda)^{-1/2}\} + \sqrt{2\pi\Lambda}]^2} - 1 \\ &\overset{\Lambda\ll 1}{\longrightarrow} \left(\frac{4}{\pi} - 1\right) - 4\Lambda = \sigma^2(0:0) - 4\Lambda.\end{aligned} \tag{3.19}$$

It is important to mention that all the results given here reproduce exactly the results discussed in [9, 10]. Finally, it should be mentioned that 2 × 2 GOE-GUE transition results were first given in [19] although the transition parameter was not identified by the authors.

3.1.2 $N \times N$ Ensemble Results for $\Sigma^2(r)$ and $\overline{\Delta}_3(r)$

Let us consider $H = H^R + i\alpha H^I$ and then $\alpha = 0$ gives GOE and $\alpha = 1$ gives GUE. The matrix elements of H satisfy the following properties (with $a = H_{ij}^R$ and $b = H_{ij}^I$),

$$\begin{aligned} \overline{H_{ij}H_{ij}} &= \overline{(a+i\alpha\, b)\,(a+i\alpha\, b)} = \overline{a^2} - \overline{\alpha^2 b^2}, \\ \overline{H_{ij}H_{ji}} &= \overline{(a+i\alpha\, b)\,(a-i\alpha\, b)} = \overline{a^2} + \overline{\alpha^2 b^2}. \end{aligned} \tag{3.20}$$

Using the normalization $v^2 d(1+\alpha^2) = 1$ $(\overline{a^2} = \overline{b^2} = v^2)$, we have

$$\begin{aligned} \overline{H_{ij}H_{ij}} &= \frac{1-\alpha^2}{d(1+\alpha^2)} = \frac{\eta}{d}, \\ \overline{H_{ij}H_{ji}} &= \frac{1}{d}, \\ \eta &= \frac{1-\alpha^2}{1+\alpha^2}. \end{aligned} \tag{3.21}$$

In the product $\overline{\langle H^\zeta\rangle\langle H^\zeta\rangle}$ there are d number of H_{ij} terms and each with its partner comes ζ times. Hence,

$$\begin{aligned} \overline{\langle H^\zeta\rangle\langle H^\zeta\rangle} &= \frac{\zeta}{d^2} d^\zeta \left[\frac{1}{d^\zeta} + \frac{\eta^\zeta}{d^\zeta}\right] = \frac{\zeta}{d^2}(1+\eta^\zeta) = A_\zeta \frac{\zeta}{d^2}; \\ A_\zeta &= 1 + \eta^\zeta . \end{aligned} \tag{3.22}$$

Equation (3.22) correctly reproduces the values for $\alpha = 0$ and $\alpha = 1$ given in Eq. (2.60). Now, extending Eq. (2.62) we have

$$\begin{aligned} S^F(x,y) &= \frac{1}{\pi^2 d^2} \sum_{\zeta=1}^{d} A_\zeta \zeta^{-1} \sin\zeta\psi(x)\, \sin\zeta\psi(y) \\ &\simeq S^F_{\mathrm{GUE}}(x,y) + \frac{1}{\pi^2 d^2} \sum_{\zeta=1}^{\infty} (\eta')^\zeta \zeta^{-1} \sin\zeta\psi(x)\, \sin\zeta\psi(y) \\ &= S^F_{\mathrm{GUE}}(x,y) + \frac{1}{4\pi^2 d^2} \ln \frac{1+(\eta')^2 - 2\eta' \cos(\psi(x)+\psi(y))}{1+(\eta')^2 - 2\eta' \cos(\psi(x)-\psi(y))} \\ &\stackrel{r \ll d}{\Rightarrow} S^F_{\mathrm{GUE}}(x,y) + \frac{1}{4\pi^2 d^2} \ln \frac{(1-\eta')^2 + 4\pi^2 \overline{\rho}^2 \eta'}{(1-\eta')^2 + r^2 \eta' / 4\pi^2 d^2 \overline{\rho}^4}; \end{aligned} \tag{3.23}$$

with

$$\begin{aligned} |x-y| &\overset{r\ll d}{\Rightarrow} r\overline{D}, \\ \eta' &\overset{r\ll d}{\Rightarrow} \eta'(x) = \eta\exp(-\tau/d\sin^2\psi) = \eta\exp(-\tau/d\pi^2\overline{\rho}^2). \end{aligned} \tag{3.24}$$

In Eq. (3.23) we have introduced an exponential cut-off in ζ in order to extend the ζ summation to ∞. The details are as follows: With a cut-off $e^{-\alpha_0\zeta}$ the ζ sum is extended to ∞ as in the case of GOE. Then $\eta^\zeta e^{-\alpha_0\zeta} = (\eta e^{-\alpha_0})^\zeta = (\eta')^\zeta$. The choice for α_0 is $\alpha_0 = \tau/d\sin^2\psi = \tau/\pi^2 d\overline{\rho}^2$. Using the simplification as it is done in the case of GOE we get step no. 3 in Eq. (3.23). Now $\cos(\psi_1+\psi_2)$ is $\cos 2\psi$ for $x\sim y$ and $\cos(\psi_1-\psi_2)$ is $1-(\delta\psi)^2/2$. Therefore,

$$\begin{aligned} \cos 2\psi &= 1-2\sin^2\psi = 1-2\pi^2\overline{\rho}^2 \\ 1-\frac{(\delta\psi)^2}{2} &= 1-\left[\frac{r}{2d\pi\overline{\rho}^2}\right]^2. \end{aligned} \tag{3.25}$$

Equation (3.25) will give the fourth equality in Eq. (3.23). Then

$$\begin{aligned} \Sigma^2_\alpha(r) &= \Sigma^2_{\mathrm{GUE}} + \frac{1}{4\pi^2}\left\{2\ln\frac{(1-\eta')^2+4\pi^2\overline{\rho}^2\eta'}{(1-\eta')^2} - 2\ln\frac{(1-\eta')^2+4\pi^2\overline{\rho}^2\eta'}{(1-\eta')^2+\frac{r^2\eta'}{4\pi^2d^2\overline{\rho}^4}}\right\} \\ &= \Sigma^2_{\mathrm{GUE}} + \frac{1}{2\pi^2}\ln\left[1+\frac{r^2\eta'}{4\pi^2d^2\overline{\rho}^4(1-\eta')^2}\right]. \end{aligned} \tag{3.26}$$

At this stage it is convenient to introduce the transition parameter

$$\Lambda(\alpha) = \alpha^2 d\overline{\rho}^2 \tag{3.27}$$

$$\begin{aligned} \eta &\overset{\alpha\ll 1}{\Rightarrow} \exp -2\alpha^2 \\ \eta' &= \exp\left(-2\alpha^2 - \frac{\tau}{d\pi^2\overline{\rho}^2}\right) \simeq 1 \\ \Rightarrow\quad 1-\eta' &\simeq 2\alpha^2 + \frac{\tau}{d\pi^2\overline{\rho}^2} = \frac{\tau+2\pi^2\Lambda(\alpha)}{d\pi^2\overline{\rho}^2}. \end{aligned} \tag{3.28}$$

Using Eq. (3.28) in Eq. (3.26) we obtain

$$\Sigma^2(r:\Lambda) \overset{\alpha^2\ll 1,\, \Lambda\ll 1}{\Longrightarrow} \Sigma^2_{\mathrm{GUE}}(r) + \frac{1}{2\pi^2}\ln\left[1+\frac{\pi^2 r^2}{4[\tau+2\pi^2\Lambda(\alpha)]^2}\right] \tag{3.29}$$

and the small Λ expansion is,

$$\Sigma^2(r:\Lambda) = \Sigma^2_{\mathrm{GUE}}(r) + \frac{1}{2\pi^2}\ln\left(1+\frac{\pi^2r^2}{4\tau^2}\right) - \frac{2\Lambda(\alpha)}{\tau[1+\frac{4\tau^2}{\pi^2r^2}]} + \cdots \tag{3.30}$$

Also note that $\Sigma^2(r:\Lambda) \to \Sigma^2_{\text{GUE}}(r)$ for $\Lambda \to \infty$. The parameter τ is fixed from GOE-GUE difference for $r=1$; $\Delta_{\Sigma^2(1)} = 0.446 - 0.344 = 0.102$ and this gives $\tau = 0.615$. Equation (3.29) was reported first in [1, 8] and later Dupuis and Montambaux [20] derived the same formula in the study of statistical behavior of the spectrum for a metallic ring pierced by a magnetic field. Here the parameter τ has a clear physical meaning. Finally we mention that an exact solution for GOE to GUE transition for $N \times N$ matrices was given by Pandey and Mehta [21].

The Δ_3 statistic for GOE to GUE transition follows by combining Eqs. (3.29) and (C.8),

$$\overline{\Delta_3}(\overline{n},\alpha) = \overline{\Delta_3}^{\text{GUE}}(\overline{n}) + \frac{1}{\pi^2\overline{n}^4}\int_0^{\overline{n}} \left(\overline{n}^3 - 2\overline{n}^2 r + r^3\right)\ln\left[1 + B(\Lambda)r^2\right] dr. \tag{3.31}$$

Here we have used Eq. (3.29) with the substitution $B(\Lambda) = \pi^2/4[\tau + 2\pi^2\Lambda(\alpha)]^2$. Solving the integral in Eq. (3.31) using MATHEMATICA gives,

$$\begin{aligned}\overline{\Delta_3}(\overline{n}:\Lambda) = \overline{\Delta_3}^{\text{GUE}}(\overline{n}) + \frac{1}{\overline{n}^4\pi^2}\Bigg[&\frac{2\,\overline{n}^3}{\sqrt{B(\Lambda)}}\tan^{-1}\left(\overline{n}\sqrt{B(\Lambda)}\right) \\ &+ \left[\frac{B^2(\Lambda)\overline{n}^4 - 1 - 4B(\Lambda)\overline{n}^2}{4B^2(\Lambda)}\ln\left(1 + B(\Lambda)\overline{n}^2\right)\right] - \overline{n}^4 \\ &+ \frac{\left(1 + B(\Lambda)\overline{n}^2\right)}{16B^4(\Lambda)}\left(6B^2(\Lambda) - 2B^3(\Lambda)\overline{n}^2 - 3\right)\Bigg].\end{aligned} \tag{3.32}$$

3.1.3 Application to TRNI in Nucleon-Nucleon Interaction

Following the fact that GOE generates stronger level repulsion compared to GOE, as seen from 2×2 $P(S)dS$, Wigner [22] suggested that this could be used to detect time reversal breaking in nuclear force. This and the close agreement between neutron resonance data (i.e. NDE) and GOE coupled with the GOE to GUE transition theory, i.e. the transition curve defined by Eq. (3.29) for $r=1$, allows us to derive a bound on time reversal non invariant (TRNI) part of the nucleon nucleon interaction. Firstly the NDE data with 1336 levels gives $\Sigma^2(1)$ value to be 0.445 and the GOE value is 0.446. Adding the sample size error on the theory value, within 3σ (99.7 % confidence), the upper bound on $\pi^2\Lambda$ is 0.145 [1]. As $\Lambda = \alpha^2 v^2/D^2$, the bound on αv is $\alpha v \simeq 0.1D$. Note that v is r.m.s. many particle nuclear matrix element for the TRI part of H. To convert this to a bound on α, i.e. TRNI in the effective nucleon-nucleon interaction, v has been determined using statistical spectroscopy methods (see Chap. 7). The deduced bound is $\alpha \leq 10^{-3}$ [23]. Recently, Morrison et al. [24] suggested that a similar analysis for T-odd, P-even interactions in atoms should be possible.

3.2 Poisson to GOE and GUE Transitions

3.2.1 2 × 2 Matrix Results for Poisson to GOE Transition

Substituting $\alpha' = 0$ in Eq. (3.6) and applying Eq. (3.7) will give the NNSD for Poisson (P) to GOE transition. To this end we use the result

$$\lim_{\alpha'\to 0}\left[\sqrt{2\pi}\,(2v\alpha')\right]^{-1}\exp[-z^2/8(v\alpha')^2] = (1/2)\delta_{z,0}$$

and the factor $1/2$ comes as we have $z \geq 0$. Then, in terms of the transition parameter

$$\Lambda = \frac{\alpha^2 v^2}{D_0^2}, \tag{3.33}$$

where the mean spacing D_0 of the unperturbed Poisson spectrum is $D_0 = 2pv$ and the mean square admixing GOE matrix element is $\alpha^2 v^2$, the NNSD for P to GOE transition, with $\hat{S} = S/D_0$, is [9, 12],

$$P_{\text{P-GOE}}(\hat{S})\,d\hat{S} = d\hat{S}\frac{\hat{S}}{4\Lambda}\exp\{-\hat{S}^2/8\Lambda\}\int_0^\infty \exp\left\{-\lambda - \frac{\lambda^2}{8\Lambda}\right\} I_0\left(\frac{\lambda\hat{S}}{4\Lambda}\right) d\lambda. \tag{3.34}$$

For $\Lambda = 0$, Eq. (3.34) gives Poisson and for $\Lambda \to \infty$ the Wigner (GOE) form. Using Eq. (3.9) with $a^2 = 1/8\Lambda$, $\mu = 2$ and $b = \lambda/4\Lambda$ (for the integral over $\hat{S}$), it is easily proved that $P_{\text{P-GOE}}(\hat{S})$ is normalized to unity.

Although we can compare $P_{\text{P-GOE}}(S)$ with $P(S)$ for various Λ values, it is more instructive to examine the $\Lambda \to 0$ and $\hat{S}$ small limit. As $\Lambda \to 0$, we can approximate $\exp -(\lambda + \lambda^2/8\Lambda)$ by $\exp -\lambda^2/8\Lambda$. Now applying Eq. (3.9) and the results given in p. 509 of [15], i.e. ${}_1F_1(\frac{1}{2}, 1, z) = [\exp z/2] I_0(z/2)$, will give

$$P_{\text{P-GOE}}(\hat{S})\,d\hat{S} = d\hat{S}\sqrt{\frac{\pi}{8}}\frac{\hat{S}}{\Lambda^{1/2}}\exp\left\{-\frac{\hat{S}^2}{16\Lambda}\right\} I_0\left(\frac{\hat{S}^2}{16\Lambda}\right). \tag{3.35}$$

Let us mention that perturbation theory also gives Eq. (3.35) for a general $N \times N$ matrix [25]. One important result that follows from Eq. (3.35) is that $P(S)$ goes to zero as S goes to zero for non-zero values of Λ (i.e. there is level repulsion as soon as GOE is switched on).

In the data analysis and applications, more useful is the variance of the NNSD, $\sigma^2(0:\Lambda) = (\overline{S^2}/\overline{S}^2) - 1$ for P to GOE transition, which defines a transition curve. Using Eq. (3.9) with $a^2 = 1/8\Lambda$, $\mu = 3$ and $b = \lambda/4\Lambda$ for the integral over $\hat{S}$ and then applying Eq. (7.628) on p. 871 in [16] will give $\overline{\hat{S}} = \sqrt{\pi}\Psi(-\frac{1}{2}, 0, 2\Lambda)$ where Ψ is Kummer's function [15]. As $\overline{\lambda^2} = 2$ for Poisson, Eq. (3.2) gives $\overline{S^2} = 8\alpha^2 v^2 + 8p^2 v^2$. Then, with $D_0 = 2pv$,

$$\sigma^2_{\text{P-GOE}}(0:\Lambda) = \frac{\overline{S^2}}{\left[\overline{\hat{S}}\right]^2 D_0^2} = \frac{8\Lambda + 2}{\pi[\Psi(-1/2, 0, 2\Lambda)]^2} - 1. \tag{3.36}$$

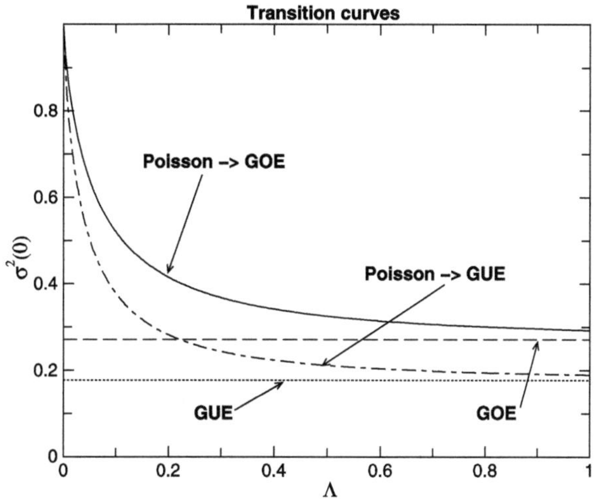

Fig. 3.1 Variance $\sigma^2(0)$ of NNSD vs transition parameter Λ for Poisson to GOE and GUE transitions. Figure is constructed using the results given in [12]. See Sect. 3.2 for details

The complete transition curve, i.e. plot of $\sigma^2(0:\Lambda)$ vs Λ is given in Fig. 3.1. It is instructive to consider small Λ expansion of $\sigma^2(0:\Lambda)$. To this end we start with the identity $\Psi(-1/2, 0, 2\Lambda) = (2\Lambda)\Psi(1/2, 2, 2\Lambda)$ and carry out small Λ expansion for $\Psi(1/2, 2, 2\Lambda)$. Using Eq. (13.1.6) on p. 504 of [15] we have,

$$\begin{aligned}\Psi(1/2,2,z) &= \frac{1}{\Gamma(-\frac{1}{2})}\left[\bigl\{1+O(z)\bigr\}\ln z + \left\{\psi\left(\frac{1}{2}\right) - \psi(1) - \psi(2)\right\} + O(z)\right] + \frac{1}{\Gamma(\frac{1}{2})z} \\ &= \frac{1}{\sqrt{\pi}\, z}\left\{1 - \frac{z\ln z}{2} - \frac{z}{2}\left[\psi\left(\frac{1}{2}\right) - \psi(1) - \psi(2)\right]\right\} + O\bigl(z^2\bigr). \end{aligned} \tag{3.37}$$

Here we used $\Gamma(-\frac{1}{2}) = -2\sqrt{\pi}$ and $\Gamma(\frac{1}{2}) = \sqrt{\pi}$. With $z = 2\Lambda$, $\psi(\frac{1}{2}) = -\gamma - 2\ln 2$, $\psi(1) = -\gamma$ and $\psi(2) = -\gamma + 1$ in Eq. (3.37), $\bar{\hat{S}}$ is

$$\bar{\hat{S}} = 2\sqrt{\pi}\Lambda\, \Psi(1/2, 2, 2\Lambda) = 1 - \Lambda\ln(2\Lambda) + \Lambda[2\ln 2 + 1 - \gamma] + O\bigl(\Lambda^2\bigr). \tag{3.38}$$

Therefore the small Λ expansion for $\sigma^2(0:\Lambda)$ is

$$\sigma^2_{\text{P-GOE}}(0:\Lambda) \xrightarrow{\Lambda\ll 1} 1 + 4\Lambda\bigl\{\ln(2\Lambda) + 1 + \gamma - 2\ln 2)\bigr\} \tag{3.39}$$

where γ is Euler's constant. Note the $\Lambda\ln\Lambda$ term also appears in the small Λ expansion for the number variance $\Sigma^2(1)$; see Eq. (3.51) ahead.

3.2.2 2 × 2 Results for Poisson to GUE Transition

Let us now consider the Poisson to GUE transition. Simplifying Eq. (3.4) after putting $\alpha = \alpha'$ will give,

$$P(S)dS = \frac{S^2 dS}{\sqrt{2\pi}(2\alpha v)^3} \exp - \frac{S^2 + 4p^2v^2\lambda^2}{8\alpha^2 v^2} \times \int_0^{\pi} d\theta \sin\theta \exp \frac{pvS\lambda\cos\theta}{2\alpha^2v^2}. \tag{3.40}$$

Now carrying the θ integration and applying Eq. (3.7), we obtain the NNSD for P to GUE and the final result is, with Λ defined in Eq. (3.33),

$$P_{\text{P-GUE}}(\hat{S})\, d\hat{S} = d\hat{S} \frac{\hat{S}}{\sqrt{2\pi}\,\Lambda^{1/2}} \exp\{-\hat{S}^2/8\Lambda\} \times \int_0^{\infty} \lambda^{-1} \exp\left\{-\lambda - \frac{\lambda^2}{8\Lambda}\right\} \sinh\left(\frac{\hat{S}\lambda}{4\Lambda}\right) d\lambda. \tag{3.41}$$

It should be noted that the mean squared GUE admixing matrix element is $2\alpha^2 v^2$ and hence in this case the transition parameter Λ, used in Eq. (3.41), is mean squared admixing GUE matrix element divided by two times the square of the mean spacing of the Poisson spectrum. Using the integrals given in p. 365 of [16], it is easy to prove that $P_{\text{P-GUE}}(\hat{S})$ is normalized to unity. Similarly, using Eq. (3.10) we have,

$$\overline{\hat{S}} = 4\Lambda \int_0^{\infty} \frac{1}{y} [\exp - \sqrt{8\Lambda}\, y] \Phi(y)\, dy + \left[\sqrt{\frac{8\Lambda}{\pi}} + [\exp 2\Lambda]\left(1 - \Phi(\sqrt{2\Lambda})\right)\right]. \tag{3.42}$$

Carrying out further simplifications using MATHEMATICA (the functions used here are ExpIntegralEi(–), HypergeometricU(–) and HypergeometricPFQ[$\{a, b\}$, $\{c, d\}$, z]), the final result is

$$\overline{\hat{S}} = X(\Lambda) = 2\Lambda \left[-Ei(2\Lambda) + 4\sqrt{2\Lambda/\pi}\, {}_2F_2\left({1/2, 1 \atop 3/2, 3/2}; \; 2\Lambda \right) \right] + \sqrt{8\Lambda/\pi} + [\exp 2\Lambda]\left[1 - \Phi(\sqrt{2\Lambda})\right]. \tag{3.43}$$

Then, the exact expression for $\sigma^2(0:\Lambda)$ is

$$\sigma^2_{\text{P-GUE}}(0:\Lambda) = \frac{12\Lambda + 2}{[X(\Lambda)]^2} - 1. \tag{3.44}$$

In Eq. (3.43) Ei is exponential integral (see p. 228 in [15]),

$$Ei(x) = \gamma + \ln x + \sum_{n=1}^{\infty} \frac{x^n}{n(n!)}.$$

Similarly ${}_2F_2$ is generalized hyper-geometric function,

$${}_2F_2\left(\begin{matrix} a, b \\ c, d \end{matrix}; \; x\right) = 1 + \frac{ab}{cd}x + \frac{a(a+1)b(b+1)}{c(c+1)d(d+1)} \frac{x^2}{2!} + \cdots.$$

The complete Poisson to GUE transition curve for $\sigma^2(0:\Lambda)$ vs Λ, from Eq. (3.44) is given in the Fig. 3.1. Once again it is instructive to consider the small Λ expansion for $\sigma^2_{\text{P-GUE}}(0:\Lambda)$. Note that,

$$\begin{aligned} X(\Lambda) &= 2\Lambda\left[\left\{-\gamma - \ln(2\Lambda) + O(\Lambda)\right\} + 4\sqrt{\frac{2\Lambda}{\pi}} + O\left(\Lambda^{3/2}\right)\right] \\ &\quad + \sqrt{\frac{8\Lambda}{\pi}} + \left[1 + 2\Lambda + O\left(\Lambda^2\right)\right]\left[1 - \frac{2}{\sqrt{\pi}}\sqrt{2\Lambda} + O\left(\Lambda^{3/2}\right)\right] \\ &= 1 + 2\Lambda[1 - \gamma - \ln 2 - \ln\Lambda] + O\left(\Lambda^{3/2}\right). \end{aligned} \tag{3.45}$$

Now Eq. (3.44) gives,

$$\sigma^2_{\text{P-GUE}}(0:\Lambda) \xrightarrow{\Lambda \ll 1} 1 + 8\Lambda\left(\ln(\Lambda) + \frac{1}{2} + \gamma + \ln 2\right). \tag{3.46}$$

Just as in the case of Poisson to GOE, here also there is the $\Lambda \ln \Lambda$ term. The approximation in Eq. (3.46) is good for $\Lambda \lesssim 0.05$.

3.2.3 *Relationship Between Λ Parameter for Poisson to GOE and the Berry-Robnik Chaos Parameter*

There are several different formulas, given by Brody [26], Berry and Robnik [27, 28], Izrailev [29, 30], Blocki [31] and many others for the NNSD $P_{\text{P-GOE}}(S)\,dS$ and $P_{\text{P-GUE}}(S)\,dS$. For example, the well known Brody (Br) distribution for Poisson to GOE transition, with the Brody parameter ω is [26]

$$P^{\text{B}}_{\text{P-GOE}}(S)\,dS = aS^{\omega}\exp\left\{-bS^{\omega+1}\right\}; \quad a = (\omega+1)b, \; b = \left\{\Gamma\left(\frac{\omega+2}{\omega+1}\right)\right\}^{\omega+1} \tag{3.47}$$

and it reduces to Poisson for $\omega = 0$ and Wigner (GOE) form for $\omega = 1$. For $0 < \omega < 1$, the distribution given by Eq. (3.47) vanishes as $S \to 0$ but has an infinite derivative at that point, an unrealistic feature. Recently a physical process that generates the Brody distribution has been identified [32] and here the Brody parameter corresponds to an appropriate fractal dimension.

The one parameter (ρ) Berry-Robnik (BR) formulas for Poisson to GOE and GUE are,

$$
\begin{aligned}
P^{BR}_{\text{P-GOE}}(S)\,dS &= (1-\rho)^2 \exp\{-(1-\rho)S\}\, \mathit{erfc}(\sqrt{\pi}\rho S/2) \\
&\quad + \big(2(1-\rho)\rho + \pi\rho^3 S/2\big) \exp\{-(1-\rho)S - \pi\rho^2 S^2/4\}, \\
P^{BR}_{\text{P-GUE}}(S)\,dS &= \big(2\rho(1-\rho) - (1-\rho)^2 \rho\, S\big) \exp\{-(1-\rho)S\}\, \mathit{erfc}\left(\frac{2}{\sqrt{\pi}}\rho\, S\right) \\
&\quad + \left(\frac{32}{\pi^2}\rho^4 S^2 + \frac{8}{\pi}(1-\rho)\rho^2 S + (1-\rho)^2\right) \\
&\quad \times \exp\left\{-(1-\rho)S - \frac{4}{\pi}\rho^2 S^2\right\}
\end{aligned}
\tag{3.48}
$$

where ρ is fractional volume, in phase space, of the chaotic region and $1-\rho$ is fractional volume of all regular regions put together. The BR forms are good when there is only one dominant chaotic region coexisting with regular regions. Note that $\rho = 0$ gives Poisson and $\rho = 1$ Wigner (GOE or GUE). Modification of BR distribution (flooding- and tunneling-improved BR) has been discussed recently in [33]. Now we will consider the relationship between the 2×2 results and the BR distribution given by Eq. (3.48) in order to give a physical meaning to the Λ parameter.

The transition curves given in Fig. 3.1 show that the Poisson to GOE and Poisson to GUE transitions are nearly complete for $\Lambda \sim \Lambda_c = 0.3$. The results in Eqs. (3.36) and (3.44) are in fact applicable to general $N \times N$ matrices (or for any interacting many particle system) through the transition parameter Λ by giving appropriate interpretations to $\alpha^2 v^2$ and D_0 in the expression for Λ; this is indeed verified by the results in Fig. 4 of [10]. With this, the results in Eqs. (3.36) and (3.44) can be applied to realistic systems. An important question is: what is the significance of the numerical value 0.3 of Λ_c for Poisson to GOE (similarly for Poisson to GUE)? Toward this end, in [12] relationship between Λ and the BR parameter ρ (ρ representing fractional volume, in phase space, of the chaotic region of a complex dynamical system) for P-GOE transition was explored. Equation (30) of [27] gives $\sigma^2(0:\rho)$ for the BR $P(S)dS$ as a function of the ρ parameter (ρ changing from 0 to 1),

$$
\sigma^2_{\text{P-GOE:BR}}(0:\rho) = \frac{2}{1-\rho}\left[1 - \exp\frac{1-\rho^2}{\pi\rho^2}\,\Phi\left(\frac{1-\rho}{\sqrt{\pi}\,\rho}\right)\right]. \tag{3.49}
$$

Say $\Lambda_{\text{BR}} = \rho/(1-\rho)$ so that Λ_{BR} changes from 0 to ∞ just as Λ. Fitting Eq. (3.49) to the curves in the Fig. 3.1, it is seen that [12],

$$
\Lambda \simeq \frac{\Lambda_{\text{BR}}}{20} = \frac{\rho}{20(1-\rho)} \quad \text{for } \Lambda \gtrsim 0.05. \tag{3.50}
$$

However for $\Lambda \lesssim 0.01$ results of Eq. (3.36) and the corresponding BR formula differ significantly. Equation (3.50) gives $\rho = 0.85$ for $\Lambda = 0.3$. Thus 85 % chaoticity can be used as a guide for deciding the marker for order (Poisson) to chaos (GOE) transition. For example, using sufficient number of energy levels near ground states or near the yrast line at high spins as the case may be in atomic nuclei (similarly in other interacting many particle systems such as atoms, molecules etc.), it is possible to deduce the corresponding $\sigma^2(0)$ values. Then from Fig. 3.1 one can read-off the value of Λ (or, depending on the sample size errors, determine a bound on Λ) for Poisson to GOE transitions in these systems. Converting this to ρ gives information about the amount of chaoticity in the system. If it is 85 % (i.e. $\Lambda \geq 0.3$), then one can argue that chaos has set in. This approach was used in deriving the order-chaos border in interacting fermion [34] and boson systems [35].

3.2.4 Poisson to GOE, GUE Transitions: N x N Ensemble Results for $\Sigma^2(r)$

Without going in details here we give the formulas, valid for $N \times N$ matrices, for $\Sigma^2(\overline{n}, \Lambda)$ for Poisson to GOE and GUE transitions. They, valid for $\overline{n} \gg \Lambda^{1/2}$, are [1]

$$
\begin{aligned}
\Sigma^2_{\text{P-GOE}}(\overline{n}, \Lambda) &\xrightarrow{\Lambda \ll 1} \overline{n} - 2\Lambda\left(\ln \frac{\overline{n}^2}{2\Lambda} + \gamma - 1 + \ln 4\right), \\
\Sigma^2_{\text{P-GUE}}(\overline{n}, \Lambda) &\xrightarrow{\Lambda \ll 1} \overline{n} - 4\Lambda\left(\ln \frac{\overline{n}^2}{2\Lambda} + \gamma\right).
\end{aligned}
\tag{3.51}
$$

More general discussion of Poisson to GUE (and GOE) transitions for $N \times N$ matrices is given in [3, 5–7].

3.2.5 Onset of Chaos at High Spins via Poisson to GOE Transition

Stephens et al. [36, 37] developed a novel technique to measure the chaoticity parameter ($\Lambda^{1/2}$) for order-chaos transition in rotational nuclei. With $\overline{D}$ giving the average spacing of the levels that are mixed and v giving the r.m.s. admixing matrix element, $\Lambda^{1/2} = v/\overline{D}$. Extending Wigner's 2×2 matrix formalism, the variance of the NNSD for Poisson to GOE transition is given by Eqs. (3.36) and (3.39). As discussed before, the Poisson to GOE transition is nearly complete for $\Lambda \sim 0.3$. and $\Lambda \simeq \frac{\rho}{20(1-\rho)}$ where ρ represents fractional volume, in phase space, of the chaotic region of a complex dynamical system. From the experiments for Yb isotopes, Stephens et al. deduced that $\Lambda^{1/2} \sim 0.15$ to 1.5. Thus at present it is not possible to make a definite statement about onset of chaoticity in the Yb isotopes.

3.3 2 x 2 Partitioned GOE

Let us consider the H matrix $H = H_0 + \alpha V$ where H_0 is a 2×2 block matrix with dimension $d = d_1 + d_2$ (d_1 is dimension of the upper block and d_2 of the lower block) and V is a d dimensional GOE(v^2). Note that GOE(v^2) stands for GOE random matrix ensemble with diagonal matrix elements of the matrices in the ensemble being $G(0, 2v^2)$ and off-diagonal matrix elements $G(0, v^2)$. We put the off-diagonal blocks of H_0 to zero and represent the upper block $\{H_{0;11}\}$ with dimension d_1 by a GOE(v_1^2) where $v_1^2 = v^2(d_1 + d_2)/d_1$ and similarly the lower block $\{H_{0;22}\}$ with dimension d_2 by a GOE(v_2^2) with $v_2^2 = v^2(d_1 + d_2)/d_2$. Thus, $\alpha = 0$ corresponds to a superposition of two GOE's and $\alpha \to \infty$ gives a single GOE. For this 2×2 partitioned GOE, binary correlation approximation gives $A_\zeta = 2(1 + [1 + \alpha^2]^{-\zeta}) \sim 2(2 - \zeta\alpha^2)$. Recall that for GOE to GUE transition we have $A_\zeta = 1 + (\frac{1-\alpha^2}{1+\alpha^2})^\zeta \sim 2(1 - \zeta\alpha^2)$. Therefore it is easy to modify the GOE-GUE derivation and derive the following result, with the transition parameter $\Lambda = \alpha^2 v^2 / \overline{D}^2$, for the number variance [1],

$$\Sigma^2_{2\times 2}(r, \Lambda) = \Sigma^2(r, \infty) + \frac{1}{\pi^2} \ln\left\{1 + \frac{\pi^2 r^2}{4(\tau + \pi^2\Lambda)^2}\right\}. \tag{3.52}$$

The cut-off parameter τ is determined using the result

$$\Sigma^2_{2\times 2}(r, 0) = \Sigma^2_{\text{GOE}}\big([d_1/d]r\big) + \Sigma^2_{\text{GOE}}\big([d_2/d]r\big)$$

and Eq. (3.52) is good for $r > 2$. Note that $\Sigma^2(r, \infty)$ is the GOE value. As discussed before, $\Sigma^2(r)$ formula gives the expression for $\overline{\Delta}_3(r)$,

$$\begin{aligned}\overline{\Delta}_{3:2\times 2}(r, \Lambda) = \overline{\Delta}_3(r, \infty) + \frac{1}{\pi^2}\bigg\{&\left[\frac{1}{2} - \frac{2}{X^2 r^2} - \frac{1}{2X^4 r^4}\right]\ln\big(1 + X^2 r^2\big) \\ &+ \frac{4}{Xr}\tan^{-1}(Xr) + \frac{1}{2X^2 r^2} - \frac{9}{4}\bigg\}; \quad X = \frac{\pi}{2(\tau + \pi^2\Lambda)}.\end{aligned} \tag{3.53}$$

A direct and good test of Eq. (3.53) came recently from experiments with two coupled flat superconducting microwave billiards [38].

3.3.1 *Isospin Breaking in ^{26}Al and ^{30}P Nuclear Levels*

Shriner and Mitchell [39, 40] considered complete spectroscopy for levels up to $\sim$8 MeV excitation in ^{26}Al and ^{30}P. For ^{26}Al, there are 75 $T = 0$ and 25 $T = 1$ levels with $J^\pi = 1^+$ to 5^+. Similarly in ^{30}P there are 69 $T = 0$ and 33 $T = 1$ levels with $J^\pi = 0^\pm$ to $5^\pm$. With Coulomb interaction breaking isospin, the appropriate random matrix model here is 2×2 GOE giving 2GOE to 1GOE transition. Analysis of data for ^{26}Al and ^{30}P is in good agreement with 2GOE to 1GOE transition. In particular, using ^{26}Al the data was analyzed using a slightly different 2×2 random matrix ensemble [41],

$$\begin{bmatrix} \mathrm{GOE}((d/d_0)v^2) & \alpha V_c(2v^2) \\ \widetilde{\alpha V_c(2v^2)} & \mathrm{GOE}((d/d_1)v^2) \end{bmatrix} \tag{3.54}$$

in $|T=0\rangle$ and $|T=1\rangle$ basis with the dimension d_0 of the $T=0$ space being $d_0 = 75$ and the dimension d_1 of the $T=1$ space being $d_1 = 25$. Then the total dimension $d = 100$. Analysis of data gave $\alpha = 0.056$ and $\overline{H_{ij}^2(c)} = \overline{\alpha^2 V_{c;ij}^2} \sim (20\ \mathrm{keV})^2$. The corresponding spreading width $\Gamma = 2\pi \overline{H_{ij}^2(c)}/\overline{D}$ is $\Gamma \sim 32$ keV. There is also an analysis of reduced transition probabilities (with about 1500 transitions) in ^{26}Al showing deviations from P-T [42]. Let us consider this in some detail.

For GOE, given the strengths $R(E_i, E_f) = |\langle E_i|\mathcal{O}|E_f\rangle|^2$, the locally renormalized transition strengths $x = R(E_i, E_f)/\overline{R(E_i, E_f)}$ are distributed according to the Porter-Thomas (P-T) law. Deviations from P-T law could be ascribed to symmetry breaking and then the questions are: (i) where to look for good data; (ii) what is the appropriate random matrix ensemble and what are its predictions. Adams et al. [42] collected data for reduced electromagnetic transition matrix elements in ^{26}Al from ground state to 8 MeV excitation. The data divides into 120 different transition sequences with each of them having about 10 matrix elements; a transition sequence is defined by initial $J_i^{\pi_i} T_i$ going to all $J_f^{\pi_f} T_f$ (with no missing transitions in between) for a given $B_T^L(E \text{ or } M)$ where L is multipole rank and $T = 0$ for isoscalar (IS) and $T = 1$ for isovector (IV) transitions. In the data set there are 211 $E1$ IS, 172 $E1$ IV, 358 $M1$ IV and 132 $E2$ IS transition matrix elements. Instead of the locally renormalized strengths x, distribution of $z = \log(x)$ is plotted by combining all the data with a proper prescription. The P-T form gives maximum at $z = 0$ while data shows the peak at ≈ -0.5. It is conjectured that this deviation is a consequence of isospin breaking. The random matrix model now consists of the 2×2 partitioned GOE for the Hamiltonian as given by Eq. (3.54) and in the same basis an independent 2×2 partitioned GOE for the transition operator $\mathcal{O}$ [43],

$$\mathcal{O} = \beta_{\mathrm{IS}} \begin{bmatrix} \mathcal{O}(0) & 0 \\ 0 & \mathcal{O}(1) \end{bmatrix} + \beta_{\mathrm{IV}} \begin{bmatrix} 0 & \mathcal{O}_c \\ \widetilde{\mathcal{O}_c} & 0 \end{bmatrix}. \tag{3.55}$$

Here, $\beta_{\mathrm{IS}} = 1$ and $\beta_{\mathrm{IV}} = 0$ for IS and $\beta_{\mathrm{IS}} = 0$ and $\beta_{\mathrm{IV}} = 1$ for IV transitions. Determining appropriately the scale parameters of the various GOE's in the H and $\mathcal{O}$ ensembles, Barbosa et al. [43] recently constructed $P(z)\,dz$ via numerical calculations by transforming the ensemble in Eq. (3.55) into $\{H\}$ basis via the unitary matrices that diagonalize H's. The random matrix model correctly predicts the shift in the peak with respect to P-T. However the data are more strongly peaked and at present there is no quantitative understanding of this feature.

3.4 Rosenzweig-Porter Model: Analysis of Atomic Levels and Nuclear 2^+ and 4^+ Levels

Rosenzweig-Porter model [44] is the appropriate random matrix model when we consider a set of levels in a spectrum $\mathscr{S}$ and the levels in $\mathscr{S}$ differ containing conserved quantum numbers which are either unknown or ignored. Then, the spectrum can be broken into r sub-spectra $\mathscr{S}_j$ of independent sequences of levels with $j = 1, 2, \ldots, r$. In the set of levels considered for the analysis of $P(S)$ (i.e. NNSD), say the fraction of levels from $\mathscr{S}_j$ is f_j. Then, $0 < f_j \leq 1$ and $\sum_{j=1}^{r} f_j = 1$. Now an appropriate random matrix model is to represent each subspace $\mathscr{S}_j$ by independent GOE's of dimension $d_j = df_j$ where d is the size of $\mathscr{S}$. NNSD for such an ensemble was first considered by Rosenzweig-Porter (RP) [44] and they showed that, with $f_j = 1/r; r \to \infty$, $P(S)$ goes to Poisson. This model has been employed in discussing LS to JJ coupling change in atomic spectra. Exact solutions for the RP model are given very recently [45, 46] but they are not useful in data analysis. Abul-Magd derived a simplified formula for $P(S)$ in terms of the chaoticity parameter $f = \sum_{j=1}^{r} f_j^2$ and it is [47],

$$P(S)dS = \left[1 - f + Q(f)\pi S/2\right] \exp\left[-(1 - f)S - Q(f)\pi S^2/4\right] \tag{3.56}$$

where $Q(f) = f(0.7 + 0.3f)$. With $f = 1/r$, $r \to \infty$, $P(S)$ goes to Poisson and $f = 1$ gives GOE. Abul-Magd, Harney, Simbel and Weidenmüller [48, 49] analyzed the NNSD of low-lying 2^+ levels (up to ∼4 MeV excitation) for Poisson to GOE transition using Eq. (3.56). They considered 1306 levels belonging to 169 nuclei (with a minimum of 5 consecutive 2^+ levels in a given nucleus). The nuclei are grouped into classes defined by the collectivity parameter $E(4_1^+)/E(2_1^+)$. In the system considered, departures from GOE arise due to the neglect of possibly good quantum numbers. Using Bayesian inference method, values of f are deduced and it is found to be small for nuclei with IBM symmetries while for the intermediate nuclei $f \sim 0.6$.

Equation (3.56) was also applied in the analysis NNSD for 2^+ levels of prolate and oblate deformed nuclei by Al-Sayed and Abul-Magd [50]. They considered 30 nuclei of oblate deformation having 246 levels and 83 nuclides of prolate deformation having 590 energy levels ranging from ^{28}Si to ^{228}Ra. Analysis showed that the chaoticity parameter f is ∼0.73 for prolate nuclei and ∼0.59 for oblate nuclei suggesting that oblate nuclei are more regular compared to prolate nuclei.

It may be useful to note that a formula for the number variance $\Sigma^2(r)$ for the RP model was given in [1] and this is not yet used in any data analysis. All the nuclear data that is analyzed so far using RMT is shown in Fig. 3.2 in the angular momentum and energy plane.

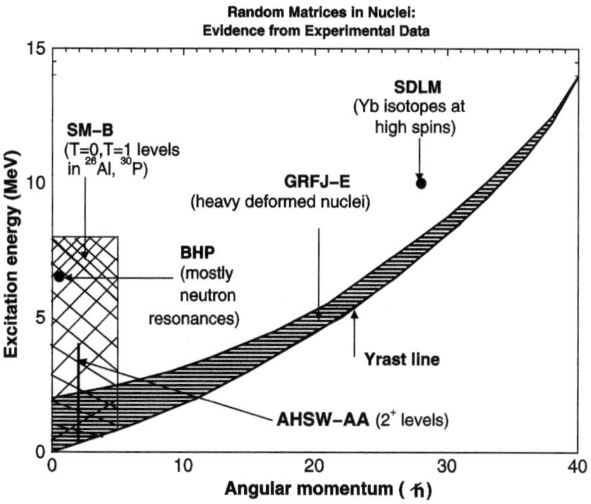

Fig. 3.2 Schematic diagram giving the regions, in the excitation energy vs angular momentum plane for nuclei, where data was analyzed for evidence for random matrices (GOE and its extensions). Details of BHP [51], SM-B [39, 40, 43], AHSW-AA [48–50], GRFJ-E [52, 53], and SDLM [36, 37] are given in the text (Color figure online)

3.5 Covariance Random Matrix Ensemble XX^T: Eigenvalue Density

Let us consider a $N \times M$ matrix X with matrix elements real and chosen to be independent $G(0, v^2)$ variables. Then the $N \times N$ random matrix ensemble $C = XX^T$, where X^T is the transpose of X, represents a GOE related covariance random matrix ensemble (GOE-CRME). It is possible to consider many other types of CRME's as discussed for example in [54–56]. The CRME's have wide ranging applications. For example: (i) they are important in multivariate statistical analysis [57]; (ii) they are used in the study of cross-correlations in financial data [58–60]; (iii) they appear in a model for mixing between distant configurations in nuclear shell model [61]; (iv) they are relevant for statistical analysis of correlations in atmospheric data [62]; (v) they determine statistical bounds on entanglement in bipartite quantum systems due to quantum chaos [63].

In this section we consider the ensemble averaged density $\rho^C(E)$ of the eigenvalues of GOE-CRME $C = XX^T$. Dyson [64, 65] first derived the result for $\rho^C(E)$ for the matrices X with $N = M$. In most applications the eigenvalue density for $N \neq M$ is needed. The result for this situation was derived using many different techniques; see [54, 55, 57, 66] and references therein. Equation (3.73) ahead gives the final result. It is indeed possible to obtain $\rho^C(E)$ using the 2×2 partitioned GOE (p-GOE:2) employed in nuclear structure studies as a statistical model for mixing between distant nuclear shell model configurations [61, 67]. This gives an easy to understand derivations of the final result [68].

3.5.1 A Simple 2 × 2 Partitioned GOE: p-GOE:2(Δ)

Let us consider two spaces #1 and #2 with dimensions d_1 and d_2 respectively . For a simple statistical model for the mixing between the spaces #1 and #2, one can assume, as a first step, that all the eigenvalues in #1 are degenerate and say their value is 0. Similarly one may assume that the eigenvalues in #2 are also degenerate with their value say Δ. More importantly, these two spaces will mix and the mixing Hamiltonian X will be a $d_1 \times d_2$ matrix. A plausible model for X is to replace it by a GOE, i.e. assume that the matrix elements of X are independent $G(0, v^2)$ variables. Then we have a 2 × 2 block structured random matrix ensemble,

$$H_\Delta = \begin{bmatrix} 0 I_1 & X \\ X^T & \Delta\, I_2 \end{bmatrix}. \tag{3.57}$$

This ensemble is called p-GOE:2(Δ). Note that the matrices I_1 and I_2 are unit matrices with dimensions d_1 and d_2 respectively and the H matrix dimension is $d = d_1 + d_2$. Now let us consider the eigenvalue density $\rho^\Delta(E)$ for the matrix H_Δ. The $\rho^\Delta(E)$ is simply,

$$\rho^\Delta(E) = \big\langle \delta(H_\Delta - E)\big\rangle^{1+2}, \tag{3.58}$$

and its decomposition into sum of the partial densities $\rho^{\Delta;1}$ and $\rho^{\Delta;2}$ defined over the spaces #1 and #2 respectively is given by,

$$\rho^{\Delta;i}(E) = \big\langle \delta(H_\Delta - E)\big\rangle^i; \qquad \rho^\Delta(E) = d^{-1}\big[d_1\rho^{\Delta;1}(E) + d_2\rho^{\Delta;2}(E)\big]. \tag{3.59}$$

As we will see ahead, the densities $\rho^{\Delta;1}(E)$ and $\rho^{\Delta;2}(E)$ differ only in a delta function. Therefore from now on we will consider only $\rho^{\Delta;1}(E)$ and also assume that $d_1 < d_2$. For mathematical simplicity, as an intermediate step, we will consider the matrix ensemble $H_{\pm\Delta'}$,

$$H_{\pm\Delta'} = \begin{bmatrix} -\Delta' I_1 & X \\ X^T & \Delta' I_2 \end{bmatrix} \tag{3.60}$$

and the corresponding $\rho^{\pm\Delta';1}(E)$. Denoting the p-th moment of this density by $M_p^{\pm\Delta';1}$, we have, with $[\frac{p}{2}]$ being the integer part of $\frac{p}{2}$,

$$M_p^{\pm\Delta';1} = (-1)^p \sum_{\nu=0}^{[\frac{p}{2}]} \binom{[\frac{p}{2}]}{\nu} (\Delta')^{p-2\nu} \big\langle (XX^T)^\nu \big\rangle^1. \tag{3.61}$$

Equation (3.61) is derived by multiplying $H_{\pm\Delta'}$ p-times and then using the first diagonal block of the resulting 2 × 2 block matrix. Similarly $M_{2\nu}^{\pm\Delta';2} = (d_1/d_2)M_{2\nu}^{\pm\Delta';1}$ and $M_{2\nu+1}^{\pm\Delta';2} = -(d_1/d_2)M_{2\nu+1}^{\pm\Delta';1}$ with $M_0^{\pm\Delta';1} = M_0^{\pm\Delta';2} = 1$. Equation (3.61) shows that there should be a generalized convolution form for

$\rho^{\pm\Delta';1}(E)$ with one of the factors being $\rho^{\Delta'=0;1}$ as the moments for a density written as a convolution of two functions follow the law $M_p(\rho_A \otimes \rho_B) = \sum \binom{p}{s} M_s(A) M_{p-s}(B)$; here $\otimes$ denotes convolution. From Eq. (3.61) we have $M_{2\nu+1}^{\pm\Delta';1} = -\Delta' M_{2\nu}^{\pm\Delta';1}$. Then $\int_{-\infty}^{\infty} E^{2\nu}(E+\Delta')\rho^{\pm\Delta';1}(E)dE = 0$ and also $\int_{-\infty}^{\infty} E^{2\nu}(E-\Delta')\rho^{\pm\Delta';1}(-E)dE = 0$. They imply that $\rho^{\pm\Delta';1}(E)$ is of the form $|\frac{E-\Delta'}{E+\Delta'}|^{1/2} f(E)$ where $f(E)$ is an even function of E. This and the fact that $\langle (XX^T)^\nu \rangle^1$ is the 2ν-th moment of $\rho^{0:1}(E)$, allow us to identify the following important result,

$$\rho^{\pm\Delta';1}(E) = \left|\frac{E-\Delta'}{E+\Delta'}\right|^{\frac{1}{2}} \rho^{\Delta'=0;1}\left(\sqrt{E^2-(\Delta')^2}\right), \quad |E| \geq \Delta'. \tag{3.62}$$

Now, putting $\Delta' = \frac{\Delta}{2}$ and shifting all the eigenvalues E by $\Delta/2$ so that $E \to (E - \frac{\Delta}{2})$, the final result for $\rho^{\Delta;1}$ is obtained,

$$\rho^{\Delta;1}(E) = \left|\frac{E-\Delta}{E}\right|^{\frac{1}{2}} \rho^{\Delta=0;1}\left(\sqrt{E(E-\Delta)}\right), \quad E \geq \Delta,\ E \leq 0. \tag{3.63}$$

Equation (3.63) was reported first in [61]. Now we will consider $\rho^{\Delta=0;1}(E)$ for p-GOE:2($\Delta = 0$).

3.5.2 Moments and the Eigenvalue Density for p-GOE:2($\Delta = 0$)

Given $H_0 = \begin{bmatrix} 0I_1 & X \\ X^T & 0I_2 \end{bmatrix}$, mathematical induction gives,

$$(H_0)^{2\nu} = \begin{bmatrix} (XX^T)^\nu & 0 \\ 0 & (X^TX)^\nu \end{bmatrix}, \qquad (H_0)^{2\nu+1} = \begin{bmatrix} 0 & (XX^T)^\nu X \\ (X^TX)^\nu X^T & 0 \end{bmatrix}. \tag{3.64}$$

Then, immediately we have $\langle\langle (H_0)^p \rangle\rangle^1 = \langle\langle (H_0)^p \rangle\rangle^2$ for $p \neq 0$ and for $p = 0$, they are d_1 and d_2 respectively. Secondly, all the odd moments of $\rho^{\Delta=0;1}(E)$ are zero. Also, for $d_1 < d_2$, $\rho^{\Delta=0;2}(E) = \frac{d_1}{d_2}\rho^{\Delta=0;1}(E) + (1 - \frac{d_1}{d_2})\delta(E)$. One way of constructing $\rho^{\Delta=0;1}(E)$ is via its moments. From Eq. (3.64) we have $M_{2\nu}(\rho^{\Delta=0;1}) = \langle (XX^T)^\nu \rangle^1$ and they can be evaluated for p-GOE:2 using BCA discussed earlier. Firstly, the ensemble averaged second moment simply is,

$$M_2(\rho^{\Delta=0;1}) = (d_1)^{-1} \sum_{i,j} \overline{X_{ij}(X^T)_{ji}} = (d_1)^{-1} \sum_{i,j} \overline{X_{ij}^2} = v^2 d_2. \tag{3.65}$$

Similarly, defining $\widetilde{M}_p = d_1 M_p = \langle\langle (H_0)^p \rangle\rangle^1$, we have

$$\widetilde{M}_4(\rho^{\Delta=0;1}) = \sum_{i,j,k,l} \overline{X_{ij}(X^T)_{jk} X_{kl} (X^T)_{li}} = \sum_{i,j,k,l} \overline{X_{ij}X_{kj}X_{kl}X_{il}}.$$

In the sum here, applying BCA, we need to consider only terms that contain pairwise correlations. Then, with $k = i$ or $l = j$,

$$
\begin{aligned}
\widetilde{M}_4\big(\rho^{\Delta=0;1}\big) &= \sum_{i,j,l}[\overline{X_{ij}X_{ij}X_{il}X_{il}}] + \sum_{i,j,k}[\overline{X_{ij}X_{kj}X_{kj}X_{ij}}] \\
&= \sum_{i,j,l}[\overline{X_{ij}X_{ij}}\;\overline{X_{il}X_{il}}] + \sum_{i,j,k}[\overline{X_{ij}X_{ij}}\;\overline{X_{kj}X_{kj}}] \\
&= v^4\big[d_1 d_2^2 + d_1^2 d_2\big].
\end{aligned}
\tag{3.66}
$$

The two terms in Eq. (3.66) can be written as $\langle \underbrace{XX^T}\,\underbrace{XX^T}\rangle$ and $\langle \underbrace{X\underbrace{X^T X}X^T}\rangle$. The terms that are dropped in Eq. (3.66) involve cross correlations, i.e. terms with odd number of matrix elements in between those that are correlated. They will be smaller by a factor of d_2 (or d_1). Thus BCA here is good if d_1 and d_2 both are large. Proceeding further we have for $\widetilde{M}_6$,

$$
\begin{aligned}
\widetilde{M}_6\big(\rho^{\Delta=0;1}\big) &= \sum_{i,j,k,l,m,n} \overline{X_{ij}X_{kj}X_{kl}X_{ml}X_{mn}X_{in}} \\
&= \sum_{i,j,l,n} \overline{X_{ij}X_{ij}X_{il}X_{il}X_{in}X_{in}} + \sum_{i,j,l,m} \overline{X_{ij}X_{ij}X_{il}X_{ml}X_{ml}X_{il}} \\
&\quad + \sum_{i,j,k,n} \overline{X_{ij}X_{kj}X_{kj}X_{ij}X_{in}X_{in}} + \sum_{i,j,k,m} \overline{X_{ij}X_{kj}X_{kj}X_{mj}X_{mj}X_{ij}} \\
&\quad + \sum_{i,j,k,l} \overline{X_{ij}X_{kj}X_{kl}X_{kl}X_{kj}X_{ij}} \\
&= v^6\big(d_1 d_2^3 + 3d_1^2 d_2^2 + d_1^3 d_2\big).
\end{aligned}
\tag{3.67}
$$

The binary correlation structure in Eq. (3.67) is clear and let us apply it to $\widetilde{M}_8$. Writing X_{ij} as X_a, symbolically $\widetilde{M}_8 = \sum \overline{X_a X_b X_c X_d X_e X_f X_g X_h}$. Now: (i) with X_a and X_b correlated, the correlations in the remaining $X_c X_d X_e X_f X_g X_h$ are same as those in $\widetilde{M}_6$; (ii) with X_a and X_d correlated, necessarily X_b and X_c must be correlated and the remaining $X_e X_f X_g X_h$ correlations are same as those in $\widetilde{M}_4$; (iii) with X_a and X_f correlated, necessarily X_g and X_h must be correlated and the remaining $X_b X_c X_d X_e$ correlations are same as those in $\widetilde{M}_4$; (iv) with X_a and X_h correlated, the correlations in the remaining $X_b X_c X_d X_e X_f X_g$ are same as those in $\widetilde{M}_6$. Then the expression for $\widetilde{M}_8$ is,

$$
\widetilde{M}_8\big(\rho^{\Delta=0;1}\big) = v^8\big[d_1^4 d_2 + 6d_1^3 d_2^2 + 6d_1^2 d_2^3 + d_1 d_2^4\big].
\tag{3.68}
$$

Continuing this will lead to a recursion formula for the moments,

$$\widetilde{M}_{2\nu}\left(\rho^{\Delta=0;1}\right) = v^2 \sum_{r=0,2,\ldots,2\nu-2} \widetilde{M}_{2\nu-2-r}\left(\rho^{\Delta=0;1}\right)\widetilde{M}_r\left(\rho^{\Delta=0;2}\right); \quad \nu \geq 1,$$
$$\widetilde{M}_{2\nu}\left(\rho^{\Delta=0;1}\right) = \widetilde{M}_{2\nu}\left(\rho^{\Delta=0;2}\right) \quad \text{for } \nu \neq 0, \qquad \widetilde{M}_0\left(\rho^{\Delta=0;1}\right) = d_1, \tag{3.69}$$
$$\widetilde{M}_0\left(\rho^{\Delta=0;2}\right) = d_2.$$

For example using Eq. (3.69) we have $\widetilde{M}_{10} = v^{10}[d_1^5 d_2 + 10 d_1^4 d_2^2 + 20 d_1^3 d_2^3 + 10 d_1^2 d_2^4 + d_1 d_2^5]$. With all the moments determined, it is possible to identify the density $\rho^{\Delta=0;1}$. Integral tables in [69], the expression for M_4 given by (3.66) and $\widetilde{M}_{2\nu}$, $\nu = 1, 2, 3, 4$ for $d_1 = d_2$ allow us to write the final solution,

$$\rho^{\Delta=0;1}(E)dE = \frac{1}{2\pi v^2 d_1} \frac{\sqrt{(R_+^2 - E^2)(E^2 - R_-^2)}}{|E|} dE, \quad R_- \leq |E| \leq R_+;$$
$$R_\pm = v(\sqrt{d_2} \pm \sqrt{d_1}). \tag{3.70}$$

Note that $\rho^{\Delta=0;1}(E) = 0$ for $|E| < R_-$ or $|E| > R_+$ and also it is a semicircle for $d_1 = d_2$. The reduced moments $\widetilde{M}_{2\nu} = M_{2\nu}/(M_2)^\nu$ of $\rho^{\Delta=0;1}(E)$ are,

$$\widetilde{M}_{2\nu}\left(\rho^{\Delta=0;1}\right) = \frac{(1+\sqrt{R_0})^{2\nu+2}}{\pi R_0} \int_{\overline{R_0}}^{1} x^{2\nu-1} \sqrt{\left(1-x^2\right)\left(x^2 - \overline{R_0}^2\right)}\, dx;$$
$$R_0 = \frac{d_1}{d_2}, \quad \overline{R_0} = \frac{1-\sqrt{R_0}}{1+\sqrt{R_0}}. \tag{3.71}$$

3.5.3 Eigenvalue Density for GOE-CRME

Our primary interest is to determine the eigenvalue density $\rho^C(E)$ for the GOE-CRME $C = XX^T$ where X is a $d_1 \times d_2$ matrix with its matrix elements being independent $G(0, v^2)$ variables; we assume $d_1 \leq d_2$. From Eq. (3.64) it is seen easily that the ν-th moment of ρ^C and the 2ν-th moment of $\rho^{\Delta=0;1}$ are simply related, $M_\nu(\rho^C) = M_{2\nu}(\rho^{\Delta=0;1})$. As $\rho^{\Delta=0;1}$ is an even function, we have

$$M_{2\nu}\left(\rho^{\Delta=0;1}\right) = 2\int_0^\infty E^{2\nu} \rho^{\Delta=0;1}(E)\, dE = \int_0^\infty y^\nu \left[\frac{\rho^{\Delta=0;1}(y^{1/2})}{y^{1/2}}\right] dy$$
$$\Longrightarrow \quad \rho^C(y) = \frac{\rho^{\Delta=0;1}(y^{1/2})}{y^{1/2}}. \tag{3.72}$$

Now, the formula for ρ^C follows simply from Eq. (3.70),

$$\rho^C(\lambda)\, d\lambda = \frac{1}{2\pi v^2 d_1} \frac{\sqrt{(\lambda_+ - \lambda)(\lambda - \lambda_-)}}{\lambda} d\lambda, \quad \lambda_- \leq \lambda \leq \lambda_+;$$

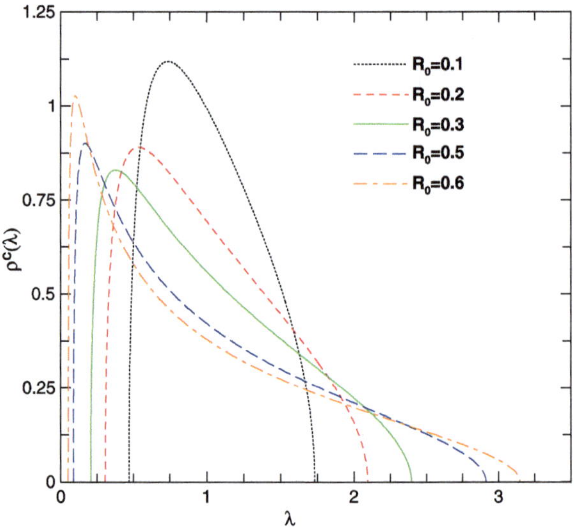

Fig. 3.3 Eigenvalue density for GOE-CRME ensemble for different values of $R_0 = d_1/d_2$. Equation (3.74) gives the formula for the eigenvalue density

$$\lambda_{\pm} = v^2(\sqrt{d_2} \pm \sqrt{d_1})^2 = \left(v^2 d_1 d_2\right)\frac{1}{d_1}\left[1 + R_0 \pm 2\sqrt{R_0}\right]. \tag{3.73}$$

With the normalization $v^2 d_2 = 1$, we have

$$\rho^C(\lambda)\,d\lambda = \frac{1}{2\pi R_0}\frac{\sqrt{(\lambda_+ - \lambda)(\lambda - \lambda_-)}}{\lambda}\,d\lambda, \quad \lambda_- \le \lambda \le \lambda_+;$$
$$\lambda_{\pm} = [1 + R_0 \pm 2\sqrt{R_0}],\ R_0 = \frac{d_1}{d_2},\ d_1 < d_2. \tag{3.74}$$

The final solution given by Eq. (3.74) is same as the result reported for example in [59] with $Q = 1/R_0$ and $\sigma^2 = 1$. Thus, $\rho^C(E)$ follows from p-GOE:2(Δ) and it is simple to deal with this ensemble. Figure 3.3 gives a plot of $\rho^C(\lambda)$ for various values of R_0 and used here is Eq. (3.74). Before going further, some comments on generalization of p-GOE will be useful.

Given $\rho(x)$, its Stieltjes transform $f(z)$ is

$$f(z) = \int_{-\infty}^{+\infty} \frac{\rho(x)}{z - x}\,dx \tag{3.75}$$

where z is a complex variable. Since $-\pi\delta(x) = \Im\langle\frac{1}{x+i0}\rangle$, we have

$$\rho(x) = -\frac{1}{\pi}\Im\left\{\left[\lim_{\varepsilon\to 0} f(x + i\varepsilon)\right]\right\}. \tag{3.76}$$

Given a general 2×2 block matrix $\left[\begin{smallmatrix} H_{11} & H_{12} \\ H_{12}^T & H_{22}\end{smallmatrix}\right]$ that is real symmetric with dimensions for the diagonal blocks being d_1 and d_2 respectively, following Eqs. (3.58) and (3.59), we have $d\rho(x) = \sum_{i=1}^{2} d_i \rho^i(x)$. Let us denote the Stieltjes transforms of ρ, ρ^1 and ρ^2 by f, f_1 and f_2 respectively. We assume that the matrix elements of H_{ij} are independent Gaussian variables with variances v_{ij}^2. Moreover, we can assume that all matrix elements are zero centered except that the diagonal matrix elements of H_{22} have centroid Δ. Then, using the moments recursion, one can prove that [4]

$$
\begin{aligned}
df_1 &= \frac{1}{z - v_{11}^2 d_1 f_1 - v_{12}^2 d_2 f_2} \\
df_2 &= \frac{1}{z - \Delta - v_{22}^2 d_2 f_2 - v_{12}^2 d_1 f_1}.
\end{aligned}
\tag{3.77}
$$

Solving these equations for f_1 with $v_{11} = v_{22} = 0$ and $\Delta = 0$ and applying Eq. (3.76) will give $\rho^{\Delta=0;1}(x)$. This will be an alternative derivation of Eq. (3.70) given earlier. However, Eq. (3.77) allows one to solve the most general 2×2 block matrix problem with $v_{11} \neq 0$, $v_{22} \neq 0$ and $\Delta \neq 0$, i.e. most general p-GOE:2 random matrix problem. Deriving an analytical form for ρ^1 (similarly for ρ^2) for the general p-GOE:2 is of considerable interest in nuclear physics [67]. Its further generalization to p-GOE:N was analyzed in [70] and the partial densities ρ^i are reduced to multiple integrals involving commuting and anticommuting variables.

3.6 Further Extensions and Applications of RMT

Here below will give a list of various extensions and applications of RMT. This list is only partial as the subject of "Random Matrices: Theory and Applications" is too vast to be covered in completeness at one place.

1. There are many new class of random matrix ensembles that are not covered in this book and some of them are: (i) β ensembles and more general random matrix ensembles related to orthogonal polynomials [71, 72]; (ii) critical random matrix ensembles [73, 74]; (iii) ensembles with non-extensive q entropy [75–77]; (iv) ensemble with super statistics [78]; (v) special constrained Gaussian ensembles [79]; (vi) Cyclic random matrix ensembles [80]; (vii) Hussein and Pato's deformed ensembles based on maximum entropy principle [81–84]; (viii) Transition ensemble for harmonic oscillators to GUE transition [85]; (ix) New versions and new applications of circular ensembles [86–90]. (x) Non-Hermitian random matrix ensembles; see [91–98] and references therein. (x) Random density matrices for entanglement related studies [99, 100].
2. There is a nice relationship between ensembles of 2×2 Hermitian matrices and Gaussian point process [101] and similarly between Poisson point process and 2×2 complex non-Hermitian random matrices [102]. Construction and applications of many other 2×2 random matrix ensembles are discussed in [76, 103, 104]. For example, introduced in [103] are 2×2 pseudo-Hermitian random matrix ensembles and in [76] introduced are ensembles based on Tsallis entropy. Thus, 2×2 ensembles have much wider relevance. As an additional

example, briefly discussed in Appendix E are 2×2 matrix results and their extensions for open quantum systems [105–112].

3. Going beyond 2×2 ensembles, recently 3×3 random matrix ensembles (first discussion on 3×3 random matrix ensembles was given in [113]) are found to be useful in deriving some new results. Using 3×3 GE, derived in [114] is the probability distribution for the ratio of consecutive level spacings (see Chap. 16). This distribution and its relatives are suggested [115–117] to be useful in understanding localization in interacting many particle systems.
4. RMT for missing levels and incomplete spectra has been discussed for example in [65, 118–120] and this is of considerable interest in data analysis and for predictions of missing levels.
5. Using the analogy between energy levels and time series, methods of time series analysis are applied to RMT spectra showing for example $1/f^2$ noise for Poisson systems and $1/f$ noise for GOE/GUE/GSE. There are several investigations in this direction as given for example in [118, 121–126].
6. There are applications of RMT for biological networks [127], neural networks [128], small world networks [129], terrace-width distributions on vicinal surfaces of vicinal crystals [130], finding words in literary texts [131] and so on.
7. There is extensive literature on results for the rate of convergence of probability distributions in RMT and on asymptotic properties of a variety of random matrices; see for example [54, 96, 132–135].
8. Extreme statistics in RMT is another important topics that is not discussed in this Section. An example is the probability distribution for the largest or the lowest eigenvalue in GOE. Tracy-Widom distribution is the starting point for all these investigations. See [134–144] and references therein.
9. New classification of random matrices, extending Dyson's 3-fold way to 10 classes, based on group theory is given in [145–147]. These new classes have applications in condensed matter physics. They also include chiral ensembles for QCD related applications [148–152].
10. Random matrix theory for random phase approximation (RPA), a widely used quantum many-body approximate method, has been introduced in [153].
11. Random matrix theory for scattering and Ericson fluctuations is an important topic that is not discussed in this section. Good references for these are [154–158].
12. Numerical methods and algorithms for constructing and analyzing random matrices of large dimensions are available for example in [159, 160].

References

1. J.B. French, V.K.B. Kota, A. Pandey, S. Tomsovic, Statistical properties of many-particle spectra V. Fluctuations and symmetries. Ann. Phys. (N.Y.) **181**, 198–234 (1988)
2. D.M. Leitner, Real symmetric random matrix ensembles of Hamiltonians with partial symmetry breaking. Phys. Rev. E **48**, 2536–2546 (1993)
3. A. Pandey, Brownian-motion model of discrete spectra. Chaos Solitons Fractals **5**, 1275–1285 (1995)

4. A. Pandey, Statistical properties of many-particle spectra IV. New ensembles by Stieltjes transform methods. Ann. Phys. (N.Y.) **134**, 110–127 (1981)
5. T. Guhr, Transitions toward quantum chaos: with supersymmetry from Poisson to Gauss. Ann. Phys. (N.Y.) **250**, 145–192 (1996)
6. K.M. Frahm, T. Guhr, A. Muller-Groeling, Between Poisson and GUE statistics: role of the Breit-Wigner width. Ann. Phys. (N.Y.) **270**, 292–327 (1998)
7. H. Kunz, B. Shapiro, Transition from Poisson to Gaussian unitary statistics: the two-point correlation function. Phys. Rev. E **58**, 400–406 (1998)
8. J.B. French, V.K.B. Kota, Statistical spectroscopy. Annu. Rev. Nucl. Part. Sci. **32**, 35–64 (1982)
9. G. Lenz, F. Haake, Reliability of small matrices for large spectra with nonuniversal fluctuations. Phys. Rev. Lett. **67**, 1–4 (1991)
10. G. Lenz, K. Zyczkowski, D. Saher, Scaling laws of the additive random-matrix model. Phys. Rev. A **44**, 8043–8050 (1991)
11. E. Caurirer, B. Grammaticos, A. Ramani, Level repulsion near integrability: a random matrix analogy. J. Phys. A **23**, 4903–4910 (1990)
12. V.K.B. Kota, S. Sumedha, Transition curves for the variance of the nearest neighbor spacing distribution for Poisson to Gaussian orthogonal and unitary ensemble transitions. Phys. Rev. E **60**, 3405–3408 (1999)
13. T. Guhr, H.A. Weidenmüller, Coexistence of collectivity and chaos in nuclei. Ann. Phys. (N.Y.) **193**, 472–489 (1989)
14. S. Schierenberg, F. Bruckmann, T. Wettig, Wigner surmise for mixed symmetry classes in random matrix theory. Phys. Rev. E **85**, 061130 (2012)
15. M. Abramowtiz, I.A. Stegun (eds.), *Handbook of Mathematical Functions*, NBS Applied Mathematics Series, vol. 55 (U.S. Govt. Printing Office, Washington, D.C., 1972)
16. I.S. Gradshteyn, I.M. Ryzhik, *Tables of Integrals, Series and Products*, 5th edn. (Academic Press, New York, 1980)
17. A. Erdelyi, W. Magnus, F. Oberhettinger, F.G. Triconi (eds.), *Higher Transcendental Functions—The Bateman Manuscript Project*, vol. I (McGraw-Hill, New York, 1953)
18. P.B. Khan, C.E. Porter, Statistical fluctuations of energy levels: the unitary ensemble. Nucl. Phys. **48**, 385–407 (1963)
19. L.D. Favro, J.F. McDonald, Possibility of detecting a small time-reversal-noninvariant term in the Hamiltonian of a complex system by measurements of energy-level spacings. Phys. Rev. Lett. **19**, 1254–1256 (1967)
20. N. Dupuis, G. Montambaux, Aharonov-Bohm flux and statistics of energy levels in metals. Phys. Rev. B **43**, 14390–14395 (1991)
21. A. Pandey, M.L. Mehta, Gaussian ensembles of random Hermitian matrices intermediate between orthogonal and unitary ones. Commun. Math. Phys. **87**, 449–468 (1983)
22. E.P. Wigner, Random matrices in physics. SIAM Rev. **9**, 1–23 (1967)
23. J.B. French, V.K.B. Kota, A. Pandey, S. Tomsovic, Statistical properties of many-particle spectra VI. Fluctuation bounds on N-N T-noninvariance. Ann. Phys. (N.Y.) **181**, 235–260 (1988)
24. M.J. Morrison, A. Derevianko, Proposed search for T-odd, P-even interactions in spectra of chaotic atoms (2012). arXiv:1206.3607 [physics.atom.ph]
25. S. Tomsovic, Bounds on the time-reversal non-invariant nucleon-nucleon interaction derived from transition-strength fluctuations, Ph.D. Thesis, University of Rochester, Rochester, New York (1986)
26. T.A. Brody, A statistical measure for the repulsion of energy levels. Lett. Nuovo Cimento **7**, 482–484 (1973)
27. M.V. Berry, M. Robnik, Semiclassical level spacings when regular and chaotic orbits coexist. J. Phys. A **17**, 2413–2422 (1984)
28. M.V. Berry, M. Robnik, Statistics of energy levels without time-reversal symmetry: Aharonov-Bohm chaotic billiards. J. Phys. A **19**, 649–668 (1986)

29. F.M. Izrailev, Intermediate statistics of the quasi-energy spectrum and quantum localisation of classical chaos. J. Phys. A **22**, 865–878 (1989)
30. G. Casati, F.M. Izrailev, L. Molinari, Scaling properties of the eigenvalue spacing distribution for band random matrices. J. Phys. A **24**, 4755–4762 (1991)
31. J.P. Blocki, A.G. Magner, Chaoticity and shell effects in the nearest-neighbor distributions for an axially-symmetric potential. Phys. Rev. C **85**, 064311 (2012)
32. J. Sakhr, J.M. Nieminen, Poisson-to-Wigner crossover transition in the nearest-neighbor statistics of random points on fractals. Phys. Rev. E **72**, 045204(R) (2005)
33. T. Rudolf, N. Mertig, S. Lock, A. Backer, Consequences of flooding on spectral statistics. Phys. Rev. E **85**, 036213 (2012)
34. M. Vyas, V.K.B. Kota, N.D. Chavda, Transitions in eigenvalue and wavefunction structure in $(1+2)$-body random matrix ensembles with spin. Phys. Rev. E **81**, 036212 (2010)
35. N.D. Chavda, V. Potbhare, V.K.B. Kota, Statistical properties of dense interacting Boson systems with one plus two-body random matrix ensembles. Phys. Lett. A **311**, 331–339 (2003)
36. F.S. Stephens et al., Order-to-chaos transition in rotational nuclei. Phys. Rev. Lett. **94**, 042501 (2005)
37. F.S. Stephens et al., Damping, motional narrowing, and chaos in rotational nuclei. Phys. Rev. C **78**, 034303 (2008)
38. H. Alt, C.I. Barbosa, H.D. Gräf, T. Guhr, H.L. Harney, R. Hofferbert, H. Rehfeld, A. Richter, Coupled microwave billiards as a model for symmetry breaking. Phys. Rev. Lett. **81**, 4847–4850 (1998)
39. G.E. Mitchell, E.G. Bilpuch, P.M. Endt, J.F. Shriner Jr., Broken symmetries and chaotic behavior in ^{26}Al. Phys. Rev. Lett. **61**, 1473–1476 (1988)
40. J.F. Shriner Jr., C.A. Grossmann, G.E. Mitchell, Level statistics and transition distributions of ^{30}P. Phys. Rev. C **62**, 054305 (2000)
41. T. Ghur, H.A. Weidenmüller, Isospin mixing and spectral fluctuation properties. Ann. Phys. (N.Y.) **199**, 412–446 (1990)
42. A.A. Adams, G.E. Mitchell, J.F. Shriner Jr., Statistical distribution of reduced transition probabilities in ^{26}Al. Phys. Lett. B **422**, 13–18 (1998)
43. C.I. Barbosa, T. Guhr, H.L. Harney, Impact of isospin on the distribution of transition probabilities. Phys. Rev. E **62**, 1936–1949 (2000)
44. N. Rosenzweig, C.E. Porter, Repulsion of energy levels in complex atomic spectra. Phys. Rev. **120**, 1698–1714 (1960)
45. N. Datta, H. Kunz, A random matrix approach to the crossover of energy-level statistics from Wigner to Poisson. J. Math. Phys. **45**, 870–886 (2004)
46. M.L. Mehta, *Random Matrices*, 3rd edn. (Elsevier, Amsterdam, 2004)
47. A. Abd El-Hady, A.Y. Abul-Magd, M.H. Simbel, Influence of symmetry breaking on the fluctuation properties of spectra. J. Phys. A **35**, 2361–2372 (2002)
48. A.Y. Abul-Magd, H.L. Harney, M.H. Simbel, H.A. Weidenmüller, Statistics of 2^+ levels in even-even nuclei. Phys. Lett. B **579**, 278–284 (2004)
49. A.Y. Abul-Magd, H.L. Harney, M.H. Simbel, H.A. Weidenmüller, Statistical analysis of composite spectra. Ann. Phys. (N.Y.) **321**, 560–580 (2006)
50. A. Al-Sayed, A.Y. Abul-Magd, Level statistics of deformed even-even nuclei. Phys. Rev. C **74**, 037301 (2006)
51. O. Bohigas, R.U. Haq, A. Pandey, Fluctuation properties of nuclear energy levels and widths: comparison of theory and experiment, in *Nuclear Data for Science and Technology*, ed. by K.H. Böckhoff (Reidel, Dordrecht, 1983), pp. 809–813
52. J.D. Garrett, J.Q. Robinson, A.J. Foglia, H.-Q. Jin, Nuclear level repulsion and order vs. chaos. Phys. Lett. B **392**, 24–29 (1997)
53. J. Enders, T. Guhr, N. Huxel, P. von Neumann-Cosel, C. Rangacharyulu, A. Richter, Level spacing distribution of scissors mode states in heavy deformed nuclei. Phys. Lett. B **486**, 273–278 (2000)
54. Z. Bai, J.W. Silverstein, *Spectral Analysis of Large Dimensional Random Matrices*, 2nd edn. (Springer, New York, 2010)

55. L. Pastur, M. Shcherbina, *Eigenvalue Distribution of Large Random Matrices*, Mathematical Surveys and Monographs, vol. 171 (Am. Math. Soc., Providence, 2011)
56. A.M. Tulino, S. Verdú, *Random Matrix Theory and Wireless Communication* (Now Publishers, Hanover, 2004)
57. I.M. Johnstone, On the distribution of the largest eigenvalue in principle components analysis. Ann. Stat. **29**, 295–327 (2001)
58. H.E. Stanley, L.A.N. Amaral, D. Canning, P. Gopikrishnan, Y. Lee, Y. Liu, Econophysics: can physics contribute to the science of economics? Physica A **269**, 156–169 (1999)
59. L. Laloux, P. Cizeau, J.P. Bouchaud, M. Potters, Noise dressing of financial correlation matrices. Phys. Rev. Lett. **83**, 1467–1470 (1999)
60. V. Plerou, P. Gopikrishnan, B. Rosenow, L.A.N. Amaral, T. Guhr, H.E. Stanley, Random matrix approach to cross correlations in financial data. Phys. Rev. E **65**, 066126 (2006)
61. J.B. French, in *Mathematical and Computational Methods in Nuclear Physics*, ed. by J.S. Dehesa, J.M.G. Gomez, A. Polls (Springer, Berlin, 1984), pp. 100–121
62. M.S. Santhanam, P.K. Patrta, Statistics of atmospheric correlations. Phys. Rev. E **64**, 016102 (2001)
63. J.N. Bandyopadhyay, A. Lakshminarayan, Testing statistical bounds on entanglement using quantum chaos. Phys. Rev. Lett. **89**, 060402 (2002)
64. F.J. Dyson, Distribution of eigenvalues for a class of real symmetric matrices. Rev. Mex. Fis. **20**, 231–237 (1971)
65. T.A. Brody, J. Flores, J.B. French, P.A. Mello, A. Pandey, S.S.M. Wong, Random matrix physics: spectrum and strength fluctuations. Rev. Mod. Phys. **53**, 385–479 (1981)
66. A.M. Sengupta, P.P. Mitra, Distributions of singular values for some random matrices. Phys. Rev. E **60**, 3389–3392 (1999)
67. V.K.B. Kota, D. Majumdar, R. Haq, R.J. Leclair, Shell model tests of the bimodal partial state densities in a 2×2 partitioned embedded random matrix ensemble. Can. J. Phys. **77**, 893–901 (1999)
68. V.K.B. Kota, Eigenvalue density for random covariance matrices from a 2×2 partitioned GOE. Adv. Stud. Theor. Phys. **2**(17), 845–854 (2008)
69. D. Bierens de Haan, *Nouvelles Tables: D'Intégrales Définies* (Hafner Publishing Company, New York, 1957), p. 111
70. Z. Pluhař, H.A. Weidenmüller, Approximation for shell-model level densities. Phys. Rev. C **38**, 1046–1057 (1988)
71. T. Nagao, P.J. Forrester, Transitive ensembles of random matrices related to orthogonal polynomials. Nucl. Phys. B **530**, 742–762 (1998)
72. I. Dumitriu, A. Edelman, Matrix models for beta ensembles. J. Math. Phys. **43**, 5830–5847 (2002)
73. K.A. Muttalib, M.E.H. Ismail, Power-law eigenvalue density, scaling and critical random-matrix ensembles. Phys. Rev. E **76**, 051105 (2007)
74. A. Ossipov, I. Rushkin, E. Cuevas, Level-number variance and spectral compressibility in a critical two-dimensional random matrix model. Phys. Rev. E **85**, 021127 (2012)
75. A.C. Bertuola, O. Bohigas, M.P. Pato, Family of generalized random matrix ensembles. Phys. Rev. E **70**, 065102(R) (2004)
76. A.Y. Abul-Magd, Non-extensive random matrix theory—a bridge connecting chaotic and regular dynamics. Phys. Lett. A **333**, 16–22 (2004)
77. A.Y. Abul-Magd, Nonextensive random matrix theory approach to mixed regular-chaotic dynamics. Phys. Rev. E **71**, 066207 (2005)
78. A.Y. Abul-Magd, Random matrix theory within superstatistics. Phys. Rev. E **72**, 066114 (2005)
79. T. Papenbrock, Z. Pluhař, H.A. Weidenmüller, Level repulsion in constrained Gaussian random-matrix ensemble. J. Phys. A **39**, 9709–9726 (2006)
80. S.R. Jain, Random cyclic matrices. Phys. Rev. E **78**, 036213 (2008)
81. M.S. Hussein, M.P. Pato, Description of chaos-order transition with random matrices within the maximum entropy principle. Phys. Rev. Lett. **70**, 1089–1092 (1993)

82. M.S. Hussein, M.P. Pato, Deformed Gaussian orthogonal ensemble description of isospin mixing and spectral fluctuation properties. Phys. Rev. C **47**, 2401–2403 (1993)
83. M.S. Hussein, M.P. Pato, Critical behavior in disordered quantum systems modified by broken time-reversal symmetry. Phys. Rev. Lett. **80**, 1003–1006 (1998)
84. J.X. de Carvalho, M.S. Hussein, M.P. Pato, A.J. Sargeant, Symmetry-breaking study with deformed ensembles. Phys. Rev. E **76**, 066212 (2007)
85. T. Guhr, T. Papenbrock, Spectral correlations in the crossover transition from a superposition of harmonic oscillators to the Gaussian unitary ensemble. Phys. Rev. E **59**, 330–336 (1999)
86. A. Pandey, P. Shukla, Eigenvalue correlations in the circular ensembles. J. Phys. A **24**, 3907–3926 (1991)
87. A. Pandey, S. Ghosh, Skew-orthogonal polynomials and universality of energy-level correlations. Phys. Rev. Lett. **87**, 024102 (2001)
88. S. Ghosh, A. Pandey, Skew-orthogonal polynomials and random-matrix ensembles. Phys. Rev. E **65**, 046221 (2002)
89. A. Pandey, S. Puri, S. Kumar, Long-range correlations in quantum-chaotic spectra. Phys. Rev. E **71**, 066210 (2005)
90. S. Kumar, A. Pandey, Nonuniform circular ensembles. Phys. Rev. E **78**, 026204 (2008)
91. J. Ginibre, Statistical ensembles of complex, quaternion, and real matrices. J. Math. Phys. **6**, 440–449 (1965)
92. H. Markum, R. Pullirsch, T. Wettig, Non-Hermitian random matrix theory and lattics QCD with chemical potential. Phys. Rev. Lett. **83**, 484–487 (1999)
93. P. Shukla, Non-Hermitian random matrices and the Calogero-Sutherland model. Phys. Rev. Lett. **87**, 194102 (2001)
94. P.J. Forrester, T. Nagao, Eigenvalues of the real Ginibre ensemble. Phys. Rev. Lett. **99**, 050603 (2007)
95. F. Haake, *Quantum Signatures of Chaos*, 3rd edn. (Springer, Heidelberg, 2010)
96. P.J. Forrester, *Log-Gases and Random Matrices* (Princeton University Press, Princeton, 2010)
97. M. Kieburg, J.J.M. Verbaarschot, S. Zafeiropoulos, Eigenvalue density of the non-Hermitian Wilson Dirac operator. Phys. Rev. Lett. **108**, 022001 (2012)
98. N. Auerbach, V. Zelevinsky, Super-radiant dynamics, doorways and resonances in nuclei and other open mesoscopic systems. Rep. Prog. Phys. **74**, 106301 (2011)
99. H. Sommers, K. Zyczkowski, Statistical properties of random density matrices. J. Phys. A, Math. Gen. **37**, 8457–8466 (2004)
100. K. Zyczkowski, K.A. Penson, I. Nechita, B. Collins, Generating random density matrices. J. Math. Phys. **52**, 062201 (2011)
101. J.M. Nieminen, Gaussian point processes and two-by-two random matrix theory. Phys. Rev. E **76**, 047202 (2007)
102. J. Sakhr, J.M. Nieminen, Wigner surmises and the two-dimensional homogeneous Poisson point process. Phys. Rev. E **73**, 047202 (2006)
103. Z. Ahmed, S.R. Jain, Gaussian ensemble of 2×2 pseudo-Hermitian random matrices. J. Phys. A, Math. Gen. **36**, 3349–3362 (2003)
104. S. Grossmann, M. Robnik, On level spacing distributions for 2D non-normal Gaussian random matrices. J. Phys. A, Math. Gen. **40**, 409–421 (2007)
105. C. Poli, G.A. Luna-Acosta, H.-J. Stöckmann, Nearest level spacing statistics in open chaotic systems: generalization of the Wigner surmise. Phys. Rev. Lett. **108**, 174101 (2012)
106. H.-J. Stöckmann, P. Sěba, The joint energy distribution function for the Hamiltonian $H = H_0 - iWW^\dagger$ for the one-channel case. J. Phys. A, Math. Gen. **31**, 3439–3449 (1998)
107. V.V. Sokolov, V.G. Zelevinsky, Dynamics and statistics of unstable quantum states. Nucl. Phys. A **504**, 562–588 (1989)
108. G. Shchedrin, V. Zelevinsky, Resonance width distribution for open quantum systems. Phys. Rev. C **86**, 044602 (2012)
109. P.E. Koehler, F. Becvar, M. Krticka, J.A. Harvey, K.H. Guber, Anomalous fluctuations of s-wave reduced neutron widths of $^{192;194}$Pt resonances. Phys. Rev. Lett. **105**, 075502 (2010)

110. E.S. Reich, Nuclear theory nudged. Nature (London) **466**, 1034 (2010)
111. G.L. Celardo, N. Auerbach, F.M. Izrailev, V.G. Zelevinsky, Distribution of resonance widths and dynamics of continuum coupling. Phys. Rev. Lett. **106**, 042501 (2011)
112. H.A. Weidenmüller, Distribution of partial neutron widths for nuclei close to a maximum of the neutron strength function. Phys. Rev. Lett. **105**, 232501 (2010)
113. C.E. Porter, *Statistical Theories of Spectra: Fluctuations* (Academic Press, New York, 1965)
114. Y.Y. Atas, E. Bogomolny, O. Giraud, G. Roux, Distribution of the ratio of consecutive level spacings in random matrix ensembles. Phys. Rev. Lett. **110**, 084101 (2013)
115. V. Oganesyan, D.A. Huse, Localization of interacting fermions at high temperature. Phys. Rev. B **75**, 155111 (2007)
116. A. Pal, D.A. Huse, Many-body localization phase transition. Phys. Rev. B **82**, 174411 (2010)
117. S. Iyer, V. Oganesyan, G. Refael, D.A. Huse, Many-body localization in a quasiperiodic system. Phys. Rev. B **87**, 134202 (2013)
118. J.M.G. Gómez, K. Kar, V.K.B. Kota, R.A. Molina, A. Relaño, J. Retamosa, Many-body quantum chaos: recent developments and applications to nuclei. Phys. Rep. **499**, 103–226 (2011)
119. O. Bohigas, M.P. Pato, Missing levels in correlated spectra. Phys. Lett. B **595**, 171–176 (2004)
120. O. Bohigas, M.P. Pato, Randomly incomplete spectra and intermediate statistics. Phys. Rev. C **74**, 036212 (2006)
121. A. Relaño, J.M.G. Gómez, R.A. Molina, J. Retamosa, E. Faleiro, Quantum chaos and $1/f$ noise. Phys. Rev. Lett. **89**, 244102 (2002)
122. E. Faleiro, J.M.G. Gómez, R.A. Molina, L. Muñoz, A. Relaño, J. Retamosa, Theoretical derivation of $1/f$ noise in quantum chaos. Phys. Rev. Lett. **93**, 244101 (2004)
123. J.M.G. Gómez, A. Relaño, J. Retamosa, E. Faleiro, L. Salasnich, M. Vranicar, M. Robnik, $1/f^{\alpha}$ noise in spectral fluctuations of quantum systems. Phys. Rev. Lett. **94**, 084101 (2005)
124. R.J. Leclair, R.U. Haq, V.K.B. Kota, N.D. Chavda, Power spectrum analysis of the average-fluctuation density separation in interacting particle systems. Phys. Lett. A **372**, 4373–4378 (2008)
125. A. Al-Sayed, Autocorrelation studies for the first 2^+ nuclear energy levels. Phys. Rev. C **85**, 037302 (2012)
126. K. Roy, B. Chakrabarti, A. Biswas, V.K.B. Kota, S.K. Haldar, Spectral fluctuation and $\frac{1}{f^{\alpha}}$ noise in the energy level statistics of interacting trapped bosons. Phys. Rev. E **85**, 061119 (2012)
127. F. Lup, J. Zhong, Y. Yang, R.H. Scheermann, J. Zhou, Application of random matrix theory to biological networks. Phys. Lett. A **357**, 420–423 (2006)
128. K. Rajan, L.F. Abbott, Eigenvalue spectra of random matrices for neural networks. Phys. Rev. Lett. **97**, 188104 (2006)
129. J.X. de Carvalho, S. Jalan, M.S. Hussein, Deformed Gaussian-orthogonal-ensemble description of small-world networks. Phys. Rev. E **79**, 056222 (2009)
130. A. Pimpinelli, H. Gebremariam, T.L. Einstein, Evolution of terrace-width distributions on vicinal surfaces: Fokker-Planck derivation of the generalized Wigner surmise. Phys. Rev. Lett. **95**, 246101 (2005)
131. P. Carpena, P. Bernaola-Galván, M. Hackenberg, A.V. Coronado, J.L. Oliver, Level statistics of words: finding keywords in literary texts and symbolic sequences. Phys. Rev. E **79**, 035102(R) (2009)
132. A. Khorunzhy, Sparse random matrices and statistics of rooted trees. Adv. Appl. Probab. **33**, 124–140 (2001)
133. O. Khorunzhiy, Rooted trees and moments of large sparse random matrices. Discrete Math. Theor. Comput. Sci. **AC**, 145–154 (2003)
134. N. El Karoui, A rate of convergence result for the largest eigenvalue of complex white Wishart matrices. Ann. Probab. **34**, 2077–2117 (2006)
135. S. Sodin, The Tracy-Widom law for some sparse random matrices. J. Stat. Phys. **136**, 834–841 (2009)

136. C.A. Tracy, H. Widom, Level-spacing distributions and the Airy kernel. Phys. Lett. B **305**, 115–118 (1993)
137. C.A. Tracy, H. Widom, Level-spacing distributions and the Airy kernel. Commun. Math. Phys. **159**, 151–174 (1994)
138. C.A. Tracy, H. Widom, On orthogonal and symplectic matrix ensembles. Commun. Math. Phys. **177**, 727–754 (1996)
139. D.S. Dean, S.N. Majumdar, Large deviations of extreme eigenvalues of random matrices. Phys. Rev. Lett. **97**, 160201 (2006)
140. S.N. Majumdar, M. Vergassola, Large deviations of the maximum eigenvalue of Wishart and Gaussian random matrices. Phys. Rev. Lett. **102**, 060601 (2009)
141. A. Lakshminarayan, S. Tomsovic, O. Bohigas, S.N. Majumdar, Extreme statistics of complex random and quantum chaotic states. Phys. Rev. Lett. **100**, 044103 (2008)
142. S.N. Majumdar, P. Vivo, Number of relevant directions in principal component analysis and Wishart random matrices. Phys. Lev. Lett. **108**, 200601 (2012)
143. M. Fridman, R. Pugatch, M. Nixon, A.A. Friesem, N. Davidson, Measuring maximal eigenvalue distribution of Wishart random matrices with coupled lasers. Phys. Rev. E **85**, 020101(R) (2012)
144. O. Bohigas, J.X. de Carvalho, M.P. Pato, Deformations of the Tracy-Widom distribution. Phys. Rev. E **79**, 031117 (2009)
145. M.R. Zirnbauer, Riemannian symmetric superspaces and their origin in random-matrix theory. J. Math. Phys. **37**, 4986–5018 (1996)
146. D.A. Ivanov, Random-matrix ensembles in p-wave vortices (2001). arXiv:cond-mat/0103089
147. P. Heinzner, A. Huckleberry, M.R. Zirnbauer, Symmetry classes of disordered fermions. Commun. Math. Phys. **257**, 725–771 (2005)
148. J.J.M. Verbaarschot, Spectrum of the QCD Dirac operator and chiral random matrix theory. Phys. Rev. Lett. **72**, 2531–2533 (1994)
149. R.G. Edwards, U.M. Heller, J. Kiskis, R. Narayanan, Quark spectra topology, and random matrix theory. Phys. Rev. Lett. **82**, 4188–4191 (1999)
150. T. Guhr, J.-Z. Ma, S. Meyer, T. Wilke, Statistical analysis and the equivalent of a Thouless energy in lattice QCD Dirac spectra. Phys. Rev. D **59**, 054501 (1999)
151. J.J.M. Verbaarschot, T. Wettig, Random matrix theory and chiral symmetry in QCD. Annu. Rev. Nucl. Part. Sci. **50**, 343–410 (2000)
152. G. Akemann, J. Bloch, L. Shifrin, T. Wettig, Individual complex Dirac eigenvalue distributions from random matrix theory and comparison to quanched lattice QCD with a quark chemical potential. Phys. Rev. Lett. **100**, 032002 (2008)
153. X. Barillier-Pertuisel, O. Bohigas, H.A. Weidenmüller, Random-metric approach to RPA equations. I. Ann. Phys. (N.Y.) **324**, 1855–1874 (2009)
154. T.E.O. Ericson, Fluctuations of nuclear cross sections in the continuum region. Phys. Rev. Lett. **5**, 430–431 (1960)
155. T.E.O. Ericson, A theory of fluctuations in nuclear cross sections. Ann. Phys. (N.Y.) **23**, 390–414 (1963)
156. T.E.O. Ericson, T. Mayer-Kuckuk, Fluctuations in nuclear reactions. Annu. Rev. Nucl. Sci. **16**, 183–206 (1966)
157. T.E.O. Ericson, Structure of amplitude correlations in open quantum systems. Phys. Rev. E **87**, 022907 (2013)
158. G.E. Mitchell, A. Richter, H.A. Weidenmüller, Random matrices and chaos in nuclear physics: nuclear reactions. Rev. Mod. Phys. **82**, 2845–2901 (2010)
159. A. Edelman, N. Raj Rao, Random matrix theory. Acta Numer. **14**, 233–297 (2005)
160. F. Mezzardi, How to generate random matrices from the classical compact groups. Not. Am. Math. Soc. **54**(5), 592–604 (2007)

Chapter 4
Embedded GOE for Spinless Fermion Systems: EGOE(2) and EGOE(k)

Matrix ensembles generated by random two-body interactions, called two-body random ensembles (TBRE), model what one may call many-body chaos or stochasticity or complexity exhibited by these systems. These ensembles are defined by representing the two-particle Hamiltonian by one of the classical ensembles (GOE or GUE or GSE) and then the $m > 2$ particle H matrix is generated by the m-particle Hilbert space geometry [1–3]. The key element here is the recognition that there is a Lie algebra that transports the information in the two-particle spaces to many-particle spaces [3–5]. Thus, in these ensembles (for many particle systems) a random matrix ensemble in two-particle spaces is embedded in the m-particle H matrix and therefore these ensembles are more generically called embedded ensembles (EE) [3, 6]. With GOE (GUE) embedding we have then EGOE(2) [EGUE(2)] with '2' denoting that in two-particles spaces the H matrix is represented by a GOE. Due to the two-body selection rules, many of the m-particle matrix elements will be zero. Figure 1.1 gives an example of a H-matrix displaying the structure due to two-body selection rules which form the basis for the EE description. Present understanding is that EE generate paradigmatic models for many-body chaos [7, 8] (one-body chaos is well understood using classical ensembles). Simplest of EE is EGOE(2) [BEGOE(2)], the embedded Gaussian orthogonal ensemble of random matrices for spinless fermion (boson) systems generated by random two-body interactions. Let us begin with EGOE for spinless fermion systems.

4.1 EGOE(2) and EGOE(k) Ensembles: Definition and Construction

The embedding algebra for EGOE(k) and EGUE(k) [also BEGOE(k) and BEGUE(k)] for a system of m spinless particles (fermions or bosons) in N single particle (sp) states with k-body interactions ($k < m$) is $SU(N)$. These ensembles are defined by the three parameters (N, m, k). The EGOE(2) ensemble for spinless fermion systems is generated by defining the two-body Hamiltonian H to be GOE

V.K.B. Kota, *Embedded Random Matrix Ensembles in Quantum Physics*,
Lecture Notes in Physics 884, DOI 10.1007/978-3-319-04567-2_4,

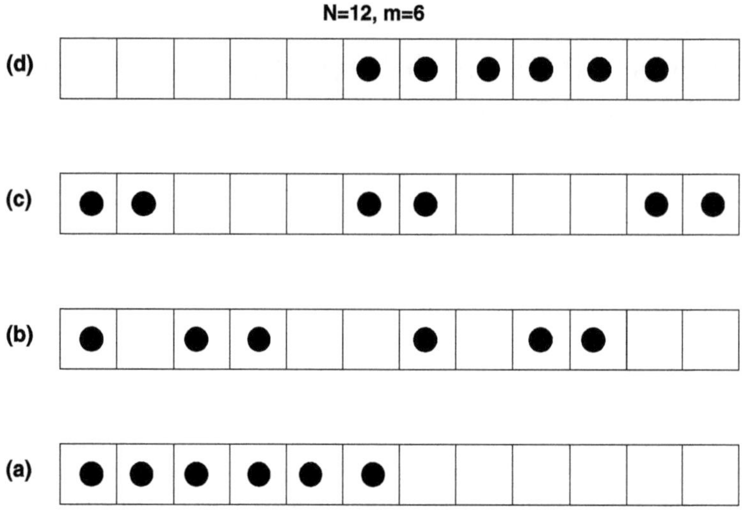

Fig. 4.1 Figure showing some configurations for the distribution of $m = 6$ spinless fermions in $N = 12$ single particle states. The m-particle configurations or basis states are similar to the distributions obtained by putting m particles in N boxes with the conditions that the occupancy of each box can be either zero or one and the total number of occupied boxes equals m. In the figure, (**a**) corresponds to the basis state $|\nu_1\nu_2\nu_3\nu_4\nu_5\nu_6\rangle$, (**b**) corresponds to the basis state $|\nu_1\nu_3\nu_4\nu_7\nu_9\nu_{10}\rangle$, (**c**) corresponds to the basis state $|\nu_1\nu_2\nu_6\nu_7\nu_{11}\nu_{12}\rangle$ and (**d**) corresponds to the basis state $|\nu_6\nu_7\nu_8\nu_9\nu_{10}\nu_{11}\rangle$

in two-particle spaces and then propagating it to many-particle spaces by using the geometry of the many-particle spaces [this is in general valid for k-body Hamiltonians, with $k < m$, generating EGOE(k)]. Let us consider a system of m spinless fermions occupying N sp states. Each possible distribution of fermions in the sp states generates a configuration or a basis state; see Fig. 4.1. Given the sp states $|\nu_i\rangle$, $i = 1, 2, \ldots, N$, EGOE(2) is defined by the Hamiltonian operator,

$$\widehat{H} = \sum_{\nu_i<\nu_j,\nu_k<\nu_\ell} \langle\nu_k\nu_\ell|\widehat{H}|\nu_i\nu_j\rangle a^\dagger_{\nu_\ell} a^\dagger_{\nu_k} a_{\nu_i} a_{\nu_j}. \tag{4.1}$$

The action of the Hamiltonian operator defined by Eq. (4.1) on the basis states $|\nu_1\nu_2\cdots\nu_m\rangle$ (Fig. 4.1 gives examples) generates the EGOE(2) ensemble in m-particle spaces. The symmetries for the antisymmetrized two-body matrix elements $\langle\nu_k\nu_\ell|\widehat{H}|\nu_i\nu_j\rangle$ are

$$\begin{aligned} \langle\nu_k\nu_\ell|\widehat{H}|\nu_j\nu_i\rangle &= -\langle\nu_k\nu_\ell|\widehat{H}|\nu_i\nu_j\rangle, \\ \langle\nu_k\nu_\ell|\widehat{H}|\nu_i\nu_j\rangle &= \langle\nu_i\nu_j|\widehat{H}|\nu_k\nu_\ell\rangle. \end{aligned} \tag{4.2}$$

Note that a_{ν_i} and $a^\dagger_{\nu_i}$ in Eq. (4.1) annihilate and create a fermion in the sp state $|\nu_i\rangle$ respectively. The Hamiltonian matrix $H(m)$ in m-particle spaces contains three different types of non-zero matrix elements (all other matrix elements are zero due to two-body selection rules) and explicit formulas for these are [7],

$$\begin{aligned}
\langle \nu_1 \nu_2 \cdots \nu_m | \widehat{H} | \nu_1 \nu_2 \cdots \nu_m \rangle &= \sum_{\nu_i < \nu_j \leq \nu_m} \langle \nu_i \nu_j | \widehat{H} | \nu_i \nu_j \rangle, \\
\langle \nu_p \nu_2 \nu_3 \cdots \nu_m | \widehat{H} | \nu_1 \nu_2 \cdots \nu_m \rangle &= \sum_{\nu_i = \nu_2}^{\nu_m} \langle \nu_p \nu_i | \widehat{H} | \nu_1 \nu_i \rangle, \\
\langle \nu_p \nu_q \nu_3 \cdots \nu_m | \widehat{H} | \nu_1 \nu_2 \nu_3 \cdots \nu_m \rangle &= \langle \nu_p \nu_q | \widehat{H} | \nu_1 \nu_2 \rangle.
\end{aligned} \tag{4.3}$$

Note that, in Eq. (4.3), the notation $|\nu_1 \nu_2 \cdots \nu_m\rangle$ denotes the orbits occupied by the m spinless fermions. The EGOE(2) is defined by Eqs. (4.2) and (4.3) with GOE(v^2) representation for $\widehat{H}$ in the two-particle spaces, i.e.,

$$\begin{aligned}
&\langle \nu_k \, \nu_\ell | \widehat{H} | \nu_i \nu_j \rangle \text{ are independent Gaussian random variables} \\
&\overline{\langle \nu_k \, \nu_\ell | \widehat{H} | \nu_i \, \nu_j \rangle} = 0, \\
&\overline{\left| \langle \nu_k \nu_\ell | \widehat{H} | \nu_i \nu_j \rangle \right|^2} = v^2 (1 + \delta_{(ij),(k\ell)}).
\end{aligned} \tag{4.4}$$

In Eq. (4.4), 'overline' indicates ensemble average and v is a constant. Now the m-fermion EGOE(2) Hamiltonian matrix ensemble is denoted by $\{H(m)\}$ where $\{\ldots\}$ denotes ensemble, with $\{H(2)\}$ being GOE. Note that, the m-particle H-matrix dimension is $d_f(N,m) = \binom{N}{m}$ and the number of independent matrix elements is $d_f(N,2)[d_f(N,2)+1]/2$; the subscript 'f' in $d_f(N,m)$ stands for 'fermions'. Computer codes for constructing EGOE(2) ensemble have been developed by many research groups; see for example [7, 9–12]. Just as the EGOE(2) ensemble, one can define k-body ($k < m$) ensembles EGOE(k) (these are also called 2-BRE, 3-BRE, ... in [13]) with GOE representation for the H matrix in k particle spaces (thus here we have random k-body interactions). It is possible to derive analytical results, using BCA, for some properties of the general EGOE(k). We will turn to these now.

4.2 Eigenvalue Density: Gaussian Form

4.2.1 Basic Results from Binary Correlation Approximation

Binary correlation theory for the moments of the eigenvalue density generated by spinless EGOE(k) has been developed by Mon and French [3, 14] and the moments given by BCA correspond to the moments in the dilute limit defined by $m \to \infty$, $N \to \infty$, $k \to \infty$ and $m/N \to 0$ and $k/m \to 0$. Alternatively one can use the condition that k is finite and $k/m \to 0$. We will describe the BCA for EGOE(k) in some detail here.

Let us begin with a k_H-body operator,

$$H(k_H) = \sum_{\alpha,\beta} v_H^{\alpha\beta} \alpha^\dagger(k_H) \beta(k_H). \tag{4.5}$$

Here, $\alpha^\dagger(k_H)$ is the k_H particle creation operator and $\beta(k_H)$ is the k_H particle annihilation operator. Similarly, $v_H^{\alpha\beta}$ are matrix elements of the operator H in k_H particle space i.e., $v_H^{\alpha\beta} = \langle k_H \beta | H | k_H \alpha \rangle$ (some authors use operators with daggers to denote annihilation operators and operators without daggers to denote creation operators). Following basic traces will be used throughout,

$$\sum_\alpha \alpha^\dagger(k)\alpha(k) = \binom{\hat{n}}{k} \quad \Rightarrow \quad \left\langle \sum_\alpha \alpha^\dagger(k)\alpha(k) \right\rangle^m = \binom{m}{k}. \tag{4.6}$$

$$\sum_\alpha \alpha(k)\alpha^\dagger(k) = \binom{N-\hat{n}}{k} \quad \Rightarrow \quad \left\langle \sum_\alpha \alpha(k)\alpha^\dagger(k) \right\rangle^m = \binom{\widetilde{m}}{k}; \quad \widetilde{m} = N - m. \tag{4.7}$$

$$\sum_\alpha \alpha^\dagger(k)B\left(k'\right)\alpha(k) = \binom{\hat{n}-k'}{k} B\left(k'\right)$$
$$\Rightarrow \quad \left\langle \sum_\alpha \alpha^\dagger(k)B\left(k'\right)\alpha(k) \right\rangle^m = \binom{m-k'}{k} B\left(k'\right). \tag{4.8}$$

$$\sum_\alpha \alpha(k)B\left(k'\right)\alpha^\dagger(k) = \binom{N-\hat{n}-k'}{k} B\left(k'\right)$$
$$\Rightarrow \quad \left\langle \sum_\alpha \alpha(k)B\left(k'\right)\alpha^\dagger(k) \right\rangle^m = \binom{\widetilde{m}-k'}{k} B\left(k'\right). \tag{4.9}$$

Equation (4.6) follows from the fact that the average should be zero for $m < k$ and one for $m = k$ and similarly, Eq. (4.7) follows from the same argument except that the particles are replaced by holes. Equation (4.8) follows first by writing the k'-body operator $B(k')$ in operator form using Eq. (4.5),

$$B\left(k'\right) = \sum_{\beta,\gamma} v_B^{\beta\gamma} \beta^\dagger\left(k'\right)\gamma\left(k'\right), \tag{4.10}$$

and then applying the commutation relations for the fermion creation and annihilation operators. This gives $\sum_{\beta,\gamma} v_B^{\beta\gamma} \beta^\dagger(k') \sum_\alpha \alpha^\dagger(k)\alpha(k)\gamma(k')$. Now applying Eq. (4.6) to the sum involving α gives Eq. (4.8). Equation (4.9) follows from the same arguments except one has to assume that $B(k')$ is a fully irreducible k'-body operator (Chap. 5 makes clear the notion of 'irreducible' operators) and therefore, it has particle-hole symmetry. For a general $B(k')$ operator, this is valid only in the $N \to \infty$ limit. Therefore, this equation has to be applied with caution.

Using the definition of the H operator in Eq. (4.5), we have

$$
\begin{aligned}
\overline{\langle H(k_H)H(k_H)\rangle^m} &= \sum_{\alpha,\beta} \overline{\{v_H^{\alpha\beta}\}^2} \langle \alpha^\dagger(k_H)\beta(k_H)\beta^\dagger(k_H)\alpha(k_H)\rangle^m \\
&= v_H^2 \left\langle \sum_\alpha \alpha^\dagger(k_H) \left\{ \sum_\beta \beta(k_H)\beta^\dagger(k_H) \right\} \alpha(k_H) \right\rangle^m \\
&= v_H^2 T(m, N, k_H).
\end{aligned} \tag{4.11}
$$

As H is taken as EGOE(k_H) with all the k_H particle matrix elements being Gaussian variables with zero center and same variance (diagonal matrix elements variance being twice that of off-diagonal matrix elements). This gives $\overline{(v_H^{\alpha\beta})^2} = v_H^2$ to be independent of the α and β labels. It is important to note that in the dilute limit, the diagonal terms [$\alpha = \beta$ in Eq. (4.11)] in the averages are neglected as they are smaller by at least one power of $1/N$ and the individual H's are irreducible K_H-body operators. These assumptions are no longer valid for finite-N systems and hence here the evaluation of averages is more complicated. In the dilute limit, we have

$$
\begin{aligned}
T(m, N, k_H) &= \left\langle \sum_\alpha \alpha^\dagger(k_H) \left\{ \sum_\beta \beta(k_H)\beta^\dagger(k_H) \right\} \alpha(k_H) \right\rangle^m \\
&= \binom{\widetilde{m} + k_H}{k_H} \left\langle \sum_\alpha \alpha^\dagger(k_H)\alpha(k_H) \right\rangle^m \\
&= \binom{\widetilde{m} + k_H}{k_H} \binom{m}{k_H}.
\end{aligned} \tag{4.12}
$$

Note that, we have used Eq. (4.7) to evaluate the summation over β and Eq. (4.6) to evaluate summation over α in Eq. (4.12). In the 'strict' $N \to \infty$ limit, we have

$$
T(m, N, k_H) \overset{N\to\infty}{\to} \binom{m}{k_H} \binom{N}{k_H}. \tag{4.13}
$$

In order to incorporate the finite-N corrections, we have to consider the contribution of the diagonal terms. Then, we have,

$$
T(m, N, k_H) = \binom{m}{k_H} \left[\binom{\widetilde{m} + k_H}{k_H} + 1 \right]. \tag{4.14}
$$

Going beyond $\langle HH \rangle$, let us consider averages involving product of four operators of the form

$$
\langle H(k_H)G(k_G)H(k_H)G(k_G) \rangle^m,
$$

where the operators H and G being of body ranks k_H and k_G respectively and they are represented by independent EGOE(k_H) and EGOE(k_G) ensembles respectively. Now, there are two possible ways of evaluating this trace. Either (a) first contract the

H operators across the G operator using Eq. (4.9) and then contract the G operators using Eq. (4.8), or (b) first contract the G operators across the H operator using Eq. (4.9) and then contract the H operators using Eq. (4.8). However, (a) and (b) give the same result only in the 'strict' $N \to \infty$ limit and also for the result incorporating finite N corrections as discussed below. In general, the final result can be expressed as,

$$\overline{\langle H(k_H)G(k_G)H(k_H)G(k_G)\rangle^m} = v_H^2\, v_G^2\, F(m, N, k_H, k_G). \tag{4.15}$$

In the 'strict' dilute limit, we have

$$\begin{aligned}
&\left(v_H^2 v_G^2\right)^{-1}\overline{\langle H(k_H)G(k_G)H(k_H)G(k_G)\rangle^m} \\
&= \sum_{\alpha,\beta,\gamma,\delta} \langle \alpha^\dagger(k_H)\beta(k_H)\gamma^\dagger(k_G)\delta(k_G)\beta^\dagger(k_H)\alpha(k_H)\delta^\dagger(k_G)\gamma(k_G)\rangle^m \\
&= \binom{\widetilde{m}-k_G+k_H}{k_H} \sum_{\alpha,\gamma,\delta} \langle \alpha^\dagger(k_H)\gamma^\dagger(k_G)\delta(k_G)\alpha(k_H)\delta^\dagger(k_G)\gamma(k_G)\rangle^m \\
&= \binom{\widetilde{m}-k_G+k_H}{k_H}\binom{m-k_G}{k_H} \sum_{\gamma,\delta} \langle \gamma^\dagger(k_G)\delta(k_G)\delta^\dagger(k_G)\gamma(k_G)\rangle^m \\
&= \binom{\widetilde{m}-k_G+k_H}{k_H}\binom{m-k_G}{k_H}\binom{\widetilde{m}+k_G}{k_G}\binom{m}{k_G}.
\end{aligned} \tag{4.16}$$

Here in the first step β and $\beta^\dagger$ are contracted using Eq. (4.9) giving $\binom{\widetilde{m}-k_G}{k_H}$ and then it is taken out of the trace. In the second step $\alpha^\dagger$ and α are contracted. Then we are left with a term that is similar to Eq. (4.12) and this gives the final result. Now in the 'strict' $N \to \infty$ limit, $F(m, N, k_H, k_G)$ is

$$\begin{aligned}
F(m, N, k_H, k_G) &= \binom{m-k_H}{k_G}\binom{m}{k_H}\binom{N}{k_H}\binom{N}{k_G} \\
&= \binom{m-k_G}{k_H}\binom{m}{k_G}\binom{N}{k_H}\binom{N}{k_G}.
\end{aligned} \tag{4.17}$$

In order to obtain correct finite-N corrections to $F(\cdots)$, we have to contract over operators whose lower symmetry parts can not be ignored. The operator $H(k_H)$ decomposes into irreducible symmetry (or tensorial) parts $\mathscr{F}(s)$ denoted by $s = 0, 1, 2, \ldots, k_H$ with respect to the unitary group $SU(N)$; see Chap. 5. For a k_H-body number conserving operator [3, 15], we have (see also Chap. 5)

$$H(k_H) = \sum_{s=0}^{k_H} \binom{m-s}{k_H-s} \mathscr{F}(s). \tag{4.18}$$

Here, the $\mathscr{F}(s)$ are orthogonal with respect to m-particle averages, i.e., $\langle \mathscr{F}(s)\mathscr{F}^\dagger(s')\rangle^m = \delta_{ss'}$. Now, $\langle H(k_H)G(k_G)H(k_H)G(k_G)\rangle^m$ will have four parts,

$$\begin{aligned}
&\overline{\left\langle H(k_H)G(k_G)H(k_H)G(k_G)\right\rangle^m} \\
&\quad= v_H^2 v_G^2 \sum_{\alpha,\beta,\gamma,\delta} \left\langle \alpha^\dagger(k_H)\beta(k_H)\gamma^\dagger(k_G)\delta(k_G)\beta^\dagger(k_H)\alpha(k_H)\delta^\dagger(k_G)\gamma(k_G)\right\rangle^m \\
&\qquad+ v_H^2 v_G^2 \sum_{\alpha,\gamma,\delta} \left\langle \alpha^\dagger(k_H)\alpha(k_H)\gamma^\dagger(k_G)\delta(k_G)\alpha^\dagger(k_H)\alpha(k_H)\delta^\dagger(k_G)\gamma(k_G)\right\rangle^m \\
&\qquad+ v_H^2 v_G^2 \sum_{\alpha,\beta,\gamma} \left\langle \alpha^\dagger(k_H)\beta(k_H)\gamma^\dagger(k_G)\gamma(k_H)\beta^\dagger(k_H)\alpha(k_H)\gamma^\dagger(k_G)\gamma(k_G)\right\rangle^m \\
&\qquad+ v_H^2 v_G^2 \sum_{\alpha,\delta} \left\langle \alpha^\dagger(k_H)\alpha(k_H)\delta^\dagger(k_G)\delta(k_G)\alpha^\dagger(k_H)\alpha(k_H)\delta^\dagger(k_G)\delta(k_G)\right\rangle^m \\
&\quad= X + Y_1 + Y_2 + Z. \qquad (4.19)
\end{aligned}$$

Note that we have decomposed each operator into diagonal and off-diagonal parts. We have used the condition that the variance of the diagonal matrix elements is twice that of the off-diagonal matrix elements in the defining spaces to convert the restricted summations into unrestricted summations appropriately to obtain the four terms in the RHS of Eq. (4.19). Following [14, 16, 17] and applying unitary decomposition to $\gamma\delta^\dagger$ (also $\delta\gamma^\dagger$) in the first two terms and $\alpha\beta^\dagger$ (also $\beta\alpha^\dagger$) in the third term we will get X, Y_1 and Y_2. To make things clear, we will discuss the derivation for X term in detail before proceeding further. Applying unitary decomposition to the operators $\gamma^\dagger(k_G)\delta(k_G)$ and $\gamma(k_G)\delta^\dagger(k_G)$ using Eq. (4.18), we have

$$X = v_H^2 v_G^2 \sum_{\alpha,\beta,\gamma,\delta} \sum_{s=0}^{k_G} \binom{m-s}{k_G-s}^2 \left\langle \alpha^\dagger(k_H)\beta(k_H)\mathscr{F}^\dagger_{\gamma\delta}(s)\beta^\dagger(k_H)\alpha(k_H)\mathscr{F}_{\gamma\delta}(s)\right\rangle^m. \qquad (4.20)$$

Contracting the operators $\beta\beta^\dagger$ across $\mathscr{F}$'s using Eq. (4.9) and operators $\alpha^\dagger\alpha$ across $\mathscr{F}$ using Eq. (4.8) gives,

$$X = v_H^2 v_G^2 \sum_{s=0}^{k_G} \binom{m-s}{k_G-s}^2 \binom{\widetilde{m}+k_H-s}{k_H} \binom{m-s}{k_H} \sum_{\gamma,\delta} \left\langle \mathscr{F}^\dagger_{\gamma\delta}(s)\mathscr{F}_{\gamma\delta}(s)\right\rangle^m. \qquad (4.21)$$

Inversion of the equation,

$$\sum_{\gamma,\delta} \left\langle \gamma^\dagger(k_G)\delta(k_G)\delta^\dagger(k_G)\gamma(k_G)\right\rangle^m = Q(m) = \sum_{s=0}^{k_G} \binom{m-s}{k_G-s}^2 \sum_{\gamma,\delta} \left\langle \mathscr{F}^\dagger_{\gamma\delta}(s)\mathscr{F}_{\gamma\delta}(s)\right\rangle^m, \qquad (4.22)$$

gives,

$$\begin{aligned}
&\binom{m-s}{k_G-s}^2 \sum_{\gamma,\delta} \langle \mathscr{F}^\dagger_{\gamma\delta}(s) \mathscr{F}_{\gamma\delta}(s)\rangle^m \\
&\quad = \binom{m-s}{k_G-s}^2 \binom{N-m}{s} \binom{m}{s} \big[(k_G-s)!s!\big]^2 \\
&\qquad \times (N-2s+1) \sum_{t=0}^{s} \frac{(-1)^{t-s}[(N-t-k_G)!]^2}{(s-t)!(N-s-t+1)!t!(N-t)!} Q(N-t). \quad (4.23)
\end{aligned}$$

For the average required in Eq. (4.22), we have

$$Q(m) = \sum_{\gamma,\delta} \langle \gamma^\dagger(k_G)\delta(k_G)\delta^\dagger(k_G)\gamma(k_G)\rangle^m = \binom{\widetilde{m}+k_G}{k_G}\binom{m}{k_G}. \quad (4.24)$$

Simplifying Eq. (4.23) using Eq. (4.24) and using the result in Eq. (4.21) along with the series summation [14]

$$\sum_{t=0}^{s} \frac{(-1)^{t-s}(N-t-k_G)!(k_G+t)!}{(s-t)!(t!)^2(N-s-t+1)!} = \frac{k_G!(N-k_G-s)!}{(N+1-s)!}\binom{k_G}{s}\binom{N+1}{s}, \quad (4.25)$$

the expression for X is,

$$\begin{aligned}
X &= v_H^2 v_G^2\, F(m,N,k_H,k_G); \\
F(m,N,k_H,k_G) &= \sum_{s=0}^{k_G} \binom{m-s}{k_G-s}^2 \binom{\widetilde{m}+k_H-s}{k_H}\binom{m-s}{k_H}\binom{\widetilde{m}}{s}\binom{m}{s}\binom{N+1}{s} \\
&\quad \times \frac{N-2s+1}{N-s+1}\binom{N-s}{k_G}^{-1}\binom{k_G}{s}^{-1}. \quad (4.26)
\end{aligned}$$

Although not obvious, X has $k_H \leftrightarrow k_G$ symmetry. This is easy to verify for $k_H, k_G \le 2$. In the large N limit, Y_1, Y_2 and Z are neglected as X will make the dominant contribution; Ref. [17] gives the formulas for Y_1, Y_2 and Z. Thus, in all the applications, we use

$$\overline{\langle H(k_H)G(k_G)H(k_H)G(k_G)\rangle^m} = X = v_H^2 v_G^2 F(m,N,k_H,k_G) \quad (4.27)$$

with Eq. (4.17) or (4.26) for $F(m,N,k_H,k_G)$ as appropriate.

4.2.2 Dilute Limit Formulas for the Fourth and Sixth Order Moments and Cumulants

In this section throughout we will use Eqs. (4.13) and (4.17) for the functions T and F respectively, i.e. we will use the strict $N \to \infty$ limit. Also in this section, we will take H to be a k-body operator. As odd order cumulants vanish for EGOE(k), the lowest two cumulants that give information about the shape of the eigenvalue density are the fourth (k_4) and sixth (k_6) order cumulants. For these we need to consider first the fourth moment and the sixth moment.

For the fourth moments given by $\overline{\langle H^4(k)\rangle^m}$, in BCA there will be three different correlation patterns that will contribute (we must correlate in pairs the operators for all moments of order >2),

$$
\begin{aligned}
\overline{\langle H^4(k)\rangle^m} &= \overline{\langle \underbrace{H(k)H(k)}\,\underbrace{H(k)H(k)}\rangle^m} \\
&+ \overline{\langle \underbrace{H(k)\underbrace{H(k)H(k)}H(k)}\rangle^m} \\
&+ \overline{\langle H(k)H(k)H(k)H(k)\rangle^m}.
\end{aligned}
\tag{4.28}
$$

In Eq. (4.28), we denote the binary correlated pairs of operators with the symbol $\underbrace{HH}$. The first two terms on the RHS of Eq. (4.28) are equal due to cyclic invariance and follow from Eq. (4.11),

$$
\begin{aligned}
\overline{\langle \underbrace{H(k)H(k)}\,\underbrace{H(k)H(k)}\rangle^m} &= \overline{\langle \underbrace{H(k)\underbrace{H(k)H(k)}H(k)}\rangle^m} \\
&= \left[\overline{\langle H^2(k)\rangle^m}\right]^2.
\end{aligned}
\tag{4.29}
$$

Similarly, the third term on the RHS of Eq. (4.28) follows from Eq. (4.27),

$$
\overline{\langle H(k)H(k)H(k)H(k)\rangle^m} = v_H^4 F(m, N, k, k). \tag{4.30}
$$

Combining Eqs. (4.28), (4.29) and (4.30), $\overline{\langle H^4(k)\rangle^m}$ is given by,

$$
\overline{\langle H^4(k)\rangle^m} = v_H^4\left[2\left\{T(m, N, k)\right\}^2 + F(m, N, k, k)\right]. \tag{4.31}
$$

Finally, fourth order cumulant k_4 in the dilute limit is

$$
\begin{aligned}
k_4 &= \gamma_2 = \left[\overline{\langle H^2(k)\rangle^m}\right]^{-2}\overline{\langle H^4(k)\rangle^m} - 1 = \binom{m-k}{k}\binom{m}{k}^{-1} - 1 \\
&\to -\frac{k^2}{m} + \frac{k^2(k-1)^2}{2m^2} + O\left(1/m^3\right).
\end{aligned}
\tag{4.32}
$$

In the last step we have used the expansion of binomials in powers of $1/m$ using,

$$\binom{m-r}{k} = \frac{m^k}{k!}\left[1 - \frac{1}{m}\left\{kr + \frac{k(k-1)}{2}\right\} + \frac{1}{m^2}\left\{\frac{k(k-1}{2}\left[r^2 + (k-1)r + \frac{(3k-1)(k-2)}{12}\right]\right\} + O\left(\frac{1}{m^3}\right)\right]. \tag{4.33}$$

Therefore, for example for a two-body operator (as in nuclei and atoms) as m increases, the excess parameter γ_2 (or k_4) goes to zero indicating that the density approaches Gaussian. We will confirm this further by deriving a formula for k_6. Before turning to this, it should be added that formulas for lower order moments for EGOE(2) were also derived by Gervios [18].

For the sixth moment $\langle (H(k))^6 \rangle^m$ there are 15 binary association diagrams and they are

$$\begin{aligned}
&\overline{\langle H^6(k)\rangle^m} \\
&\quad = \overline{\langle H(k)H(k)H(k)H(k)H(k)H(k)\rangle^m} \\
&\quad\quad \oplus \overline{\langle H(k)H(k)H(k)H(k)H(k)H(k)\rangle^m} \oplus \overline{\langle H(k)H(k)H(k)H(k)H(k)H(k)\rangle^m} \\
&\quad\quad \oplus \overline{\langle H(k)H(k)H(k)H(k)H(k)H(k)\rangle^m} \oplus \overline{\langle H(k)H(k)H(k)H(k)H(k)H(k)\rangle^m} \\
&\quad\quad \oplus \overline{\langle H(k)H(k)H(k)H(k)H(k)H(k)\rangle^m} \oplus \overline{\langle H(k)H(k)H(k)H(k)H(k)H(k)\rangle^m} \\
&\quad\quad \oplus \overline{\langle H(k)H(k)H(k)H(k)H(k)H(k)\rangle^m} \oplus \overline{\langle H(k)H(k)H(k)H(k)H(k)H(k)\rangle^m} \\
&\quad\quad \oplus \overline{\langle H(k)H(k)H(k)H(k)H(k)H(k)\rangle^m} \oplus \overline{\langle H(k)H(k)H(k)H(k)H(k)H(k)\rangle^m} \\
&\quad\quad \oplus \overline{\langle H(k)H(k)H(k)H(k)H(k)H(k)\rangle^m} \oplus \overline{\langle H(k)H(k)H(k)H(k)H(k)H(k)\rangle^m} \\
&\quad\quad \oplus \overline{\langle H(k)H(k)H(k)H(k)H(k)H(k)\rangle^m} \oplus \overline{\langle H(k)H(k)H(k)H(k)H(k)H(k)\rangle^m}.
\end{aligned} \tag{4.34}$$

As all the correlated H's in Eq. (4.34) are dummy operators, it is easy to see that the first five terms on RHS of Eq. (4.34) are all same. Similarly, the next six terms and also the following three terms are same. This gives,

$$
\begin{aligned}
&\overline{\langle H^6(k)\rangle^m} \\
&\quad = 5\overline{\langle H(k)H(k)H(k)H(k)H(k)H(k)\rangle^m} \oplus 6\overline{\langle H(k)H(k)H(k)H(k)H(k)H(k)\rangle^m} \\
&\qquad \oplus 3\overline{\langle H(k)H(k)H(k)H(k)H(k)H(k)\rangle^m} \\
&\qquad \oplus \overline{\langle H(k)H(k)H(k)H(k)H(k)H(k)\rangle^m}.
\end{aligned} \tag{4.35}
$$

The first correlation diagram in Eq. (4.35) is simply $\{\overline{\langle H^2(k)\rangle^m}\}^3$. With the normalization, which we will use from now onwards, $v_H^2\binom{N}{k} = 1$, this gives $\binom{m}{k}^3$. The second correlation diagram is also simple as we can take out the two directly correlated H's outside the average and then we are left with $\overline{\langle HGHG\rangle^m}$ type term. This gives $\binom{m-k}{k}\binom{m}{k}^2$. For the third correlation diagram, we can use the rule, that follows from Eqs. (4.8) and (4.9),

$$
\begin{aligned}
\alpha^\dagger(k)\beta(k)H(k)\beta^\dagger(k)\alpha(k) &= v^2\binom{N}{k}\alpha^\dagger(k)H(k)\alpha(k) \\
&= v^2\binom{N}{k}\binom{m-k}{k}H(k) = \binom{m-k}{k}H(k).
\end{aligned} \tag{4.36}
$$

By contracting the first and third correlated H's and similarly the fourth and the sixth H's in the average gives the third term to be $\binom{m-k}{k}^2\binom{m}{k}$. In the last correlation diagram, we have to necessarily contract across two H's i.e., we have to contract two H's across an effectively $2k$-body operator. Then, first contracting the first and the fourth correlated H's, we are left with $\overline{\langle HGHG\rangle^m}$ type term. This gives $\binom{m-2k}{k}\binom{m-k}{k}\binom{m}{k}$. Substituting these results, Eq. (4.35) gives

$$
\begin{aligned}
\overline{\langle H^6(k)\rangle^m} &= 5\binom{m}{k}^3 + 6\binom{m-k}{k}\binom{m}{k}^2 \\
&\quad + 3\binom{m-k}{k}^2\binom{m}{k} + \binom{m-2k}{k}\binom{m-k}{k}\binom{m}{k}.
\end{aligned} \tag{4.37}
$$

First converting the sixth order moment into sixth order cumulant k_6 using Eq. (B.5) gives,

$$
k_6 = 5 - 9\binom{m}{k}^{-1}\binom{m-k}{k} + 3\binom{m}{k}^{-2}\binom{m-k}{k}^2 + \binom{m}{k}^{-2}\binom{m-k}{k}\binom{m-2k}{k}. \tag{4.38}
$$

Now, expanding the binomials in Eq. (4.38) in powers of $1/m$ using Eq. (4.33), we have

$$
k_6 = \frac{k^3(6k-1)}{m^2} + O\left(\frac{1}{m^3}\right). \tag{4.39}
$$

Similarly, for the eight order cumulant [7] we have

$$k_8 = \frac{-4k^5(23k-9)}{m^3} + O\left(\frac{1}{m^4}\right). \tag{4.40}$$

Equations (4.32), (4.39) and (4.40) clearly show that in the dilute limit, as m increases (from $m = k$) the density approaches Gaussian as the cumulants k_r approach zero. In fact if we neglect all the cross correlated terms in the moment expressions, clearly we have $\mu_{2r} = (2r-1)!!$ and they are the reduced central moments of a Gaussian. Although this result is derived for the dilute limit, in practice Gaussian form is seen even when the stringent dilute limit conditions are not valid (see Chap. 5 for examples). Thus the eigenvalue density tends to Gaussian form for EGOE(k).

For $m = k$, EGOE (EGUE) reduces to GOE (GUE) and the state density then is a semi-circle. For fixed k as we increase m starting from k (or vice verse) there will be semi circle to Gaussian transition in state densities. Numerically this was studied in the past [6] but the transition point was not known. Simplifying Eq. (4.32) for fixed (m,k) and $N \to \infty$, it is seen that $\gamma_2 \to -1$ for $m < 2k$. This is suggestive that $m = 2k$ is the transition point. To prove this conclusively, Benet et al. [19, 20] solved EGUE(k) [it is possible to solve EGOE(k) also] using super symmetry (SUSY) method and showed that the density is semi-circle for $m < 2k$. It is also proved that there will be non-vanishing corrections to the semi-circle shape for $m \geq 2k$. However the SUSY method fails for $m > 2k$ and therefore SUSY method could not be used to prove that for $m \gg 2k$ the eigenvalue density takes Gaussian form. In conclusion, as m increases from k, state densities exhibit semi-circle to Gaussian transition with $m = 2k$ being the transition point.

4.3 Average-Fluctuation Separation and Lower-Order Moments of the Two-Point Function

4.3.1 Level Motion in Embedded Ensembles

Given a normalized state density $\rho(E)$, it is possible to expand it in terms of its asymptotic (or smoothed) form $\overline{\rho}(E)$ and the orthonormal polynomials $P_\mu(E)$ defined by the asymptotic density. For EGOE ensembles $\overline{\rho}(E)$ is a Gaussian, i.e. $\overline{\rho}(E) = \rho_{\mathscr{G}}(E) = (\sqrt{2\pi}\sigma)^{-1}\exp[-(E-E_c)^2/2\sigma^2]$. Then the Gram-Charlier (GC) expansion [21] gives,

$$\rho(E) = \rho_{\mathscr{G}}(E)\left\{1 + \sum_{\zeta \geq 3} (\zeta!)^{-1} S_\zeta He_\zeta(\widehat{E})\right\}. \tag{4.41}$$

In Eq. (4.41), $\widehat{E} = (E - E_c)/\sigma$ is the standardized E. The centroid $E_c = \langle H \rangle^m$ and the variance $\sigma^2 = \langle H^2 \rangle^m - E_c^2$ of the Gaussian $\rho_{\mathscr{G}}$ are same as that of ρ. He_ζ are Hermite polynomials and S_ζ are, in principle, related to higher moments of the state density $\rho(E)$. One can apply Eq. (4.41) to EGOE(2) and BEGOE(2) by noting that for fermions in the dilute limit and for bosons in the dense limit (see Chap. 9), $\overline{\rho}(E) = \rho_{\mathscr{G}}(E)$. Thus, at this stage distinction between boson and fermion systems is not important. We will not consider boson systems here and return to them in Chap. 9. Since S_ζ's change from member to member of the EGOE(2) ensemble, one can treat them as independent random variables with zero center,

$$\overline{S_\zeta} = 0, \qquad \overline{S_\zeta S_{\zeta'}} = 0 \quad \text{for } \zeta \neq \zeta'. \tag{4.42}$$

This is consistent with the result $\overline{\rho}(E) = \rho_{\mathscr{G}}(E)$ where the 'bar' denotes ensemble average. Each ζ term in Eq. (4.41) represents an excitation 'mode' and the wavelength of the modes is proportional to ζ^{-1}. Therefore small ζ terms are long wavelength modes and large ζ are short wavelength modes. The distribution function $F(E)$, the integrated version of $\rho(E)$, is $F(E) = d\int_{\infty}^{E} \rho(E')dE'$ where d is the dimensionality. Deviation of a given level with energy E from its smoothed (with respect to the ensemble) counter part $\overline{E}$ gives the level motion. In terms of $F(E)$ and the local mean spacing $\overline{D(E)}$, we have $\delta E = E - \overline{E} = [F(E) - \overline{F(E)}]\overline{D(E)}$. Then, the variance of the level motion is given by the ensemble average of $\frac{(\delta E)^2}{\overline{D(E)}^2}$. Using Eq. (4.41) we have easily,

$$\begin{aligned}
\overline{\frac{(\delta E)^2}{\overline{D(E)}^2}} &= \overline{\left[F(x) - \overline{F(x)}\right]^2} \\
&= d^2\sigma^2\left[\rho_{\mathscr{G}}(E)\right]^2\left\{\sum_{\zeta\geq 3}(\zeta!)^{-2}\overline{S_\zeta^2}\left[He_{\zeta-1}(\widehat{E})\right]^2\right\}.
\end{aligned} \tag{4.43}$$

By adding centroid and variance fluctuations, the summation in Eq. (4.43) extends to $\zeta \geq 1$. Then,

$$\overline{\frac{(\delta E)^2}{\overline{D(E)}^2}} = d^2\sigma^2\left[\rho_{\mathscr{G}}(E)\right]^2\left\{\sum_{\zeta\geq 1}(\zeta!)^{-2}\overline{S_\zeta^2}\left[He_{\zeta-1}(\widehat{E})\right]^2\right\}. \tag{4.44}$$

Thus we need $\overline{S_\zeta^2}$ for EGOE and BEGOE and we will address this now.

4.3.2 $\overline{S_\zeta^2}$ in Binary Correlation Approximation

Definition of the co-variances $\Sigma_{p,q}$ and an expression for them in terms of $\overline{S_\zeta^2}$ are,

$$\begin{aligned}
\Sigma_{p,q} &= \overline{\langle H^p\rangle\langle H^q\rangle} - \overline{\langle H^p\rangle}\,\overline{\langle H^q\rangle} \\
&= \sum_{\zeta\geq 1}\overline{S_\zeta^2}(\sigma)^{p+q}\binom{p}{\zeta}\binom{q}{\zeta}(p-\zeta-1)!!(q-\zeta-1)!!.
\end{aligned} \tag{4.45}$$

The above relation follows from Eq. (4.41) as

$$\langle H^p \rangle = \overline{\langle H^p \rangle} + \sum_{\zeta \geq 3} (\zeta!)^{-1} S_\zeta \int E^p \rho_{\mathscr{G}}(E) He_\zeta(\widehat{E})\, dE. \tag{4.46}$$

We have used in Eq. (4.45) the fact that $\sigma^p \binom{p}{\zeta}(p-\zeta-1)!!$ is the pth (central) moment of $\rho(E)He_\zeta(\widehat{E})$. Note that $E_c = \overline{E_c} = 0$. On the other hand, using BCA we have [3]

$$\begin{aligned} \Sigma_{p,q} &= \overline{\langle H^p \rangle \langle H^q \rangle} - \overline{\langle H^p \rangle}\, \overline{\langle H^q \rangle} \\ &= \sum_{\zeta=0}^{\infty} \binom{p}{\zeta}\binom{q}{\zeta} \overline{\{\langle H^\zeta \rangle \overline{\langle H^{p-\zeta} \rangle}\}\ \{\langle H^\zeta \rangle \overline{\langle H^{q-\zeta} \rangle}\}} - \overline{\langle H^p \rangle}\, \overline{\langle H^q \rangle}. \end{aligned} \tag{4.47}$$

The last term of Eq. (4.47) will cancel with the $\zeta = 0$ term of the first term. Then we have,

$$\Sigma_{p,q} = \sum_{\zeta \geq 1} \binom{p}{\zeta}\binom{q}{\zeta} \overline{\langle H^{p-\zeta} \rangle}\, \overline{\langle H^{q-\zeta} \rangle}\, \overline{\underbrace{\langle H^\zeta \rangle \langle H^\zeta \rangle}}. \tag{4.48}$$

The Gaussian moments of $\overline{\langle H^{p-\zeta} \rangle}$ are $(p-\zeta-1)!!(\sigma)^{p-\zeta}$. Therefore,

$$\Sigma_{p,q} = \sum_{\zeta \geq 1} \binom{p}{\zeta}\binom{q}{\zeta} (p-\zeta-1)!!(q-\zeta-1)!!(\sigma)^{p+q-2\zeta}\, \overline{\underbrace{\langle H^\zeta \rangle \langle H^\zeta \rangle}}. \tag{4.49}$$

Comparing Eqs. (4.49) and (4.45) will give the important relation,

$$\overline{S_\zeta^2} = \overline{\underbrace{\langle H^\zeta \rangle \langle H^\zeta \rangle}}(\sigma)^{-2\zeta} = (\sigma)^{-2\zeta}\, \tilde{\Sigma}_{\zeta\zeta}. \tag{4.50}$$

Thus for studying $\frac{\overline{(\delta E)^2}}{\overline{D}^2}$ via (4.44), all we need to evaluate is $\overline{\underbrace{\langle H^\zeta \rangle \langle H^\zeta \rangle}}$.

4.3.3 Average-Fluctuations Separation in the Spectra of Dilute Fermion Systems: Results for EGOE(1) and EGOE(2)

For one body interactions as discussed by Bloch in 1969 [22], fluctuations are of Poisson type. The argument is that without interactions there are many conserved symmetries. An example is $U(N_1) \oplus U(N_2) \oplus --$, where $N_i = 2j+1$ for a nuclear or atomic shell model j-orbit. Note that the nearest neighbor spacing S_n for the n'th level is $S_n = E_{n+1} - E_n$ where $E_{n+1} = \sum_{i=1}^m \varepsilon_i'$ and $E_n = \sum_{i=1}^m \varepsilon_i''$. Here for example ε_i' are the energies of the single particle states that are occupied by the m fermions for generating the $(n+1)$-th state. Similarly ε_i'' generate the n-th state.

Then, obviously S_n's will be uncorrelated giving Poisson fluctuations. In [23] (see also [8])), the authors argued that there will be effects in the lowing part of the many particle spectrum that depend explicitly on the structure of the single particle spectrum. These specific effects are not yet verified in any data analysis. However, after a critical excitation strength Poisson fluctuations set in. Thus generically, for correlations and hence for the level repulsion we require k-body interactions with $k \geq 2$. Now, we will consider $k = 2$ and the results extend to any $k > 2$ [3, 6].

In the dilute limit $H = H(2)$ will be effectively an irreducible two-body operator. Chapter 5 gives details of the decomposition of $H(2)$ into irreducible zero, one and two-body operators. Then, using trace propagation results discussed in Chap. 5, we have $\sigma^2(m) = \langle H^2(2)\rangle^m \overset{\text{dilute limit}}{\Rightarrow} \binom{m}{2}\langle H^2(2)\rangle^2$. Here on-wards we will use the normalization $\langle H^2(2)\rangle^2 = 1$. Then, $\sigma^2(m) = \langle H^2(2)\rangle^m = \binom{m}{2}$. Also, in the dilute limit, as $H(2)$ is an irreducible two-body operator, the propagation equation for $\langle H^p\rangle^m$ is

$$
\begin{aligned}
\langle H^p\rangle^m &= \frac{m(m-1)(N-m)(N-m-1)}{N(N-1)(N-2)(N-3)} x_2^{(p)} \\
&\quad + \frac{m(m-1)(m-2)(N-m)(N-m-1)(N-m-2)}{N(N-1)(N-2)(N-3)(N-4)(N-5)} x_3^{(p)} \\
&\quad + \cdots \\
&\xrightarrow[m/N\to 0]{m\to\infty, N\to\infty} \frac{m^2}{N^2} x_2^{(p)} + \frac{m^3}{N^3} x_3^{(p)} + \cdots \\
&\to \frac{m^2}{N^2} x_2^{(p)} = \frac{m^2}{2}\binom{N^2}{2}^{-1} x_2^{(p)} = \binom{m}{2}\langle H^p\rangle^2 .
\end{aligned} \tag{4.51}
$$

Equation (4.51) gives the correct result for $p = 2$. Now the cross correlated trace is,

$$
\begin{aligned}
\overline{\langle H^\zeta(2)\rangle^m \langle H^\zeta(2)\rangle^m} &= \binom{m}{2}^2 \overline{\langle H^\zeta(2)\rangle^2 \langle H^\zeta(2)\rangle^2} \\
&= 2\zeta \binom{m}{2}^2 \binom{N}{2}^{-2} .
\end{aligned} \tag{4.52}
$$

Here, as H in 2-particle spaces is a GOE, we used the GOE result for $\overline{\langle H^\zeta(2)\rangle^2 \langle H^\zeta(2)\rangle^2}$ given by Eq. (2.60). Now, Eqs. (4.50) and (4.52) along with $\sigma^2(m) = \binom{m}{2}$ will give the important result

$$
\overline{S_\zeta^2} = 2\zeta \binom{m}{2}^{2-\zeta} \binom{N}{2}^{-2} . \tag{4.53}
$$

Substituting Eq. (4.53) in Eq. (4.44) will give the final result for level motion in EGOE(2),

$$\begin{aligned}
\frac{\overline{(\delta E)^2}}{\overline{D(E)}^2} &= \binom{N}{m}^2 \binom{m}{2}^2 \left[\rho_{\mathscr{G}}(E)\right]^2 \\
&\quad \times \left\{ \sum_{\zeta \geq 1} (\zeta!)^{-2} 2\zeta \binom{m}{2}^{2-\zeta} \binom{N}{2}^{-2} \left[He_{\zeta-1}(\widehat{E})\right]^2 \right\} \\
&\xrightarrow{\hat{E}=0} \frac{1}{\pi} \binom{N}{m}^2 \binom{m}{2} \binom{N}{2}^{-2} \\
&\quad \times \left\{ 1 + \frac{1}{12} \binom{m}{2}^{-2} + \frac{1}{320} \binom{m}{2}^{-4} + \cdots \right\}.
\end{aligned} \tag{4.54}$$

Thus, as ζ increases, deviations in $\overline{(\delta E)^2}$ from the leading term rapidly go to zero due to the $\binom{m}{2}^{-2r}$, $r = 1, 2, \ldots$ terms in Eq. (4.54). There will be no change until $\zeta \sim m/2$, thereby defining separation. Beyond this, for $\zeta \gg m/2$ the deviations grow, i.e. fluctuations set in and they will tend to that of GOE [the GOE nature of fluctuations is seen in large number of numerical calculations and therefore it is conjectured in [3, 6] that the EGOE fluctuations in energy levels and strengths will follow GOE—however there is no analytical proof]. Note that for GOE, from Eq. (2.67), we have

$$\frac{\overline{(\delta E)^2}}{\overline{D(E)}^2} \xrightarrow{\widehat{E}=0} \frac{\gamma}{\pi^2} \ln 2d, \tag{4.55}$$

where γ is Euler constant and d is m-particle H matrix dimension. It is important to stress that the BCA for EGOE(2), that gave Eq. (4.54) fails for $\zeta > m/2$. However, before this limit is reached separation sets in. An important consequence of the separation is that the only a few long wavelength modes are required to define the averages. Thus we need a few lower order moments for spectral averages and they can be calculated using trace propagation equations without recourse to H matrix construction and diagonalization. The separation and the GOE nature of fluctuations (then they will be small) form the basis for statistical spectroscopy (SS) [24]. We will discuss this further in Chaps. 5 and 7.

4.3.4 Lower-Order Moments of the Two-Point Function and Cross Correlations in EGOE

Unlike GOE, for EGOE's with N the number of single particle states fixed, two-point function involves in general the two energies drawn from the spectra for two different particle numbers say m_1 and m_2. It is important to note that the GOE in the defining space will be same for the systems with m_1 fermions and m_2 fermions as N is fixed. The two-point function $S^{\rho:m_1,m_2,N}(x_1,x_2)$ is defined by

$$S^{\rho:m_1,m_2,N}(x_1,x_2) = \overline{\langle\delta(H-x_1)\rangle_N^{m_1}\langle\delta(H-x_2)\rangle_N^{m_2}} - \overline{\langle\delta(H-x_1)\rangle_N^{m_1}}\;\overline{\langle\delta(H-x_2)\rangle_N^{m_1}}. \tag{4.56}$$

Here, in the densities we have also shown N explicitly to stress that N is same in all the densities. In general we have $m_1 = m_2$ or $m_1 \neq m_2$. The bivariate moments $\Sigma_{p,q}$ in Eq. (4.45) are the moments for the two-point function with $m_1 = m_2$. Similarly the level motion, discussed in the previous subsections, for a (m,N) system derives from $S^{\rho:m,m,N}(x_1,x_2)$. More importantly, Eq. (4.56) shows that EGOE generates cross correlations, that is correlations between spectra with different particle numbers, as the bivariate moments

$$\Sigma_{p,q}(m_1,m_2,N) = \overline{\langle H^p\rangle_N^{m_1}\langle H^q\rangle_N^{m_2}} - \overline{\langle H^p\rangle_N^{m_1}}\;\overline{\langle H^q\rangle_N^{m_2}} \tag{4.57}$$

will be in general non-zero for $m_1 \neq m_2$. It is important to stress that so far all attempts to derive the form of $S^{\rho:m_1,m_2,N}(x_1,x_2)$ for EGOE have failed; see for example [3, 25, 26]. However, it is possible to derive the formulas for the lower order bivariate moments, i.e. $\Sigma_{p,q}(m_1,m_2,N)$ with $p+q \leq 4$. These give some information about cross correlations generated by EGOE. We will discuss this important aspect in later chapters and in detail in Chap. 12.

4.4 Transition Strength Density: Bivariate Gaussian Form

The strength $R(E_i,E_f)$ generated by a transition operator $\mathcal{O}$ in the H-diagonal basis is $R(E_i,E_f) = |\langle E_f \mid \mathcal{O} \mid E_i\rangle|^2$. Correspondingly, the bivariate strength density $I_{biv;\mathcal{O}}(E_i,E_f)$ or $\rho_{biv;\mathcal{O}}(E_i,E_f)$ which is positive definite and normalized to unity is defined by

$$\begin{aligned} I_{biv;\mathcal{O}}(E_i,E_f) &= \langle\langle \mathcal{O}^\dagger\delta(H-E_f)\mathcal{O}\delta(H-E_i)\rangle\rangle \\ &= I^f(E_f)\big|\langle E_f|\mathcal{O}|E_i\rangle\big|^2 I^i(E_i) \\ &= \langle\langle \mathcal{O}^\dagger\mathcal{O}\rangle\rangle \rho_{biv;\mathcal{O}}(E_i,E_f). \end{aligned} \tag{4.58}$$

With ε_i and ε_f being the centroids and σ_i^2 and σ_f^2 being the variances of the marginal densities $\rho_{i;\mathcal{O}}(E_i)$ and $\rho_{f;\mathcal{O}}(E_f)$ respectively of the bivariate density $\rho_{biv;\mathcal{O}}$, the bivariate reduced central moments of are $\mu_{pq} = \langle \mathcal{O}^\dagger(\frac{H-\varepsilon_f}{\sigma_f})^q \mathcal{O}(\frac{H-\varepsilon_i}{\sigma_i})^p\rangle / \langle \mathcal{O}^\dagger\mathcal{O}\rangle$ and $\zeta = \mu_{11}$ is the bivariate correlation coefficient. In order to obtain the asymptotic form of $\rho_{biv;\mathcal{O}}$ for EGOE, formulas for μ_{pq} with $p+q=4$ and 6 are derived using BCA and thereby the reduced cumulants k_{pq} with $p+q=4$ and 6.

Firstly, H is represented by EGOE(k). Given the transition operator $\mathcal{O}$ of body rank t, we can decompose it into a part that is correlated with H and represent the remaining part say R by a EGOE(t) independent of EGOE(k) representing H.

Then $\mathcal{O} = \alpha H + R$ and the αH term generates the expectation values or the diagonal matrix elements $\langle E|\mathcal{O}|E\rangle$ where E are H eigenvalues. Note that, as H and R are independent, $\alpha = \langle \mathcal{O}H\rangle / \langle H^2\rangle$. Therefore R generates the off-diagonal, in the H diagonal basis, transition matrix elements $|\langle E_f|\mathcal{O}|E_i\rangle|^2$, $E_i \neq E_f$. Thus, by removing the diagonal or expectation value producing part of $\mathcal{O}$, we can assume that H and the part R of $\mathcal{O}$ can be represented by EGOE(k) and EGOE(t) respectively and further they can be assumed to be independent. Once we remove the αH part from $\mathcal{O}$, we need not to make a distinction between $\mathcal{O}$ and R and hence from now on we use only $\mathcal{O}$. Thus, the theory for transition strengths should be applied only to the off-diagonal matrix elements. Now, we proceed to derive formulas for the bivariate moments μ_{pq} using BCA with independent EGOE(k) and EGOE(t) representations for H and $\mathcal{O}$ respectively [16, 27]. The matrix elements variances v_H^2 and $v_{\mathcal{O}}^2$ respectively in the defining space will be in general different for EGOE(k) and EGOE(t). However they will not appear in the formulas for μ_{pq} as these are reduced moments. It is useful to point out that the correlations in μ_{pq} arise due to the non-commutability of H and $\mathcal{O}$ operators. Firstly it is seen that all μ_{pq} with $p+q$ odd will vanish on ensemble average and also $\mu_{pq} = \mu_{qp}$. Moreover $\sigma_i^2 = \langle \mathcal{O}^\dagger \mathcal{O} H^2\rangle^m / \langle \mathcal{O}^\dagger \mathcal{O}\rangle^m = \langle H^2\rangle$ and $\sigma_f^2 = \sigma_i^2$. Thus the first non-trivial moment is μ_{11} and it is given by,

$$\zeta = \mu_{11} = \left\{\overline{\langle O^\dagger(t) O(t)\rangle^m}\; \overline{\langle H(k)H(k)\rangle^m}\right\}^{-1} \overline{\langle O^\dagger(t)H(k)O(t)H(k)\rangle^m}. \tag{4.59}$$

Applying Eqs. (4.11), (4.13), (4.15) and (4.17) will then give,

$$\zeta = \binom{m}{k}^{-1}\binom{m-t}{k} = 1 - \frac{kt}{m} + \frac{k(k-1)t(t-1)}{2m^2} + O\left(\frac{1}{m^3}\right). \tag{4.60}$$

In the cases with $p+q=4$, the moments to be evaluated are $\mu_{40} = \mu_{04}$, $\mu_{31} = \mu_{13}$ and μ_{22}. The diagrams for these follow by putting $\mathcal{O}^\dagger$ and $\mathcal{O}$ at appropriate places in the $\langle H^4\rangle$ diagrams in Eq. (4.28). Firstly, μ_{04} is given by

$$\begin{aligned}
\mu_{04} &= \left[\overline{\langle \mathcal{O}^\dagger \mathcal{O}\rangle^m}\left(\overline{\langle H^2\rangle^m}\right)^2\right]^{-1} \overline{\langle \mathcal{O}^\dagger H^4(k)\mathcal{O}\rangle^m} \\
&= \left[\overline{\langle \mathcal{O}^\dagger \mathcal{O}\rangle^m}\left(\overline{\langle H^2\rangle^m}\right)^2\right]^{-1} \overline{\langle \mathcal{O}^\dagger \mathcal{O}\rangle^m}\;\overline{\langle H^4(k)\rangle^m} \\
&= 2 + \binom{m}{k}^{-1}\binom{m-k}{k} = \mu_{40}.
\end{aligned} \tag{4.61}$$

Here we have used independence of $\mathcal{O}$ and H ensembles and used BCA that led to Eq. (4.32). Similarly,

$$\begin{aligned}
\mu_{13} = &\left[\overline{\langle \mathcal{O}^\dagger \mathcal{O}\rangle^m}\left(\overline{\langle H^2\rangle^m}\right)^2\right]^{-1}\left[\overline{\langle \mathcal{O}^\dagger H(k)H(k)H(k)\mathcal{O}H(k)\rangle^m}\right. \\
&\oplus \overline{\langle \mathcal{O}^\dagger H(k)H(k)H(k)\mathcal{O}H(k)\rangle^m} \\
&\left.\oplus \overline{\langle \mathcal{O}^\dagger H(k)H(k)H(k)\mathcal{O}H(k)\rangle^m}\right].
\end{aligned} \tag{4.62}$$

The first two terms in Eq. (4.62) are equal and the directly correlated H–H pair can be removed from the trace giving $\binom{m}{k}$. Then we are left with $\langle \mathcal{O}^\dagger H \mathcal{O} H\rangle^m$ term that gives $\binom{m-t}{k}\binom{m}{t}$. In the last term, we have to first contract the first and third H's across the second H giving $\binom{m-k}{k}$ factor. Then we are left with $\langle \mathcal{O}^\dagger H \mathcal{O} H\rangle^m$ term that gives $\binom{m-t}{k}\binom{m}{t}$ using Eq. (4.36) for contracting H's across the $\mathcal{O}$ operator. Combining all these, we have

$$\mu_{13} = \binom{m}{k}^{-2}\left[2\binom{m-t}{k}\binom{m}{k} + \binom{m-t}{k}\binom{m-k}{k}\right] = \zeta\,\mu_{04}. \tag{4.63}$$

Alternatively, it is possible to consider μ_{31} and this gives immediately Eq. (4.63). Note that μ_{31} involves $\langle \mathcal{O}^\dagger H \mathcal{O} H^3\rangle^m$ with $\mathcal{O}^\dagger$ and $\mathcal{O}$ correlated and $[\binom{m}{t}]^{-1}\binom{m-k}{t} = [\binom{m}{k}]^{-1}\binom{m-t}{k}$. This proof also gives immediately that $\mu_{15} = \mu_{51} = \zeta\,\mu_{06}$. Now, we will consider μ_{22} where

$$\begin{aligned}
\mu_{22} = {} & \left[\overline{\langle \mathcal{O}^\dagger \mathcal{O}\rangle^m}\left(\overline{\langle H^2\rangle^m}\right)^2\right]^{-1}\Big[\overline{\langle \mathcal{O}^\dagger H(k)H(k)\mathcal{O}H(k)H(k)\rangle^m} \\
& \oplus \overline{\langle \mathcal{O}^\dagger H(k)H(k)\mathcal{O}H(k)H(k)\rangle^m} \\
& \oplus \overline{\langle \mathcal{O}^\dagger H(k)H(k)\mathcal{O}H(k)H(k)\rangle^m}\Big] \\
= {} & \binom{m}{k}^{-2}\left[\binom{m}{k}^2 + \binom{m-t}{k}^2 + \binom{m-k-t}{k}\binom{m-t}{k}\right].
\end{aligned} \tag{4.64}$$

The first term in Eq. (4.64) is simple as we can take out the correlated pairs of H's from the trace. The second term follows by applying Eq. (4.36) twice for the contraction of H's across $\mathcal{O}$. The third term follows by first contracting two H's across $H\mathcal{O}$ operator (effective body rank $k+t$) and then we are left with the $\overline{\langle \mathcal{O}^\dagger H \mathcal{O} H\rangle^m}$ term. Using (4.60), (4.61), (4.63) and (4.64), formulas for the 4th order cumulants are obtained and they are

$$\begin{aligned}
k_{04} = k_{40} = \mu_{04} - 3 &= \binom{m}{k}^{-1}\binom{m-k}{k} - 1 \\
&= -\frac{k^2}{m} + \frac{k^2(k-1)^2}{2m^2} + O\left(\frac{1}{m^3}\right), \\
k_{13} = k_{31} = \mu_{13} - 3\mu_{11} = \zeta\,k_{04} &= -\frac{k^2}{m} + \frac{k^2[(k-1)^2 + 2kt]}{2m^2} + O\left(\frac{1}{m^3}\right), \\
k_{22} = \mu_{22} - 2\mu_{11}^2 - 1 &= \zeta^2\left\{\binom{m-k-t}{k}\binom{m-t}{k}^{-1} - 1\right\} \\
&= -\frac{k^2}{m} + \frac{k^2[(k-1)^2 + 4kt - 2t]}{2m^2} + O\left(\frac{1}{m^3}\right).
\end{aligned} \tag{4.65}$$

In order to establish the structure of the bivariate cumulants, the cumulants to order $p+q=6$ are also derived starting with the 15 diagrams in Eq. (4.34). Following the μ_{04} and μ_{13} derivations, we have simply,

$$\begin{aligned}
\mu_{06} &= \left[\left(\overline{\langle H^2\rangle^m}\right)^3\right]^{-1}\overline{\langle H^6\rangle^m} \\
&= 5 + 6\binom{m}{k}^{-1}\binom{m-k}{k} + 3\binom{m}{k}^{-2}\binom{m-k}{k}^2 \\
&\quad + \binom{m}{k}^{-2}\binom{m-2k}{k}\binom{m-k}{k}, \\
\mu_{15} &= \zeta\,\mu_{06}.
\end{aligned} \tag{4.66}$$

Here we have used Eq. (4.37) and ζ is given by Eq. (4.60). Now, we will consider μ_{24} and it is given by,

$$\begin{aligned}
\mu_{24} = {} & \left[\overline{\langle \mathcal{O}^\dagger \mathcal{O}\rangle^m}\left(\overline{\langle H^2\rangle^m}\right)^3\right]^{-1} \\
& \times \Big[\langle\{\mathcal{O}^\dagger H(k)H(k)\mathcal{O}H(k)H(k)H(k)H(k) \\
& \oplus \mathcal{O}^\dagger H(k)H(k)\mathcal{O}H(k)H(k)H(k)H(k)\} \\
& \oplus \{\mathcal{O}^\dagger H(k)H(k)\mathcal{O}H(k)H(k)H(k)H(k) \\
& \oplus \mathcal{O}^\dagger H(k)H(k)\mathcal{O}H(k)H(k)H(k)H(k) \\
& \oplus \mathcal{O}^\dagger H(k)H(k)\mathcal{O}H(k)H(k)H(k)H(k)\} \\
& \oplus \{\mathcal{O}^\dagger H(k)H(k)\mathcal{O}H(k)H(k)H(k)H(k) \\
& \oplus \mathcal{O}^\dagger H(k)H(k)\mathcal{O}H(k)H(k)H(k)H(k) \\
& \oplus \mathcal{O}^\dagger H(k)H(k)\mathcal{O}H(k)H(k)H(k)H(k)\} \\
& \oplus \{\mathcal{O}^\dagger H(k)H(k)\mathcal{O}H(k)H(k)H(k)H(k) \\
& \oplus \mathcal{O}^\dagger H(k)H(k)\mathcal{O}H(k)H(k)H(k)H(k)\} \\
& \oplus \{\mathcal{O}^\dagger H(k)H(k)\mathcal{O}H(k)H(k)H(k)H(k)
\end{aligned}$$

Table 4.1 Diagrams for the bivariate reduced moment μ_{24} and the corresponding BCA formulas. In the table, $X = [\overline{\langle \mathcal{O}^\dagger \mathcal{O}\rangle^m}\ (\overline{\langle H^2\rangle^m})^3]$

Correlation diagram	Formula in BCA
$X^{-1}2\overline{\langle \mathcal{O}^\dagger AA\mathcal{O}CCBB\rangle^m}$	2
$X^{-1}3\overline{\langle \mathcal{O}^\dagger AC\mathcal{O}CABB\rangle^m}$	$3\binom{m}{k}^{-2}\binom{m-t}{k}^{2}$
$X^{-1}3\overline{\langle \mathcal{O}^\dagger AB\mathcal{O}ABCC\rangle^m}$	$3\binom{m}{k}^{-2}\binom{m-t}{k}\binom{m-t-k}{k}$
$X^{-1}2\overline{\langle \mathcal{O}^\dagger AC\mathcal{O}BCBA\rangle^m}$	$2\binom{m}{k}^{-3}\binom{m-k}{k}\binom{m-t}{k}^{2}$
$X^{-1}2\overline{\langle \mathcal{O}^\dagger AC\mathcal{O}BABC\rangle^m}$	$2\binom{m}{k}^{-3}\binom{m-k}{k}\binom{m-t}{k}\binom{m-t-k}{k}$
$X^{-1}\overline{\langle \mathcal{O}^\dagger AA\mathcal{O}CBCB\rangle^m}$	$\binom{m}{k}^{-1}\binom{m-k}{k}$
$X^{-1}\overline{\langle \mathcal{O}^\dagger AC\mathcal{O}BACB\rangle^m}$	$\binom{m}{k}^{-3}\binom{m-2k}{k}\binom{m-t-k}{k}\binom{m-t}{k}$
$X^{-1}\overline{\langle \mathcal{O}^\dagger AC\mathcal{O}BCAB\rangle^m}$	$\binom{m}{k}^{-3}\binom{m-t}{k}\sum_{\nu=k}^{2k}\binom{m-\nu}{k}\binom{m-k-t}{\nu-k}\binom{k}{2k-\nu}$

$$\oplus\, \mathcal{O}^\dagger H(k)H(k)\mathcal{O}H(k)H(k)H(k)H(k)\}$$

$$\oplus\, \mathcal{O}^\dagger H(k)H(k)\mathcal{O}H(k)H(k)H(k)H(k)$$

$$\oplus\, \mathcal{O}^\dagger H(k)H(k)\mathcal{O}H(k)H(k)H(k)H(k)$$

$$\oplus\, \mathcal{O}^\dagger H(k)H(k)\mathcal{O}H(k)H(k)H(k)H(k)\rangle^m]. \tag{4.67}$$

For simplicity, the 'overline' symbol is dropped in Eq. (4.67). All the terms in '{}' brackets are equal and we show in Table 4.1, using the same alphabet for correlated pairs of H's, the diagrams and the formula for them in BCA. The first seven terms in Table 4.1 are easy to recognize following the results already given before using BCA. The last term is special as we need to contract over two operators that are correlated in a different way than in all the other diagrams we have considered so far. Therefore, this needs special treatment as discussed in the context of the 8th moment of the eigenvalue density in [3]. Finally, in BCA μ_{33} can be written as follows (again here also the 'overline' symbol is dropped everywhere),

$$
\begin{aligned}
\mu_{33} = {} & \left[\langle \mathcal{O}^\dagger \mathcal{O} \rangle^m \left(\langle H^2 \rangle^m\right)^3\right]^{-1} \\
& \times \Big[\big\langle \big\{\mathcal{O}^\dagger H(k)H(k)H(k)\mathcal{O}H(k)H(k)H(k) \\
& \oplus \mathcal{O}^\dagger H(k)H(k)H(k)\mathcal{O}H(k)H(k)H(k) \\
& \oplus \mathcal{O}^\dagger H(k)H(k)H(k)\mathcal{O}H(k)H(k)H(k) \\
& \oplus \mathcal{O}^\dagger H(k)H(k)H(k)\mathcal{O}H(k)H(k)H(k)\big\} \\
& \oplus \big\{\mathcal{O}^\dagger H(k)H(k)H(k)\mathcal{O}H(k)H(k)H(k) \\
& \oplus \mathcal{O}^\dagger H(k)H(k)H(k)\mathcal{O}H(k)H(k)H(k) \\
& \oplus \mathcal{O}^\dagger H(k)H(k)H(k)\mathcal{O}H(k)H(k)H(k) \\
& \oplus \mathcal{O}^\dagger H(k)H(k)H(k)\mathcal{O}H(k)H(k)H(k)\big\} \\
& \oplus \mathcal{O}^\dagger H(k)H(k)H(k)\mathcal{O}H(k)H(k)H(k) \\
& \oplus \big\{\mathcal{O}^\dagger H(k)H(k)H(k)\mathcal{O}H(k)H(k)H(k) \\
& \oplus \mathcal{O}^\dagger H(k)H(k)H(k)\mathcal{O}H(k)H(k)H(k)\big\} \\
& \oplus \mathcal{O}^\dagger H(k)H(k)H(k)\mathcal{O}H(k)H(k)H(k) \\
& \oplus \big\{\mathcal{O}^\dagger H(k)H(k)H(k)\mathcal{O}H(k)H(k)H(k) \\
& \oplus \mathcal{O}^\dagger H(k)H(k)H(k)\mathcal{O}H(k)H(k)H(k)\big\} \\
& \oplus \mathcal{O}^\dagger H(k)H(k)H(k)\mathcal{O}H(k)H(k)H(k)\big\rangle^m\Big].
\end{aligned} \tag{4.68}
$$

Just as for μ_{24}, we can write Eq. (4.68) as a sum of seven terms by recognizing that the terms in a given '{}' will give the same result. In Table 4.2 given are the BCA formulas for these terms.

From the previous discussion, it is easy to derive all the formulas given in Table 4.2. Using the formulas given in Appendix B, all the bivariate reduced moments can be converted into bivariate cumulants and then the $1/m$ expansions for the 6th

Table 4.2 Diagrams for the bivariate reduced moment μ_{33} and the corresponding BCA formulas. In the table, $X = [\overline{\langle \mathcal{O}^\dagger \mathcal{O} \rangle^m}\,(\overline{\langle H^2 \rangle^m})^3]$

Correlation diagram	Formula in BCA
$X^{-1} 4 \overline{\langle \mathcal{O}^\dagger ACC\mathcal{O}ABB \rangle^m}$	$4\binom{m}{k}^{-1}\binom{m-t}{k}$
$X^{-1} 4 \overline{\langle \mathcal{O}^\dagger ABA\mathcal{O}BCC \rangle^m}$	$4\binom{m}{k}^{-2}\binom{m-k}{k}\binom{m-t}{k}$
$X^{-1} \overline{\langle \mathcal{O}^\dagger ABC\mathcal{O}CBA \rangle^m}$	$\binom{m}{k}^{-3}\binom{m-t}{k}^3$
$X^{-1} 2 \overline{\langle \mathcal{O}^\dagger ACB\mathcal{O}CAB \rangle^m}$	$2\binom{m}{k}^{-3}\binom{m-t}{k}\binom{m-t-k}{k}^2$
$X^{-1} \overline{\langle \mathcal{O}^\dagger ABA\mathcal{O}CBC \rangle^m}$	$\binom{m}{k}^{-3}\binom{m-k}{k}^2\binom{m-t}{k}$
$X^{-1} 2 \overline{\langle \mathcal{O}^\dagger ABC\mathcal{O}CAB \rangle^m}$	$2\binom{m}{k}^{-3}\binom{m-t}{k}^2\binom{m-k-t}{k}$
$X^{-1} \overline{\langle \mathcal{O}^\dagger ACB\mathcal{O}ACB \rangle^m}$	$\binom{m}{k}^{-3}\binom{m-2k-t}{k}\binom{m-t-k}{k}\binom{m-t}{k}$

order cumulants are,

$$
\begin{aligned}
k_{06} = k_{60} &= \mu_{06} - 15\mu_{04} + 30 \\
&= \frac{k^3(6k-1)}{m^2} - \frac{k^3(k-1)^2(7k-1)}{m^3} + O\left(\frac{1}{m^4}\right), \\
k_{15} = k_{51} &= \zeta k_{06} = \frac{k^3(6k-1)}{m^2} - \frac{k^3[(k-1)^2(7k-1) + kt(6k-1)]}{m^3} + O\left(\frac{1}{m^4}\right), \\
k_{24} = k_{42} &= \mu_{24} - \mu_{04} - 8\zeta\mu_{13} - 6\mu_{22} + 24\zeta^2 + 6 \\
&= \frac{k^3(6k-1)}{m^2} - \frac{k^3[(k-1)^2(7k-1) + t(12k^2 - 6k - 1)]}{m^3} + O\left(\frac{1}{m^4}\right), \\
k_{33} &= \mu_{33} - 6\mu_{13} - 9\zeta\mu_{22} + 12\zeta^3 + 18\zeta \\
&= \frac{k^3(6k-1)}{m^2} - \frac{k^3[(k-1)^2(7k-1) + t(16k^2 - 13k + 2)]}{m^3} + O\left(\frac{1}{m^4}\right).
\end{aligned}
\tag{4.69}
$$

As discussed in Appendix B, for a bivariate Gaussian all cumulants k_{pq} with $p + q \geq 3$ should be zero. Therefore, using Eqs. (4.65) and (4.69), it is seen that in the dilute limit (just as in the case of state densities, here also one needs $k^2/m \to 0$), the transition strength densities approach bivariate Gaussian form. Thus, we have

$$
\begin{aligned}
&\rho_{biv;\mathcal{O}}(E_i, E_f) \\
&\xrightarrow{EGOE} \overline{\rho_{biv;\mathcal{O}}(E_i, E_f)} = \rho_{biv-\mathcal{G};\mathcal{O}}(E_i, E_f\,;\,\varepsilon_i, \varepsilon_f, \sigma_i, \sigma_f, \zeta) \\
&= \frac{1}{2\pi\sigma_i\sigma_f\sqrt{1-\zeta^2}} \\
&\quad \times \exp\left\{-\frac{1}{2(1-\zeta^2)}\left[\left(\frac{E_i-\varepsilon_i}{\sigma_i}\right)^2 - 2\zeta\left(\frac{E_i-\varepsilon_i}{\sigma_i}\right)\left(\frac{E_f-\varepsilon_f}{\sigma_f}\right)\right.\right. \\
&\quad \left.\left. + \left(\frac{E_f-\varepsilon_f}{\sigma_f}\right)^2\right]\right\}.
\end{aligned} \tag{4.70}
$$

However, for the strict validity of the Gaussian form, $k_{pq} = 0$ for $p + q \geq 3$ should be valid for any rotation of the (E_i, E_f) variables. To examine this, we convert the bivariate moments μ_{pq} given above in the (E_i, E_f) variables into those defined for the sum and difference variables $(E_i + E_f, E_i - E_f)$. Reduced moments and cumulants defined by these new variables will be denoted by μ'_{pq} and k'_{pq} respectively. For example, denoting E_i by x_1 and E_f by x_2, we have (without loss of generality, we assume (x_1, x_2) are standardized variables)

$$
\begin{aligned}
\mu'_{20} &= \left\langle (x_1+x_2)^2\right\rangle^m = \left\langle 2x_1^2 + 2x_1x_2\right\rangle^m = 2(1+\zeta), \\
\mu'_{02} &= \left\langle (x_1-x_2)^2\right\rangle^m = \left\langle 2x_1^2 - 2x_1x_2\right\rangle^m = 2(1-\zeta), \\
\mu'_{40} &= \left[4(1+\zeta)^2\right]^{-1}\left\langle (x_1+x_2)^4\right\rangle^m \\
&= \left[4(1+\zeta)^2\right]^{-1}\left\langle 2x_1^4 + 6x_1^2x_2^2 + 8x_1x_2^3\right\rangle^m \\
&= \left[4(1+\zeta)^2\right]^{-1}(2\mu_{40} + 6\mu_{22} + 8\mu_{31}), \\
\mu'_{04} &= \left[4(1-\zeta)^2\right]^{-1}\left\langle (x_1-x_2)^4\right\rangle^m \\
&= \left[4(1-\zeta)^2\right]^{-1}\left\langle 2x_1^4 + 6x_1^2x_2^2 - 8x_1x_2^3\right\rangle^m \\
&= \left[4(1-\zeta)^2\right]^{-1}(2\mu_{40} + 6\mu_{22} - 8\mu_{31}).
\end{aligned} \tag{4.71}
$$

Here, we have used the results $\langle x_1^2\rangle = \langle x_2^2\rangle$ and $\langle x_i^2 x_j\rangle = \langle x_j^2 x_i\rangle$. Converting the moments μ_{pq} into cumulants k_{pq}, we obtain (it should be noted that $\zeta' = 0$) using Eq. (4.65),

$$
\begin{aligned}
k'_{40} &= \left[2(1+\zeta)^2\right]^{-1}(k_{40} + 3k_{22} + 4k_{31}) = -\frac{k^2}{m} + O\left(\frac{1}{m^2}\right), \\
k'_{04} &= \left[2(1-\zeta)^2\right]^{-1}(k_{40} + 3k_{22} - 4k_{31}) = \frac{k-3/2}{t} + O\left(\frac{1}{m}\right).
\end{aligned} \tag{4.72}
$$

Similarly, it is easy to see that $\mu'_{13} = \mu'_{31} = 0$ and $k'_{13} = k'_{31} = 0$. The μ'_{22} and k'_{22} are given by

$$\begin{aligned}
\mu'_{22} &= \left[4\left(1-\zeta^2\right)\right]^{-1}\left\langle (x_1+x_2)^2(x_1-x_2)^2\right\rangle^m \\
&= \left[4\left(1-\zeta^2\right)\right]^{-1}(2\mu_{40}-2\mu_{22}), \\
k'_{22} &= \left[2\left(1-\zeta^2\right)\right]^{-1}(k_{40}-k_{22}) = \frac{k(1-2k)}{4m} + O\left(\frac{1}{m^2}\right).
\end{aligned} \tag{4.73}$$

From these equations, it is clearly seen that k_{04} in the difference variable will not approach zero even if m is large although all the other cumulants approach zero as $m \to \infty$. Therefore, even in the dilute limit, EGOE will not generate a strict bivariate Gaussian. To further confirm this result, sixth order cumulants k'_{pq} with $p+q=6$ are considered. Following the same procedure as in Eq. (4.72) for the sixth order cumulants, we get the following results using Eqs. (4.65) and (4.69),

$$\begin{aligned}
k'_{60} &= \left[4(1+\zeta)^3\right]^{-1}[k_{60}+6k_{51}+15k_{42}+10k_{33}] \\
&= \frac{k^3(6k-1)}{m^2} + O\left(\frac{1}{m^3}\right), \\
k'_{51} &= k'_{15} = 0, \qquad k'_{33} = 0, \\
k'_{42} &= \left[4(1-\zeta)(1+\zeta)^2\right]^{-1}[k_{60}+2k_{51}-k_{42}-2k_{33}] \\
&= \frac{k^2(32k^2-30k+3)}{16m^2} + O\left(\frac{1}{m^3}\right), \\
k'_{24} &= \left[4(1+\zeta)(1-\zeta)^2\right]^{-1}[k_{60}-2k_{51}-k_{42}+2k_{33}] \\
&= \frac{k(-8k^2+18k-5)}{8mt} + O\left(\frac{1}{m^2}\right), \\
k'_{06} &= \left[4(1-\zeta)^3\right]^{-1}[k_{60}-6k_{51}+15k_{42}-10k_{33}] \\
&= \frac{16k^2-46k+35}{4t^2} + O\left(\frac{1}{m}\right).
\end{aligned} \tag{4.74}$$

It is seen from Eq. (4.74) that the cumulants k'_{24} and k'_{06} will not approach zero even if m is large. Thus, in practice one has to apply the bivariate Edgeworth corrections (given in Appendix B) to the bivariate Gaussian form of the transition strength density.

The peculiar behavior of k'_{rs} is a result of the behavior of the bivariate correlation coefficient ζ in the original (E_i, E_f) variables. It is seen from Eq. (4.60) that $\zeta \to 1$ as $m \to \infty$ (with $k/m \to 0$ and $t/m \to 0$). This implies that as m increases, the strength density will become narrower. The value $\zeta = 1$ is unphysical as this implies H and $\mathcal{O}$ commute. In practice, $\zeta = 0.6$–0.8 and it will not be very close to 1. Note that $\zeta = 0$ implies that the strengths are constant, i.e. the system reduces to a GOE representation. An expansion for the strengths that starts from the GOE result can be obtained by expanding the delta functions in Eq. (4.58) in terms of polynomials defined by the H eigenvalue density. Given the eigenvalue density $\rho(E)$ and the corresponding orthonormal polynomials $P_\mu(E)$, we have [28]

$$\delta(H-E) = \rho(E) \sum_\mu P_\mu(H) P_\mu(E). \tag{4.75}$$

Given the moments of $\rho(E)$, we can write the polynomials $P_\mu(E)$. As odd moments vanish for EGOE, the lowest four polynomials, in terms of standardized variables x, are

$$P_0(x) = 1, \qquad P_1(x) = x, \qquad P_2(x) = \frac{x^2-1}{\sqrt{\mu_4 - 1}}, \qquad P_3(x) = \frac{x^3 - \mu_4 x}{\sqrt{\mu_6 - \mu_4^2}}. \tag{4.76}$$

Substituting in Eq. (4.58) the delta function expansion given by Eq. (4.75), we obtain [29]

$$\left|\langle m_f, E_f | \mathcal{O} | m_i E_i \rangle\right|^2 = \sum_{\mu,\nu} \left\langle \mathcal{O}^\dagger P_\mu^{m_f}(H) \mathcal{O} P_\nu^{m_i}(H) \right\rangle^{m_i} P_\mu^{m_f}(E_f) P_\nu^{m_i}(E_i). \tag{4.77}$$

For simplicity we assume that $m_i = m_f = m$. Now, using the results for μ_{pq} given before one can write down formulas using BCA for $g_{\mu\nu} = [\langle \mathcal{O}^\dagger \mathcal{O} \rangle^m]^{-1} \langle \mathcal{O}^\dagger P_\mu(H) \times \mathcal{O} P_\nu(H) \rangle^m$, $\mu + \nu \leq 6$. Then,

$$\begin{aligned} &g_{00} = 1, \qquad g_{11} = \zeta, \qquad g_{22} = \zeta^2 \left[1 - \frac{k^2 t}{2m^2} + O\left(\frac{1}{m^3}\right)\right], \\ &g_{33} = \zeta^3 \left[1 - 3\frac{k^2 t}{2m^2} + O\left(\frac{1}{m^3}\right)\right]. \end{aligned} \tag{4.78}$$

All other $g_{\mu\nu} = 0$ or at least $O(\frac{1}{m^3})$. For example,

$$g_{24} = g_{42} = \frac{k^3 t(2k-1)}{4\sqrt{3} m^3} + O\left(\frac{1}{m^4}\right). \tag{4.79}$$

Generalizing the results in Eq. (4.78) we have in the dilute limit

$$\begin{aligned} &\left[\langle \mathcal{O}^\dagger \mathcal{O} \rangle^m\right]^{-1} \left\langle \mathcal{O}^\dagger P_\mu(H) \mathcal{O} P_\nu(H) \right\rangle^m = \delta_{\mu\nu} (\zeta)^\mu \\ &\quad \Rightarrow \quad \rho_{biv:\mathcal{O}}(E_i, E_f) = \rho_1(E_i) \rho_2(E_f) \sum_{\mu=0}^{\infty} (\zeta)^\mu P_\mu(E_f) P_\mu(E_i) \\ &\qquad\qquad = \rho_{biv-\mathcal{G};\mathcal{O}}(E_i, E_f). \end{aligned} \tag{4.80}$$

For EGOE the eigenvalue densities are Gaussians and hence the polynomials are Hermite polynomials. Then the sum over the polynomials gives exactly

$\rho_{biv-\mathcal{G};\mathcal{O}}(E_i, E_f)$ with correlation coefficient ζ. Therefore the polynomial expansion has to be summed to very high orders to recover the bivariate Gaussian form. This implies that larger the ζ value, slower will be the convergence of the polynomial expansion for transition strengths. For EGOE, the correlation coefficient $\zeta = \binom{m}{k}^{-1}\binom{m-t}{k}$ and this will be closer to unity. Therefore, expansions for transition strength densities starting with a bivariate Gaussian form will be appropriate. In practice, it is important to employ the bivariate Edgeworth expansion given by Eq. (B.15) incorporating k_{rs}, $r+s=3,4$ corrections.

4.5 Strength Sums and Expectation Values: Ratio of Gaussians

Given a operator $\mathcal{O}$ acting on an eigenstate with energy E_i, the transition strength sum, sum of the strengths going to all states with energies E_f, is $\sum_{E_f} |\langle E_f|\mathcal{O}|E_i\rangle|^2$ and this is nothing but the expectation value $\langle \mathcal{O}^\dagger \mathcal{O}\rangle^{E_i} = \langle E_i|\mathcal{O}^\dagger \mathcal{O}|E_i\rangle$. However, taking degeneracies into account, one has to deal with strength sum or expectation value densities. Given a positive definite operator $K = \mathcal{O}^\dagger \mathcal{O}$, the expectation density $I_K(E) = I_{\mathcal{O}^\dagger\mathcal{O}}(E)$ and its normalized version $\rho_K(E)$ are

$$\begin{aligned} I_K^m(E) &= \langle m, E|K|m, E\rangle I^m(E) = \langle K\rangle^{m,E} I^m(E) \\ &= \langle\!\langle K\delta(H-E)\rangle\!\rangle^m; \\ \rho_k(E) &= \frac{\langle K\delta(H-E)\rangle^m}{\langle K\rangle^m}. \end{aligned} \tag{4.81}$$

Clearly, expectation value will be the ratio of expectation value density and state density. More importantly, strength sum density [for this $K = \mathcal{O}^\dagger\mathcal{O}$ in Eq. (4.81)] will be a marginal density of the bivariate strength density. For EGOE(k), as the bivariate strength density is a bivariate Gaussian, the strength sum density will be a Gaussian and strength sum will be a ratio of Gaussians [27],

$$\langle \mathcal{O}^\dagger\mathcal{O}\rangle^E = \frac{I^m_{\mathcal{O}^\dagger\mathcal{O}}(E)}{I^m(E)} \xrightarrow{\text{EGOE}} \frac{I^m_{\mathcal{O}^\dagger\mathcal{O}:\mathcal{G}}(E)}{I^m_{\mathcal{G}}(E)} = \langle \mathcal{O}^\dagger\mathcal{O}\rangle^m \frac{\rho^m_{\mathcal{O}^\dagger\mathcal{O}:\mathcal{G}}(E)}{\rho^m_{\mathcal{G}}(E)}. \tag{4.82}$$

Moments of the strength sum density are

$$M_p(\mathcal{O}^\dagger\mathcal{O}) = \frac{\langle \mathcal{O}^\dagger\mathcal{O}H^p\rangle^m}{\langle \mathcal{O}^\dagger\mathcal{O}\rangle^m}. \tag{4.83}$$

Using the moments to fourth order it is possible to add Edgeworth corrections to the Gaussian densities in Eq. (4.82). With $\mathcal{O} = a_i$, Eq. (4.82) gives expectation values of the number operator $\hat{n}_i$ or the occupancies of the sp states $|i\rangle$. Similarly $\mathcal{O}$ is GT operator gives GT strength sums [30] in nuclei and dipole operator gives dipole strength sums in atoms [31].

4.6 Level Fluctuations

In this section, we will briefly discuss the various attempts made in literature to derive the two-point correlation function in energy levels for EGOE(k) and similarly for EGUE(k).

French [3, 6] has conjectured in early 70's, as already stated in Sect. 4.3, that the level and strength fluctuations for EGOE(k) follow GOE. This inference came from many numerical examples (with unfolding of each member of the ensemble with Edgeworth corrected Gaussian defined by the moments generated by individual members, i.e. using spectral unfolding rather than ensemble unfolding) both from EGOE(2) and random two-body interactions in nuclear shell model. These showed that the NNSD is close to Wigner form, $\overline{\Delta_3(\overline{n})}$ fits Dyson-Mehta formula and strength fluctuations follow P-T law. See for example Figs. 2.2, 2.4, 5.3 and [6, 7]. However, the two-point correlation function could not be derived as the BCA fails here.

In 1984, Verbaarschot and Zirnbauer [32] used the replica trick, developed in statistical mechanics for the study of spin glasses and Anderson localization, to derive the two-point function for EGOE(k). However their attempted was not successful. Later in 2000, Weidenmüller's group made another attempt [19, 20]. They have used two extreme models, one called EGE$_{min}(k)$ where all the k-particle matrix elements are assumed to be same. Thus it will have only one independent variable. The other ensemble is called EGE$_{max}(k)$ where all matrix elements, in the m particle space H matrix, allowed by symmetries are assumed to be independent Gaussian random variable and the rest are put to zero. Clearly EGE$_{min}(k)$ represents an integrable system and therefore follows Poisson. Similarly, it was shown explicitly using the SUSY method that EGE$_{max}(k)$ follows GOE. Then, using the sparsity of EGOE(k) ensemble it is argued that EGOE(k) fluctuations should be in between Poisson and GOE. However, explicit form of the two-point correlation function could not be derived [25, 26]. More recently Papenbrock et al. [33], made another attempt to establish the nature of fluctuations generated by EGOE(k). They have, motivated by the analogy to metal-insulator transition (MIT) and a special power-law random band matrix (PLRBM) that simulates the critical statistic at the MIT, constructed a random matrix ensemble called scaffolding ensemble (ScE) having properties: (i) ScE is more sparse than EGOE(k) ensemble; (ii) ScE spectral fluctuations are those of the critical ensemble. Using arguments based on a combination of analytical results, numerical examples and application of a criterion due to Levitov [34], it is argued that EGOE(k) H matrices (with $k \geq 2$) lie on the delocalized side of the MIT and is therefore chaotic or equivalently EGOE(k) fluctuations follow GOE.

It is important and also of interest to understand ergodicity and universality of embedded ensembles. Width of the fluctuations in energy centroids and spectral variances, discussed in detail in Chaps. 11 and 12, clearly indicate that in the dilute limit (for boson systems in the dense limit) EE will be ergodic. However there is not yet an explicit analytical derivation of the result that EE are ergodic. Larger variety of EE described in Chaps. 5–11 and 13 also show that EE have universality—their results apply to a variety of physical systems. Finally, there are some attempts to study fluctuations in energy levels near the ground state in EE. For example, Bohigas and Flores [35] compared the properties of the low-lying part of the spectrum generated by random interactions in shell model (called TBRE—see Chap. 13) and showed that the widths of the positions of individual eigenvalues were much larger for the TBRE than for the GOE. Cota et al. [36, 37] analyzed NNSD and obtained for the Brody parameter the value ∼0.8. More recent results by Flores et al. [38] show that the semi-Poisson distribution gives a better fit than the Brody distribution, if spectral unfolding is used.

4.7 Summary

In summary, EGOE(k) [similarly EGUE(k) discussed in Chap. 11] generates for $m \gg k$ Gaussian form for state densities with $\gamma_2 \to -k^2/m$ and this is established using BCA. In fact, as m increases from $m = k$, state densities exhibit semi-circle to Gaussian transition with $m = 2k$ being the transition point. The semi-circle form for $m < 2k$ has been proved using SUSY method and the result beyond this follows from the BCA method. Thus, the one-point function for EGOE(k) differs from that of GOE. Secondly, using BCA it is established that the smoothed transition strength densities will take close to a bivariate Gaussian form. Then smoothed transition strength sums, being marginal densities divided by the state density, will be ratio of two Gaussians. Thirdly, EGOE(k) exhibits average-fluctuation separation (as m increases) and also non-zero cross correlations between spectra with different particle numbers (Chap. 12 gives details). Finally, it is seen (from transition strengths and level fluctuations with both being essentially two-point in nature) that there are important differences between GOE and EGOE in the two-point functions.

References

1. J.B. French, S.S.M. Wong, Validity of random matrix theories for many-particle systems. Phys. Lett. B **33**, 449–452 (1970)
2. O. Bohigas, J. Flores, Two-body random Hamiltonian and level density. Phys. Lett. B **34**, 261–263 (1971)
3. K.K. Mon, J.B. French, Statistical properties of many-particle spectra. Ann. Phys. (N.Y.) **95**, 90–111 (1975)
4. V.K.B. Kota, SU(N) Wigner-Racah algebra for the matrix of second moments of embedded Gaussian unitary ensemble of random matrices. J. Math. Phys. **46**, 033514 (2005)
5. Z. Pluhař, H.A. Weidenmüller, Symmetry properties of the k-body embedded unitary Gaussian ensemble of random matrices. Ann. Phys. (N.Y) **297**, 344–362 (2002)
6. T.A. Brody, J. Flores, J.B. French, P.A. Mello, A. Pandey, S.S.M. Wong, Random matrix physics: spectrum and strength fluctuations. Rev. Mod. Phys. **53**, 385–479 (1981)
7. V.K.B. Kota, Embedded random matrix ensembles for complexity and chaos in finite interacting particle systems. Phys. Rep. **347**, 223–288 (2001)
8. J.M.G. Gómez, K. Kar, V.K.B. Kota, R.A. Molina, A. Relaño, J. Retamosa, Many-body quantum chaos: recent developments and applications to nuclei. Phys. Rep. **499**, 103–226 (2011)

9. Ph. Jacquod, D.L. Shepelyansky, Emergence of quantum chaos in finite interacting Fermi systems. Phys. Rev. Lett. **79**, 1837–1840 (1997)
10. V.V. Flambaum, F.M. Izrailev, Statistical theory of finite Fermi systems based on the structure of chaotic eigenstates. Phys. Rev. E **56**, 5144–5159 (1997)
11. Y. Alhassid, Statistical theory of quantum dots. Rev. Mod. Phys. **72**, 895–968 (2000)
12. A. Relaño, R.A. Molina, J. Retamosa, $1/f$ noise in the two-body random ensemble. Phys. Rev. E **70**, 017201 (2004)
13. A. Volya, Emergence of symmetry from random n-body interactions. Phys. Rev. Lett. **100**, 162501 (2008)
14. K.K. Mon, Ensemble-averaged eigenvalue distributions for k-body interactions in many-particle systems, B.A. Dissertation, Princeton University (1973)
15. F.S. Chang, J.B. French, T.H. Thio, Distribution methods for nuclear energies, level densities and excitation strengths. Ann. Phys. (N.Y.) **66**, 137–188 (1971)
16. S. Tomsovic, Bounds on the time-reversal non-invariant nucleon-nucleon interaction derived from transition-strength fluctuations, Ph.D. Thesis, University of Rochester, Rochester, New York (1986)
17. M. Vyas, Some studies on two-body random matrix ensembles, Ph.D. Thesis, M.S. University of Baroda, India (2012)
18. A. Gervois, Level densities for random one- and two-body potentials. Nucl. Phys. A **184**, 507–532 (1972)
19. L. Benet, T. Rupp, H.A. Weidenmüller, Nonuniversal behavior of the k-body embedded Gaussian unitary ensemble of random matrices. Phys. Rev. Lett. **87**, 010601 (2001)
20. L. Benet, T. Rupp, H.A. Weidenmüller, Spectral properties of the k-body embedded Gaussian ensembles of random matrices. Ann. Phys. (N.Y.) **292**, 67–94 (2001)
21. H. Cramer, *Mathematical Methods of Statistics* (Princeton University Press, Princeton, 1974)
22. C. Bloch, Statistical nuclear theory, in *Physique Nucléaire*, ed. by C. Dewitt, V. Gillet (Gordon and Breach, New York, 1969), pp. 303–412
23. L. Muñoz, E. Faleiro, R.A. Molina, A. Relaño, J. Retamosa, Spectral statistics in noninteracting many-particle systems. Phys. Rev. E **73**, 036202 (2006)
24. V.K.B. Kota, R.U. Haq, *Spectral Distributions in Nuclei and Statistical Spectroscopy* (World Scientific, Singapore, 2010)
25. L. Benet, H.A. Weidenmüller, Review of the k-body embedded ensembles of Gaussian random matrices. J. Phys. A **36**, 3569–3594 (2003)
26. M. Srednicki, Spectral statistics of the k-body random interaction model. Phys. Rev. E **66**, 046138 (2002)
27. J.B. French, V.K.B. Kota, A. Pandey, S. Tomsovic, Statistical properties of many-particle spectra VI. Fluctuation bounds on N-N T-noninvariance. Ann. Phys. (N.Y.) **181**, 235–260 (1988)
28. G. Szegő, *Orthogonal Polynomials*, Colloquium Publications, vol. 23 (Am. Math. Soc., Providence, 2003)
29. J.P. Draayer, J.B. French, S.S.M. Wong, Spectral distributions and statistical spectroscopy: I General theory. Ann. Phys. (N.Y.) **106**, 472–502 (1977)
30. V.K.B. Kota, D. Majumdar, Bivariate distributions in statistical spectroscopy studies: IV. Interacting particle Gamow-Teller strength densities and β-decay rates of fp-shell nuclei for presupernova stars. Z. Phys. A **351**, 377–383 (1995)
31. V.V. Flambaum, A.A. Gribakina, G.F. Gribakin, Statistics of electromagnetic transitions as a signature of chaos in many-electron atoms. Phys. Rev. A **58**, 230–237 (1998)
32. J.J.M. Verbaarschot, M.R. Zirnbauer, Replica variables, loop expansion, and spectral rigidity of random-matrix ensembles. Ann. Phys. (N.Y.) **158**, 78–119 (1984)
33. T. Papenbrock, Z. Pluhař, J. Tithof, H.A. Weidenmüller, Chaos in fermionic many-body systems and the metal-insulator transition. Phys. Rev. E **83**, 031130 (2011)
34. L.S. Levitov, Delocalization of vibrational modes caused by electric dipole interaction. Phys. Rev. Lett. **64**, 547–550 (1990)

35. O. Bohigas, J. Flores, Spacing and individual eigenvalue distributions of two-body random Hamiltonians. Phys. Lett. B **35**, 383–387 (1971)
36. E. Cota, J. Flores, P.A. Mello, E. Yépez, Level repulsion in the ground-state region of nuclei. Phys. Lett. B **53**, 32–34 (1974)
37. T.A. Brody, E. Cota, J. Flores, P.A. Mello, Level fluctuations: a general properties of spectra. Nucl. Phys. A **259**, 87–98 (1976)
38. J. Flores, M. Horoi, M. Mueller, T.H. Seligman, Spectral statistics of the two-body random ensemble revisited. Phys. Rev. E **63**, 026204 (2000)

Chapter 5
Random Two-Body Interactions in Presence of Mean-Field: EGOE(1 + 2)

5.1 EGOE(1 + 2): Definition and Construction

Hamiltonian for realistic systems such as nuclei and atoms consists of a mean-field one-body (defined by a finite set of single particle states) plus a complexity generating two-body interaction. Then, the appropriate random matrix ensemble, studied first by Flambaum et al. [1], is EGOE(1 + 2) defined by the ensemble of H operators

$$\{\widehat{H}\} = \widehat{h}(1) + \lambda\{\widehat{V}(2)\}, \tag{5.1}$$

where { } denotes an ensemble. The mean-field one-body Hamiltonian $\widehat{h}(1) = \sum_i \varepsilon_i n_i$ is a fixed one-body operator defined by the sp energies ε_i with average spacing Δ (note that n_i is the number operator for the sp state $|\nu_i\rangle$). In general one can choose ε_i's to form an ensemble. The $\{\widehat{V}(2)\}$ ensemble in two-particle spaces is a GOE(1) and λ is the strength of the two-body interaction. Thus, EGOE(1 + 2) is defined by the three parameters (m, N, λ) and without loss of generality we put $\Delta = 1$ so that λ is in units of Δ. From now on 'hat' over H, h and V is dropped when there is no confusion. Construction of EGOE(1 + 2) in the occupation number basis defined by $|\nu_1 \nu_2 \cdots \nu_m\rangle$ follows easily from Eq. (4.3). It should be noted that $h(1)$ contributes only to the diagonal matrix elements and for a given $|\nu_1 \nu_2 \cdots \nu_m\rangle$, the $h(1)$ contribution is simply $\sum_{i=1}^{m} \varepsilon_{\nu_i}$. Before proceeding further let us point out that many different choices for the sp energies have been adopted in literature. For example, EGOE(1 + 2) with $h(1)$ a fixed Hamiltonian (usually generating a uniform sp spectrum) has been used by Flambaum and Izrailev [1] and Kota and collaborators [2]. Similarly, Alhassid et al. [3, 4] used sp energies drawn from the eigenvalues around the center of the semicircle density of a GOE (or a GUE). Alternatively, Jacquod et al. [5, 6] considered sp energies to be random such that $\varepsilon_i = \Delta + \delta_i$ where δ_i are uniform random variables. At the outset, it should be clear that EGOE(1 + 2) reduces to EGOE(2) as $\lambda \to \infty$ and it is seen ahead that in practice λ need not be very large for the approach to EGOE(2). In addition, for EGOE(1 + 2) also, just as with EGOE(2), the embedding is generated by $U(N)$.

V.K.B. Kota, *Embedded Random Matrix Ensembles in Quantum Physics*,
Lecture Notes in Physics 884, DOI 10.1007/978-3-319-04567-2_5,

Before discussing some of the important properties of EGOE(1 + 2), we will digress and consider tensorial decomposition of the Hamiltonian and propagation of energy centroids and spectral variances. These will play important role in the results presented ahead in this chapter and in the other chapters to follow.

5.2 Unitary Decomposition and Trace Propagation

5.2.1 *Unitary or $U(N)$ Decomposition of the Hamiltonian Operator*

General references here are [2, 7–12]. Let us consider a system of m spinless fermions in N sp states with a (1 + 2)-body Hamiltonian $H = h(1) + V(2)$ where $h(1) = \sum_i \varepsilon_i \hat{n}_i$ and $V(2)$ is defined by the two-body matrix elements $V_{ijkl} = \langle kl|V(2)|ij\rangle$. The embedding $U(N)$ algebra is generated by the N^2 number of operators $a_i^\dagger a_j$. Similarly the corresponding $SU(N)$ algebra is generated by the $N^2 - 1$ independent operators $a_i^\dagger a_j - \frac{1}{N}[\sum_k a_k^\dagger a_k]\delta_{i,j}$. With respect to the $U(N)$ group, tensorial or unitary decomposition of H can be obtained as follows. Firstly the irreducible representations (irreps) of $U(N)$ are denoted, in Young tableaux notation, by $\{\lambda_1, \lambda_2, \ldots, \lambda_N\}$ where the λ_i are positive integers with $\lambda_1 \geq \lambda_2 \geq \cdots \geq \lambda_N \geq 0$. In the Young tableaux picture of the irrep, one has λ_1 boxes in the first row, λ_2 boxes in the second row and so on with λ_N boxes in the Nth row; Fig. 5.1 shows several examples. Given a $U(N)$ irrep, the corresponding $SU(N)$ irrep is $\{\lambda_1 - \lambda_N, \lambda_2 - \lambda_N, \ldots, \lambda_{N-1} - \lambda_N\}$. A one body operator contains terms of the type $a_i^\dagger a_j$. The $a_i^\dagger$ being a creation operator for a fermion in the ith sp state, it will transform as the $U(N)$ irrep $\{1\}$; note that here $\lambda_2 = \lambda_3 = \cdots = \lambda_N = 0$. Similarly, a_j being a hole creation operator it will transform as $\{1^{N-1}\}$. With these, clearly $h(1)$ will transform as the Kronecker product $\{1\} \times \{1^{N-1}\}$. This product can be reduced to a direct sum ($\oplus$) of the $U(N)$ irreps $\{1^N\} \oplus \{21^{N-2}\}$ or equivalently $SU(N)$ irreps $\{0\} \oplus \{21^{N-2}\}$. The first part $\{1^N\}$ or $\{0\}$ is a scalar with respect to $U(N)$. From now on we will not make a distinction between $U(N)$ and $SU(N)$ unless specifically needed. As the number operator $\hat{n}$ with eigenvalue m is the only scalar available, this piece, called $\nu = 0$ part, will be proportional to $\hat{n}$. Thus, the second part $\{21^{N-2}\}$ is the irreducible one-body part or the $\nu = 1$ part of $h(1)$. Therefore we have, with respect to $SU(N)$

$$h(1) = h^{\{0\}} + h^{\{21^{N-2}\}} = h^{\nu=0} + h^{\nu=1}. \tag{5.2}$$

Proceeding to a two-body operator $V(2)$, it is a sum of the pieces of the form $a_i^\dagger a_j^\dagger a_k a_l$. As $a_i^\dagger a_j^\dagger$ creates a two fermion state, it will transform as $\{1^2\}$. Similarly, the two hole creation operator $a_k a_l$ transforms as $\{1^{N-2}\}$. Therefore $V(2)$ will transform as the Kronecker product $\{1^2\} \times \{1^{N-2}\}$ and its reduction gives,

$$V(2) = V^{\{0\}} + V^{\{21^{N-2}\}} + V^{\{2^2 1^{N-4}\}} = V^{\nu=0} + V^{\nu=1} + V^{\nu=2}. \tag{5.3}$$

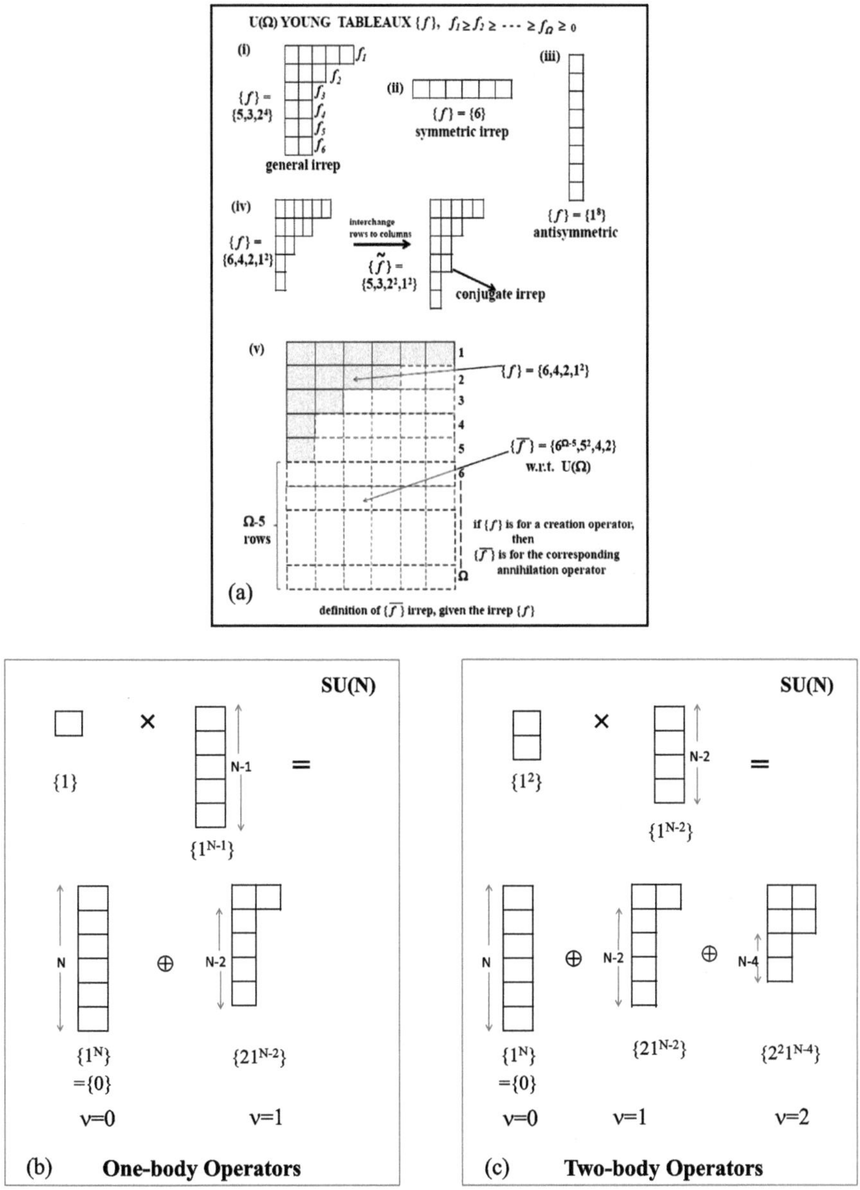

Fig. 5.1 (**a**) Young tableaux representation of the irreps of $U(\Omega)$ and shown are: (*i*) a general irrep $\{f\}$; (*ii*) symmetric irrep $\{m\}$; (*iii*) antisymmetric irrep $\{1^m\}$; (*iv*) conjugate irrep $\{\tilde{f}\}$ that corresponds to a given $\{f\}$; (*v*) irrep $\overline{\{f\}}$ that corresponds to a given $\{f\}$. *Shaded part* in (*v*) is the irrep $\{f\}$ and the remaining Young tableaux read bottom to top is $\overline{\{f\}}$. Note the importance of Ω in defining $\overline{\{f\}}$ and this is used in Chap. 11 ahead. (**b**) Young tableaux for various tensor parts of a one-body operators with respect to $SU(N)$. (**c**) Same as (**b**) but for two-body operators. See [9, 13] for details regarding (**b**) and (**c**) (Color figure online)

Thus, $V(2)$ will have a scalar ($\nu = 0$) part with respect to $U(N)$ and obviously this will be proportional to $\binom{\hat{n}}{2}$. Similarly there will be an effective one-body ($\nu = 1$) part (Hartree-Fock like) and an irreducible two-body ($\nu = 2$) part. Given the sp energies ε_i and the two-particle matrix elements V_{ijkl}, it is easy to identify the $\nu = 0$ parts of $h(1)$ and $V(2)$,

$$\begin{aligned} h^{\nu=0} &= \hat{n}\overline{\varepsilon}; \qquad \overline{\varepsilon} = \frac{1}{N}\sum_{i=1}^{N} \varepsilon_i, \\ V^{\nu=0} &= \binom{\hat{n}}{2} V_0; \qquad V_0 = \binom{N}{2}^{-1} \sum_{i<j} V_{ijij}. \end{aligned} \tag{5.4}$$

Subtraction of $h^{\nu=0}$ from h gives $h^{\nu=1}$,

$$h^{\nu=1} = \sum_i \varepsilon_i^1 \hat{n}_i, \qquad \varepsilon_i^1 = \varepsilon_i - \overline{\varepsilon}. \tag{5.5}$$

The $V^{\nu=1}$ part, as it is derived from a two-body Hamiltonian, should be of the form $(a\hat{n} + b)\sum \zeta_{i,j} a_i^\dagger a_j$. Also, as this operator has to vanish in zero and one particle spaces, we have $b = -a$. In addition, the ζ matrix should be traceless (as $\nu = 0$ part of V is removed). Choosing $a = 1/(N-2)$ and applying contraction of one index in V_{ijkl} gives [2, 9],

$$V^{\nu=1} = \frac{\hat{n}-1}{N-2}\sum_{i,j} \zeta_{i,j} a_i^\dagger a_j; \qquad \zeta_{i,j} = \left[\sum_k V_{kikj}\right] - \left[(N)^{-1}\sum_{r,s} V_{rsrs}\right]\delta_{i,j}. \tag{5.6}$$

Now, the V_{ijkl} matrix elements defining $V^{\nu=2}$ follow from simple subtraction and this gives,

$$\begin{aligned} V^{\nu=2} &= V - V^{\nu=0} - V^{\nu=1} \quad \Longleftrightarrow \quad V_{ijkl}^{\nu=2}; \\ V_{ijij}^{\nu=2} &= V_{ijij} - V_0 - (N-2)^{-1}(\zeta_{i,i} + \zeta_{j,j}), \\ V_{ijik}^{\nu=2} &= V_{ijik} - (N-2)^{-1}\zeta_{j,k} \quad \text{for } j \neq k, \\ V_{ijkl}^{\nu=2} &= V_{ijkl} \quad \text{for all other cases.} \end{aligned} \tag{5.7}$$

Figures 5.1b and c show the $SU(N)$ tensorial decomposition of one and two-body operators in Young tableaux notation.

5.2.2 Propagation Equations for Energy Centroids and Spectral Variances

One very important property of the $U(N)$ decomposition is that the various ν parts of H will be orthogonal with respect to m-particle space averages,

$$\langle H^2\rangle^m = \sum_{\nu=0,1,2} \langle [H^\nu]^2\rangle^m . \tag{5.8}$$

This result follows from the facts: (i) for any m-particle average of an operator only the scalar part with respect to $U(N)$ will contribute; (ii) Kronecker product of two operators with unitary ranks ν and ν' will give a scalar ($\nu''=0$) term only if $\nu=\nu'$. Therefore, it is possible to define a $U(N)$ geometry with norms for an operator $\mathcal{O}$ defined by

$$\|\mathcal{O}\|_m = \left\{\langle [\mathcal{O} - \mathcal{O}^{\nu=0}]^2\rangle^m\right\}^{1/2} . \tag{5.9}$$

Thus, the square of the norm of $\mathcal{O}$ will be the sum of squares of the norms of its various ν parts (with $\mathcal{O}^{\nu=0}$ dropped).

The m-particle space averages follow from the simple rule that the average of a k-body operator will be a polynomial in m of order k. Then the propagation equation for the m-particle energy centroids is given by,

$$E_c(m) = \langle H\rangle^m = \langle H^{\nu=0}\rangle^m = \langle h(1)\rangle^m + \langle V(2)\rangle^m = m\overline{\varepsilon} + \binom{m}{2} V_0 . \tag{5.10}$$

Similarly, the spectral variances

$$\sigma^2(m) = \langle H^2\rangle^m - \left[E_c(m)\right]^2 = \langle [H^{\nu=1}]^2\rangle^m + \langle [H^{\nu=2}]^2\rangle^m \tag{5.11}$$

also propagate simply. Propagation equations for the $\nu=1$ and $\nu=2$ parts are,

$$\begin{aligned} \langle [H^{\nu=1}]^2\rangle^m &= \frac{m(N-m)}{N(N-1)} \sum_{i,j} \left\{\varepsilon_i^1 \delta_{i,j} + \frac{m-1}{N-2}\zeta_{i,j}\right\}^2 , \\ \langle [H^{\nu=2}]^2\rangle^m &= \frac{m(m-1)(N-m)(N-m-1)}{2(N-2)(N-3)} \langle (V^{\nu=2})^2\rangle^2 . \end{aligned} \tag{5.12}$$

For later discussion it is useful to consider ensemble averaged spectral variances generated by $V(2)$. For these first we need $\overline{\langle [V^\nu]^2\rangle^2}$. We have easily, with variance of the two-particle V matrix elements being unity (diagonal matrix elements variance is 2),

$$\begin{aligned}
\overline{\langle [V^{\nu=0}]^2\rangle^2} &= 2\binom{N}{2}^{-1}, \\
\overline{\langle [V^{\nu=1}]^2\rangle^2} &= \frac{2}{N(N-1)(N-2)}\sum_{i,j}\overline{\zeta_{i,j}^2}, \\
\overline{\langle V^2\rangle^2} &= \binom{N}{2}+1, \\
\overline{\langle [V^{\nu=2}]^2\rangle^2} &= \overline{\langle V^2\rangle^2} - \overline{\langle [V^{\nu=0}]^2\rangle^2} - \overline{\langle [V^{\nu=1}]^2\rangle^2}.
\end{aligned} \tag{5.13}$$

The only unknown is now $\sum_{i,j}\overline{\zeta_{i,j}^2}$ and it is evaluated as

$$\begin{aligned}
\sum_{i,j}\overline{\zeta_{i,j}^2} &= \sum_{i,j,k}\overline{V_{kikj}^2} - \frac{1}{N}\overline{\left(\sum_{i,j} V_{ijij}^2\right)^2} \\
&= \left[N(N-1)(N-1)+N(N-1)\right] - 4(N-1) \\
&= (N-1)(N-2)(N+2).
\end{aligned} \tag{5.14}$$

Combining Eqs. (5.13) and (5.14) we obtain,

$$\overline{\langle [V^{\nu=2}]^2\rangle^2} = \frac{(N-3)(N^2+N+2)}{2(N-1)}. \tag{5.15}$$

5.3 Chaos Markers Generated by EGOE(1 + 2)

Most significant aspect of EGOE(1 + 2) is that as λ changes, in terms of state density, level fluctuations, strength functions, entropy and occupancies, the ensemble admits three chaos markers. We will turn to this now.

Firstly, the state densities $\rho(E)$ take Gaussian form, for large enough m, for all λ values. This follows from the fact that, as discussed in detail in the previous section, EGOE(2) gives in general Gaussian state densities and also in general the $h(1)$'s produce Gaussian densities. The later follows easily from the result that m-fermion state density will be essentially a m-fold convolution of the single particle density generated by $h(1)$. Except for singular $h(1)$'s, as discussed for example in [12], this leads to Gaussian form for large enough m. It should be added that the fluctuations in $\rho(E)$ will be large for $H = h(1)$ but we will not discuss this further here. Figure 5.2 shows an example for the Gaussian densities generated by EGOE(1 + 2).

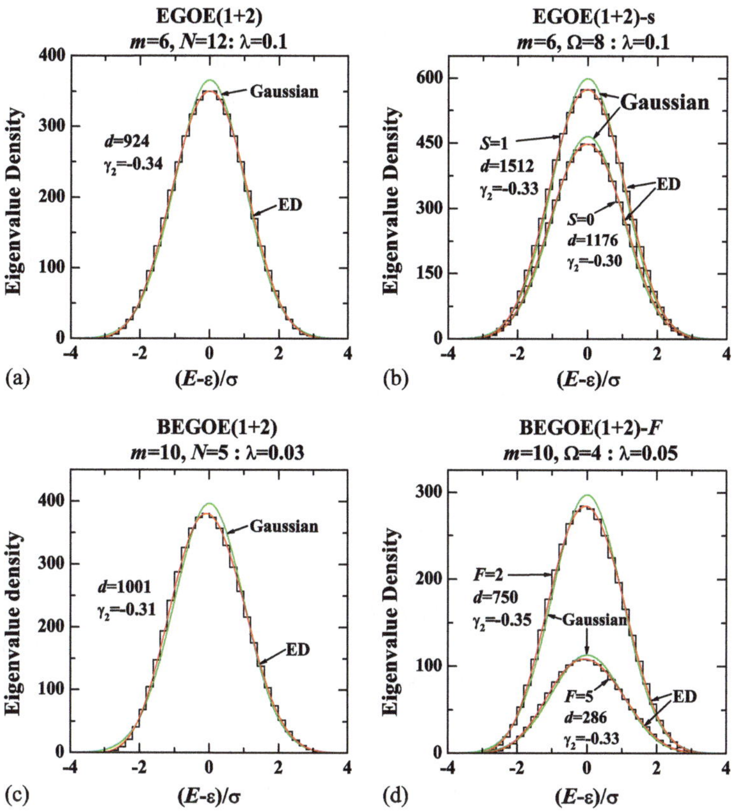

Fig. 5.2 Typical examples for the eigenvalue density given by various embedded ensembles. (**a**) EGOE(1 + 2) ensemble for $m = 6$ and $N = 12$ system with interaction strength $\lambda = 0.1$ in Eq. (5.1). (**b**) EGOE(1 + 2)-s ensemble for $m = 6$, $\Omega = 8$ ($N = 16$) and $S = 0$ and 1 systems with $\lambda_0 = \lambda_1 = 0.1$ in Eq. (6.1). (**c**) BEGOE(1 + 2) ensemble for $m = 10$ and $N = 5$ system with $\lambda = 0.03$ in Eq. (9.5). (**d**) BEGOE(1 + 2)-F ensemble for $m = 10$, $\Omega = 4$ and $F = 2$ and 5 systems with $\lambda_0 = \lambda_1 = \lambda = 0.05$ in Eq. (10.3). In all the examples used are 500 member ensembles and the sp energies are taken as $\varepsilon = i + 1/i$; $i = 1, 2, \ldots, N$ for EGOE(1 + 2) and BEGOE(1 + 2) and $i = 1, 2, \ldots, \Omega$ for EGOE(1 + 2)-s and BEGOE(1 + 2)-F ensembles. Shown in the figure are the dimensions and the values of ensemble averaged γ_2; ensemble averaged $\gamma_1 \sim 0$ in all the examples. See Chaps. 6, 9 and 10 for details regarding the last three ensembles. All the four ensembles generate Gaussian densities and the ED representation given by Eq. (B.7) is excellent. Note that $(E - \varepsilon)/\sigma$ is standardized variable

5.3.1 *Chaos Marker λ_c*

With λ increasing from zero value, there is a chaos marker λ_c such that for $\lambda \geq \lambda_c$ the level fluctuations follow GOE, i.e. λ_c marks the transition in the nearest-neighbor spacing distribution from Poisson to Wigner form. This transition occurs when the interaction strength λ is of the order of the spacing Δ_c between the states that are directly coupled by the two-body interaction. This definition came out of nuclear structure calculations by Aberg [14]. However it was proposed independently much later by Jacquod and Shepelyansky [5] in analyzing EGOE$(1+2)$. Therefore from now we refer to this as AJS criterion. Given a typical many particle configuration, the action of $V(2)$ will change the occupancy at two places or one place or none. Therefore, given a m particle configuration, the number K of states directly coupled by $V(2)$ is

$$K = 1 + m(N-m) + \frac{1}{4}m(m-1)(N-m)(N-m-1) \stackrel{dilute-limit}{\longrightarrow} m^2N^2/4. \quad (5.16)$$

Similarly, action of $V(2)$ on a configuration changes the energy of the configuration by Δ_c and this spreads over the K states directly coupled by $V(2)$. The value of Δ_c can be estimated using the $h(1)$ spectrum. The energy of the lowest two particle state is $\varepsilon_1 + \varepsilon_2 \sim \Delta$. Similarly the highest two particle state energy is $(2N-3)\Delta$. Therefore $\Delta_c \sim 2N\Delta$. Then, for the Poisson to Wigner transition chaos marker, AJS gives [5]

$$\lambda_c = \Delta_c/K \propto 1/m^2N. \quad (5.17)$$

In practice, to determine λ_c from NNSD calculated as a function of λ, we need a criterion for defining the onset of GOE NNSD. There are several recipes for this [5, 15, 16]. For example, in [16], $\sigma^2(0:\lambda)$ vs λ is calculated and λ_c is determined using the condition, following the discussion in Sect. 3.2,

$$\sigma^2(0:\lambda_c) = 0.37. \quad (5.18)$$

Figure 5.3 gives an example for the Poisson to GOE transition in NNSD generated by EGOE$(1+2)$. As the sp energies used are $\varepsilon_i = i + 1/i$, it is easy to see that in the $\lambda \sim 0$ limit, majority of many-body eigenvalues approach a perturbed picket-fence spectrum. Away from this, the spectrum is not picket-fence but deviates from Poisson as can be seen from Fig. 5.3. However, if we had used sp energies drawn from the center of a GOE or from the eigenvalues of an irregular system, the fluctuations will be generically Poisson for $\lambda \sim 0$. Thus, strictly speaking, the results in Fig. 5.3 show transition from Poisson like to GOE and generically this corresponds to Poisson to GOE transition. This aspect is also used in Chaps. 6, 9 and 10.

5.3.2 *Chaos Marker* λ_F

As λ increases further from λ_c, the strength functions change from Breit-Wigner (BW) [18] to Gaussian form and the transition point is denoted by λ_F. Note that the strength functions $F_k(E)$ are defined by Eq. (2.81) and here k denotes a m-fermion mean-field basis state. Similarly, Appendix D gives the standard model derivation of the BW form for strength functions. Now, the BW to Gaussian chaos marker λ_F can be understood as follows. Firstly there are two scales in EGOE$(1+2)$ with the first one being Δ_c and the other being the m-particle level spacing Δ_m. As the

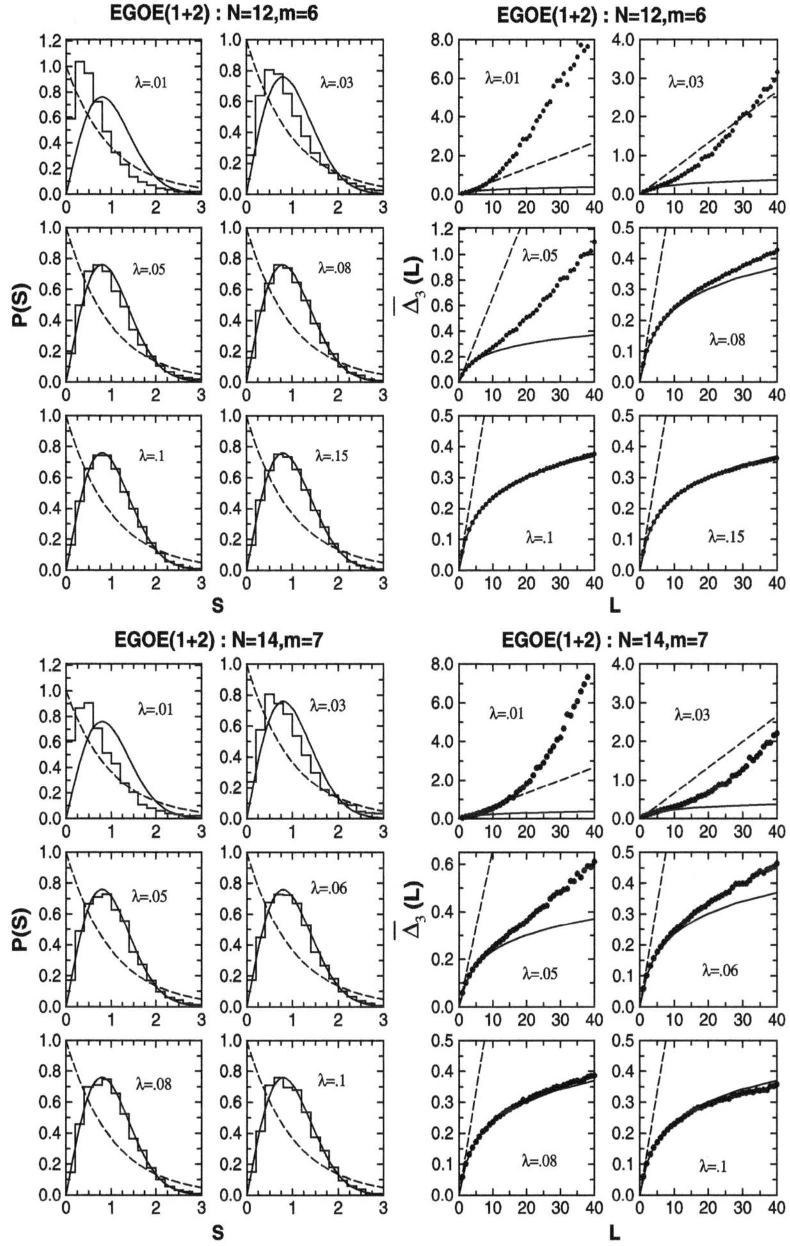

Fig. 5.3 Nearest neighbor spacing distribution $P(S)$ vs S and Dyson-Mehta $\overline{\Delta_3}(L)$ statistic for $0 \leq L \leq 40$ for various values of the interaction strength λ in EGOE$(1+2)$. For $P(S)$, histograms are EGOE$(1+2)$ results, *dashed curves* are Poisson and *continuous curves* correspond to Wigner distribution. For $\overline{\Delta_3}(L)$, *filled circles* are EGOE$(1+2)$ results, *dashed curves* are Poisson and *continuous curves* are for GOE. Results are shown for six and seven particle examples with 20 and 1 member respectively. Figure is from [17]

m-particle spectrum span, estimated using $h(1)$ spectrum, $B_m = m(N-m) \simeq mN$, we have $\Delta_m = mN/d_f(N,m)$. Now the Fermi golden rules gives the spreading width to be $\Gamma \propto \lambda^2/\Delta_c \sim m^2 N \lambda^2$ [18]. Then the participation ratio $\xi \propto \Gamma/\Delta_m = \lambda^2 m\, d_f(N,m)$. In the BW domain $\Gamma < B_m/f_0$, where $f_0 > 1$ and $\xi \gg 1$. This gives $\frac{1}{\sqrt{m\, df(N,M)}} \ll \lambda_F < \frac{1}{\sqrt{f_0 m}}$ [19, 20]. As $d_f(N,m)$ is usually large, the BW form sets in fast and

$$\lambda_F \propto 1/\sqrt{m}. \tag{5.19}$$

The $\lambda_c \le \lambda \le \lambda_F$ region is called the *BW domain*, with strength functions close to BW form and fluctuations following GOE. Similarly, the $\lambda > \lambda_F$ region is called the *Gaussian domain* with strength functions close to Gaussian form. In fact the BW form starts in a region below λ_c [there is a λ_0 such that below λ_0 the strength functions are close to δ-function form and for $\lambda > \lambda_0$ there is onset of BW form but the fluctuations here will be close to Poisson for $\lambda < \lambda_c$]. The BW to Gaussian transition was first recognized by Frazier et al. in ^{24}Mg shell-model results [21]. Similarly, first example for BW to Gaussian transition in atoms was given in [22]. More importantly, these were shown to be a feature of EGOE(1 + 2), for the first time, by Kota and Sahu [17]. Figure 5.4 shows an example for the BW to Gaussian transition in strength functions. With the basis state energies defined as $E_k = \langle k|H|k\rangle$, in the calculations E and E_k are zero centered for each member and scaled by the spectrum (E's) width σ so that $\hat{E}_k = (E_k - \varepsilon)/\sigma$ and $\hat{E} = (E - \varepsilon)/\sigma$. Then, th results in the figure are for the k states with $\hat{E}_k = 0 \pm \Delta$. The value of Δ is chosen to be 0.025 for $\lambda \le 0.1$ and beyond this $\Delta = 0.1$ for the results in the figure.

An important aspect is that in the Gaussian regime and in a limited manner in BW to Gaussian regime, it is possible to derive analytical formulas for ensemble averaged NPC and S^{info} for EGOE(1 + 2) wavefunctions expanded in the mean-field [$h(1)$] basis. It is also possible to derive formulas for occupancies and entropy defined by occupancies. First we will give the results for: (i) NPC and S^{info}; (ii) occupancies. Next, we will use these results in the discussion of a third chaos marker λ_t.

5.3.3 *NPC and S^{info} in* EGOE(1 + 2)

Given the mean-field basis states $|k\rangle$ defined by energies $E_k = \langle k|H|k\rangle$, one can assume that E_k are generated by a Hamiltonian H_k. Taking degeneracies of E and E_k energies into account we have,

$$\begin{aligned}
\rho_{biv}(E, E_k) &= \big\langle \delta(H-E)\delta(H_k - E_k)\big\rangle \\
&= (1/d) \sum_{\alpha \in k, \beta \in E} \big|C_{k,\alpha}^{E,\beta}\big|^2 \\
&= (1/d)\overline{\big|C_k^E\big|^2}\big[d\rho^H(E)\big]\big[d\rho^{H_k}(E_k)\big]
\end{aligned} \tag{5.20}$$

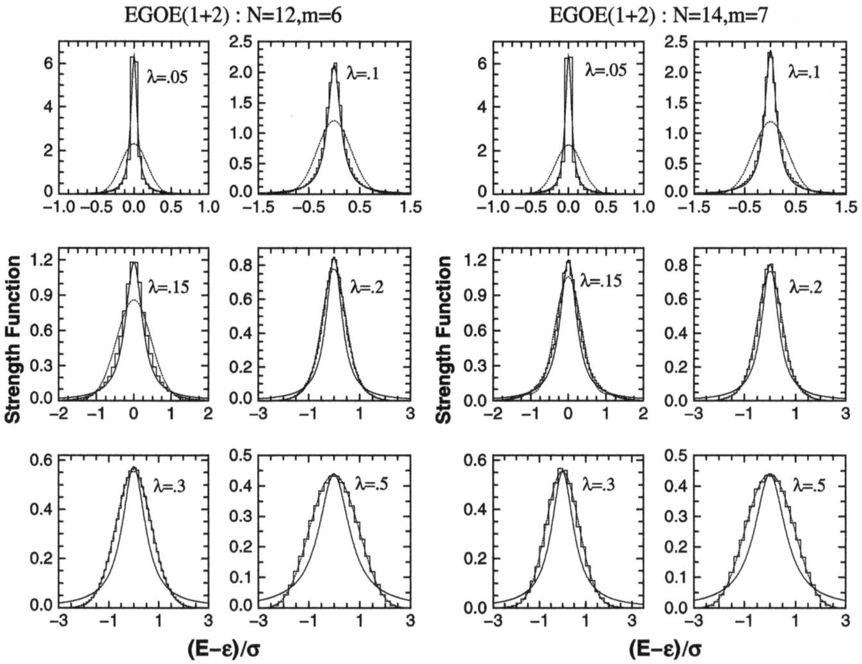

Fig. 5.4 Strength functions for EGOE(1 + 2) for various values of the interaction strength λ: (i) for a system of 6 fermions in 12 single particle states with 25 members; (ii) for a system of 7 fermions in 14 single particle states with one member. In the figure, the histograms are EGOE(1 + 2) results and *continuous curves* are BW fit. For the 6 fermions case, the *dotted curves* are Gaussians for λ ≤ 0.15 and Edgeworth corrected Gaussians (ED) for λ > 0.15. Similarly for the 7 fermions case, the *dotted curves* are Gaussians for λ ≤ 0.1 and Edgeworth corrected Gaussians (ED) for λ > 0.1. Figure is from [17]

$$\Longrightarrow \quad F_k(E) = \rho_{biv}(E, E_k)/\rho^{H_k}(E_k),$$
$$\overline{\left|C_k^E\right|^2} = \rho_{biv}(E, E_k)\big/\big[d\rho^H(E)\rho^{H_k}(E_k)\big].$$

In Eq. (5.20), $\overline{|C_k^E|^2}$ is the average of $|C_k^E|^2$ over all the degenerate states and $d = d_f(N, m)$. Let us now examine the structure of H_k and $\rho_{biv}(E, E_k)$. Firstly it should be noted that the two-body interaction $V(2)$ can be decomposed, just as the $U(N)$ decomposition discussed in Sect. 5.2.1, into three parts $V(2) = V^{[0]} + V^{[1]} + \boldsymbol{V}$ so that $h(1) + V^{[0]}$ generates the E_k energies (diagonal matrix elements of H in the m-particle mean-field basis states). As given Sect. 5.2.1, $V^{[0]}$ decomposes into a scalar part $V^{[0],0}$, an effective one-body part $V^{[0],1}$ and an irreducible 2-body part $V^{[0],2}$. Adding $V^{[0],0} + V^{[0],1}$ to $h(1)$ gives an effective one-body part $\boldsymbol{h}$ of H, $\boldsymbol{h} = h(1) + V^{[0],0} + V^{[0],1}$. The important point now being that, with respect to the $U(N)$ norm, the size of $V^{[0],2}$ is usually very small compared to the size of $\boldsymbol{h}$ in the m-particle spaces and similarly the norm of $V^{[1]}$ is small compared to the norm of $\boldsymbol{V}$. With this, $H = \boldsymbol{h} + \boldsymbol{V}$ and then the H_k is nothing but $\boldsymbol{h}$. The piece $\boldsymbol{V}$ generates the widths and other shape parameters of $F_k(E)$. It should be added that with respect to the $U(N)$ norm $\boldsymbol{h}$ and $\boldsymbol{V}$ are orthogonal and therefore $\sigma_H^2(m) = \sigma_{\boldsymbol{h}}^2(m) + \sigma_{\boldsymbol{V}}^2(m)$. For EGOE(1 + 2), it is well known that the widths of $F_k(E)$ are in general constant (this is seen in many numerical calculations and there is some analytical understanding as discussed in Chaps. 11, 12 and 14). Then, the average variance of $F_k(E)$'s is given simply by

$$\overline{\sigma_k^2} = \sigma_{\boldsymbol{V}}^2 = (d)^{-1} \sum_{\alpha \neq \beta} \left| \langle \alpha | H | \beta \rangle \right|^2$$

where α and β are the indices for the m-particle mean-field basis state. In the following EGOE(1 + 2) discussion we assume that $\boldsymbol{h}$ is $h(1)$ and $\boldsymbol{V}$ is $V(2)$, i.e. $H = \boldsymbol{h} + \lambda \boldsymbol{V} \to h(1) + \lambda V(2)$.

In the chaotic domain with $\lambda > \lambda_F$ we have: (i) E_k are generated by $H_k = h(1)$, therefore the variance of $\rho^{H_k}(E_k)$ is $\sigma_{\boldsymbol{h}}^2$; (ii) widths of the strength functions are constant and they are generated by $V(2)$ with the average variance $\overline{\sigma_k^2} = \sigma_{\boldsymbol{V}}^2$; (iii) $F_k(E)$'s are Gaussian in form; (iv) $F_k(E)$ is a conditional density of the bivariate Gaussian $\rho_{biv:\mathscr{G}}(E, E_k)$. The correlation coefficient ζ of $\rho_{biv:\mathscr{G}}(E, E_k)$ is given by,

$$\zeta(m) = \frac{\langle (H - \varepsilon_H)(H_k - \varepsilon_H) \rangle^m}{\sqrt{\langle (H - \varepsilon_H)^2 \rangle^m \langle (H_k - \varepsilon_H)^2 \rangle^m}} = \sqrt{\left(1 - \frac{\overline{\sigma_k^2}}{\sigma_H^2(m)}\right)} = \frac{\sigma_{\boldsymbol{h}}(m)}{\sigma_H(m)}. \tag{5.21}$$

Note that the centroids of the E and E_k energies are both given by $\varepsilon_H = \langle H \rangle$. Similarly, ζ^2 is nothing but the variance of E_k's (the centroids of $F_k(E)$) normalized by the state density variance. The $\rho_{biv:\mathscr{G}}(E, E_k)$, which takes into account the fluctuations in the centroids of $F_k(E)$ and assumes that the variances are constant, is used to derive formulas for $\xi_2(E)$ and localization length $\ell_H(E)$ in the wavefunctions. Note that Eq. (2.78) gives the definitions for $\xi_2(E)$ and $\ell_H(E)$. We will also give the results that take into account variance fluctuations.

In terms of the locally renormalized amplitudes $\mathscr{C}_k^E = C_k^E / \sqrt{\overline{|C_k^E|^2}}$ where the bar denotes ensemble average with respect to EGOE(1 + 2), $\sum_k |C_k^E|^4 = \sum_k |\mathscr{C}_k^E|^4 (\overline{|C_k^E|^2})^2$. Now,

$$\begin{aligned}
\overline{\sum_k |C_k^E|^4} \overset{\text{EGOE(1+2)}}{\longrightarrow} & \sum_k \overline{|\mathscr{C}_k^E|^4} \left(\overline{|C_k^E|^2}\right)^2 \\
= & \; 3 \sum_k \left(\overline{|C_k^E|^2}\right)^2 \\
= & \; \frac{(3/d)}{[\rho_{\mathscr{G}}^H(E)]^2} \int dE_k \frac{[\rho_{biv:\mathscr{G}}(E, E_k)]^2}{\rho_{\mathscr{G}}^{H_k}(E_k)} \\
= & \; \frac{(3/d)}{[\rho_{\mathscr{G}}^H(E)]^2} \int dE_k \rho_{\mathscr{G}}^{H_k}(E_k) \left[F_{k:\mathscr{G}}(E)\right]^2.
\end{aligned} \tag{5.22}$$

Then, the formula for NPC or the ensemble averaged $\xi_2(E)$ is [23],

$$\xi_2(E) = \xi_{\text{GOE}}\sqrt{1-\zeta^4}\exp-\left\{\frac{\zeta^2}{1+\zeta^2}\widehat{E}^2\right\}; \quad \xi_{\text{GOE}} = d/3. \tag{5.23}$$

Note that $\widehat{E} = (E - \varepsilon_H)/\sigma_H$. In the first step in Eq. (5.22) the fact that EGOE exhibits average fluctuations separation (see Chap. 4) with little communication between the two is used. This allows one to carry out $|\mathscr{C}_k^E|^4$ ensemble average independent of the other smoothed (average) term. In the second line the Porter-Thomas form of local strength fluctuations is used and then $\overline{|\mathscr{C}_k^E|^4} = 3$, a GOE result. In the third step the result in Eq. (5.20) is used. Then, the Gaussian forms, valid in the chaotic domain ($\lambda > \lambda_{F_k}$), of all the densities for EGOE(1 + 2) give the final formula. As seen from Eq. (5.23), NPC for EGOE(1 + 2) is entirely determined by the correlation coefficient ζ. In order to estimate ζ we will use the last equality in Eq. (5.21). In addition we will consider $\{H\} = \alpha h(1) + \lambda\{V(2)\}$. Neglecting the contributions of $V(2)$ to $\sigma_{\boldsymbol{h}}$ and assuming a uniform sp spectrum, one gets $\sigma_{\boldsymbol{h}}^2(m) \sim (mN^2/12)\alpha^2$. Similarly $\sigma_{\boldsymbol{V}}^2 \sim \binom{m}{2}\binom{N}{2}\lambda^2$. Here, trace propagation equations in Eq. (5.12) are used. Therefore, $\zeta^2 = [1 + 3m(\lambda/\alpha)^2]^{-1}$ and this expression gives 0.51 and 0.76 for the $\alpha = 0.5$ and 1 cases in Fig. 5.5. They compare well with the exact numbers 0.59 and 0.82 respectively. However this estimate fails in the $\alpha \to 0$ and in this limit, the $\boldsymbol{h}$ has to be replaced by $V^{[0]}$. Then the E_k energies are a sum of $\binom{m}{2}$ zero centered Gaussian variables each with variance λ^2. This together with the $\sigma_{\boldsymbol{V}}^2$ expression, gives $\zeta^2 \sim \binom{N}{2}^{-1}$ for $\alpha \sim 0$. The number quoted for the $\alpha = 0$ case in Fig. 5.5 is close to this estimate.

Correction to NPC due to fluctuations in σ_k, i.e. for $\delta\sigma_k^2 = \sigma_k^2 - \overline{\sigma_k^2} \neq 0$, is obtained by using, for small $|\delta\sigma_k^2|$, the Hermite polynomial expansion which gives [24], $F_{k:\mathscr{G}}(E) \to F_{k:\mathscr{G}}(E)\{1 + c_2(\mathscr{E}_k^2 - 1)\}$ where $c_2 = \delta\sigma_k^2/2\overline{\sigma_k^2}$ and $\mathscr{E}_k = (E - E_k)/\sqrt{\overline{\sigma_k^2}}$. This corrected $F_k(E)$ is used in the integral form with $F_k(E)$ in Eq. (5.22) by treating $(\delta\sigma_k^2)$'s as random. Keeping only the terms that are quadratic in $(\delta\sigma_k^2)$ in the integral form for NPC gives [23],

$$\begin{aligned}
\xi_2(E) &= \frac{(3/d)}{[\rho_{\mathscr{G}}^H(E)]^2}\int dE_k\rho_{\mathscr{G}}^{H_k}(E_k)\left[F_{k:\mathscr{G}}(E)\right]^2\left(1+\frac{(\delta\sigma^2)}{2\overline{\sigma_k^2}}(\mathscr{E}_k^2-1)\right)^2 \\
&= \frac{d}{3}\sqrt{1-\zeta^4}\exp-\left\{\frac{\zeta^2}{1+\zeta^2}\widehat{E}^2\right\}\left\{1+\frac{1}{4}\left[\frac{(\delta\sigma^2)}{\sigma_H^2}\right]^2 X(E)\right\}^{-1}; \\
X(E) &= \frac{1}{(1+\zeta^2)^4}\left[\widehat{E}^4 - 2\frac{(1+\zeta^2)(1-2\zeta^2)}{1-\zeta^2}\widehat{E}^2 + \left(\frac{1+\zeta^2}{1-\zeta^2}\right)^2(1+2\zeta^4)\right].
\end{aligned} \tag{5.24}$$

The $\delta\sigma^2$ correction term above is valid only when the fluctuations in the variances of $F_k(E)$'s are small (this is in general always true). An estimate for $[(\delta\sigma^2)/\sigma_H^2]^2$ is obtained from Eq. (5.12) by noting that $\sigma_{\boldsymbol{V}}^2$ is a sum of $K \sim \binom{m}{2}\binom{N}{2}$ number of χ^2

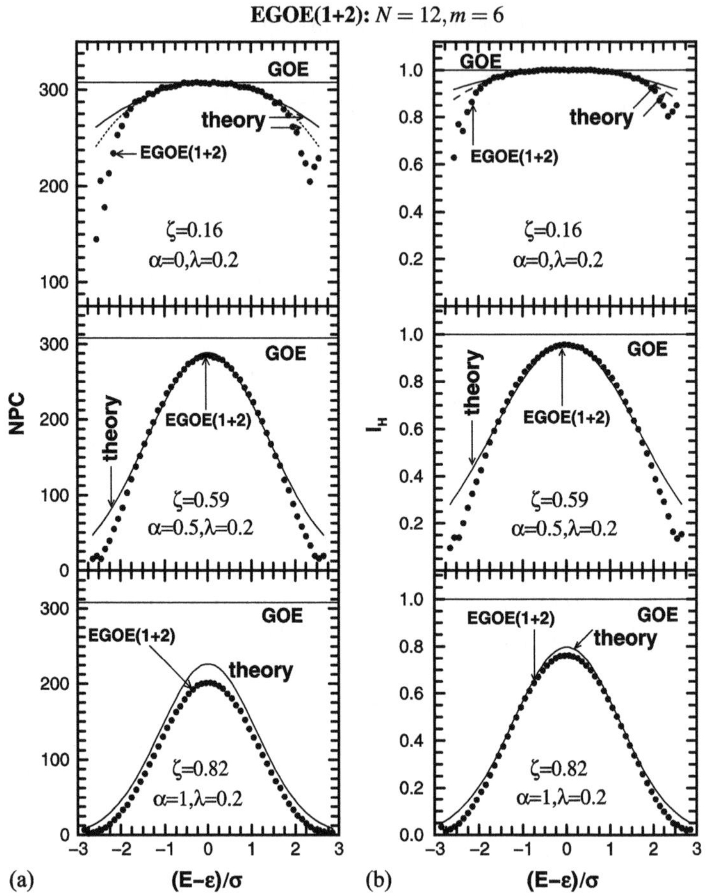

Fig. 5.5 (**a**) NPC and (**b**) ℓ_H vs E for a 20 member EGOE(1 + 2) with $N = 12, m = 6$. Results are shown, as discussed in the text, for the ensemble $\{H\} = \alpha h(1) + \lambda\{V(2)\}$ and the sp energies defining $h(1)$ are taken as $\varepsilon_i = i + 1/i$. For the three values ($\alpha = 0, \lambda = 0.2$), ($\alpha = 0.5, \lambda = 0.2$) and ($\alpha = 1, \lambda = 0.2$) the corresponding value for the correlation coefficient ζ is shown in the figure. For NPC, Eq. (5.23) and for ℓ_H, Eq. (5.25) give the theoretical formulas respectively. The figures are taken from [23] with permission from American Physical Society

variables and therefore $[(\delta\sigma^2)/\sigma_V^2]^2 = 2/K$ as given first in [1]. Then,

$$\left[(\delta\sigma^2)/\sigma_H^2\right]^2 \sim 2(1-\zeta^2)\bigg/\binom{m}{2}\binom{N}{2}.$$

For finite N, the correlation coefficient and the variance corrections are small but non zero and in the large N limit they are zero giving the GOE result. As we add the mean-field part to the EGOE(2), ζ increases and at the same time the correction due to variance fluctuations decreases. Thus the formula Eq. (5.24) with the $(\delta\sigma^2)$ term is important only for small ζ (this equation was also derived by Kaplan and Papenbrock [25] but using some what a different approach). Equation (5.23) is accurate for reasonably large ζ ($\zeta \geq 0.3$). All these results are well tested in Fig. 5.5.

Proceeding exactly as above, expression for ℓ_H in wavefunctions is obtained and the result is [23],

$$\begin{aligned} \ell_H(E) &\xrightarrow{\text{EGOE(1+2)}} -\int dE_k \frac{\rho_{biv:\mathscr{G}}(E,E_k)}{\rho_{\mathscr{G}}^{H}(E)} \ln\left\{ \frac{\rho_{biv:\mathscr{G}}(E,E_k)}{\rho_{\mathscr{G}}^{H_k}(E_k)\,\rho_{\mathscr{G}}^{H}(E)} \right\} \\ &= \sqrt{1-\zeta^2}\exp\left(\frac{\zeta^2}{2}\right)\exp-\left(\frac{\zeta^2\widehat{E}^2}{2}\right). \end{aligned} \tag{5.25}$$

Thus $\mathbf{l}_H(E)$ or the $S^{info}(E)$ is determined completely by the correlation coefficient ζ. Figure 5.5 gives a numerical test of the EGOE(1 + 2) formula given by Eq. (5.25). By rewriting the integral in Eq. (5.25) in terms of $F_k(E)$ and making small $(\delta\sigma^2)$ expansion just as in the case of NPC, gives

$$\begin{aligned} \ell_H(E) &= \sqrt{1-\zeta^2}\exp\left(\frac{\zeta^2}{2}\right)\exp-\left(\frac{\zeta^2\widehat{E}^2}{2}\right)\left(1-\frac{1}{8}\left[\frac{(\delta\sigma^2)}{\sigma_H^2}\right]^2 Y(E)\right); \\ Y(E) &= \frac{1}{(1-\zeta^2)^2}\left\{\left(1-\zeta^2\right)^2\left(\widehat{E}^2-1\right)^2+4\zeta^2\left(1-\zeta^2\right)\widehat{E}^2+2\zeta^4\right\}. \end{aligned} \tag{5.26}$$

5.3.3.1 NPC in BW to Gaussian Interpolating Region

In the BW to Gaussian transition regime (i.e. in the $\lambda_c \leq \lambda \leq \lambda_F$ region) of EGOE(1 + 2), there is no analytical method available to solve for $F_k(E)$. However it is possible to use for example a linear interpolation form [17, 26] $F_k(E) = (1-\alpha)F_{k:BW}(E)+\alpha F_{k:\mathscr{G}}(E)$. Then $\alpha = 0$ gives BW and $\alpha = 1$ the Gaussian forms. However, a non-linear form, a priori better as argued in [22], is to use an extended t-distribution,

$$F_{k:BW-\mathscr{G}}(E:\alpha,\beta) = \frac{(\alpha\beta)^{\alpha-\frac{1}{2}}\Gamma(\alpha)}{\sqrt{\pi}\,\Gamma(\alpha-\frac{1}{2})}\frac{1}{((E-E_k)^2+\alpha\beta)^{\alpha}}, \quad \alpha \geq 1. \tag{5.27}$$

Here $\Gamma(-)$ are Γ-functions and α and β are parameters. Eq. (5.27) gives BW for $\alpha = 1$ and Gaussian for $\alpha \to \infty$ (this can be easily checked using Stirling's approximation). As required, it is normalized to unity for any positive value of the continuous parameter α. For $2\alpha - 1$ an integer, $F_{k:BW-\mathscr{G}}(E:\alpha,\beta)$ gives the so called *Student's* t-distribution [24], which is well known in statistics. The parameter α is sensitive to shape changes, while the parameter β supplies the energy scale over which $F_{k:BW-\mathscr{G}}(E:\alpha,\beta)$ extends. Since we focus on the shape transformations, α is the significant parameter. It is easy to see that $F_{k:BW-\mathscr{G}}(E:\alpha,\beta)$ is an even function of $E-E_k$, so that all of its finite odd cumulants vanish (strictly speaking, the centroid is E_k only for $\alpha > 1$). The variance σ_k^2 is

$$\sigma_k^2 = \frac{\alpha}{2\alpha-3}\beta; \quad \alpha > 3/2 \tag{5.28}$$

and this can be used to eliminate the parameter β. For $\alpha \leq 3/2$, it is the spreading width Γ (this is well defined for all α values) that is useful for fixing the β value. There is no simple expression for Γ as a function of α and β. Numerical EGOE(1+2) calculations showed that Eq. (5.27) describes quite well $F_k(E)$ as λ changes from λ_c to λ_F and beyond. The fits of $F_k(E)$, constructed for various λ values, to $F_{k:BW-\mathscr{G}}(E:\alpha,\beta)$ give λ vs α (note that σ_k eliminates the parameter β). It is seen that $\alpha = 4$ gives the chaos marker $\lambda_F = 0.2$ for the examples in Fig. 5.4.

Substituting Eq. (5.27) in the last equality in Eq. (5.22), assuming that the parameters α and β to be k independent, using Eq. (5.28) to eliminate β, simplifying all the variances to ζ^2 using Eq. (5.21) and then carrying out the integral for $\widehat{E} = 0$ gives (for $\alpha > 3/2$) [22],

$$\xi_2(E=0)/\xi_2^{\mathrm{GOE}} = \left\{\sqrt{\frac{2}{(2\alpha-3)}\frac{\Gamma^2(\alpha)}{\Gamma^2(\alpha-\frac{1}{2})}}\frac{1}{\sqrt{\zeta^2(1-\zeta^2)}}U\left(\frac{1}{2},\frac{3}{2}-2\alpha,\frac{(2\alpha-3)(1-\zeta^2)}{2\zeta^2}\right)\right\}^{-1} \tag{5.29}$$

where $U(---)$ is hyper-geometric-U function [27].

5.3.4 Occupancies and Single Particle Entropy in Gaussian Region

In the Gaussian domain, EGOE(1 + 2) is effectively EGOE(2) and therefore Eq. (4.82) with $\mathscr{O} = a_i$, gives a theory for occupancies except that all the parameters defining the Gaussians and higher cumulants should be calculated using EGOE(1+2) Hamiltonian. Figure 5.6 shows an example. An alternative, often quite good and easier for analytical treatment, is to consider the linear response of $\rho^H(E)$ under the deformation $H \to H_\alpha = H + \alpha\hat{n}_i$. Then, it is easily seen that [28]

$$\langle n_i\rangle^E = -\left[\rho^H(E)\right]^{-1}\lim_{\alpha\to 0}\int_{-\infty}^{E}\frac{\partial\rho^{H_\alpha}(x)}{\partial\alpha}\,dx. \tag{5.30}$$

Under $H \to H_\alpha$, the single particle energy $\varepsilon_i \to \varepsilon_i + \alpha$. With H represented by EGOE(1 + 2), H_α for α small is also represented by EGOE(1 + 2) and therefore the shape of $\rho^H(E)$ will be unchanged from the Gaussian form under the α deformation. Using this and applying Eq. (5.30) one gets,

$$\langle\hat{n}_i\rangle^E \xrightarrow{\mathrm{EGOE}(1+2)} \langle\hat{n}_i\rangle^m + \big\langle\hat{n}_i\big(H-\varepsilon_H(m)\big)\big\rangle^m\big(E-\varepsilon_H(m)\big)/\sigma_H^2(m). \tag{5.31}$$

The linear form given by Eq. (5.31) is close to the numerical results given in Fig. 5.6. For the present purpose more relevant is to consider entropy defined by occupancies [called single particle entropy $S^{sp}(E)$],

$$S^{sp}(E) = -\sum_i\left\{\langle\hat{n}_i\rangle^E\ln\big(\langle\hat{n}_i\rangle^E\big) + \big(1-\langle\hat{n}_i\rangle^E\big)\ln\big(1-\langle\hat{n}_i\rangle^E\big)\right\}. \tag{5.32}$$

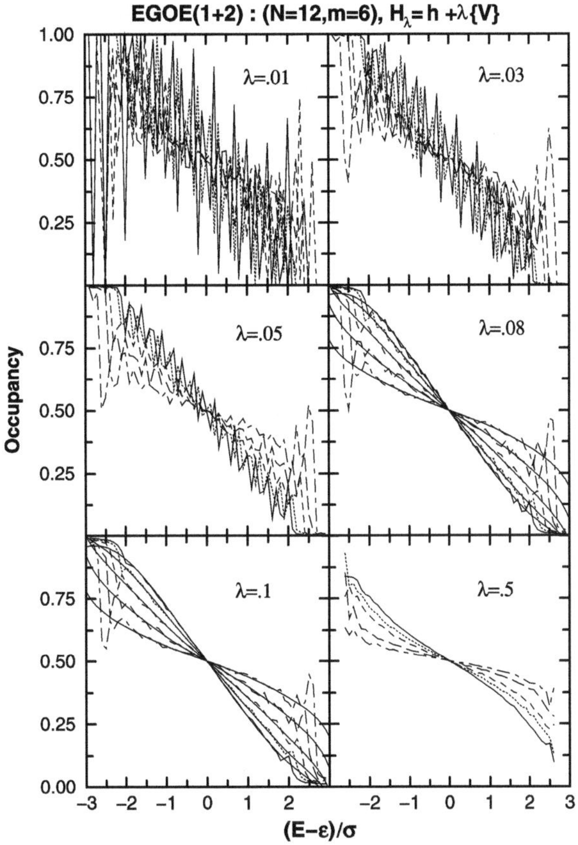

Fig. 5.6 Occupation numbers for a 25 member 924 dimensional EGOE(1 + 2) ensemble. Note that the sp energies used are $\varepsilon_i = i + 1/i$. Results are shown for the lowest 5 single particle states and for six values of the interpolating parameter λ. The estimate from Eq. (5.18) gives $\lambda_c \sim 0.05$ for order-chaos border in the present EGOE(1 + 2) example. It is clearly seen that once chaos sets in, the occupation numbers take stable smoothed forms. For $\lambda = 0.08$ and 0.1, the EGOE(1 + 2) results are compared with the EGOE smoothed form given by Eq. (4.82) which is a ratio of Gaussians (smoothed curves in the figure and here Edgeworth corrections are added). Figure is taken from [2] with permission from Elsevier

Let us assume that the sp spectrum is a uniform spectrum with level spacing Δ. Then the sum in Eq. (5.32) can be replaced by the integral, to good approximation, $\sum_i \cdots = \int \cdots \rho(\varepsilon) d\varepsilon = [\Delta]^{-1} \int \cdots d\varepsilon$. Using this and substituting Eq. (5.31) for $\langle \hat{n}_i \rangle^E$, gives [29], when truncated to $\widehat{E}^2$ term

$$\exp\bigl(S^{sp}(E) - S^{sp}_{max}\bigr) = \exp -\frac{1}{2}\zeta^2 \widehat{E}^2. \tag{5.33}$$

It is important to note that the correlation coefficient ζ in Eq. (5.33) is the same as the one that enters into EGOE(1 + 2) formulas given before for NPC and S^{info}. A numerical test of Eq. (5.33) is shown in Fig. 5.7 and the agreement is good for a wider range of ζ values.

5.3.5 Chaos Marker λ_t

A very important question for isolated finite interacting particle systems is the following [30]: in the chaotic domain will there be a point or a region where ther-

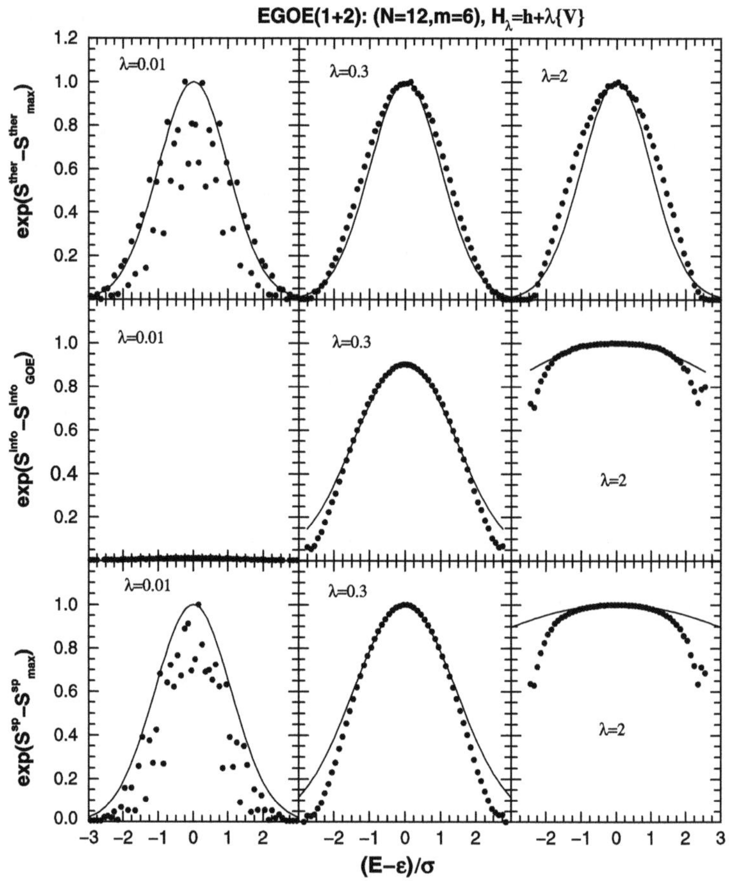

Fig. 5.7 Thermodynamic entropy $\exp(S^{ther} - S^{ther}_{max})$, information entropy $\exp(S^{info} - S^{info}_{\mathrm{GOE}})$ and single-particle entropy $\exp(S^{sp} - S^{sp}_{max})$ vs $(E - \varepsilon)/\sigma$ for EGOE(1 + 2) for three values of λ. The same $N = 12$ and $m = 6$ systems as in Fig. 5.6 has been employed in the calculations. The *filled circles* are EGOE(1 + 2) results and the *continuous curves* are the theoretical EGOE(1 + 2) predictions as given by Eqs. (5.25), (5.33) and (5.34). Figure is taken from [29] with permission from American Physical Society

malization occurs, i.e. will there be a region where different definitions of entropy, temperature, specific heat and other thermodynamic variables give the same results (as valid for infinite systems)? Obviously this has to happen beyond λ_F and this gives the third chaos marker λ_t. To understand this marker, in the Gaussian domain of EGOE(1 + 2), three different entropies are considered: thermodynamic (S^{ther}), information (S^{info}) and single particle (S^{sp}) entropies; note that $(S^{ther})_E = \ln \rho^{H,m}(E)$. Trivially, the EGOE(1 + 2) formula for S^{ther} is

$$\exp\bigl[\bigl(S^{ther}\bigr)_E - \bigl(S^{ther}\bigr)_{max}\bigr] \longrightarrow \exp -\frac{1}{2}\bigl(\widehat{E}^2\bigr) \tag{5.34}$$

and the other two entropies are given by Eqs. (5.25) and (5.33). These results are compared with numerical EGOE(1 + 2) calculations in Fig. 5.7 and they are understood as follows. With $H = h(1) + V(2)$ (here we consider $h(1)$ with single-particle level spacing Δ and $V(2)$ with matrix elements variance λ^2), there are two natural basis defined by h and V respectively. Then for thermodynamic considerations to apply, the entropy measures should be independent of the chosen basis. Firstly, in the dilute limit h and V will be orthogonal. The variance of h in m-particle spaces is $\sigma_h^2(m) = [(mN^2)/12]\Delta^2 = f^2\Delta^2$. Similarly the variance of V is $\sigma_V^2(m) \sim [(m^2N^2)/4]\lambda^2 = g^2\lambda^2$. The S^{info} and S^{sp} are basically determined by ζ and for strength functions expanded in the $h(1)$ basis, $\zeta_0(\lambda) = \sqrt{(f^2\Delta^2)/(f^2\Delta^2 + g^2\lambda^2)}$. Similarly for strength functions expanded in the V(2) basis, $\zeta_\infty(\lambda) = \sqrt{(g^2\lambda^2)/(f^2\Delta^2 + g^2\lambda^2)}$. Now the following is clear: when $\lambda \to \infty$, ζ_0 gets close to zero. Similarly when $\Delta \to \infty$, ζ_∞ gets close to zero. In both these situations S^{info} takes GOE values and S^{sp} approaches its maximum value. The condition $\zeta_0(\lambda_t) = \zeta_\infty(\lambda_t)$ gives $\lambda_t = |\Delta f/g|$ and here $\zeta^2 = 0.5$. Also note that, with $\Delta = 1$ [22],

$$\lambda_t \sim \frac{1}{\sqrt{m}}; \qquad \zeta^2 = 0.5. \tag{5.35}$$

With λ_t defined, it is easily seen that $\zeta_\infty(\lambda) = \zeta_0(\lambda_t^2/\lambda)$, thus there is a duality in EGOE(1 + 2); duality in EGOE(1 + 2) was first discussed by Jacquod and Varga [20]. At the duality point $\lambda = \lambda_t$, the entropies are basis independent. Thus $\lambda \sim \lambda_t$ with $\zeta^2 \sim 0.5$ defines the thermodynamic region for interacting particle systems. In this region, as stated in [31]: *the thermodynamic entropy defined via the global level density or in terms of occupation numbers behaves similar to the information entropy*. Comparing Fig. 5.7 with the shell-model calculations due to Horoi et al. [31] for ^{28}Si nucleus and due to Kota and Sahu [29] for the ^{24}Mg nucleus, it is seen that atomic nuclei in general will be in the thermodynamic regime (i.e. $\lambda \sim \lambda_t$).

5.4 Transition Strengths in EGOE(1 + 2)

5.4.1 Bivariate t-Distribution Interpolating Bivariate Gaussian and BW Forms

In the Gaussian domain of EGOE(1+2), just as the strength functions, it is plausible to argue that the transition strength densities, generated by a transition operator take bivariate Gaussian form with correlation coefficient ζ defined as in Sect. 4.4 except H is now the EGOE(1 + 2) Hamiltonian. However in the BW domain, for $\lambda_c \le \lambda < \lambda_F$, a form interpolating bivariate BW and bivariate Gaussian is appropriate. Then, as argued in [32], it is possible to represent the transition strength density by the bivariate t-distribution with a parameter ν_t,

$$\begin{aligned}
&\rho_{biv-t}(E_i, E_f; \varepsilon_i, \varepsilon_f, \sigma_1, \sigma_2, \zeta; \nu_t) \\
&\quad = \frac{1}{2\pi\sigma_1\sigma_2\sqrt{1-\zeta^2}} \\
&\qquad \times \left[1 + \frac{1}{\nu_t(1-\zeta^2)} \left\{ \left(\frac{E_i - \varepsilon_i}{\sigma_1}\right)^2 - 2\zeta \left(\frac{E_i - \varepsilon_i}{\sigma_1}\right)\left(\frac{E_f - \varepsilon_f}{\sigma_2}\right) \right.\right. \\
&\qquad \left.\left. + \left(\frac{E_f - \varepsilon_f}{\sigma_2}\right)^2 \right\}\right]^{-\frac{\nu_t+2}{2}}, \quad \nu_t \geq 1.
\end{aligned} \tag{5.36}$$

For $\nu_t = 1$, ρ_{biv-t} gives a bivariate Cauchy (i.e. bivariate BW) distribution and as $\nu_t \to \infty$, ρ_{biv-t} becomes a bivariate Gaussian. Secondly, the marginal distributions of ρ_{biv-t} are easily seen to be univariate t-distributions, with ν_t degrees of freedom, independent of ζ with univariate Cauchy (BW) distribution for $\nu_t = 1$ and Gaussian as $\nu_t \to \infty$. Thus, as the parameter ν (when there is no confusion, we will write ν_t as ν) changes from 1 to ∞, ρ_{biv-t} changes from bivariate BW to bivariate Gaussian. In Eq. (5.36), ε_i and ε_f are the centroids of the two marginals of ρ_{biv-t} and ζ is the bivariate correlation coefficient. However, σ_1 and σ_2 will approach the marginal widths σ_i and σ_f only in the limit $\nu \to \infty$, i.e. for the bivariate Gaussian. In fact, the marginal variances are $\sigma_i^2 = \frac{\nu}{\nu-2}\sigma_1^2$ and $\sigma_f^2 = \frac{\nu}{\nu-2}\sigma_2^2$ for $\nu > 2$; For $\nu \leq 2$, the spreading widths of the marginal densities define σ_1 and σ_2. Figure 5.8 shows some EGOE(1 + 2) examples.

5.4.2 NPC and S^{info} in Transition Strengths

Given a transition operator $\mathcal{O}$ and the corresponding transition strengths $|\langle E_f|\mathcal{O}|E\rangle|^2$, the normalized strength $\mathcal{R}$, ensemble averaged (smoothed) normalized strength $\overline{\mathcal{R}}$ and locally re-normalized strength $\hat{R}$ are defined by

$$\begin{aligned}
\mathcal{R}(E, E_f) &= \left\{\langle E|\mathcal{O}^\dagger\mathcal{O}|E\rangle\right\}^{-1} \left|\langle E_f|\mathcal{O}|E\rangle\right|^2, \\
\overline{\mathcal{R}(E, E_f)} &= \left\{\overline{\langle E|\mathcal{O}^\dagger\mathcal{O}|E\rangle}\right\}^{-1} \overline{\left|\langle E_f|\mathcal{O}|E\rangle\right|^2}, \\
\hat{R}(E, E_f) &= \left\{\overline{\left|\langle E_f|\mathcal{O}|E\rangle\right|^2}\right\}^{-1} \left|\langle E_f|\mathcal{O}|E\rangle\right|^2.
\end{aligned} \tag{5.37}$$

Here bar denotes average over the EGOE(1 + 2) ensemble. Then the measures NPC or $\xi_2^{(s)}$ and $S^{info:s}$ for transition strengths are

$$\begin{aligned}
\xi_2^{(s)}(E) &= \left\{\sum_{E_f} \left\{\mathcal{R}(E, E_f)\right\}^2\right\}^{-1}, \\
\left(S^{info:s}\right)_E &= -\sum_{E_f} \mathcal{R}(E, E_f) \ln \mathcal{R}(E, E_f).
\end{aligned} \tag{5.38}$$

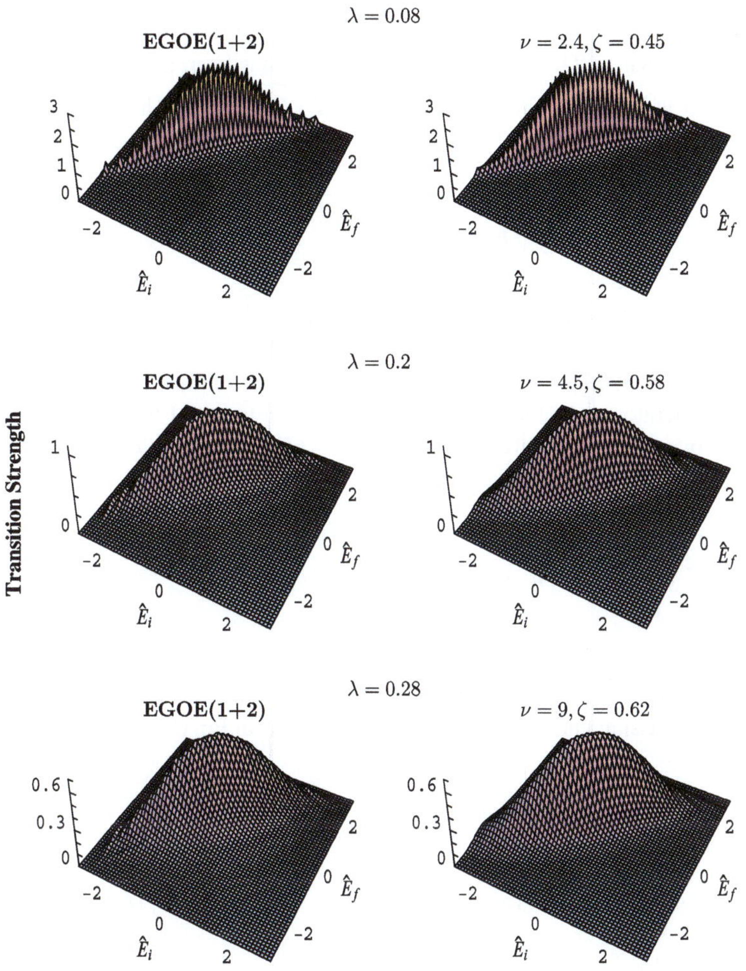

Fig. 5.8 Transition strength $|\langle E_f|\mathcal{O}|E_i\rangle|^2$ vs (E_i, E_f) for $\lambda = 0.08$, 0.2 and 0.28. In the figure $\widehat{E}_i$ and $\widehat{E}_f$ are standardized E_i and E_f. In all the figures, the ensemble-averaged strengths in the window $\widehat{E}_i \pm \delta/2$ and $\widehat{E}_f \pm \delta/2$ are summed and plotted at $(\widehat{E}_i, \widehat{E}_f)$; δ is chosen to be 0.1. The EGOE(1 + 2) system is same as that was used in Fig. 5.6 with $N = 12$ and $m = 6$ and the one-body transition operator $\mathcal{O} = a_2^\dagger a_9$. For this system, the total strength is 252. As λ changes from 0.08 to 0.28, the ν value changes from 2.4 to 9 and the bivariate correlation coefficient ξ changes from 0.45 to 0.62. Note the change in the scales of the vertical axes in the figures. Figures for various λ values are take from [32] with permission from American Physical Society

Here after $\overline{\xi_2^{(2)}(E)}$ and $\overline{S^{info:s}(E)}$ correspond to averages over the EGOE(1 + 2) ensemble in the Gaussian domain. The EGOE formula for $\xi_2^{(s)}(E)$ is derived by first writing $\overline{\xi_2^{(s)}(E)}$ in terms of $\overline{(\hat{R}^2)}$ and $(\overline{\mathscr{R}})^2$ using Eqs. (5.37) and (5.38). In the second step used is the fact that there is average-fluctuation separation in transition strengths (i.e. assuming that the results in Sect. 4.3 extend to transition strengths).

Then we can evaluate $\overline{\{\hat{R}(E, E_f)\}^2}$ separately. To this end the numerically observed result that EGOE fluctuations follow GOE has been applied, i.e. $\hat{R}(E, E_f)$ distribution is Porter-Thomas. This gives $\overline{\{\hat{R}(E, E_f)\}^2} = 3$. Thus we are left with $\sum_{E_f} [\overline{\mathscr{R}(E, E_f)}]^2$ giving,

$$\{\xi_2^{(s)}(E)\}^{-1} = 3 \frac{\sum_{E_f} [\overline{|\langle E_f|\mathscr{O}|E\rangle|^2}]^2}{[\overline{\langle E|\mathscr{O}^\dagger \mathscr{O}|E\rangle}]^2}. \tag{5.39}$$

To proceed further we consider the bivariate transition strength densities $I_{\mathscr{O}}^{m,m_f}(E, E_f) = I^{m_f}(E_f)|\langle E_f \mid \mathscr{O} \mid E\rangle|^2 I^m(E)$ and they take, for EGOE(1 + 2) in the Gaussian domain, bivariate Gaussian form with normalization $\langle\langle \mathscr{O}^\dagger \mathscr{O}\rangle\rangle^m$. Writing the numerator and denominator in Eq. (5.39) in terms of the I's and replacing the sums over E_f by the integral $\int(- - -)I^{m_f}(E_f)\,dE_f$ will lead to the form,

$$\begin{aligned} &\{\xi_2^{(s)}(E)\}^{-1} \\ &\quad \xrightarrow{\text{EGOE}} 3\left\{\int I_{\mathscr{O}}^{m,m_f}(E, E_f)\,dE_f\right\}^{-2} \int \left[I^{m_f}(E_f)\right]^{-1}\left[I_{\mathscr{O}}^{m,m_f}(E, E_f)\right]^2 dE_f. \end{aligned} \tag{5.40}$$

Now replacing $I_{\mathscr{O}}^{m,m_f}(E, E_f)$ by the EGOE bivariate Gaussian and $I^{m_f}(E_f)$ by univariate Gaussian and carrying out the integrations in Eq. (5.40) will give the final result [33],

$$\begin{aligned} &\xi_2^{(s)}(E) \xrightarrow{\text{EGOE}} \frac{d_f}{3}\left\{\hat{\sigma}\sqrt{1-\zeta^2}X \exp -\left(\frac{\hat{\sigma}\zeta\widehat{E} + \hat{\Delta}}{X}\right)^2\right\}; \\ &\hat{\sigma} = \sigma_2/\sigma_f, \qquad \hat{\Delta} = (\varepsilon_2 - \varepsilon_f)/\sigma_f, \qquad \widehat{E} = (E - \varepsilon_1)/\sigma_1, \\ &X = \left[2 - (\hat{\sigma})^2(1-\zeta^2)\right]^{1/2}. \end{aligned} \tag{5.41}$$

Here ε_f and σ_f^2 are the centroid and variance of $I^{m_f}(E_f)$ and d_f is the dimension of the E_f space. Similarly, $(\varepsilon_1, \varepsilon_2)$ and (σ_1^2, σ_2^2) are the centroids and variances of the marginal densities of the bivariate transition strength density. The formula for $S^{info:s}(E)$ is,

$$\exp\left[S^{info:s}(E)\right] = 0.48 d_f \left[\hat{\sigma}\sqrt{1-\zeta^2} \exp \frac{1 - \hat{\sigma}^2(1-\zeta^2)}{2} \exp -\frac{(\hat{\sigma}\zeta\widehat{E} + \hat{\Delta})^2}{2}\right]. \tag{5.42}$$

The crucial factor that determines the EGOE structure of $(\text{NPC})_E$ is the bivariate correlation coefficient ζ. Also, it is easy to see that there is a close relationship between NPC and S^{info} in wavefunctions and in transition strengths except that the meaning of the correlation coefficient ζ is different for these two. For further details and discussion see [23, 33].

5.5 Simple Applications of NPC in Transition Strengths

Comparing Eq. (5.41), valid for EGOE(1 + 2) in the $\lambda > \lambda_F$ region, with the GOE result $d/3$ for NPC (see Sect. 2.3), it is possible to define effective dimension d_{eff} so that

$$\xi_2^{(s)}(E) = d_{eff}(E)/3. \tag{5.43}$$

Assuming $\hat{\sigma} = 1$ and $\hat{\Delta} = 0$ and putting $d = d_f$, Eq. (5.41) gives

$$\begin{aligned} d_{eff}(E) &= \frac{\sigma}{\overline{D(E)} f_\zeta(E)}; \\ f_\zeta(E) &= \sqrt{2\pi\left(1-\zeta^4\right)} \exp \frac{1-\zeta^2}{2(1+\zeta^2)} \widehat{E}^2, \\ \left\{\overline{D(E)}\right\}^{-1} = \overline{I(E)} &= \frac{d}{\sqrt{2\pi}\,\sigma} \exp -\frac{\widehat{E}^2}{2}. \end{aligned} \tag{5.44}$$

Now consider the transition parameter Λ for TRNI in nucleon-nucleon interaction as discussed in Sect. 3.1.3. Let us denote the TRI part of the Hamiltonian as H_R. Say, TRNI to TRI ratio in the interaction is α and v^2 is the square of the size of the H_R matrix elements in the neutron resonance region. Then (with the energies of the resonances $\sim E$),

$$\Lambda = \frac{\alpha^2 v^2(E)}{\{\overline{D(E)}\}^2} = \frac{\alpha^2 \sigma^2}{d_{eff}(E)\{\overline{D(E)}\}^2}. \tag{5.45}$$

With σ in MeV and $\overline{D}$ in eV units, Eqs. (5.44) and (5.45) will give

$$\alpha = \sqrt{\frac{f_\zeta(E)}{\sigma}} \Lambda^{1/2} \overline{D(E)}^{1/2} 10^{-3}. \tag{5.46}$$

In Eq. (5.46) one can assume, though we have used spinless EGOE ensemble, that all quantities are in J spaces as required for nuclei. Substituting typical values $\widehat{E} = 3-4$, $\zeta = 0.8-0.9$ and $\sigma \sim 2.25$, finally gives the result

$$\alpha \sim 1.5 \times 10^{-3} \Lambda^{1/2} \overline{D(E)}^{1/2}. \tag{5.47}$$

This compares quite well with Eq. (30) of [34]. Note that in [34], Eq. (5.47) was derived by using the results of detailed calculations for v^2 for large number of nuclei in the neutron resonance region.

In another application, we will show that the width of the fluctuations in strength sums is given by d_{eff}. Let us denote the strength sum, generated by a transition operator $\mathcal{O}$ acting on a state with energy E by $M_0(E) = \langle E|\mathcal{O}^\dagger \mathcal{O}|E\rangle$ and the locally averaged strength sum by $\overline{M_0(E)}$. The bar over $M_0(E)$ and other quantities denote local average which is equivalent to ensemble averaging. Assuming that the fluctuations in the locally renormalized strengths follow P-T (the locally averaged strengths taking EGOE bivariate Gaussian strength density form), it is seen that the transition matrix elements $\langle E'|\mathcal{O}|E\rangle$ are locally independent and they are zero centered Gaussian variables. The relationship between $\overline{\Sigma^2(E)}$, the mean square deviation in the strength sums (from the averages) normalized by the square of the average strength sum $\overline{[M_0(E)]^2}$ and NPC or $\xi_2^{(s)}(E)$ in transition strengths originating from $|E\rangle$ is derived as follows,

$$
\begin{aligned}
\overline{[M_0(E)]^2} &= \overline{\left[\sum_{E'} |\langle E'|\mathcal{O}|E\rangle|^2\right]^2} \\
&= \overline{\left[\sum_{E'} |\langle E'|\mathcal{O}|E\rangle|^4\right]} + \overline{\sum_{E'\neq E''} |\langle E'|\mathcal{O}|E\rangle|^2 |\langle E''|\mathcal{O}|E\rangle|^2} \\
&= \overline{\left[\sum_{E'} |\langle E'|\mathcal{O}|E\rangle|^4\right]} + \sum_{E'\neq E''} \overline{|\langle E'|\mathcal{O}|E\rangle|^2}\;\overline{|\langle E''|\mathcal{O}|E\rangle|^2} \\
&= \sum_{E'} [\overline{|\langle E'|\mathcal{O}|E\rangle|^4} - \{\overline{|\langle E'|\mathcal{O}|E\rangle|^2}\}^2] + (\overline{M_0(E)})^2 \\
\Rightarrow \quad \frac{\overline{\Sigma^2(E)}}{(\overline{M_0(E)})^2} &= \frac{2\sum_{E'} [\overline{|\langle E'|\mathcal{O}|E\rangle|^2}]^2}{(\overline{M_0(E)})^2} \\
&= \frac{2}{d'} \times \left\{ \left(\int dE' \overline{\rho_{biv;\mathcal{O}}(E,E')} \right)^2 \right\}^{-1} \int dE' \frac{(\overline{\rho_{biv;\mathcal{O}}(E,E')})^2}{\overline{\rho'(E')}} \\
&= \frac{2}{3\,\xi_{2:\mathcal{O}}^{(s)}(E)} = \frac{2}{d_{eff}(E)}.
\end{aligned}
\tag{5.48}
$$

The second step to the third in Eq. (5.48) follows from the independence of the strengths and the third reduces to the fourth step by adding and subtracting terms with $E' = E''$. The fourth step is simplified using the result that for a zero centered Gaussian variable x one has $\overline{x^4} = 3(\overline{x^2})^2$. Note that $\overline{\Sigma^2(E)}$ is the mean square deviation in the strength sum from the average and in relating it to NPC in transition strengths, results in Sect. 5.4.2 are used. Also, in $\xi_{2:\mathcal{O}}^{(s)}$ we have shown explicitly the transition operator $\mathcal{O}$. The final result in Eq. (5.48) was given first in [28] and its relation to NPC was pointed out in [2]. The results of Eq. (5.48) are compared with nuclear shell model results in [2, 35, 36].

References

1. V.V. Flambaum, G.F. Gribakin, F.M. Izrailev, Correlations within eigenvectors and transition amplitudes in the two-body random interaction model. Phys. Rev. E **53**, 5729–5741 (1996)
2. V.K.B. Kota, Embedded random matrix ensembles for complexity and chaos in finite interacting particle systems. Phys. Rep. **347**, 223–288 (2001)

3. Y. Alhassid, Statistical theory of quantum dots. Rev. Mod. Phys. **72**, 895–968 (2000)
4. Y. Alhassid, H.A. Weidenmüller, A. Wobst, Disordered mesoscopic systems with interaction: induced two-body ensembles and the Hartree-Fock approach. Phys. Rev. B **72**, 045318 (2005)
5. Ph. Jacquod, D.L. Shepelyansky, Emergence of quantum chaos in finite interacting Fermi systems. Phys. Rev. Lett. **79**, 1837–1840 (1997)
6. Ph. Jacquod, A.D. Stone, Ground state magnetization for interacting fermions in a disordered potential: kinetic energy, exchange interaction, and off-diagonal fluctuations. Phys. Rev. B **64**, 214416 (2001)
7. M. Hamermesh, *Group Theory and Its Application to Physical Problems* (Addison-Wesley, New York, 1962)
8. B.G. Wybourne, *Symmetry Principles and Atomic Spectroscopy* (Wiley, New York, 1970)
9. F.S. Chang, J.B. French, T.H. Thio, Distribution methods for nuclear energies, level densities and excitation strengths. Ann. Phys. (N.Y.) **66**, 137–188 (1971)
10. B.G. Wybourne, *Classical Groups for Physicists* (Wiley, New York, 1974)
11. J.C. Parikh, *Group Symmetries in Nuclear Structure* (Plenum, New York, 1978)
12. V.K.B. Kota, R.U. Haq, *Spectral Distributions in Nuclei and Statistical Spectroscopy* (World Scientific, Singapore, 2010)
13. C.M. Vincent, Group classification of many-body interactions. Phys. Rev. **163**, 1044–1050 (1967)
14. S. Aberg, Onset of chaos in rapidly rotating nuclei. Phys. Rev. Lett. **64**, 3119–3122 (1990)
15. R. Berkovits, Y. Avishai, Localization in Fock space: a finite-energy scaling hypothesis for many-particle excitation statistics. Phys. Rev. Lett. **80**, 568–571 (1998)
16. N.D. Chavda, V. Potbhare, V.K.B. Kota, Statistical properties of dense interacting Boson systems with one plus two-body random matrix ensembles. Phys. Lett. A **311**, 331–339 (2003)
17. V.K.B. Kota, R. Sahu, Breit-Wigner to Gaussian transition in strength functions, arXiv:nucl-th/0006079
18. B. Georgeot, D.L. Shepelyansky, Breit-Wigner width and inverse participation ratio in finite interacting Fermi systems. Phys. Rev. Lett. **79**, 4365–4368 (1997)
19. V.V. Flambaum, F.M. Izrailev, Statistical theory of finite Fermi systems based on the structure of chaotic eigenstates. Phys. Rev. E **56**, 5144–5159 (1997)
20. Ph. Jacquod, I. Varga, Duality between the weak and strong interaction limits of deformed fermionic two-body random ensembles. Phys. Rev. Lett. **89**, 134101 (2002)
21. N. Frazier, B.A. Brown, V. Zelevinsky, Strength functions and spreading widths of simple shell model configurations. Phys. Rev. C **54**, 1665–1674 (1996)
22. D. Angom, S. Ghosh, V.K.B. Kota, Strength functions, entropies and duality in weakly to strongly interacting fermion systems. Phys. Rev. E **70**, 016209 (2004)
23. V.K.B. Kota, R. Sahu, Structure of wavefunctions in $(1+2)$-body random matrix ensembles. Phys. Rev. E **64**, 016219 (2001)
24. A. Stuart, J.K. Ord, *Kendall's Advanced Theory of Statistics: Distribution Theory* (Oxford University Press, New York, 1987)
25. L. Kaplan, T. Papenbrock, Wave function structure in two-body random matrix ensembles. Phys. Rev. Lett. **84**, 4553–4556 (2000)
26. N.D. Chavda, V. Potbhare, V.K.B. Kota, Strength functions for interacting bosons in a mean-field with random two-body interactions. Phys. Lett. A **326**, 47–54 (2004)
27. M. Abramowtiz, I.A. Stegun (eds.), *Handbook of Mathematical Functions*, NBS Applied Mathematics Series, vol. 55 (U.S. Govt. Printing Office, Washington, D.C., 1972)
28. J.P. Draayer, J.B. French, S.S.M. Wong, Spectral distributions and statistical spectroscopy: I General theory. Ann. Phys. (N.Y.) **106**, 472–502 (1977)
29. V.K.B. Kota, R. Sahu, Single particle entropy in $(1+2)$-body random matrix ensembles. Phys. Rev. E **66**, 037103 (2002)
30. M. Rigol, V. Dunjko, M. Olshanii, Thermalization and its mechanism for generic isolated quantum systems. Nature (London) **452**, 854–858 (2008)

31. M. Horoi, V. Zelevinsky, B.A. Brown, Chaos vs thermalization in the nuclear shell model. Phys. Rev. Lett. **74**, 5194–5197 (1995)
32. V.K.B. Kota, N.D. Chavda, R. Sahu, Bivariate t-distribution for transition matrix elements in Breit-Wigner to Gaussian domains of interacting particle systems. Phys. Rev. E **73**, 047203 (2006)
33. V.K.B. Kota, R. Sahu, Information entropy and number of principal components in shell model transition strength distributions. Phys. Lett. B **429**, 1–6 (1998)
34. J.B. French, V.K.B. Kota, A. Pandey, S. Tomsovic, Statistical properties of many-particle spectra VI. Fluctuation bounds on N-N T-noninvariance. Ann. Phys. (N.Y.) **181**, 235–260 (1988)
35. J.P. Draayer, J.B. French, S.S.M. Wong, Spectral distributions and statistical spectroscopy: II Shell-model comparisons. Ann. Phys. (N.Y.) **106**, 503–524 (1977)
36. J.M.G. Gómez, K. Kar, V.K.B. Kota, R.A. Molina, J. Retamosa, Number of principal components and localization length in E2 and M1 transition strengths in ^{46}V. Phys. Rev. C **69**, 057302 (2004)

Chapter 6
One Plus Two-Body Random Matrix Ensembles for Fermions with Spin Degree of Freedom: EGOE(1 + 2)-s

First non-trivial but at the same time very important (from the point of view of its applications) embedded ensemble is the embedded Gaussian orthogonal ensemble of one plus two-body interactions with spin degree of freedom [EGOE(1 + 2)-**s**] for a system of interacting fermions. This ensemble is directly applicable, as spin degree of freedom is explicitly included, to mesoscopic systems such as quantum dots and small metallic grains. Spin degree of freedom allows for inclusion of both exchange interaction and pairing interaction in the Hamiltonian. Secondly EGOE(1 + 2)-**s** ensemble exhibits three chaos markers, just as the EGOE(1 + 2) for spinless fermion systems, with the markers depending on the total m fermion spin S. The spin dependent chaos markers provide a much stronger basis for statistical (nuclear and atomic) spectroscopy [1]. Also,thermalization in generic isolated quantum systems has applications in QIS as emphasized in some recent papers [2–6] and with the chaos markers, EGOE(1 + 2) and EGOE(1 + 2)-**s** ensembles allow us to study thermalization in finite quantum systems [7]. In addition, as recognized recently, entanglement and strength functions essentially capture the same information about eigenvector structure and therefore the change in the form (δ-function to BW to Gaussian) of the strength functions in different regimes defined by the chaos markers determines entanglement properties in multi-qubit systems [8–11]. Hence, EGOE(1 + 2) and EGOE(1 + 2)-**s** ensembles will be useful in multi-qubit entanglement studies as emphasized in [11].

In this chapter we will present some of the general properties of EGOE(1 + 2)-**s** and in the next chapter applications are given.

6.1 EGOE(1 + 2)-s: Definition and Construction

Let us begin with a system of m ($m > 2$) fermions distributed say in Ω number of single particle orbits each with spin $\mathbf{s} = \frac{1}{2}$ so that the number of single particle states $N = 2\Omega$. Single particle states are denoted by $|i, m_s = \pm\frac{1}{2}\rangle$ with $i = 1, 2, \ldots, \Omega$ and similarly two particle antisymmetric stats are denoted by

V.K.B. Kota, *Embedded Random Matrix Ensembles in Quantum Physics*,
Lecture Notes in Physics 884, DOI 10.1007/978-3-319-04567-2_6,

$|(ij)s, m_s\rangle_a$ with $s = 0$ or 1. For one plus two-body Hamiltonians preserving m particle spin S, the one-body Hamiltonian is $\widehat{h}(1) = \sum_{i=1,2,\ldots,\Omega} \varepsilon_i \hat{n}_i$ where the orbits i are doubly degenerate, $\hat{n}_i$ are number operators and ε_i are single particle energies [it is in principle possible to consider $\widehat{h}(1)$ with off-diagonal energies ε_{ij}]. Similarly the two-body Hamiltonian $\widehat{V}(2)$ is defined by the two-body matrix elements $V^s_{ijkl} = {}_a\langle(kl)s, m_s|\widehat{V}(2)|(ij)s, m_s\rangle_a$ with the two-particle spin $s = 0, 1$ and they are independent of the m_s quantum number; note that for $s = 1$, only $i \neq j$ and $k \neq l$ matrix elements exist. Thus $\widehat{V}(2) = \lambda_0 \widehat{V}^{s=0}(2) + \lambda_1 \widehat{V}^{s=1}(2)$; the sum here is a direct sum. Now, EGOE(2)-**s** for a given (m, S) system is generated by defining the two parts of the two-body Hamiltonian to be independent GOE's [one for $\widehat{V}^{s=0}(2)$ and other for $\widehat{V}^{s=1}(2)$] in the 2-particle spaces and then propagating the $V(2)$ ensemble $\{\widehat{V}(2)\} = \lambda_0\{\widehat{V}^{s=0}(2)\} + \lambda_1\{\widehat{V}^{s=1}(2)\}$ to the m-particle spaces with a given spin S by using the geometry (direct product structure) of the m-particle spaces; here $\{\}$ denotes ensemble. Then EGOE(1 + 2)-**s** is defined by

$$\{\widehat{H}\}_{\text{EGOE(1+2)-s}} = \widehat{h}(1) + \lambda_0\big\{\widehat{V}^{s=0}(2)\big\} + \lambda_1\big\{\widehat{V}^{s=1}(2)\big\}, \tag{6.1}$$

where $\{\widehat{V}^{s=0}(2)\}$ and $\{\widehat{V}^{s=1}(2)\}$ in two-particle spaces are GOE(1) and λ_0 and λ_1 are the strengths of the $s = 0$ and $s = 1$ parts of $\widehat{V}(2)$, respectively. From now onwards we drop the "hat" symbol over H, h and V operators when there is no confusion. The mean-field one-body Hamiltonian $h(1)$ in Eq. (6.1) is a fixed one-body operator defined by the single particle energies ε_i with average spacing Δ [it is possible to draw the ε_i's from the eigenvalues of a random ensemble just as in EGOE(1+2)]. Thus, EGOE(1 + 2)-**s** is defined by the five parameters $(\Omega, m, S, \lambda_0, \lambda_1)$; without loss of generality we put $\Delta = 1$ so that λ_0 an λ_1 are in units of Δ. The $\widehat{H}$ matrix structure in the defining space is shown in Fig. 6.1a.

Starting with Eq. (6.1), matrix for $\widehat{H}$ in m-particle spaces can be constructed as described ahead. As $\widehat{H}$ preserves S, the m particle matrix will be a direct sum of matrices in each (m, S) spaces as shown in Fig. 6.1b. It is useful to note that a formula for the H matrix dimension $d_f(\Omega, m, S)$ in a given (m, S) space, i.e. number of levels in the (m, S) space with each of them being $(2S + 1)$-fold degenerate, with the fermions in Ω number of sp levels is easy to write down. Given the S_z quantum number M_S, for fixed (m, M_S) we have $m_1 = (m + 2M_S)/2$ and $m_2 = (m - 2M_S)/2$. Then the (m, M_S) space dimension $\mathcal{D}(\Omega, m, M_S) = \binom{\Omega}{m_1}\binom{\Omega}{m_2}$. Now the simple rule $d_f(\Omega, m, S) = \mathcal{D}(\Omega, m, S) - \mathcal{D}(\Omega, m, S + 1)$ gives,

$$d_f(\Omega, m, S) = \frac{(2S + 1)}{(\Omega + 1)}\binom{\Omega + 1}{m/2 + S + 1}\binom{\Omega + 1}{m/2 - S}. \tag{6.2}$$

They satisfy the sum rule $\sum_S (2S + 1) d_f(\Omega, m, S) = \binom{N}{m}$. For example for $m = \Omega = 8$, the dimensions are 1764, 2352, 720, 63 and 1 for $S = 0$, 1, 2, 3 and 4 respectively. Similarly, for $\Omega = m = 12$, they are 226512, 382239, 196625, 44044, 4214, 143, and 1 for $S = 0$–6. Often we will drop the suffix 'f' in $d_f()$ when there is no confusion. It is useful to note that for the EGOE(1 + 2)-**s** ensemble three group structures are relevant and they are $U(\Omega) \otimes SU(2)$, $\sum_{S=0,1} O(N_{2,S}) \oplus$ and

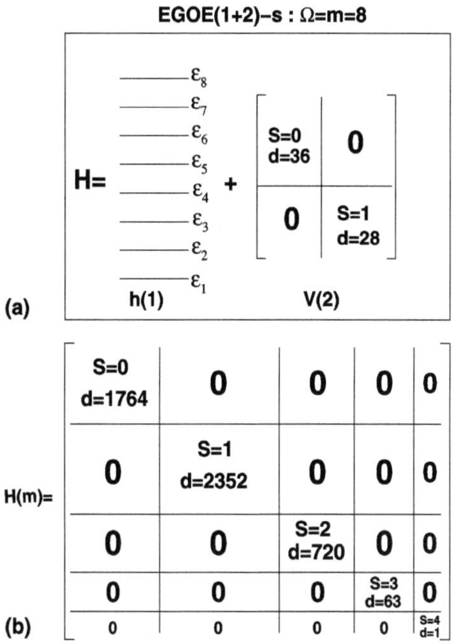

Fig. 6.1 (**a**) Hamiltonian generating EGOE(1 + 2)-**s** ensemble for 8 fermions in 8 sp levels. Shown are the sp spectrum defining the mean-filed part $h(1)$ and the $V(2)$ matrix in two-particle spaces. Note that each sp level is doubly degenerate. (**b**) Decomposition of the m particle space H matrix into direct sum of matrices with fixed spin S value and there is a EGOE(1 + 2)-**s** ensemble in each (m, S) space corresponding to each diagonal block in the figure

$\sum_S O(N_{m,S})\oplus$, $m > 2$. Here $N_{m,S} = d_f(\Omega, m, S)$, the symbol $\oplus$ stands for direct sum and $O(r)$ is the orthogonal group in r dimensions. The $U(\Omega) \otimes SU(2)$ algebra defines the embedding. The EGOE(2) ensemble has orthogonal invariance with respect to the $\sum_{S=0,1} O(N_{2,S})\oplus$ group acting in two-particle spaces. However it is not invariant under the $\sum_S O(N_{m,S})\oplus$ group for $m > 2$. This group is appropriate if GOE representation for fixed-(m, S) H matrices is employed; i.e., there is an independent GOE for each (m, S) subspace.

In order to construct the many particle Hamiltonian matrix for a given (m, S), one approach is as follows [12]. Consider the single particle states $|i, m_s = \pm\frac{1}{2}\rangle$ and arrange them in such a way that the first Ω state have $m_s = \frac{1}{2}$ and the remaining Ω states have $m_s = -\frac{1}{2}$ so that a state $|r\rangle = |i = r, m_s = \frac{1}{2}\rangle$ for $r \leq \Omega$ and $|r\rangle = |i = r - \Omega, m_s = -\frac{1}{2}\rangle$ for $r > \Omega$. Now the m-particle configurations $\mathbf{m}$ and the corresponding M_S values are,

$$\mathbf{m} = (m_1, m_2, \ldots, m_\Omega, m_{\Omega+1}, m_{\Omega+2}, \ldots, m_{2\Omega}), \qquad m_r = 0 \quad \text{or} \quad 1,$$
$$M_S = \frac{1}{2}\left[\sum_{r=1}^{\Omega} m_r - \sum_{r'=\Omega+1}^{2\Omega} m_{r'}\right]. \tag{6.3}$$

Two examples for $\mathbf{m}$ for a ($\Omega = 6$, $m = 6$) system are shown in Fig. 6.2. It is important to note that the $\mathbf{m}$'s with $M_S = 0$ will contain states with all S values for even m and similarly with $M_S = \frac{1}{2}$ for odd m. Therefore, we construct the m particle Hamiltonian matrix using the basis defined by $\mathbf{m}$'s with

$M_S = 0$ for even m and $M_S = \frac{1}{2}$ for odd m. The dimension of this basis space is $\mathscr{D}(\Omega, m, M_S^{min}) = \sum_S d_f(\Omega, m, S)$. For example, $\mathscr{D}(8, 8, 0) = 4900$, $\mathscr{D}(8, 6, 0) = 3136$ and $\mathscr{D}(10, 10, 0) = 63404$. To proceed further, the $(1 + 2)$-body Hamiltonian defined by $(\varepsilon_i, V_{ijkl}^{s=0,1})$'s should be converted into the $|i, m_s = \pm\frac{1}{2}\rangle$ basis. Then ε_i change to ε_r with the index r defined as above and $V_{ijkl}^{s=0,1}$ change to $V_{im_i, jm_j, km_k, lm_l}$ where,

$$
\begin{aligned}
V_{i\frac{1}{2}, j\frac{1}{2}, k\frac{1}{2}, l\frac{1}{2}} &= V_{ijkl}^{s=1} \\
V_{i-\frac{1}{2}, j-\frac{1}{2}, k-\frac{1}{2}, l-\frac{1}{2}} &= V_{ijkl}^{s=1} \\
V_{i\frac{1}{2}, j-\frac{1}{2}, k\frac{1}{2}, l-\frac{1}{2}} &= \frac{\sqrt{(1+\delta_{ij})(1+\delta_{kl})}}{2}\left\{V_{ijkl}^{s=1} + V_{ijkl}^{s=0}\right\}
\end{aligned}
\tag{6.4}
$$

with all other matrix elements being zero except for the symmetries,

$$
\begin{aligned}
V_{im_i, jm_j, km_k, lm_l} &= -V_{im_i, jm_j, lm_l, km_k} \\
V_{im_i, jm_j, km_k, lm_l} &= V_{km_k, lm_l, im_i, jm_j}.
\end{aligned}
\tag{6.5}
$$

Using $(\varepsilon_r, V_{im_i, jm_j, km_k, lm_l})$'s, construction of the m particle H matrix in the basis defined by Eq. (6.3) reduces to the problem of EGOE(1 + 2) for spinless fermion systems and hence Eq. (4.3) will apply. For the S^2 operator, $\varepsilon_i = 3/4$ independent of i, $V_{ijij}^{s=0} = -3/2$ and $V_{ijij}^{s=1} = 1/2$ independent of (ij) and all other V_{ijkl}^s are zero. Using these for the S^2 operator, the m particle matrix with $M_S = 0$ for even m ($M_S = \frac{1}{2}$ for odd m) is constructed and diagonalized. This gives a direct sum of unitary matrices and the unitary matrix that corresponds to a given S is identified by the eigenvalue $S(S + 1)$. Applying the unitary transformation defined by this unitary matrix, the m particle H matrix with $M_S = 0$ for even m ($M_S = \frac{1}{2}$ for odd m) is transformed to the basis with good S values. Alternative method of construction is to directly construct the H matrix in a good S basis using angular-momentum algebra [13]. Employing good-M_S basis is equivalent to employing the algebra $U(2\Omega) \supset U(\Omega) \oplus U(\Omega)$ and the good spin basis corresponds to $U(2\Omega) \supset U(\Omega) \otimes SU(2)$. For the construction of EGOE(1 + 2)-**s**, besides Kota et al. [12, 14], computer codes were also written by Jaquod [15], Papenbrock [16] and Alhassid [13] groups.

6.2 Fixed-S Eigenvalue Densities, NPC and Information Entropy

6.2.1 Eigenvalue Densities

Using the EGOE(1 + 2)-**s** codes, several groups have [12, 14–16], in large number of examples, numerically constructed the H matrix and by diagonalizing them obtained the ensemble averaged eigenvalue (level) densities $\overline{\rho^{m,S}(E)} =$

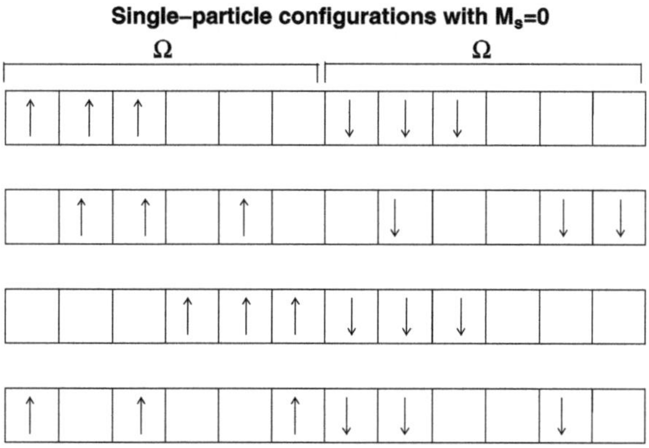

Fig. 6.2 Examples of single particle configurations with $M_S = 0$ for a $\Omega = 6$ and $m = 6$ system. Note that the number of sp states $N = 12$. In the figure, first Ω number of sp states correspond to spin up ($m_s = 1/2$) and the remaining Ω number of sp states correspond to spin down ($m_s = -1/2$) states

$\overline{\langle \delta(H-E)\rangle}^{m,S}$. Note that the trace of an operator $\mathscr{O}$ over a fixed-(m,S) space is defined by $\langle\langle \mathscr{O}\rangle\rangle^{m,S} = (2S+1)^{-1}\sum_\alpha \langle m,S,\alpha|\mathscr{O}|m,S,\alpha\rangle$ and similarly (m,S) space average is $\langle \mathscr{O}\rangle^{m,S} = [(2S+1)d_f(\Omega,m,S)]^{-1}\sum_\alpha \langle m,S,\alpha|\mathscr{O}|m,S,\alpha\rangle$. From now onwards, we drop the 'bar' over ρ when there is no confusion. As an example, results for $\Omega = 8$ and $m = 6$ system are shown in Fig. 5.2. In this example, $\lambda_0 = \lambda_1 = \lambda = 0.1$. Also the sp energies taken to be $\varepsilon_i = i + 1/i$ with $i = 1, 2, \ldots, \Omega$. To construct the eigenvalue density, the centroids $E_c(m,S)$ of all the members of the ensemble are made to be zero and variance $\sigma^2(m,S)$ unity i.e. for each member we change the eigenvalues E to the standardized variables $\widehat{E} = [E - E_c(m,S)]/\sigma(m,S)$. Note that the parameters $E_c(m,S)$ and $\sigma^2(m,S)$ depend also on Ω. But for convenience, we will drop Ω in $E_c(m,S)$ and $\sigma^2(m,S)$ throughout. Then, using a bin-size $\Delta\widehat{E} = 0.2$, histograms for $\rho^{m,S}(E)$ are generated. The calculated results are compared with both the Gaussian ($\rho_{\mathscr{G}}$) and Edgeworth (ED) corrected Gaussian (ρ_{ED}) with γ_1 and γ_2 corrections. From the results in Fig. 5.2, it is seen that the agreement between the exact and ED corrected Gaussians is excellent. It has been well established, as discussed in Chaps. 4 and 5, that the ensemble averaged eigenvalue density takes Gaussian form in the case of EGOE(1+2). Combining this with the numerical results for the fixed-(m,S) level densities, it can be concluded that the Gaussian form is generic for the embedded ensembles extending to those with good quantum numbers. This is further substantiated by the analytical results for the ensemble averaged $\gamma_2(m,S)$ for EGOE(2)-**s** extracted from [17] (see also [18–20]) and for a general $h(1)$ Hamiltonian given in [14]; they give to lowest order $\gamma_2(m,S) \sim \frac{C_0}{m} + \frac{C_1}{\Omega}[1 + \frac{4S(S+1)}{m^2}]$ where C_0 and C_1 are constants.

6.2.2 NPC and S^{info}

Basis states with good S used in the previous section for constructing H matrix are also eigenstates of the mean-field Hamiltonian $h(1)$ [this is ensured by a further diagonalization of $h(1)$ in the basis with good S]. Simple expressions for NPC or $\xi_2(E)$ and $S^{info}(E)$ in $h(1)$ basis for spinless EGOE(1 + 2) in the Gaussian regime are given by Eqs. (5.23) and (5.25). Extending these results to EGOE(1 + 2)-**s** by replacing the fixed-m variances in this expression by fixed-(m, S) variances will give,

$$
\begin{aligned}
&\xi_2(E,S)/\xi_2^{GOE} \stackrel{\text{EGOE(1+2)-s}}{\longrightarrow} \sqrt{1-\left[\zeta(m,S)\right]^4}\exp-\frac{[\zeta(m,S)]^2\widehat{E}^2}{1+[\zeta(m,S)]^2},\\
&\ell_H(E,S) \stackrel{\text{EGOE(1+2)-s}}{\longrightarrow} \sqrt{1-\left[\zeta(m,S)\right]^2}\exp\left(\frac{[\zeta(m,S)]^2}{2}\right)\exp-\left(\frac{[\zeta(m,S)]^2\widehat{E}^2}{2}\right);\\
&\left[\zeta(m,S)\right]^2 = 1-\frac{\sigma^2_{off\text{-}diagonal}(m,S)}{\sigma^2(m,S)} \sim \frac{\sigma^2_{h(1)}(m,S)}{\sigma^2_{h(1)}(m,S)+\sigma^2_{V(2)}(m,S)}.
\end{aligned}
\tag{6.6}
$$

Note that $\xi_2^{GOE} = d(m, S)/3$ independent of E and also $\widehat{E}$ is zero centered and scaled to unit width. These results, expected to be good for reasonably large values for λ_0 and λ_1 (this will be clear in Sect. 6.4), have been tested in many examples [12, 14]. It is important to stress here that a formula for $\sigma^2(m, S)$ is available as given ahead by Eq. (6.18). This also gives $\sigma^2_{off\text{-}diagonal}(m, S)$ by dropping the first three terms and putting $\lambda_{i,i} = 0$ in the next three terms in this equation. Therefore, the correlation coefficient ζ can be calculated without constructing the H matrices in (m, S) spaces. A seen from Fig. 6.3, $\xi_2(E)$ and $S^{info}(E)$ calculated using Eq. (6.6) describe the numerical matrix diagonalization results rather well. Thus, in general, the spectral variance given by Eq. (6.18) ahead together with Eq. (6.6) can be used to predict $\xi_2(E)$ and $S^{info}(E)$, in the Gaussian domain (defined in Sect. 6.4) for any (m, Ω) system with fixed S. Finally, from the agreements seen in Figs. 5.2 and 6.3, we can conclude that to a good approximation many of the results of EGOE(1 + 2) extend to EGOE(1 + 2)-**s** with the parameters calculated in (m, S) spaces.

6.3 Fixed-Spin Energy Centroids and Variances

Let us start with the fixed-(m, S) energy centroids $E_c(m, S) = \langle H\rangle^{m,S}$ for a one plus two-body Hamiltonian $H = h(1) + V(2) = h(1) + [\lambda_0 V^{s=0}(2) + \lambda_1 V^{s=1}(2)]$. The operator generating $\langle H\rangle^{m,S}$ will be a polynomial, in the scalar operators $\hat{n}$ and $\hat{S}^2$, of maximum body rank 2. A two-body operator is said to be of body rank 2, a three-body operator of body rank 3 and so on [21]. Then, $E_c(m, S) = a_0 + a_1 m + a_2 m^2 + a_3 S(S + 1)$. Solving for the a_i's in terms of E_c for $m \le 2$, we obtain [22]

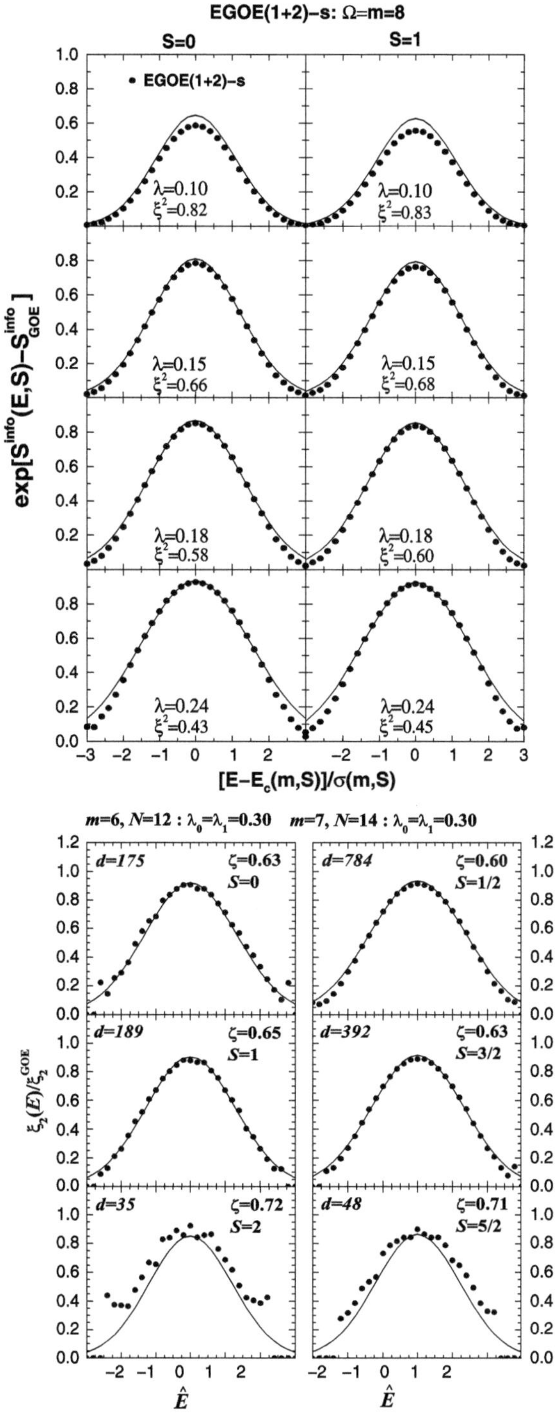

Fig. 6.3 Results for information entropy S^{info} and number of principle components (ξ_2). For S^{info}, results are shown for $\exp[S^{info}(E,S) - S^{info}_{\mathrm{GOE}}]$ for a 20 member EGOE(1 + 2)-s ensemble with $\Omega = m = 8$ and $S = 0$ and 1. The *continuous curves* correspond to Eq. (6.6). Similarly, results for ξ_2 are shown for a 20 member EGOE(1 + 2)-s with $\Omega = m = 6$ and $S = 0$, 1 and 2. Also shown are results for a 5 member ensemble with $\Omega = m = 7$ and $S = 1/2, 3/2$ and $5/2$. Here again, *continuous curves* are from Eq. (6.6). All the ensemble results in the figures are averaged over a bin-size of 0.2 for $\widehat{E} = [E - \varepsilon(m,S)]/\sigma(m,S)$ and they are shown as *filled circles*. Some of the results for ξ_2 are given in [12]. The S^{info} figure is taken from [14] with permission from American Physical Society

$$
\begin{aligned}
E_c(m,S) &= \big[\langle h(1)\rangle^{1,\frac{1}{2}}\big]m \\
&\quad + \lambda_0 \langle\!\langle V^{s=0}(2)\rangle\!\rangle^{2,0}\frac{P^0(m,S)}{4\Omega(\Omega+1)} \\
&\quad + \lambda_1 \langle\!\langle V^{s=1}(2)\rangle\!\rangle^{2,1}\frac{P^1(m,S)}{4\Omega(\Omega-1)}; \\
P^0(m,S) &= \big[m(m+2)-4S(S+1)\big], \\
P^1(m,S) &= \big[3m(m-2)+4S(S+1)\big], \\
\langle h(1)\rangle^{1,\frac{1}{2}} &= \overline{\varepsilon} = \Omega^{-1}\sum_{i=1}^{\Omega}\varepsilon_i, \\
\langle\!\langle V^{s=0}(2)\rangle\!\rangle^{2,0} &= \sum_{i\le j} V^{s=0}_{ijij}, \qquad \langle\!\langle V^{s=1}(2)\rangle\!\rangle^{2,1} = \sum_{i<j} V^{s=1}_{ijij}.
\end{aligned}
\tag{6.7}
$$

Trivially the ensemble average of E_c from the $V(2)$ part will be zero. However the covariances in the energy centroids generated by the two-body part $H(2) = V(2)$ of H are non-zero,

$$
\begin{aligned}
&\overline{\langle H(2)\rangle^{m,S}\langle H(2)\rangle^{m',S'}} \\
&\quad = \frac{\lambda_0^2}{16\Omega(\Omega+1)}P^0(m,S)P^0(m',S') + \frac{\lambda_1^2}{16\Omega(\Omega-1)}P^1(m,S)P^1(m',S').
\end{aligned}
\tag{6.8}
$$

The spectral variances $\sigma^2(m,S) = \langle H^2\rangle^{m,S} - [\langle H\rangle^{m,S}]^2$ are generated by an operator that is a polynomial, in the scalar operators $\hat{n}$ and $\hat{S}^2$, of maximum body rank 4. This gives $\sigma^2(m,S) = \sum_{p=0}^{4} a_p m^p + \sum_{q=0}^{2} b_q m^q S(S+1) + c_0[S(S+1)]^2$. The nine parameters (a_i, b_i, c_i) can be written in terms of ε_i and the two-body matrix elements $V^{s=0,1}_{ijkl}$ using the embedding algebra $U(N) \supset U(\Omega)\otimes SU(2)$. With respect to this algebra, as pointed out in [23, 24], $h(1)$ decomposes into a scalar $\nu = 0$ part [given by the first term in the first equation in Eq. (6.7)] and an irreducible one-body part with $\nu = 1$. The $\nu = 0$ and $\nu = 1$ parts transform, in Young tableaux notation [25], as the irreps [0] and $[21^{\Omega-2}]$ respectively of $U(\Omega)$. Similarly $V^s(2)$, $s = 0, 1$ decompose into $\nu = 0, 1$ and 2 parts. The scalar parts $V^{\nu=0:s=0,1}$ can be identified from Eq. (6.7) and they will not contribute to the variances. The effective one-body parts $V^{\nu=1:s=0,1}$, generated by $V^{s=0,1}_{ijkl}$, are defined by the induced single particle energies $\lambda_{i,j}(s)$ given ahead in Eq. (6.9). The diagonal induced energies $\lambda_{i,i}(s)$ are identified for the first time in [23]. However for EGOE(1 + 2)-**s** it is possible to have $\lambda_{i,j}(s)$, $i \neq j$. Now the irreducible two-body part $V^{\nu=2:s=0} = V - V^{\nu=0:s=0} - V^{\nu=1:s=0}$ and similarly $V^{\nu=2:s=1}$ is defined. It should be noted that the two $\nu = 0$ parts of $V(2)$ transform as the $U(\Omega)$ irrep $[0]$ and the two $\nu = 1$ parts of $V(2)$ transform as the irrep $[21^{\Omega-2}]$. Similarly $V^{\nu=2:s=0}$ transforms as the irrep $[42^{\Omega-2}]$ and the $V^{\nu=2:s=1}$ as the irrep $[2^2 1^{\Omega-4}]$. Figures 5.1c and 9.2 ahead show the corresponding Young tableaux; note that N in these figures should be replaced by Ω for EGOE(1 + 2)-**s**. Using these and the group theory of $U(N) \supset U(\Omega) \otimes SU(2)$ algebra as given by Hecht and Draayer [26], a compact and easy to understand expression for fixed-S variances emerges, with $\mathcal{S}^2 = S(S+1)$, $m^x = \Omega - m/2$, $X(m,S) = m(m+2) - 4S(S+1)$ and $Y(m,S) = m(m-2) - 4S(S+1)$,

$$
\begin{aligned}
\sigma^2_{H=h(1)+V(2)}(m,S) = {} & \frac{(\Omega+2)mm^x - 2\Omega\mathcal{S}^2}{\Omega(\Omega-1)(\Omega+1)} \sum_i \tilde{\varepsilon}_i^2 \\
& + \frac{m^x X(m,S)}{2\Omega(\Omega-1)(\Omega+1)} \sum_i \tilde{\varepsilon}_i \lambda_{i,i}(0) \\
& + \frac{(\Omega+2)m^x[3Y(m,S) + 16\mathcal{S}^2] - 8\Omega(m-1)\mathcal{S}^2}{2\Omega(\Omega-1)(\Omega+1)(\Omega-2)} \\
& \times \sum_i \tilde{\varepsilon}_i \lambda_{i,i}(1) \\
& + \frac{[(m+2)m^x/2 + \mathcal{S}^2]X(m,S)}{8\Omega(\Omega-1)(\Omega+1)(\Omega+2)} \sum_{i,j} \lambda_{i,j}^2(0) \\
& + \frac{1}{8\Omega(\Omega-1)(\Omega+1)(\Omega-2)^2} \\
& \times \big\{8\Omega(m-1)(\Omega-2m+4)\mathcal{S}^2 \\
& + (\Omega+2)\big[3(m-2)m^x/2 - \mathcal{S}^2\big]\big[3Y(m,S) + 8\mathcal{S}^2\big]\big\} \\
& \times \sum_{i,j} \lambda_{i,j}^2(1) \\
& + \frac{[3(m-2)m^x/2 - \mathcal{S}^2]X(m,S)}{4\Omega(\Omega-1)(\Omega+1)(\Omega-2)} \sum_{i,j} \lambda_{i,j}(0)\lambda_{i,j}(1) \\
& + P_2^0(m,S)\big\langle\big(V^{\nu=2,s=0}\big)^2\big\rangle^{2,0} + P_2^1(m,S)\big\langle\big(V^{\nu=2,s=1}\big)^2\big\rangle^{2,1};
\end{aligned}
\tag{6.9}
$$

$$
P_2^0(m,S) = \frac{[m^x(m^x+1) - \mathcal{S}^2]X(m,S)}{8\Omega(\Omega-1)}, \tag{6.10}
$$

$$
\begin{aligned}
P_2^1(m,S) = {} & \frac{1}{\Omega(\Omega+1)(\Omega-2)(\Omega-3)} \\
& \times \big\{\big(\mathcal{S}^2\big)^2\big(3\Omega^2 - 7\Omega + 6\big)/2 + 3m(m-2)m^x\big(m^x - 1\big)
\end{aligned}
$$

$$\times (\Omega+1)(\Omega+2)/8 - \mathscr{S}^2\big[(5\Omega-3)(\Omega+2)m^x m + \Omega(\Omega-1)(\Omega+1)(\Omega+6)\big]/2\big\}, \tag{6.11}$$

with

$$\begin{aligned}
&\tilde{\varepsilon}_i = \varepsilon_i - \overline{\varepsilon},\\
&\lambda_{i,i}(s) = \sum_j V^s_{ijij}(1+\delta_{ij}) - (\Omega)^{-1}\sum_{k,l} V^s_{klkl}(1+\delta_{kl}),\\
&\lambda_{i,j}(s) = \sum_k \sqrt{(1+\delta_{ki})(1+\delta_{kj})}V^s_{kikj} \quad \text{for } i \neq j,\\
&V^{\nu=2,s}_{ijij} = V^s_{ijij} - \big[\langle V(2)\rangle^{2,s} + \big(\lambda_{i,i}(s)+\lambda_{j,j}(s)\big)\big(\Omega+2(-1)^s\big)^{-1}\big],\\
&V^{\nu=2,s}_{kikj} = V^s_{kikj} - \big(\Omega+2(-1)^s\big)^{-1}\sqrt{(1+\delta_{ki})(1+\delta_{kj})}\lambda^s_{i,j} \quad \text{for } i \neq j,\\
&V^{\nu=2,s}_{ijkl} = V^s_{ijkl} \quad \text{for all other cases.}
\end{aligned} \tag{6.12}$$

Equation (6.9) along with Eq. (6.12) applies to individual members of the EGOE(1 + 2)-**s** ensemble. Using these, formula for the ensemble averaged variance is obtained as follows. With $h(1)$ and $V(2)$ being independent gives,

$$\overline{\sigma^2_H(m,S)} = \sigma^2_{h(1)}(m,S) + \overline{\sigma^2_{V(2)}(m,S)}. \tag{6.13}$$

The propagation formula for $\sigma^2_{h(1)}$ is the first term in Eq. (6.9).

$$\sigma^2_{h(1)}(m,S) = \frac{(\Omega+2)m(\Omega-m/2) - 2\Omega S(S+1)}{(\Omega-1)(\Omega+1)}\sigma^2_{h(1)}\left(1,\frac{1}{2}\right). \tag{6.14}$$

Similarly,

$$\overline{\sigma^2_{V(2)}(m,S)} = \sum_{s=0,1}\lambda^2_s \sum_{\nu=1,2}\overline{\big\langle\big[V^{s,\nu}(2)\big]^2\big\rangle^{m,S}} \tag{6.15}$$

and the four terms here correspond to terms 4, 5, 7 and 8 in Eq. (6.9). For evaluating $\overline{\langle[V^{s,\nu=1}(2)]^2\rangle^{m,S}}$ we need $\sum_{i,j}\overline{\lambda^2_{i,j}(s)}$ and similarly, for evaluating $\overline{\langle[V^{s,\nu=2}(2)]^2\rangle^{m,S}}$ we need $\overline{\langle[V^{s,\nu=2}(2)]^2\rangle^{2,s}}$. Firstly, applying the fact that the V^s matrix elements are independent Gaussian random variables with zero center and variance unity (except for the diagonal matrix elements it is 2) and simplifying using Eq. (6.12), we obtain

$$\sum_{i,j} \overline{\lambda^2_{i,j}(0)} = (\Omega - 1)(\Omega + 2)^2,$$
$$\sum_{i,j} \overline{\lambda^2_{i,j}(1)} = (\Omega - 1)(\Omega - 2)(\Omega + 2). \tag{6.16}$$

Also, $\overline{\langle [V^s(2)]^2\rangle^{2,s}} = [d(m,s) + 1]$. This along with Eq. (6.9) and Eqs. (6.15) and (6.16) will give $\overline{\langle [V^{s,\nu=2}(2)]^2\rangle^{2,s}}$,

$$\overline{\left\langle \left[V^{s=0,\nu=2}(2)\right]^2\right\rangle^{2,0}} = \frac{1}{2}(\Omega - 1)(\Omega + 2),$$
$$\overline{\left\langle \left[V^{s=1,\nu=2}(2)\right]^2\right\rangle^{2,1}} = \frac{(\Omega - 3)(\Omega^2 + \Omega + 2)}{2(\Omega - 1)}. \tag{6.17}$$

Substituting Eqs. (6.16) and (6.17) in Eq. (6.9) gives the final result,

$$\begin{aligned}
&\overline{\sigma^2_{V(2)}(m,S)} \\
&\quad= \frac{\lambda_0^2}{\Omega(\Omega+1)/2}\left[\frac{\Omega+2}{\Omega+1} Q^1(\{2\}:m,S) + \frac{\Omega^2+3\Omega+2}{\Omega^2+3\Omega} Q^2(\{2\}:m,S)\right] \\
&\qquad+ \frac{\lambda_1^2}{\Omega(\Omega-1)/2}\left[\frac{\Omega+2}{\Omega+1} Q^1(\{1^2\}:m,S)\right. \\
&\qquad\left.+ \frac{\Omega^2+\Omega+2}{\Omega^2+\Omega} Q^2(\{1^2\}:m,S)\right]; \\
&Q^1(\{2\}:m,S) = \left[(\Omega+1)P^0(m,S)/16\right]\left[m^x(m+2)/2 + \mathscr{S}^2\right], \\
&Q^2(\{2\}:m,S) = \left[\Omega(\Omega+3)P^0(m,S)/32\right]\left[m^x(m^x+1) - \mathscr{S}^2\right], \\
&Q^1(\{1^2\}:m,S) = \frac{(\Omega-1)}{16(\Omega-2)}\left[(\Omega+2)P^1(m,S)P^2(m,S)\right. \\
&\qquad\qquad\left.+ 8\Omega(m-1)(\Omega-2m+4)\mathscr{S}^2\right], \\
&Q^2(\{1^2\}:m,S) = \frac{\Omega}{8(\Omega-2)}\left[(3\Omega^2-7\Omega+6)\left(\mathscr{S}^2\right)^2\right. \\
&\qquad\qquad + 3m(m-2)m^x(m^x-1)(\Omega+1)(\Omega+2)/4 \\
&\qquad\qquad + \mathscr{S}^2\left\{-mm^x(5\Omega-3)(\Omega+2)\right. \\
&\qquad\qquad\left.\left.+ \Omega(\Omega-1)(\Omega+1)(\Omega+6)\right\}\right], \\
&P^2(m,S) = 3m^x(m-2)/2 - \mathscr{S}^2, \quad m^x = \left(\Omega - \frac{m}{2}\right).
\end{aligned} \tag{6.18}$$

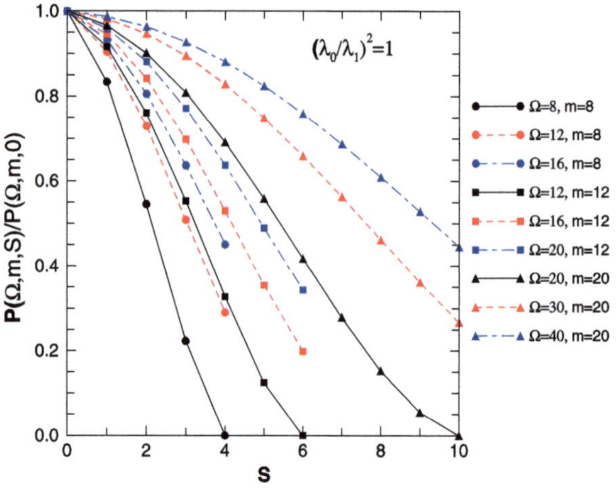

Fig. 6.4 Variance propagator $P(\Omega, m, S)$ vs S for different values of Ω and m. Equation (6.19) gives the formula for $P(\Omega, m, S)$. Figure is taken from [14] with permission from American Physical Society

Note that the $\nu = 1$ terms (they correspond to the Q^1's) are $1/\Omega^2$ times smaller as compared to the $\nu = 2$ terms (they correspond to the Q^2's). Therefore in the dilute limit defined by $\Omega \to \infty$, $m \to \infty$, $m/\Omega \to 0$ and $m >> S$, the $V^{s=0,1:\nu=2}$ parts determine the variances $\sigma_H^2(m, S)$ as given in [24]. Let us add that for $\lambda_0 = \lambda_1 = \lambda$, the $\overline{\sigma^2_{V(2)}(m, S)}$ takes the following simpler form,

$$\overline{\sigma^2_{V(2)}(m, S)} \xrightarrow{\lambda_0=\lambda_1=\lambda} \lambda^2 P(\Omega, m, S);$$

$$\begin{aligned} &P(\Omega, m, S) \\ &\quad = \frac{1}{\Omega(\Omega+1)/2}\left[\frac{\Omega+2}{\Omega+1} Q^1(\{2\} : m, S) + \frac{\Omega^2+3\Omega+2}{\Omega^2+3\Omega} Q^2(\{2\} : m, S)\right] \\ &\qquad + \frac{1}{\Omega(\Omega-1)/2}\left[\frac{\Omega+2}{\Omega+1} Q^1(\{1^2\} : m, S) + \frac{\Omega^2+\Omega+2}{\Omega^2+\Omega} Q^2(\{1^2\} : m, S)\right]. \end{aligned} \tag{6.19}$$

A plot of $P(\Omega, m, S)/P(\Omega, m, 0)$ vs S is shown in Fig. 6.4. It is seen that $P(\Omega, m, S)$ decreases with spin and this plays an important role in understanding the behavior of the chaos markers generated by EGOE(1 + 2)-**s**.

6.4 Chaos Markers and Their Spin Dependence

6.4.1 Poisson to GOE Transition in Level Fluctuations: $\lambda_c(S)$ Marker

Fluctuations in the eigenvalues of a fixed-(m, S) spectrum derive from the two and higher point correlation functions. For example, the two-point function, in a fixed-

(m, S) space, is

$$S^{\rho:m,S}(E_i, E_f) = \overline{\rho^{m,S}(E_i)\rho^{m,S}(E_f)} - \overline{\rho^{m,S}(E_i)}\,\overline{\rho^{m,S}(E_f)}. \tag{6.20}$$

The Dyson-Mehta Δ_3 statistic is an exact two-point measure while variance $\sigma^2(0)$ of the nearest neighbor spacing distribution (NNSD) is essentially a two-point measure as discussed in Chap. 2. In all the discussion in this section and all other remaining sections in this and in the next chapter, results are discussed for $\lambda_0 = \lambda_1 = \lambda$ (some results for $\lambda_0 \neq \lambda_1$ are given in [14]). In this situation, the EGOE(1 + 2)-**s** Hamiltonian is

$$H_\lambda = h(1) + \lambda\big[V^{s=0}(2) + V^{s=1}(2)\big]. \tag{6.21}$$

The NNSD and $\overline{\Delta}_3$ statistics show Poisson character in general for very small values of λ due to the presence of many good quantum numbers defined by $h(1)$. As the value of λ increases, there is delocalization in the Fock space, i.e. the eigenstates spread over all the basis states leading to complete mixing of the basis states giving GOE behavior for large λ values. For a 20 member EGOE(1 + 2)-**s** ensemble with $\Omega = m = 8$ and spins $S = 0$ and 2, NNSD for various λ values changing from 0.01 to 0.2 are constructed [14] and the results are shown in Fig. 6.5. Reference [14] also gives results for the Δ_3 statistic. As we increase λ, NNSD changes rapidly from a form close to Poisson to a form close to that of GOE (Wigner distribution) as seen from Fig. 6.5. For a given λ, $\sigma^2(0)$ gives the transition parameter Λ introduced in Sect. 3.2 and then λ_c corresponds to $\Lambda = 0.3$ and equivalently one can use Eq. (5.18). In Fig. 6.5, the values of the Λ parameter are given for different λ values and it is seen that the transition point λ_c is 0.028 and 0.047 for $S = 0$ and 2 respectively. For a qualitative understanding of the variation of λ_c with spin S, it is plausible to employ the same arguments used, based on lowest order perturbation theory, for EGOE(1 + 2), i.e. AJS criterion discussed in Sect. 5.3.1. Then, Poisson to GOE transition occurs when λ is of the order of the spacing Δ_c between the m particle states that are directly coupled by the two-body interaction. Given the two particle spectrum span to be B_2 and the number of fixed-(m, S) states directly coupled by the two-body interaction to be $K(\Omega, m, S)$, we have $\Delta_c(\Omega, m, S) \propto B_2/K(\Omega, m, S)$ and therefore, $\lambda_c \propto B_2/K(\Omega, m, S)$. Using the $h(1)$ spectrum, it is easy to see that $B_2 \propto \Omega$. Assuming that the spectral variance generated by $V(2)$ spreads uniformly over the directly connected states, we have $\sigma^2_{V(2)}(m, S) \approx \lambda^2 K(\Omega, m, S)$. Then, Eq. (6.19) gives $K(\Omega, m, S) \approx P(\Omega, m, S)$. With this, we have

$$\lambda_c(S) \propto \frac{\Omega}{P(\Omega, m, S)}. \tag{6.22}$$

For $\Omega = m = 8$, Eq. (6.22) and the formula for $P(\Omega, m, S)$ gives $P(8, 8, S = 1)/P(8, 8, S = 0) = 0.834$ and $P(8, 8, S = 2)/P(8, 8, S = 0) = 0.55$. These and the result $\lambda_c(S = 0) = 0.028$ from Fig. 6.5 will give $\lambda_c(S = 1) = 0.034$ and $\lambda_c(S = 2) = 0.05$. This prediction is close to the numerical results as shown for $S = 2$ in Fig. 6.5. Therefore Eq. (6.22) gives a good qualitative understanding of

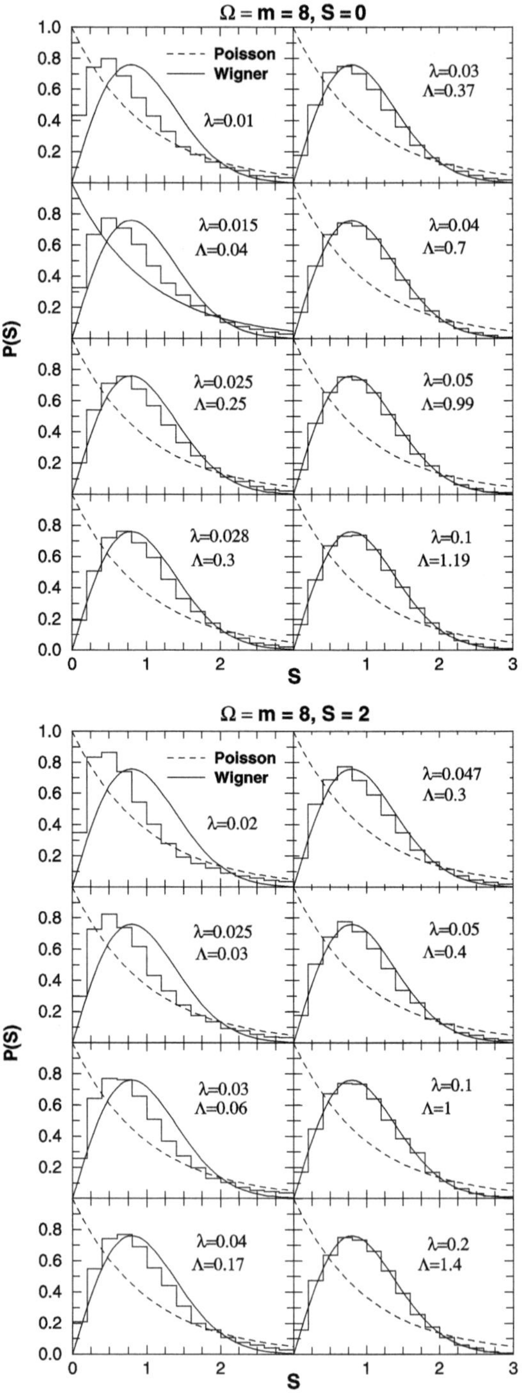

Fig. 6.5 NNSD for a 20 member EGOE(1 + 2)-**s** ensemble with $\Omega = m = 8$ and spins $S = 0$ and 2. Calculated NNSD are compared to the Poisson and Wigner (GOE) forms. Values of the interaction strength λ and the transition parameter Λ are given in the figure. The chaos marker λ_c corresponds to $\Lambda = 0.3$. Bin-size for the histograms is 0.2. Figures for $S = 0$ and 2 are taken from [14] with permission from American Physical Society

the $\lambda_c(S)$ variation with S. In the dilute limit (equivalent to asymptotic limit), as defined just after Eq. (6.18), it is easily seen that $P(\Omega, m, S) \to m^2\Omega^2$ and hence $\lambda_c \to 1/m^2\Omega$. Thus we recover Eq. (5.17) for spinless fermion systems as a limiting case.

6.4.2 Breit-Wigner to Gaussian Transition in Strength Functions: $\lambda_F(S)$ Marker

Given the mean field $h(1)$ basis states (denoted by $|k\rangle$) expanded in the H eigenvalue (E) basis,

$$|k, S, M_S\rangle = \sum_E C_{k,S}^{E,S} |E, S, M_S\rangle, \tag{6.23}$$

the fixed-S strength functions $F_{k,S}(E, S)$, extending Eq. (2.81), are defined by

$$F_{k,S}(E, S) = \sum_{E'} \left|C_{k,S}^{E',S}\right|^2 \delta\left(E - E'\right) = \left|\mathscr{C}_{k,S}^{E,S}\right|^2 d(m, S)\rho^{m,S}(E). \tag{6.24}$$

Here $|\mathscr{C}_{k,S}^{E,S}|^2$ denotes the average of $|C_{k,S}^{E,S}|^2$ over the eigenstates with the same energy E. In Eq. (6.23), $M_S = 0$ for even m and $M_S = 1/2$ for odd m. From now on we will drop M_S. Trivially, for $\lambda = 0$, the strength functions will be δ-functions at the $h(1)$ eigenvalues. As λ increases from zero, the strength functions first change from δ-function form to BW form at $\lambda = \lambda_\delta$ where λ_δ is very small; see Eq. (6.25) ahead. With further increase of λ, just as in EGOE(1 + 2), the BW form changes to Gaussian form. Figure 6.6 shows strength functions as a function of λ for a $\Omega = m = 8$ system with spins $S = 0$, 1 and 2. In the calculations, E and the basis state energies E_k are zero centered for each member and scaled by the width of the eigenvalue spectrum. The new energies are called $\widehat{E}$ and $\widehat{E}_k$ respectively. For each member $|C_{k,S}^{E,S}|^2$ are summed over the basis states in the energy window $\widehat{E}_k \pm \Delta_k$ and then the ensemble averaged $F_k(\widehat{E}, S)$ vs $\widehat{E}$ are constructed as histograms. For the results in the figure, $\Delta_k = 0.025$ for $\lambda < 0.1$ and beyond this $\Delta_k = 0.1$. For each λ value, the strength functions are fitted to the t-distribution given by Eq. (5.27) and deduced the value of the shape parameter α; note that $\beta = \sigma_{F_k}^2(2\alpha - 3)/\alpha$ for $\alpha > 1.5$ and the spreading width determines the parameter β for $\alpha \le 1.5$. As seen from Fig. 6.6, the fits are excellent over a wide range of λ values. The parameter α rises slowly up to λ_F, then it increases sharply (for $\alpha > 16$ the curves are indistinguishable from Gaussian). As pointed out in Sect. 5.4, the criterion $\alpha \sim 4$ defines the transition point λ_F. From the results in Fig. 6.6 it is seen that the transition point λ_F is 0.15, 0.16 and 0.19 for $S = 0$, 1 and 2 respectively.

For a qualitative understanding of the variation of λ_F with spin S, we will follow the same procedure used in Sect. 5.3.2 and for this, the spreading width $\Gamma(S)$ and the inverse participation ratio (IPR) $\xi_2(S)$ need to be estimated. Firstly,

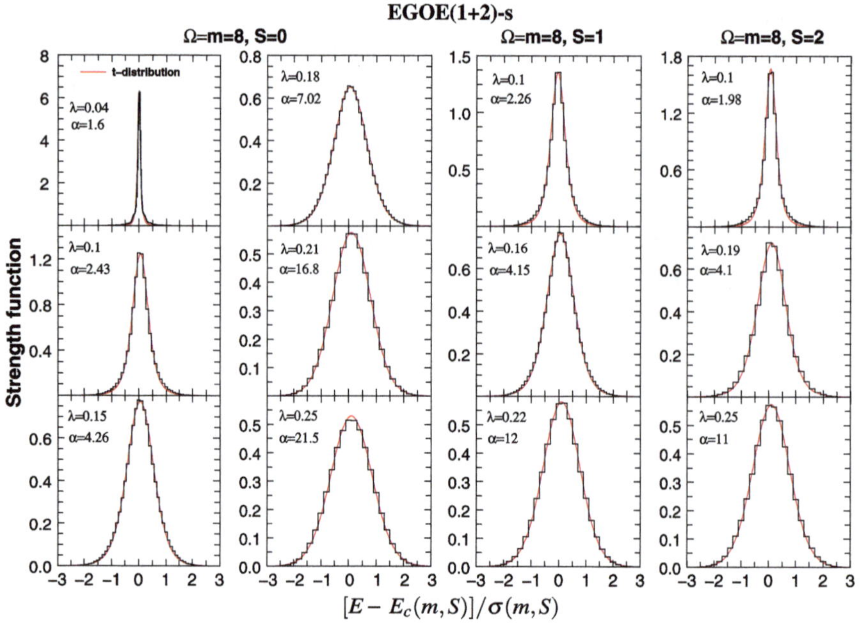

Fig. 6.6 Strength functions $F_k(\widehat{E}, S)$, for $\widehat{E}_k = 0$, as a function of λ for a 20 member EGOE(1 + 2)-s ensemble. Calculations (histograms) are for a $\Omega = m = 8$ system with spins $S = 0$, 1 and 2. Note that the widths $\sigma_{F_k}(m, S)$ of the strength functions are different from the spectral widths $\sigma(m, S)$. *Continuous curves* in the figures correspond to the t-distribution given by Eq. (5.27). In the plots $\int F_k(\widehat{E}, S) d\widehat{E} = 1$. See text for further details. Figure is taken from [14] with permission from American Physical Society

Fermi golden rule gives $\Gamma(S) = 2\pi\lambda^2/\overline{D(S)}$ with $\overline{D(S)} = \Delta_c(\Omega, m, S)$ as established in [27]. Therefore, using Eq. (6.22) gives $\Gamma(S) \propto 2\pi\lambda^2 P(\Omega, m, S)/\Omega$. Similarly, $\xi_2(S) \sim \Gamma(S)/\Delta_m(S)$ with $\Delta_m(S)$ being the average spacing of the m particle fixed-S spectrum. The total spectrum span considering only $h(1)$ is $B_m \propto m\Omega$ and therefore $\Delta_m(S) \propto m\Omega/d_f(\Omega, m, S)$. In the BW domain, $\Gamma(S)$ and $\xi_2(S)$ should be such that (i) $\Gamma(S) < f_0 B_m$ and (ii) $\xi_2(S) \gg 1$ where $f_0 < 1$. Condition (i) gives, $\lambda^2 < C_0 m\Omega^2/P(\Omega, m, S)$ and condition (ii) gives, $\lambda^2 \gg B_0 m\Omega^2/P(\Omega, m, S)\, d_f(\Omega, m, S)$. Note that the constants C_0 and B_0 are positive. Therefore,

$$\begin{gathered}\sqrt{\frac{B_0 m\Omega^2}{P(\Omega, m, S) d_f(\Omega, m, S)}} \ll \lambda < \sqrt{\frac{C_0 m\Omega^2}{P(\Omega, m, S)}} \\ \Rightarrow \quad \lambda_F(S) \propto \sqrt{\frac{m\Omega^2}{P(\Omega, m, S)}}.\end{gathered} \tag{6.25}$$

This equation shows that just as λ_c, the marker λ_F is essentially determined by the variance propagator $P(\Omega, m, S)$. Also as λ increases from zero, the BW form sets in fast as $d_f(\Omega, m, S)$ is usually very large. From the results in Fig. 6.6, it is clear that

λ_F should increase with S. Equation (6.25) along with the result $\lambda_F(S=0)=0.15$ gives $\lambda_F(S=1)=0.16$ and $\lambda_F(S=2)=0.2$. All these are in close agreement with the numerical results. In the dilute limit with $P(\Omega, m, S) \to m^2\Omega^2$, we have $\lambda_F \to 1/\sqrt{m}$ and thus reducing to Eq. (5.19) for spinless fermion systems.

6.4.3 Thermodynamic Region: $\lambda_t(S)$ Marker

Following the EGOE(1 + 2) analysis, let us compare, for different λ values, the thermodynamic entropy

$$S^{ther}(E) = \ln \rho^{m,S}(E), \tag{6.26}$$

the information entropy,

$$S^{info}(E,S) = -\frac{1}{d(m,S)\rho^{m,S}(E)} \sum_{E'} \sum_{k} \left|C_{k,S}^{E',S}\right|^2 \ln\left|C_{k,S}^{E',S}\right|^2 \delta(E-E'), \tag{6.27}$$

and the sp entropy

$$S^{sp}(E,S) = -\sum_i 2\{f_i(E,S)\ln f_i(E,S) + [1-f_i(E,S)]\ln[1-f_i(E,S)]\}, \tag{6.28}$$

where the fractional occupation probabilities $f_i(E,S) = \frac{1}{2}\langle n_i\rangle^{m,S,E}$. As already discussed in Sect. 6.2, Eq. (6.6) describes S^{info} and similarly, Eq. (5.33) for S^{sp} extends to fixed-(m,S) spaces by replacing ζ by $\zeta(m,S)$. For $\Omega = m = 8$ and $S = 0$ system with 20 members, results for $\lambda = \lambda_t(S) = 0.21$, $\lambda = 0.01 \ll \lambda_t(S)$ and $\lambda = 2 \gg \lambda_t(S)$ are shown in Fig. 6.7. Note that $\exp[S^{ther}(E,S) - S^{ther}_{max}] \longrightarrow \exp -\frac{1}{2}\widehat{E}^2$ for all λ values as the eigenvalue density is a Gaussian essentially independent of λ. For the examples in Fig. 6.7, $\zeta^2 = 0.998$, 0.5 and 0.039 for $\lambda = 0.01$, 0.21 and 2 respectively. It is clearly seen from Fig. 6.7 that the three entropies differ as we go away from $\lambda = \lambda_t$ and at $\lambda = \lambda_t$ they all look similar. Therefore, $\lambda = \lambda_t$ region can be interpreted as the thermodynamic region in the sense that all different definitions of entropy coincide in this region. Then as in EGOE(1 + 2), in the $\lambda \sim \lambda_t$ region all quantities are expected to be basis independent. This, if we consider the h and V basis, reduces to the criterion that the spreadings produced by $h(1)$ and $V(2)$ should be equal at $\lambda = \lambda_t$, i.e. $\sigma^2_{h(1)}(m,S) = \sigma^2_{V(2)}(m,S)$. To determine $\sigma^2_{h(1)}(m,S)$, we consider a uniform spectrum with $\Delta = 1$. Then, $\sigma^2_{h(1)}(1,\frac{1}{2}) = (\Omega^2-1)/12$ and using this in Eq. (6.14) we have,

$$\sigma^2_{h(1)}(m,S) = \mathscr{H}(\Omega,m,S) = \frac{1}{12}\left[m(\Omega+2)(\Omega-m/2) - 2\Omega S(S+1)\right]. \tag{6.29}$$

Combining this with Eq. (6.19) will give finally

$$\lambda_t(S) \propto \sqrt{\frac{\mathscr{H}(\Omega,m,S)}{P(\Omega,m,S)}}. \tag{6.30}$$

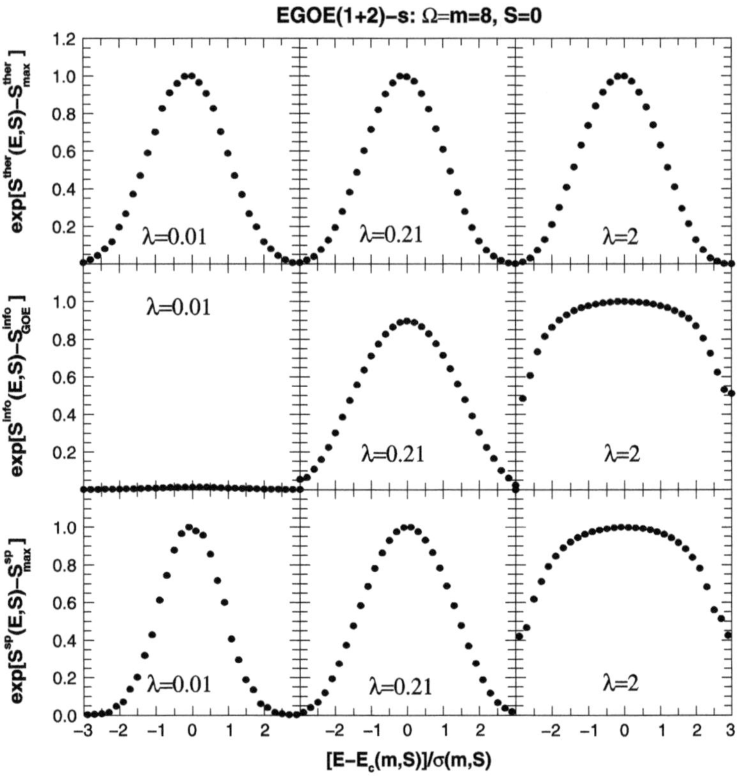

Fig. 6.7 Thermodynamic entropy $\exp[S^{ther}(E,S) - S^{ther}_{max}]$, information entropy $\exp[S^{info}(E,S) - S^{info}_{\mathrm{GOE}}]$ and single-particle entropy $\exp[S^{sp}(E,S) - S^{sp}_{max}]$ vs $\widehat{E} = [E - E_c(m,S)]/\sigma(m,S)$ for a 20 member EGOE(1 + 2)-s ensemble with $\Omega = m = 8$ and $S = 0$ for different λ values. Entropies averaged over bin-size 0.2 are shown as *filled circles*. Note that for $\lambda = 0.01$, $\exp[S^{info}(E,S) - S^{info}_{\mathrm{GOE}}]$ is close to zero for all $\widehat{E}$ values. Figure is taken from [14] with permission from American Physical Society

For the numerical results shown in Fig. 6.7, $\lambda_t(S=0) = 0.21$. Then, Eq. (6.30) gives $\lambda_t(S=1) = 0.22$ and $\lambda_t(S=2) = 0.24$. In the dilute limit, simplifying the $\mathscr{H}$ and P factors, we have $\lambda_t \to 1/\sqrt{m}$ and this is same as Eq. (5.35) for spinless fermion systems. This also shows that in the dilute limit λ_t and λ_F have same scale. However these scales differ parametrically as m approaches Ω (for $m > \Omega$ one has to consider holes) and $S \gtrsim m/4$. In this situation $\lambda_t(S)/\lambda_F(S) \propto \sqrt{\frac{[m(\Omega+2)(\Omega-m/2)-2\Omega S(S+1)]}{m\Omega^2}}$. Thus the variance propagator determines the behavior of the three transition markers $\lambda_c(S)$, $\lambda_F(S)$ and $\lambda_t(S)$.

6.5 Pairing and Exchange Interactions in EGOE(1 + 2)-s Space

6.5.1 *Pairing Hamiltonian and Pairing Symmetry*

With spin degree of freedom for the fermions, it is possible to introduce pairing in the (m, S) spaces of EGOE(1 + 2)-**s**. General discussion of pairing algebras for fermion systems for example is given in [28–35]. Results in this section follow from [33]. Let us start with the single particle states $a^{\dagger}_{i,\frac{1}{2},m_{\mathbf{s}}}|0\rangle = |i, m_{\mathbf{s}} = \pm\frac{1}{2}\rangle$ with $i = 1, 2, \ldots, \Omega$ and define coupled (in spin space) one-body operators $u^r_{\mu}(i, j)$,

$$u^r_{\mu}(i, j) = \left(a^{\dagger}_i \tilde{a}_j\right)^r_{\mu}; \quad r = 0, 1. \tag{6.31}$$

These $4\Omega^2$ number of operators generate the $U(2\Omega)$ algebra. They satisfy the following commutation relations,

$$\begin{aligned}\left[u^r_{\mu}(i, j), u^{r'}_{\mu'}(k, l)\right]_- &= \sum_{r''}(-1)^{r+r'}\left\langle r\ \mu\ r'\ \mu' \,|\, r''\ \mu''\right\rangle\sqrt{(2r+1)(2r'+1)} \\ &\quad \times \begin{Bmatrix} r & r' & r'' \\ 1/2 & 1/2 & 1/2 \end{Bmatrix} \\ &\quad \times \left[u^{r''}_{\mu''}(k, j)\delta_{il} - (-1)^{r+r'+r''} u^{r''}_{\mu''}(i, l)\delta_{jk}\right].\end{aligned} \tag{6.32}$$

Note that, from now on we do not include, for obvious reasons, the index μ in Eq. (6.31) for $r = 0$. For m fermions, all states belong to the $U(2\Omega)$ totally antisymmetric irrep $\{1^m\}$ and therefore uniquely represented by the particle number m. A simple subalgebra of $U(2\Omega)$ is generated by the space × spin decomposition and this corresponds $U(2\Omega) \supset U(\Omega) \otimes SU(2)$ with $SU(2)$ generating spin S and the 'space' part $U(\Omega)$ corresponds to the sp levels $i = 1, 2, \ldots, \Omega$. It is easily seen that the operators $u^0_{ij} = (a^{\dagger}_i \tilde{a}_j)^0$, which are Ω^2 in number, generate $U(\Omega)$ algebra. Similarly, the operators $C_{ij} = u^0_{ij} - u^0_{ji}$, $i > j$, which are $\Omega(\Omega - 1)/2$ in number, generate the $SO(\Omega)$ sub-algebra of $U(\Omega)$. The spin operator $\hat{S} = S^1_{\mu}$, the number operator $\hat{n}$ and the quadratic Casimir operators C_2's of $U(\Omega)$ and $SO(\Omega)$ are

$$\begin{aligned} S^1_{\mu} &= \frac{1}{\sqrt{2}}\sum_{i=1}^{\Omega} u^1_{ii;\mu}, \\ \hat{n} &= \sum_i n_i, \quad n_i = \sqrt{2}u^0_{ii}, \\ C_2\big(U(\Omega)\big) &= 2\sum_{i,j} u^0_{ij}u^0_{ji}, \\ C_2\big(SO(\Omega)\big) &= 2\sum_{i>j} C_{ij}C_{ji}. \end{aligned} \tag{6.33}$$

The structure of $C_2(U(\Omega))$ in terms of the number operator and the $\hat{S} \cdot \hat{S} = \hat{S}^2$ operator is,

$$\begin{aligned} C_2\big(U(\Omega)\big) &= \hat{n}\left(\Omega + 2 - \frac{\hat{n}}{2}\right) - 2\hat{S}^2, \\ \big\langle C_2\big(U(\Omega)\big)\big\rangle^{m,S} &= m\left(\Omega + 2 - \frac{m}{2}\right) - 2S(S+1). \end{aligned} \tag{6.34}$$

Note that $\langle C_2(U(\Omega))\rangle^{\{f\}} = \sum_i f_i(f_i + \Omega + 1 - 2i)$. As $U(2\Omega) \supset U(\Omega) \otimes SU(2)$ with the $SU(2)$ algebra generating total spin S, the $U(\Omega)$ irreps are labeled by two column irreps $\{2^p 1^q\}$ with $m = 2p + q$, $S = q/2$. As a consequence, the $SO(\Omega)$ irreps are also of two column type and we will denote them by $[2^{v_1} 1^{v_2}]$. Here, $v_S = 2v_1 + v_2$ is called seniority and $\tilde{s} = v_2/2$ is called reduced spin. We also have

$$\begin{aligned} \big\langle C_2\big(SO(\Omega)\big)\big\rangle^{\langle\omega\rangle} &= \sum_i \omega_i(\omega_i + \Omega - 2i) \\ \Rightarrow \quad \big\langle C_2\big(SO(\Omega)\big)\big\rangle^{\langle 2^{v_1} 1^{v_2}\rangle} &= v_S\left(\Omega + 1 - \frac{v_S}{2}\right) - 2\tilde{s}(\tilde{s}+1). \end{aligned} \tag{6.35}$$

Significance of $SO(\Omega)$ follows by defining the pairing Hamiltonian H_p where,

$$\begin{aligned} &H_p = P^2 = PP^\dagger, \quad P = \frac{1}{\sqrt{2}} \sum_i \big(a_i^\dagger a_i^\dagger\big)^0 = \sum_i P_i, \\ &\big\langle (k\ell)s \big| H_p \big| (ij)s \big\rangle = \delta_{s,0}\delta_{i,j}\delta_{k,\ell}. \end{aligned} \tag{6.36}$$

Note that P is the generalized spin $S = 0$ pair creation operator and P_i is the pair creation operator for each orbit i. After some commutator algebra it can be shown that [36],

$$\begin{aligned} 2H_p &= -C_2\big(SO(\Omega)\big) + \hat{n}\left(\Omega + 1 - \frac{\hat{n}}{2}\right) - 2\hat{S}^2, \\ \langle H_p\rangle^{(m,S,v_S,\tilde{s})} &= \frac{1}{4}(m - v_S)(2\Omega + 2 - m - v_S) + \big[\tilde{s}(\tilde{s}+1) - S(S+1)\big]. \end{aligned} \tag{6.37}$$

To proceed further, classification of $U(2\Omega) \supset [U(\Omega) \supset SO(\Omega)] \otimes SU(2)$ states defined by $(m, S, v_S, \tilde{s})$ quantum numbers is needed. This problem, i.e. $(m, S) \rightarrow (v_S, \tilde{s})$ reductions, is solved in principle by group theory. Using the tabulations in [37], results are given in Tables 6.1 and 6.2 for: (i) $m \leq 4, \Omega \geq 4$; (ii) $m = 5 - 8, \Omega = 6, 8$.

A much simpler approach to derive the eigenvalue formula for the pairing operator, the pairing quantum numbers and the irrep reductions is to use [38] the group-subgroup chain $U(2\Omega) \supset Sp(2\Omega) \supset SO(\Omega) \otimes SU_S(2)$. This shows that pairing is defined by a complementary $SU(2)$ algebra. Firstly, the $2\Omega(\Omega-1)$ number of operators $V^r_\mu(i,j)$ along with 3Ω number of operators $u^1_\mu(i,i)$ form the $Sp(2\Omega)$ algebra

Table 6.1 $(m, S) \to (v_S, \tilde{s})$ reductions for $m \leq 4$ and $\Omega \geq 4$

(m, S)	$(v_S, \tilde{s})$
$(0, 0)$	$(0, 0)$
$(1, \frac{1}{2})$	$(1, \frac{1}{2})$
$(2, 0)$	$(2, 0)$, $(0, 0)$
$(2, 1)$	$(2, 1)$
$(3, \frac{1}{2})$	$(3, \frac{1}{2})$, $(1, \frac{1}{2})$
$(3, \frac{3}{2})$	$\{(1, \frac{1}{2})_{\Omega=4};\ (2, 1)_{\Omega=5};\ (3, \frac{3}{2})_{\Omega\geq 6}\}$
$(4, 0)$	$(4, 0)$, $(2, 0)$, $(0, 0)$
$(4, 1)$	$\{(2, 0)_{\Omega=4};\ (3, \frac{1}{2})_{\Omega=5};\ (4, 1)_{\Omega\geq 6}\}$, $(2, 1)$
$(4, 2)$	$\{(0, 0)_{\Omega=4};\ (1, \frac{1}{2})_{\Omega=5};\ (2, 1)_{\Omega=6},\ (3, \frac{3}{2})_{\Omega=7};\ (4, 2)_{\Omega\geq 8}\}$

Table 6.2 $(m, S) \to (v_S, \tilde{s})$ irrep reductions for $(\Omega = 6; m = 6)$ and $(\Omega = 8; m = 5 - 8)$. Note that the dimensions of the irreps are given as subscripts

Ω	$(m, S)_{D(m,S)}$	$(v_S, \tilde{s})_{D(v_S,\tilde{s})}$
6	$(6, 0)_{175}$	$(6, 0)_{70}$, $(4, 0)_{84}$, $(2, 0)_{20}$, $(0, 0)_1$
	$(6, 1)_{189}$	$(4, 1)_{90}$, $(4, 0)_{84}$, $(2, 1)_{15}$
	$(6, 2)_{35}$	$(2, 1)_{15}$, $(2, 0)_{20}$
	$(6, 3)_1$	$(0, 0)_1$
8	$(5, \frac{1}{2})_{1008}$	$(5, \frac{1}{2})_{840}$, $(3, \frac{1}{2})_{160}$, $(1, \frac{1}{2})_8$
	$(5, \frac{3}{2})_{504}$	$(5, \frac{3}{2})_{448}$, $(3, \frac{3}{2})_{56}$
	$(5, \frac{5}{2})_{56}$	$(3, \frac{3}{2})_{56}$
	$(6, 0)_{1176}$	$(6, 0)_{840}$, $(4, 0)_{300}$, $(2, 0)_{35}$, $(0, 0)_1$
	$(6, 1)_{1512}$	$(6, 1)_{1134}$, $(4, 1)_{350}$, $(2, 1)_{28}$
	$(6, 2)_{420}$	$(4, 1)_{350}$, $(4, 2)_{70}$
	$(6, 3)_{28}$	$(2, 1)_{28}$
	$(7, \frac{1}{2})_{2352}$	$(7, \frac{1}{2})_{1344}$, $(5, \frac{1}{2})_{840}$, $(3, \frac{1}{2})_{160}$, $(1, \frac{1}{2})_8$
	$(7, \frac{3}{2})_{1344}$	$(5, \frac{1}{2})_{840}$, $(5, \frac{3}{2})_{448}$, $(3, \frac{3}{2})_{56}$
	$(7, \frac{5}{2})_{216}$	$(3, \frac{1}{2})_{160}$, $(3, \frac{3}{2})_{56}$
	$(7, \frac{7}{2})_8$	$(1, \frac{1}{2})_8$
	$(8, 0)_{1764}$	$(8, 0)_{588}$, $(6, 0)_{840}$, $(4, 0)_{300}$, $(2, 0)_{35}$, $(0, 0)_1$
	$(8, 1)_{2352}$	$(6, 0)_{840}$, $(6, 1)_{1134}$, $(4, 1)_{350}$, $(2, 1)_{28}$
	$(8, 2)_{720}$	$(4, 0)_{300}$, $(4, 1)_{350}$, $(4, 2)_{70}$
	$(8, 3)_{63}$	$(2, 0)_{35}$, $(2, 1)_{28}$
	$(8, 4)_1$	$(0, 0)_1$

[total number of generators $= 2\Omega(\Omega - 1) + 3\Omega = \Omega(2\Omega + 1)$] where $V^r_\mu(i, j)$ are,

$$V^r_\mu(i, j) = \sqrt{(-1)^{r+1}}\left[u^r_\mu(i, j) - (-1)^r u^r_\mu(j, i)\right]; \quad i > j,\ r = 0, 1. \tag{6.38}$$

The quadratic Casimir operators of the $U(2\Omega)$ and $Sp(2\Omega)$ algebras are [33],

$$
\begin{aligned}
C_2\big[U(2\Omega)\big] &= \sum_{i,j,r} u^r(i,j)\cdot u^r(j,i), \\
C_2\big[Sp(2\Omega)\big] &= 2\sum_i u^1(i,i)\cdot u^1(i,i) + \sum_{i>j,r} V^r(i,j)\cdot V^r(i,j).
\end{aligned}
\tag{6.39}
$$

Simplifying these, using angular momentum algebra and the anti-commutation relations for fermion creation and annihilation operators, will give

$$
\begin{aligned}
C_2\big[U(2\Omega)\big] &= 2\hat{n}\Omega - 2\sum_i P_i P_i^\dagger \\
&\quad - \sum_{i\neq j,s} \sqrt{2s+1}\big[s(s+1)-1\big]\big[\big(a_i^\dagger a_j^\dagger\big)^s(\tilde{a}_j\tilde{a}_i)^s\big]^0, \\
C_2\big[Sp(2\Omega)\big] &= (2\Omega+1)\hat{n} - 6\sum_i P_i P_i^\dagger - 4\sum_{i>j}\big(P_i P_j^\dagger + P_j P_i^\dagger\big) \\
&\quad - \sum_{i\neq j,s} \sqrt{2s+1}\big[s(s+1)-1\big]\big[\big(a_i^\dagger a_j^\dagger\big)^s(\tilde{a}_j\tilde{a}_i)^s\big]^0.
\end{aligned}
\tag{6.40}
$$

Therefore,

$$
C_2\big[U(2\Omega)\big] - C_2\big[Sp(2\Omega)\big] = 4PP^\dagger - \hat{n}. \tag{6.41}
$$

Here, a crucial point is that the operators P, $P^\dagger$ and P_0 form a $SU(2)$ algebra,

$$
\big[P, P^\dagger\big] = \hat{n} - \Omega = 2P_0, \qquad [P_0, P] = P, \qquad \big[P_0, P^\dagger\big] = -P^\dagger \tag{6.42}
$$

with $P_0 = (\hat{n} - \Omega)/2$. The spin that corresponds to this $SU(2)$ is called quasi-spin Q. Then the Q_z operator is nothing but P_0 and its eigenvalues are $M_Q = (m - \Omega)/2$. This then gives $Q = (\Omega - v)/2$ and, for $m \le \Omega$, v take values $v = m, m-2, \ldots, 0$ or 1. The situation here is same as identical particle pairing discussed extensively in nuclear structure [28]. The quasi-spin $SU(2)$ algebra easily gives the eigenvalues of the pairing Hamiltonian $H_P = P\,P^\dagger$,

$$
E_p(m,v,S) = \langle H_P\rangle^{m,v,S} = \big\langle PP^\dagger\big\rangle^{m,v,S} = \frac{1}{4}(m-v)(2\Omega+2-m-v). \tag{6.43}
$$

Also, as all the m nucleon states behave as basis states of the totally anti-symmetric irrep $\{1^m\}$ with respect to $U(2\Omega)$ algebra, we have

$$
\big\langle C_2\big[U(2\Omega)\big]\big\rangle^{\{1^m\}} = m(2\Omega+1-m). \tag{6.44}
$$

Combining this with Eq. (6.41) will give,

$$
C_2\big[Sp(2\Omega)\big] = 2v\left(\Omega + 1 - \frac{v}{2}\right). \tag{6.45}
$$

Table 6.3 Classification of states in the $U(2\Omega) \supset Sp(2\Omega) \supset SO(\Omega) \otimes SU_S(2)$ limit for ($\Omega = 6$, $m = 6$). Given are (m, v, S) labels, the corresponding dimensions $D(m, v, S)$ and the pairing Hamiltonian H_P eigenvalues

m	S	v	$D(\Omega, m, v, S)$	$\langle H_P \rangle^{m,v,S}$
6	0	6	70	0
		4	84	2
		2	20	6
		0	1	12
	1	6	84	0
		4	90	2
		2	15	6
	2	6	20	0
		4	15	2
	3	6	1	0

Therefore, seniority quantum number v corresponds to totally anti-symmetric irrep $\langle 1^v \rangle$ of $Sp(2\Omega)$. Thus $Sp(2\Omega)$ corresponds to $SU(2)$ algebra generated by (P, $P^\dagger$, P_0). Also, v uniquely fixes $SO(\Omega)$ irrep for fixed (m, S). From $SU(2)$ quasi-spin algebra it is easy to write the structure of H_p eigenfunctions,

$$|m, v, S, \alpha\rangle = \sqrt{\frac{(\Omega - v - p)!}{(\Omega - v)!p!}} P^{(m-v)/2} |m = v, v, S, \alpha\rangle; \quad p = \frac{m - v}{2}. \tag{6.46}$$

The spin S is generated by the v free particles and therefore $v \geq 2S$. Then, given m and S we have (for $m \leq \Omega$),

$$v = m, m - 2, \ldots, 2S. \tag{6.47}$$

Similarly, the dimensions of the (m, v, S) irreps are given by

$$D(\Omega, m, v, S) = d_f(\Omega, m = v, S) - d_f(\Omega, m = v - 2, S) \tag{6.48}$$

and Eq. (6.2) gives the formula for $d_f(\Omega, m, S)$. Note that $\sum_{v,S}(2S + 1) \times D(\Omega, m, v, S) = \binom{2\Omega}{m}$ and $\sum_v D(\Omega, m, v, S) = d_f(\Omega, m, S)$.

It is also seen that both S^1_μ and C_{ij} are in the $Sp(2\Omega)$ algebra and therefore $Sp(2\Omega) \supset SO(\Omega) \otimes SU(2)$. Comparing the results in Tables 6.1 and 6.2 with those in Tables 6.3 and 6.4 also the irrep reductions, it is clearly seen that the $Sp(2\Omega)$ irreps uniquely define the $SO(\Omega)$ irrep for a given (m, S). Therefore, as it is much simpler, one can use $U(2\Omega) \supset Sp(2\Omega) \supset SO(\Omega) \otimes SU_S(2)$ symmetry scheme for pairing in EGOE(1 + 2)-s.

Table 6.4 Classification of states in the $U(2\Omega) \supset Sp(2\Omega) \supset SO(\Omega) \otimes SU_S(2)$ limit for ($\Omega = 8$, $m = 5, 6, 7, 8$). Given are (m, v, S) labels, the corresponding dimensions $D(\Omega, m, v, S)$ and the pairing Hamiltonian H_P eigenvalues

m	S	v	$D(\Omega, m, v, S)$	$\langle H_P \rangle^{m,v,S}$
5	$\frac{1}{2}$	5	840	0
		3	160	5
		1	8	12
	$\frac{3}{2}$	5	448	0
		3	56	5
	$\frac{5}{2}$	5	56	0
6	0	6	840	0
		4	300	4
		2	35	10
		0	1	18
	1	6	1134	0
		4	350	4
		2	28	10
	2	6	350	0
		4	70	4
	3	6	28	0
7	$\frac{1}{2}$	7	1344	0
		5	840	3
		3	160	8
		1	8	15
	$\frac{3}{2}$	7	840	0
		5	448	3
		3	56	8
	$\frac{5}{2}$	7	160	0
		5	56	3
	$\frac{7}{2}$	7	8	0
8	0	8	588	0
		6	840	2
		4	300	6
		2	35	12
		0	1	20
	1	8	840	0
		6	1134	2
		4	350	6
		2	28	12
	2	8	300	0
		6	350	2
		4	70	6
	3	8	35	0
		6	28	2
	4	8	1	0

6.5.2 Fixed-(m, v, S) Partial Densities

Expansion of the eigenstates in the $|m, v, S, \alpha\rangle$ basis with the expansion coefficients being $C_E^{m,v,S,\alpha}$ will allow us to define fixed-(m, v, S) partial densities $\rho^{m,v,S}(E)$,

$$\begin{aligned}\rho^{m,v,S}(E) &= \big\langle\delta(H-E)\big\rangle^{m,v,S} = \frac{1}{D(\Omega, m, v, S)}\sum_\alpha \big|C_E^{m,v,S,\alpha}\big|^2 \\ \Rightarrow \quad I^{m,v,S}(E) &= D(\Omega, m, v, S)\rho^{m,v,S}(E) = \sum_\alpha \big|C_E^{m,v,S,\alpha}\big|^2.\end{aligned} \tag{6.49}$$

It is important to note that fixed-S density of states $\rho^{m,S}(E)$ decomposes into a sum of fixed-(m, v, S) partial densities,

$$\begin{aligned}\rho^{m,S}(E) &= \sum_v \frac{D(\Omega, m, v, S)}{d_f(\Omega, m, S)}\rho^{m,v,S}(E) \\ \Rightarrow \quad I^{m,S}(E) &= \sum_v I^{m,v,S}(E).\end{aligned} \tag{6.50}$$

The partial densities $\rho^{m,v,S}(E)$ are defined over broken symmetry subspaces and also they are sums of strength functions (strength functions are defined for each basis state). For EGOE(1 + 2) and EGOE(1 + 2)-**s** we have already shown that the strength functions are Gaussian for $\lambda > \lambda_F$ and by an extension this, it is to be expected that the partial densities $\rho^{m,v,S}(E)$ will take Gaussian form in the Gaussian domain with $\lambda > \lambda_F$. On the other hand, Eqs. (6.47) and (6.48) clearly show that the state density generated by the pairing Hamiltonian $H = -H_p$ will be a highly skewed distribution and therefore, one may expect that the fixed-(m, v, S) partial densities may be highly skewed. Therefore it is important to establish the shape of $\rho^{m,v,S}(E)$. Numerical examples shown in Fig. 6.8 and in Ref. [38] confirm that the partial densities $\rho^{m,v,S}(E)$ indeed take Gaussian form for $\lambda \gg \lambda_F$. Thus the fixed-(m, v, S) partial densities take Gaussian form in the Gaussian domain defined by $\lambda > \lambda_F$ for EGOE(1 + 2)-**s**. Extension of this result for EGOE(1 + 2)-J ensemble (see Chap. 13 for the definition of this ensemble) with subspaces defined by the pairing Hamiltonian, i.e. for fixed-(m, v, J) partial densities, is often used in statistical nuclear spectroscopy [39–41] without proof.

For constructing Gaussian partial (m, v, S) densities, we need fixed-(m, v, S) centroids $E_c(m, v, S) = \langle H\rangle^{m,v,S}$ and variances $\sigma^2(m, v, S) = \langle H^2\rangle^{m,v,S} - [E_c(m, v, S)]^2$. Simple (Casimir) propagation equations for these are possible. From Table 6.1 one can see that the number of (m, v, S) irreps Λ_i is 5 for m up to 2 and there are 5 simple scalar operators $\widehat{C}_i$ of maximum body rank 2, $\widehat{C}_i = 1, \hat{n}, \binom{\hat{n}}{2}, H_p$ and $\hat{S}^2$ for $i = 1$–5 respectively. Note that $\langle H_p\rangle^{m,v,S}$ and $\langle\hat{S}^2\rangle^{m,\mathrm{v},S}$ are $E_p(m, v, S)$ [see Eq. (6.43)] and $S(S + 1)$ respectively. More remarkable is that, for $m \le 4$, the

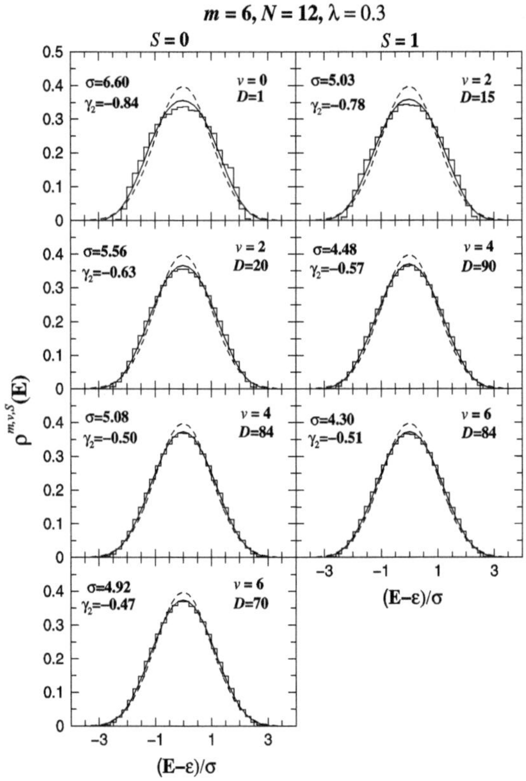

Fig. 6.8 Partial densities $\rho^{m,v,S}(E)$ vs E for a 20 member EGOE(1 + 2)-**s** ensemble for $\Omega = m = 6$ and $\lambda_0 = \lambda_1 = \lambda = 0.3$ in Eq. (6.1). The values of (v, S), dimension D, width σ and γ_2 for the densities are given in the figure. Note that $\gamma_1 \sim 0$ in all cases. The energies E are zero centered with respect to the centroid ε and scaled with the width σ of $\rho^{m,v,S}(E)$. The histograms (with 0.2 bin size) are exact results, *dashed curves* are Gaussians and the *continuous curves* are Edgeworth corrected Gaussians. Similar results for $\Omega = m = 8$ are reported in [38]

number of (m, v, S) irreps Υ_i is 14 as seen from Table 6.1 and also the available simple scalars $\widehat{\mathscr{C}_i}$ of maximum body rank 4 are exactly 14. These are $\widehat{\mathscr{C}_i} = 1, \hat{n}, \binom{\hat{n}}{2}, \binom{\hat{n}}{3}, \binom{\hat{n}}{4}, H_p, \hat{n}H_p, \binom{\hat{n}}{2}H_p, (H_p)^2, H_p\hat{S}^2, \hat{S}^2, \hat{n}\hat{S}^2, \binom{\hat{n}}{2}\hat{S}^2$ and $(\hat{S}^2)^2$ for $i = 1$–14 respectively. For any $\Gamma = (m, \mathrm{v}, S)$, propagation equation for the energy centroids is easy to write in terms of the row matrices $[C(\Gamma)]$ and $[\mathscr{E}]$ with 5 elements and the 5×5 matrix $[X]$, where

$$\begin{aligned}
&\langle H\rangle^{\Gamma} = \big[C(\Gamma)\big][X]^{-1}\widetilde{[\mathscr{E}]};\\
&\big[C(\Gamma)\big] \quad \Leftrightarrow \quad C_i(\Gamma) = \langle\widehat{C}_i\rangle^{\Gamma},\\
&[\mathscr{E}] \quad \Leftrightarrow \quad \mathscr{E}_i = \langle H\rangle^{\Lambda_i},\\
&[X] \quad \Leftrightarrow \quad X_{ij} = \langle\widehat{C}_j\rangle^{\Lambda_i}.
\end{aligned} \tag{6.51}$$

Similarly, the propagation equation for $\langle H^2\rangle^{\Gamma}$ in terms of the row matrices $[\mathscr{C}(\Gamma)]$ and $[\mathscr{S}]$ with 14 elements and the 14×14 matrix $[Y]$, is

$$
\begin{aligned}
& \langle H^2\rangle^{\Gamma} = [\mathscr{C}(\Gamma)][Y]^{-1}[\widetilde{\mathscr{S}}]; \\
& [\mathscr{C}(\Gamma)] \quad \Leftrightarrow \quad \mathscr{C}_i(\Gamma) = \langle\widehat{\mathscr{C}_i}\rangle^{\Gamma}, \\
& [\mathscr{S}] \quad \Leftrightarrow \quad \mathscr{S}_i = \langle H^2\rangle^{\Upsilon_i}, \\
& [Y] \quad \Leftrightarrow \quad Y_{ij} = \langle\widehat{\mathscr{C}_j}\rangle^{\Upsilon_i}.
\end{aligned}
\tag{6.52}
$$

Using EGOE(1 + 2)-**s** computer codes, it is easy to construct, even for large Ω values, the H matrices for $m \leq 4$ and using them, it is easy to obtain the input matrices $[\mathscr{E}]$ and $[\mathscr{S}]$ for centroids and variances propagation.

6.5.3 Exchange Interaction

Space exchange or the Majorana operator M that exchanges the spatial coordinates of the particles and leaves the spin unchanged is defined by

$$M|i,\alpha;\, j,\beta\rangle = |j,\alpha;\, i,\beta\rangle. \tag{6.53}$$

In Eq. (6.53), labels $(i,\ j)$ and (α,β) denote the spatial and spin labels respectively. As $|i,\alpha;\, j,\beta\rangle = (a^{\dagger}_{i,\alpha}a^{\dagger}_{j,\beta})|0\rangle$, we have

$$
\begin{aligned}
M &= \sum_{i,j,\alpha,\beta} \big(a^{\dagger}_{j,\alpha}a^{\dagger}_{i,\beta}\big)\big(a^{\dagger}_{i,\alpha}a^{\dagger}_{j,\beta}\big)^{\dagger} \\
&= \frac{1}{2}\big(C_2\big[U(\Omega)\big] - \Omega\hat{n}\big).
\end{aligned}
\tag{6.54}
$$

In Eq. (6.54), $C_2[U(\Omega)] = \sum_{i,j,\alpha,\beta} a^{\dagger}_{i,\alpha}a_{j,\alpha}a^{\dagger}_{j,\beta}a_{i,\beta}$ is the quadratic Casimir invariant of the $U(\Omega)$ and simple algebra gives

$$C_2\big[U(\Omega)\big] = \hat{n}(\Omega+2) - \frac{\hat{n}^2}{2} - 2\hat{S}^2. \tag{6.55}$$

Finally, combining Eqs. (6.54) and (6.55) will give,

$$M = -\hat{S}^2 - \hat{n}\left(\frac{\hat{n}}{4} - 1\right). \tag{6.56}$$

Therefore, the interaction generated by the $\hat{S}^2$ operator is the exchange interaction with a number dependent term. This number dependent term becomes important when the particle number m changes. Applications of EGOE(1 + 2)-**s** and its extensions including the pairing Hamiltonian and the Majorana operator will be discussed in the next chapter.

References

1. V.K.B. Kota, R.U. Haq, *Spectral Distributions in Nuclei and Statistical Spectroscopy* (World Scientific, Singapore, 2010)
2. J.M. Deutsch, Quantum statistical mechanics in a closed system. Phys. Rev. A **43**, 2046–2049 (1991)
3. M. Srednicki, Chaos and quantum thermalization. Phys. Rev. E **50**, 888–901 (1994)
4. M. Rigol, V. Dunjko, M. Olshanii, Thermalization and its mechanism for generic isolated quantum systems. Nature (London) **452**, 854–858 (2008)
5. L.F. Santos, M. Rigol, Onset of quantum chaos in one dimensional bosonic and fermionic systems and its relation to thermalization. Phys. Rev. E **81**, 036206 (2010)
6. L.F. Santos, M. Rigol, Localization and the effects of symmetries in the thermalization properties of one-dimensional quantum systems. Phys. Rev. E **82**, 031130 (2010)
7. V.K.B. Kota, A. Relaño, J. Retamosa, M. Vyas, Thermalization in the two-body random ensemble, J. Stat. Mech. P10028 (2011)
8. C. Mejía-Monasterio, G. Benenti, G.G. Carlo, G. Casati, Entanglement across a transition to quantum chaos. Phys. Rev. A **71**, 062324 (2005)
9. S. Montangero, L. Viola, Multipartite entanglement generation and fidelity decay in disordered qubit systems. Phys. Rev. A **73**, 040302(R) (2006)
10. I. Pižorn, T. Prosen, T.H. Seligman, Loschmidt echoes in two-body random matrix ensembles. Phys. Rev. B **76**, 035122 (2007)
11. W.G. Brown, L.F. Santos, D.J. Starling, L. Viola, Quantum chaos, delocalization and entanglement in disordered Heisenberg models. Phys. Rev. E **77**, 021106 (2008)
12. V.K.B. Kota, N.D. Chavda, R. Sahu, One plus two-body random matrix ensemble with spin: analysis using spectral variances. Phys. Lett. A **359**, 381–389 (2006)
13. H.E. Türeci, Y. Alhassid, Spin-orbit interaction in quantum dots in the presence of exchange correlations: an approach based on a good-spin basis of the universal Hamiltonian. Phys. Rev. B **74**, 165333 (2006)
14. M. Vyas, V.K.B. Kota, N.D. Chavda, Transitions in eigenvalue and wavefunction structure in $(1+2)$-body random matrix ensembles with spin. Phys. Rev. E **81**, 036212 (2010)
15. Ph. Jacquod, A.D. Stone, Ground state magnetization for interacting fermions in a disordered potential: kinetic energy, exchange interaction, and off-diagonal fluctuations. Phys. Rev. B **64**, 214416 (2001)
16. L. Kaplan, T. Papenbrock, C.W. Johnson, Spin structure of many-body systems with two-body random interactions. Phys. Rev. C **63**, 014307 (2000)
17. J. Planelles, F. Rajadell, J. Karwowski, Spectral density distribution moments of N-electron Hamiltonians in the low-density limit. J. Phys. A **30**, 2181–2196 (1997)
18. J. Karwowski, Statistical theory of spectra. Int. J. Quant. Chem. **51**, 425–437 (1994)
19. J. Karwowski, F. Rajadell, J. Planelles, V. Mas, The first four moments of spectral density distribution of an N-electron Hamiltonian matrix defined in an antisymmetric and spin-adapted model space. At. Data Nucl. Data Tables **61**, 177–232 (1995)
20. J. Planelles, F. Rajadell, J. Karwowski, V. Mas, A diagrammatic approach to statistical spectroscopy of many-fermion Hamiltonians. Phys. Rep. **267**, 161–194 (1996)
21. K.K. Mon, J.B. French, Statistical properties of many-particle spectra. Ann. Phys. (N.Y.) **95**, 90–111 (1975)
22. J.C. Parikh, *Group Symmetries in Nuclear Structure* (Plenum, New York, 1978)
23. V.K.B. Kota, Sizes of effective single particle fields for s-d shell effective interactions. Phys. Rev. C **20**, 347–356 (1979)
24. V.K.B. Kota, K. Kar, Group symmetries in two-body random matrix ensembles generating order out of complexity. Phys. Rev. E **65**, 026130 (2002)
25. K.T. Hecht, Summation relation for $U(N)$ Racah coefficients. J. Math. Phys. **15**, 2148–2156 (1974)
26. K.T. Hecht, J.P. Draayer, Spectral distributions and the breaking of isospin and supermultiplet symmetries in nuclei. Nucl. Phys. A **223**, 285–319 (1974)

27. B. Georgeot, D.L. Shepelyansky, Breit-Wigner width and inverse participation ratio in finite interacting Fermi systems. Phys. Rev. Lett. **79**, 4365–4368 (1997)
28. I. Talmi, *Simple Models of Complex Nuclei: The Shell Model and Interacting Boson Model* (Harwood Academic Publishers, Chur, 1993)
29. K.T. Hecht, Some simple R_5 Wigner coefficients and their application. Nucl. Phys. **63**, 177–213 (1965)
30. K.T. Hecht, Five-dimensional quasi-spin the n, T dependence of shell-model matrix elements in the seniority scheme. Nucl. Phys. A **102**, 11–80 (1967)
31. K.T. Hecht, J.P. Elliott, Coherent-state theory for the proton-neutron quasispin group. Nucl. Phys. A **438**, 29–40 (1985)
32. K.T. Hecht, Wigner coefficients for the proton-neutron quasispin group: an application of vector coherent state techniques. Nucl. Phys. A **493**, 29–60 (1989)
33. V.K.B. Kota, J.A. Castilho Alcarás, Classification of states in $SO(8)$ proton-neutron pairing model. Nucl. Phys. A **764**, 181–204 (2006)
34. D.J. Rowe, J.L. Wood, *Fundamentals of Nuclear Models: Foundational Models* (World Scientific, Singapore, 2010)
35. M.A. Caprio, J.H. Skrabacz, F. Iachello, Dual algebraic structures for the two-level pairing model. J. Phys. A, Math. Theor. **44**, 075303 (2011)
36. V.K.B. Kota, Two-body ensembles with group symmetries for chaos and regular structures. Int. J. Mod. Phys. E **15**, 1869–1883 (2006)
37. B.G. Wybourne, *Symmetry Principles and Atomic Spectroscopy* (Wiley, New York, 1970)
38. M. Vyas, V.K.B. Kota, N.D. Chavda, One-plus two-body random matrix ensembles with spin: results for pairing correlations. Phys. Lett. A **373**, 1434–1443 (2009)
39. C. Quesne, S. Spitz, Spectral distributions of mixed configurations of identical nucleons in the seniority scheme I. Generalized seniority scheme. Ann. Phys. (N.Y.) **85**, 115–151 (1974)
40. C. Quesne, S. Spitz, Spectral distributions of mixed configurations of identical nucleons in the seniority scheme II. Configuration-seniority scheme. Ann. Phys. (N.Y.) **112**, 304–327 (1978)
41. B.J. Dalton, S.M. Grimes, J.P. Vary, S.A. Williams (eds.), *Moment Methods in Many Fermion Systems* (Plenum, New York, 1980)

Chapter 7
Applications of EGOE(1 + 2) and EGOE(1 + 2)-s

7.1 Mesoscopic Systems: Quantum Dots and Small Metallic Grains

Quantum dots and small metallic grains, being mesoscopic systems whose transport properties can be measured [1, 2], are generic systems for exploring physics of small coherent structures [3–5]. As the electron phase is preserved in mesoscopic systems [the phase coherence length increases rapidly with decreasing temperature and for system size $\sim$100 μm, the system becomes mesoscopic below $\sim$100 mK], these are ideal to observe new phenomenon governed by the laws of quantum mechanics not observed in macroscopic conductors. Also, the transport properties of mesoscopic systems are readily measured with almost all system parameters (like the shape and size of the system, number of electrons in the system and the strength of coupling with the leads) under experimental control.

Quantum dots are artificial devices obtained by confining a finite number of electrons to regions with diameter $\sim$100 nm by electrostatic potentials. Typically it consists of 10^9 real atoms but the number of mobile electrons is much lower, $\sim$100. Their level separation is $\sim 10^{-4}$ eV. In isolated or closed quantum dots, the coupling to leads is weak and conductance occurs only by tunneling. Also the charge on the closed dot is quantized and they have discrete excitation spectrum. The tunneling of an electron into the dot is usually blocked by the classical Coulomb repulsion of the electrons already in the dot. This phenomenon is called Coulomb blockade. This repulsion can be overcome by changing the gate voltage. At appropriate gate voltage, the charge on the dot will fluctuate between m and $m+1$ electrons giving rise to a peak in the conductance. The oscillations in conductance as a function of gate voltage are called Coulomb blockade oscillations and at sufficiently low temperatures, these oscillations turn into sharp peaks. In Coulomb blockade regime $kT \ll \Delta \ll E_c$ where T is the temperature, Δ is the mean single particle level spacing and E_c is the charging energy. The quantum limits of electrical conduction are revealed in quantum dots and conductivity exhibits statistical properties which reflect the presence of one-body chaos, quantum interference and electron-electron interaction.

V.K.B. Kota, *Embedded Random Matrix Ensembles in Quantum Physics*, Lecture Notes in Physics 884, DOI 10.1007/978-3-319-04567-2_7,

Ultra-small metallic grains are small pieces of metals of size ~2–10 nm. The level separation for nm-size metallic grains is smaller than in quantum dots of similar size and thus experiments can easily probe the Coulomb blockade regime in metallic grains. Also, some of the phenomena observed in nm-size metallic grains are strikingly similar to those seen in quantum dots.

Mesoscopic fluctuations are universal dictated only by a few basic symmetries of the system. Random matrix theory describes the statistical fluctuations in the universal regime i.e. at energy scales below the Thouless energy $E = g\Delta$, g is the Thouless conductance. In this universal regime, random matrix theory addresses questions about statistical behavior of eigenvalues and eigenfunctions rather than their individual description. A closed mesoscopic system (quantum dot or small metallic grain) with chaotic single particle dynamics and with large Thouless conductance g is described by an effective Hamiltonian which comprises of a mean field and two-body interactions preserving spin quantum number. For chaotic isolated mesoscopic systems, randomness of single particle energies leads to randomness in effective interactions that are two-body in nature. Hence, it is appropriate to invoke the ideas of embedded ensembles generated by random two-body interactions to understand and also predict properties of these systems theoretically [4, 5]. A realistic Hamiltonian for isolated mesoscopic systems conserves total spin S and includes a mean field one-body part, (random) two-body interaction, pairing interaction H_p and exchange interaction $\hat{S}^2$. Thus, an appropriate Hamiltonian is (with λ_p and λ_S being positive),

$$\left\{\widehat{H}(\lambda_0, \lambda_1, \lambda_p, \lambda_S)\right\} = \widehat{h}(1) + \lambda_0 \left\{\widehat{V}^{s=0}(2)\right\} + \lambda_1 \left\{\widehat{V}^{s=1}(2)\right\} - \lambda_p H_p - \lambda_S \hat{S}^2. \tag{7.1}$$

The constant part arising due to the charging energy that depends on the number of fermions in the system can be easily incorporated in the Hamiltonian when required.

7.1.1 Delay in Stoner Instability

Standard Stoner picture of ferromagnetism in itinerant systems is based on the competition between the one-body kinetic energy [generated by $h(1)$ in Eq. (7.1)] and the exchange interaction ($\hat{S}^2$). One-body kinetic energy (Pauli principle as applied to the distribution of fermions in sp levels) favors demagnetized ground states while sufficiently strong repulsive exchange interaction ($-\hat{S}^2$) favors maximum spin to be ground state. However, random interactions also disfavor magnetized ground states. In order to understand this, first we need a prescription for determining the ground state (gs) energies. The eigenvalue density of a system modeled by EGOE(1 + 2)-s is a Gaussian and therefore the gs energies are largely determined by the widths of the corresponding Gaussians. With a Gaussian density of states, gs energy E_{gs} can be obtained by inverting the equation $1/2 = \int_{-\infty}^{E_{gs}} I_{\mathcal{G}}(x)\, dx$; as $\overline{E_c(m, S)} = 0$, $I_{\mathcal{G}}(E)$ is zero centered. The inversion gives, using the results in [6] and ignoring the dimension effects as they will be logarithmic, the following simple relation

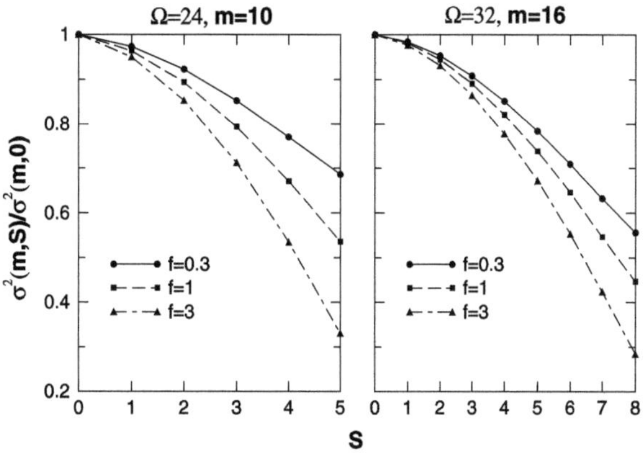

Fig. 7.1 Comparison of spectral variance for various spin sectors for $\Omega = 24, m = 10$ and $\Omega = 32, m = 16$ with three different values for $f = \lambda_0^2/\lambda_1^2$ for EGOE(2)-**s**. As seen from the figures, spectral variances decrease with increasing spin and there by implying that random interaction favor minimal gs spin

$$E_{gs}(m, S) \propto -\beta \overline{\sigma(\Omega, m, S, \lambda_0, \lambda_1)}. \tag{7.2}$$

Here, β is a positive constant. In Eq. (7.2) we have shown, for later use, that σ depends on Ω, λ_0 and λ_1. It is possible to incorporate in Eq. (7.2) the effects due to the deviations of the spectral shape from exact Gaussian form [7, 8]. Though this is well known in nuclear physics [9, 10], it was advocated in mesoscopic physics in the context of EGOE(1 + 2)-**s** by Jacquod and Stone [11] and hence we call it JS prescription from now on. Therefore, applying the JS criterion and comparing the ensemble averaged spectral variances generated by a random interactions [given by Eq. (6.18)] for different spins S, for fixed Ω and m values, will give the S value for the absolute gs of the system. It is seen from Fig. 7.1, constructed using Eq. (6.18), that the variances decrease with increasing spin and this behavior is independent of the ratio $f = \lambda_0^2/\lambda_1^2$. Therefore, random interactions generate gs with minimum value for S. Thus, the result that follows easily from Fig. 6.4 extends to the situations with $\lambda_0 \neq \lambda_1$ and explain in a simple way that in general there will be preponderance of gs with spin $S_{min} = 0$ for m even or $S_{min} = \frac{1}{2}$ (m odd) for mesoscopic systems. As minimum spin ground states are favored by random interactions, the Stoner transition will be delayed in presence of a strong random two-body part in the Hamiltonian. For a better understanding of these results, numerical calculations are carried out [12] for $\Omega = m = 8$ using $H(\lambda, \lambda, 0, \lambda_S)$ in Eq. (7.1) and the probability $P(S > 0)$ for the gs to be with $S > 0$ (for m even) is shown as a function of λ and λ_S in Fig. 7.2. Similar calculations for smaller systems with $\Omega = m = 6$ were given in [11, 13, 14]. It is seen from the results in Fig. 7.2 that the probability $P(S > 0)$ for ground state to have $S > 0$ is very small when $\lambda > \lambda_S$ and it increases with increasing λ_S. Figure also gives for a fixed λ value, the minimum λ_S needed for ground states to have $S > 0$ with 100 % probability. These numerical results clearly bring out the demagnetizing effect of random interaction. Thus the model given by the Hamiltonian $\{\widehat{H}(\lambda_0, \lambda_1, 0, \lambda_S)\}$ explains the strong bias for low-spin ground states and the delayed ground state magnetization by random two-body interactions.

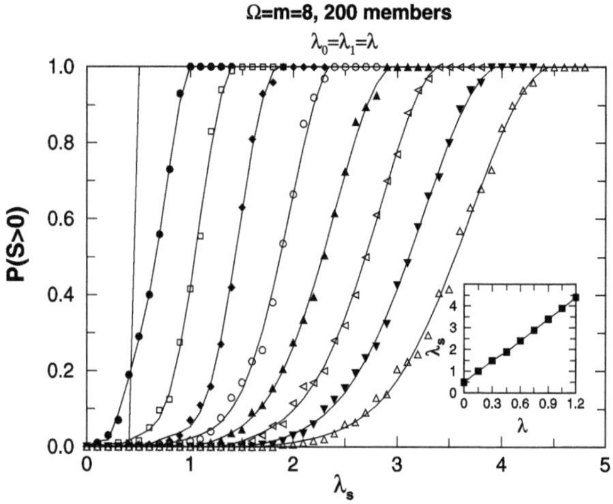

Fig. 7.2 Probability $P(S > 0)$ for ground states to have $S > 0$ as a function of exchange interaction strength λ_S for $\lambda = 0$ to 1.2 in steps of 0.15; used here is $\widehat{H}(\lambda, \lambda, 0, \lambda_S)$ defined by Eq. (7.1). The calculations are for 200 member EGOE(2)-**s** ensemble with $\Omega = m = 8$. *Inset* of figure shows the minimum exchange interaction strength λ_S required for the ground states to have $S > 0$ with 100 % probability as a function of λ. It is seen from the results that the probability $P(S > 0)$ for gs to have $S > 0$ is very small when $\lambda > \lambda_S$ and it increases with increasing λ_S. The results clearly bring out the demagnetizing effect of random interaction. Similar calculations have been performed in the past for smaller systems with $\Omega = m = 6$ [13, 14]. Figure is taken from [12, 15]

7.1.2 Odd-Even Staggering in Small Metallic Grains

For nm-scale Al particles (5–13 nm in radius), odd-even staggering is observed in gs energies measured using electron tunneling [16]. This phenomenon is normally associated with pairing interaction effects. Surprisingly, it can also arise from random two-body interactions as pointed out first in [17]. Odd-even staggering implies that the gs energy of even particle system is larger than the arithmetic mean of its odd number members. Then,a good staggering indicator is

$$\Delta_2(m) = E_{gs}(m + 1, S_{min}) + E_{gs}\big(m - 1, S'_{min}\big) - 2E_{gs}\big(m, S''_{min}\big) \tag{7.3}$$

and this is the second derivative of the gs energy with particle number m. Firstly, it is easy to see that $\widehat{H} = h(1)$ will generate odd-even staggering. Given the sp energies ε_i, $i = 1, 2, \ldots, \Omega$, we have $\Delta_2(m) = \varepsilon_{m/2+1} - \varepsilon_{m/2}$ for m even and $\Delta_2(m) = 0$ for m odd showing odd-even staggering. Going to the strong interaction regime, one can use EGOE(2)-**s** in the Gaussian domain. In this situation, staggering effect generated by random interactions can be understood by employing the JS criterion. For EGOE(2)-**s** using Eq. (6.18), spectral variance $\sigma^2(m, S)$ can be written as $\lambda_1^2\sigma^2(\Omega, m, S, f = \lambda_0^2/\lambda_1^2)$. Then, Eqs. (7.2) and (7.3) will give

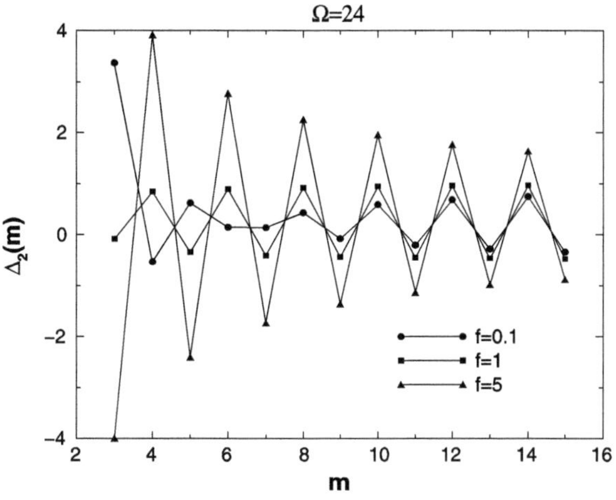

Fig. 7.3 Figure showing staggering in ground state energies with random two-body interactions. Calculations are repeated for various combinations of λ_0 and λ_1 values and it is found that the effect is independent of the ratio $f = \lambda_0^2/\lambda_1^2$. See text for details

$$\begin{aligned}\Delta_2(m) = \frac{-|\beta\lambda_1|}{2}\big[&\sigma(\Omega, m+1, S_{min}, f) + \sigma(\Omega, m-1, S_{min}, f) \\ &- 2\sigma\big(\Omega, m, S'_{min}, f\big)\big]\end{aligned} \tag{7.4}$$

where $S_{min} = 0$ for even number of particles and $\frac{1}{2}$ for odd particle number. Note that β is a constant and $f = \lambda_0^2/\lambda_1^2$. In Fig. 7.3, the staggering indicator $\Delta_2(m)$ with $|\beta\lambda_1| = 1$ is shown as a function of particle number m for $\Omega = 24$ and three different values of f. Results in the figure confirm that random interactions generate odd-even staggering in gs energies even when $f \neq 1$. The $S(S+1)$ and $[S(S+1)]^2$ terms in the variance propagators P's given by Eq. (6.18) [similarly, for $\lambda_0 = \lambda_1$ the propagator is given by Eq. (6.19)] are responsible for the staggering effect.

Odd-even staggering in the transition region between pure mean-field limit and strong two-body interaction has been studied numerically by Papenbrock et al. [17] using a 200 member EGOE(1 + 2)-**s** ensemble with $\Omega = 10$ and $m = 3, 4, \ldots, 17$. The Hamiltonian adopted was a variant of Eq. (6.1),

$$\{\widehat{H}\} = \cos\phi\,\widehat{h}(1) + \sin\phi\big[\lambda_0\big\{\widehat{V}^{s=0}(2)\big\} + \lambda_1\big\{\widehat{V}^{s=1}(2)\big\}\big] \tag{7.5}$$

so that $\phi = 0$ gives the mean-field limit and $\phi = \pi/2$ gives strong interaction limit. Varying ϕ, calculations are carried out and the largest matrix in these calculation has the dimension $d = 63504$. It is important to mention that even with the best available computing facilities, it is not yet feasible to numerically study the properties of large systems ($\Omega \gg 10$) modeled by Eq. (7.5). Results for three different choices of $\lambda's$ and ϕ's are shown in Fig. 7.4. Clearly, there is staggering effect in the transition domain. Results in the figures show that the interaction induced staggering can be clearly distinguished from mean-field effects. Also, random interaction induced staggering is a smooth function of particle number.

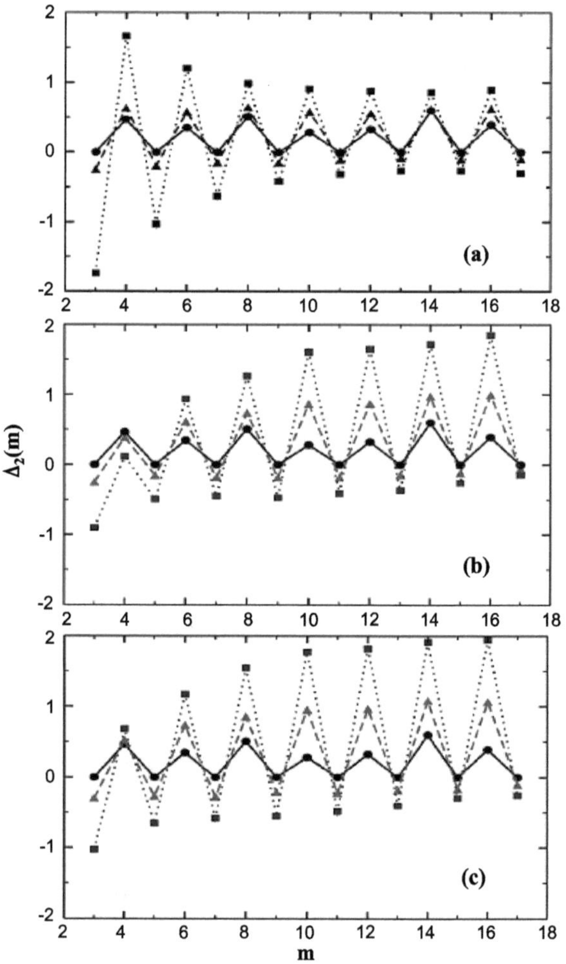

Fig. 7.4 Figure showing staggering in ground state energies generated by a Hamiltonian consisting of mean-field plus random two-body interactions defined by Eq. (7.5): (**a**) $\lambda_0 = 1$ and $\lambda_1 = 0$; (**b**) $\lambda_0 = 0$ and $\lambda_1 = 1$; (**c**) $\lambda_0 = \lambda_1 = 1$. In (**a**), (**b**) and (**c**) results are shown for $\phi = 0$ (*full line*), $\phi = \pi/12$ (*dashed line*) and $\phi = \pi/2$ (*dotted line*—scaled by a factor 1/2 for display purpose). Note that $\Delta_2(m)$ is dimensionless and the energy scale in the calculations is set by the overall energy scale of the Hamiltonian. Also used in the calculations are sp energies drawn from the center of a GOE. Figures (**a**), (**b**) and (**c**) are taken from [17] with permission from American Physical Society

7.1.3 Conductance Peak Spacings

Coulomb blockade oscillations yield detailed information about the energy and wavefunction statistics of mesoscopic systems. Spacing between two neighboring conductance peaks, as a function of the gate voltage for temperatures less than the average level spacing, is simply given by $\Delta_2(m)$ defined by Eq. (7.3). Therefore, the peak spacing distribution is the distribution $P(\Delta_2)$ of the spacings Δ_2. The $P(\Delta_2)$ has been used in the study of the distribution of conductance peak spacings in chaotic quantum dots [4, 18] and small metallic grains [19] using chaotic sp dynamics. Alhassid and collaborators have demonstrated for the first time [4, 20] that EGOE but not GOE describes the experimental results for $P(\Delta_2)$.

Let us first consider non-interacting spinless finite Fermi systems, i.e. $H = h(1)$ with no spin and say the sp energies are ε_i; $i = 1, 2, \ldots, N$. Now the ground state energy $E_{gs}^{(m-1)}$ for $m-1$ particles is obtained by filling the sp states from bottom by applying Pauli principle. Addition of one particle in the system results in the gs energy $E_{gs}^{(m)} = E_{gs}^{(m-1)} + \varepsilon_m$ and similarly $E_{gs}^{(m+1)} = E_{gs}^{(m-1)} + \varepsilon_m + \varepsilon_{m+1}$, by Pauli principle. Then, $\Delta_2 = \varepsilon_{m+1} - \varepsilon_m$, irrespective of whether m is even or odd. For mesoscopic systems, it is possible to consider sp energies drawn from GOE eigenvalues [4, 18] with the assumption that the single particle motion will be chaotic. Therefore $P(\Delta_2)$ corresponds to GOE spacing distribution $P_W(\Delta)$—the Wigner distribution. However, experiments showed that $P(\Delta_2)$ is a Gaussian in many situations [21] as shown in Fig. 7.5a. This calls for inclusion of two-body interaction and hence the importance of EGOE(1 + 2) in the study of conductance fluctuations in mesoscopic systems [4, 20]. Alhassid et al. [20] considered m spinless fermions in N sp states with $H = h(1) + \lambda V(2)$ and $h(1)$ in one particle space is chosen to be $N \times N$ GOE with average spacing Δ_s between the levels in one particle space. Similarly $V(2)$ in two-particle space is a GOE with unit variance for the matrix elements. Using this EGOE(1 + 2) (called RIMM in [20]), calculations are carried out for a system with $N = 12$ and $m = 4$. Results are shown in Fig. 7.5b for the distribution of $\widetilde{\Delta_2} = (\Delta_2 - \langle \Delta_2 \rangle)/\Delta_s$ for several different values of λ/Δ_s. Clearly it is seen that with strong enough interaction, peak spacing distribution approaches Gaussian form as seen from Fig. 7.5b. Although the results in Fig. 7.5, with a unimodal form for the peak spacing distribution, appear to indicate that the spin degree of freedom of electrons (hence pairing) is not important, later experiments with appropriate system parameters did show effects due to spin degree of freedom. Now we will turn to this.

With electron carrying spin degree of freedom, pairing effects are expected to be seen in conductance peak spacings distributions. With the system Hamiltonians conserving total spin S, it is important to consider sp levels that are doubly degenerate. Thus, more generally EGOE(1 + 2)-**s** is relevant. Again, we start with non-interacting finite Fermi systems with sp energies ε_i, $i = 1, 2 \ldots, \Omega$ and drawn from a GOE; total number of sp states $N = 2\Omega$. In this scenario Δ_2 depends on whether m is odd or even. For m odd, say $m = 2k + 1$, the $(m - 1)$ fermion ground state energy $E_{gs}^{(m-1)} = 2\sum_{i=1}^{k} \varepsilon_i$, $E_{gs}^{(m)} = E_{gs}^{(m-1)} + \varepsilon_{k+1}$ and $E_{gs}^{(m+1)} = E_{gs}^{(m-1)} + 2\varepsilon_{k+1}$ resulting in $\Delta_2 = 0$. Similar analysis for even $m = 2k$ yields $\Delta_2 = \varepsilon_{k+1} - \varepsilon_k$; note that $E_{gs}^{(m)} = 2\sum_{i=1}^{k} \varepsilon_i$, $E_{gs}^{(m-1)} = E_{gs}^{(m)} - \varepsilon_k$ and $E_{gs}^{(m+1)} = E_{gs}^{(m)} + \varepsilon_{k+1}$. For odd m, Δ_2 corresponds to even-odd-even transition and $P(\Delta_2)$ is a delta function. For even m, we have odd-even-odd transitions with $P(\Delta_2)$ following Wigner distribution. As we need to include, for real systems, both these transitions, inclusion of spin degree of freedom gives bimodal distribution for $P(\Delta_2)$,

$$P(\Delta_2) = \frac{1}{2}\big[\delta(\Delta_2) + P_W(\Delta_2)\big]. \tag{7.6}$$

Convolution of this bimodal form with a Gaussian has been used in the analysis of data for quantum dots obtained for situations that correspond to weak interactions [22]. This analysis showed that spin degree of freedom and pairing correlations are important for mesoscopic systems. Note that pairing correlations (H_p) favor

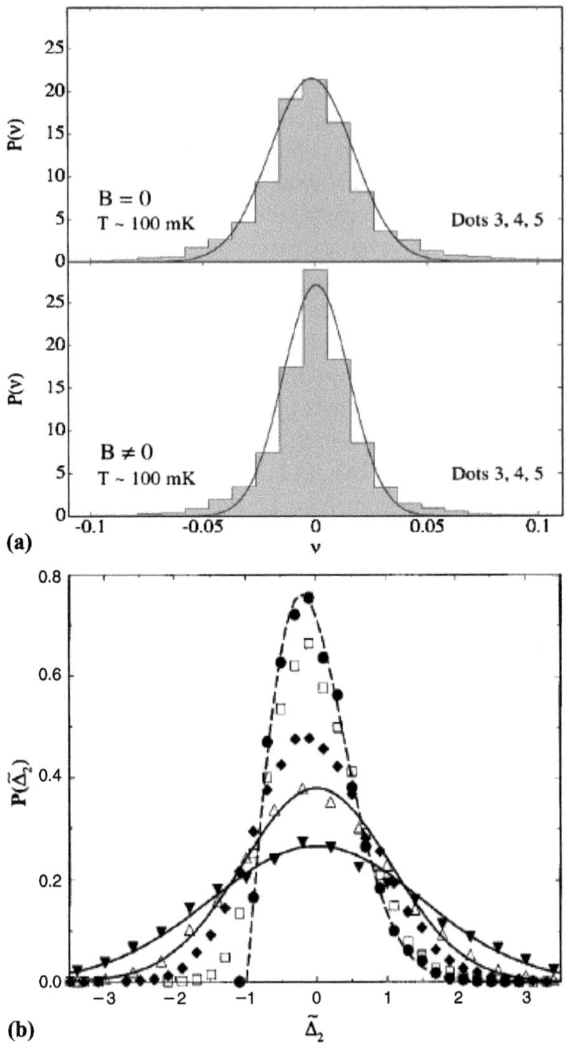

Fig. 7.5 (**a**) Conducting peak spacing distributions for GaAs quantum dots. Histograms of normalized peak spacing (ν) distributions with magnetic field $B = 0$ and $B \neq 0$ from three devices. Shown also are the best fits to normalized Gaussians. Given the spacings Δ, $\nu = \Delta/\langle\Delta\rangle - 1$. Figure is taken from [21], by removing the insect figures in the original figure, with permission from American Physical Society. (**b**) Peak spacing distributions for EGOE(1 + 2) model described in the text for a $N = 12$, $m = 4$ system with 10000 members; $\widetilde{\Delta_2}$ is defined in the text. Results are shown for $\lambda/\Delta_s = 0$ (*solid circles*), 0.35 (*open circles*), 0.7 (*solid diamonds*), 1.1 (*open triangles*) and 1.8 (*solid triangles*). For $\lambda = 0$, the distribution is Wigner-Dyson like and it is Gaussian like for $\lambda/\Delta_s > 1$. Figure is taken from [20] with permission from American Physical Society

minimum spin ground state whereas the exchange interaction ($-\hat{S}^2$) tends to maximize the ground state spin. Competition between pairing and exchange interaction is equivalent to competition between ferromagnetism and superconductivity [19]. Hence, it is imperative to study $P(\Delta_2)$ with a Hamiltonian that includes mean field one-body part, (random) two-body interaction, exchange interaction and pairing (defined by H_p). For small metallic grains, using a microscopic model with pairing interaction, it was shown in [19] that $P(\Delta_2)$ is bimodal when pairing interaction is dominant whereas it is unimodal for strong exchange interaction. Same result is expected to follow from the extended EGOE(1+2)-s with $\widehat{H} = \widehat{H}(\lambda, \lambda, \lambda_p, \lambda_S)$ given by Eq. (7.1). As this ensemble is not analytically tractable to derive $P(\Delta_2)$, numer-

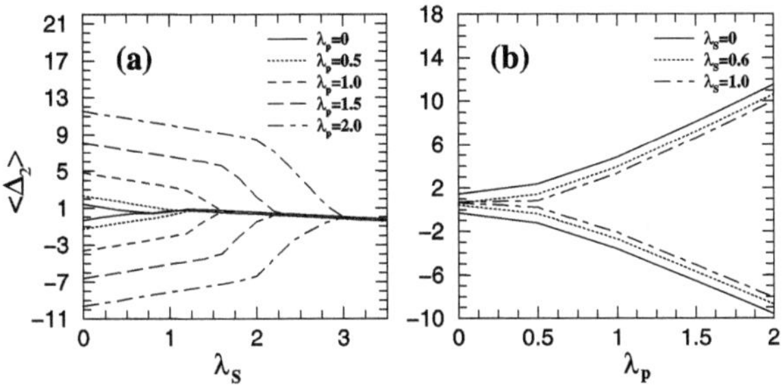

Fig. 7.6 Average peak spacing $\langle \Delta_2 \rangle$ (**a**) as a function of exchange interaction strength λ_S for several values of pairing strength λ_p and (**b**) as a function of λ_p for several values of λ_S, for a 1000 member EGOE(1 + 2)-s ensemble, defined by Eq. (7.1), with $\Omega = 6$. The *curves* in the *upper part* correspond to $m = 4$ $(3 \to 4 \to 5)$ and those in the *lower part* to $m = 5$ $(4 \to 5 \to 6)$ in (7.3). Figure is taken [12, 15]

ical calculations are carried out in [23] with focus on the strong interaction regime using $\lambda_0 = \lambda_1 = \lambda \geq 0.3$ in Eq. (7.1) and employing a fixed set of sp energies.

Figure 7.6a shows the variation of average peak spacing with exchange interaction strength λ_S for several λ_p values. The curves in the upper part correspond to $m = 4$ and those in the lower part to $m = 5$. As the exchange strength increases, the average peak spacing $\langle \Delta_2 \rangle$ is almost same for odd-even-odd and even-odd-even transitions. Value of average peak spacing and its variation with λ_S is different for odd-even-odd and even-odd-even transitions when pairing correlations are strong. The curve for fixed value of λ_p can be divided into two linear regions whose slopes can be determined considering only exchange interactions, i.e. $E_{gs} = C_0 - \lambda_S S(S+1)$. For weak exchange interaction strength, ground state spin is 0(1/2) for m even(odd) and thus for this linear region, $\langle \Delta_2 \rangle / \lambda_S \propto -3/2(3/2)$. The linear region where exchange interactions are dominant, $\langle \Delta_2 \rangle / \lambda_S \propto -1/2$ as ground state spin is $m/2$. Figure 7.6b shows the variation of average peak spacing with pairing strength for several λ_S values. It clearly shows that the separation between the two modes of the distributions becomes larger with increasing λ_p. These results are in good agreement with the numerically obtained results for the $P(\Delta_2)$ variation as a function of λ_p and λ_S as shown in Fig. 7.7. Similar results were reported for small metallic grains in [19] where a microscopic model is employed. The model with H defined in Eq. (7.1) thus explains the interplay between exchange (favoring ferromagnetism) and pairing (favoring superconductivity) interaction in the Gaussian domain and can be used for investigating in more detail various transport properties of mesoscopic systems.

7.1.4 Induced Two-Body Ensembles

In the previous Sects. 7.1.1–7.1.3, we have shown that EE give generic description of properties of mesoscopic systems where interaction effects are important. However, this description is good only in the $g \to \infty$ (g is the dimensionless Thouless

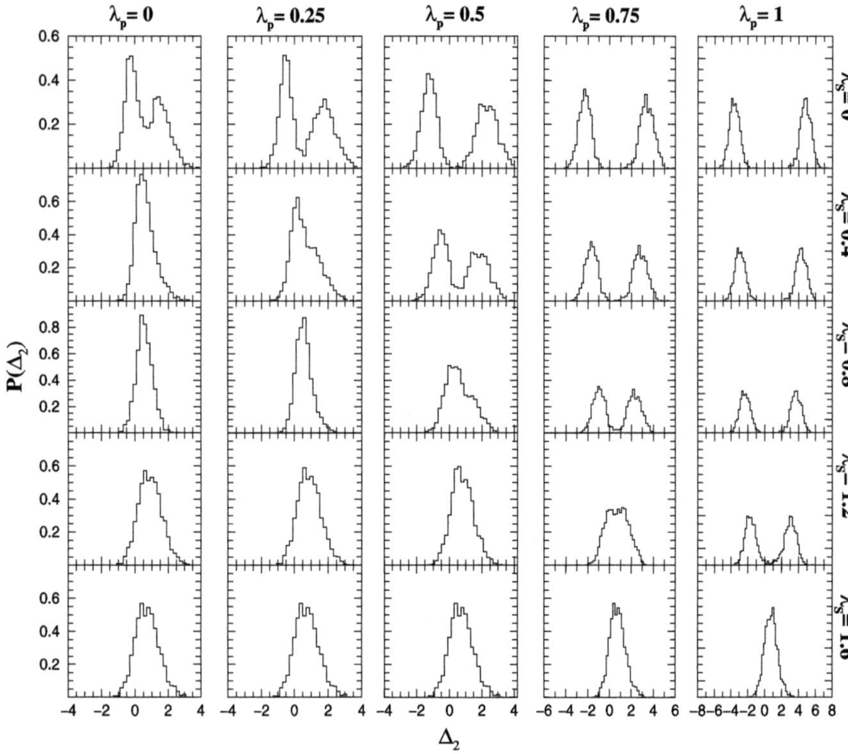

Fig. 7.7 $P(\Delta_2)$ vs Δ_2 for various values of the pairing strength λ_p and exchange interaction strength λ_S for the same EGOE(1 + 2)-**s** system used in Fig. 7.6. The distributions $P(\Delta_2)$ are constructed (with bin size 0.2) by combining the results for Δ_2 with $m = 4$ and 5. See text for further details. Figure is taken from [23] with permission from Elsevier

conductance) limit. For (almost) closed diffusive or chaotic dots for example, g may be large but finite. In this situation, strictly speaking one has to employ induced two-body ensembles rather than EE(2) for the two-body part of the interaction. We will discuss this aspect briefly in this section.

In general, for a disordered system (diffusive or chaotic quantum dot being an example), $H = h(1) + V(2)$ where the one-body part $h(1) = \sum_{i,j} h_{ij} a_i^\dagger a_j$ is due to the kinetic energy and the disordered potential and the $V(2)$ matrix elements V_{ijkl} are due to the screened Coulomb interaction. One can model the h_{ij} matrix by one of the classical random matrix ensemble (GOE or GUE or GSE) depending on the symmetries. Diagonalizing each member of the ensemble gives the eigenbasis say $|\alpha\rangle$ for that member. By expressing $V(2)$ in the $h(1)$ eigenbasis, we will have an ensemble representation for $V(2)$ with $V_{\alpha\beta\gamma\delta}$ random two-body matrix elements. Thus, we have a induced two-body ensemble and adding to this the one-body part $h'(1) = \sum_\alpha \varepsilon_\alpha \hat{n}_\alpha$, we have an induced EE(1 + 2). Statistical properties of $V_{\alpha\beta\gamma\delta}$ (or $V^s_{\alpha\beta\gamma\delta}$ for fermions with spin) will depend on (i) the invariance properties of the h_{ij} matrix (orthogonal or unitary or symplectic), (ii) the form of the two-body interaction and (iii) whether we have spinless fermions or fermions with spin degree of freedom. For various choices of (i), (ii) and (iii), formulas, valid to order $1/g^2$, are derived by Alhassid et al. [18] for the average and variance of $V_{\alpha\beta\gamma\delta}$. Higher order ($k$-th order, $k > 2$) cumulants are expected to be $\sim 1/g^k$. As Alhassid et al. [18] add, "whenever an observable contains in addition to $V_{\alpha\beta\gamma\delta}$'s other expressions which depend upon the $|\alpha\rangle$'s, the ensemble averages of this observable taken over the induced two-body ensemble and over the full two-body ensemble will differ". Although, obviously induced EE$(1+2)$ are more appropriate for mesoscopic systems, they are not yet explored in any detail due to inherent difficulties with these ensembles [18, 24].

7.2 Statistical Spectroscopy: Spectral Averages for Nuclei

EGOE$(1+2)$ with or without spin generates three chaos markers as discussed in Chaps. 5 and 6. These chaos markers provide the basis for statistical spectroscopy where the forms generated by EGOE$(1+2)$ for various smoothed densities, ignoring fluctuations, are used to give a theory for calculating spectroscopic quantities such as level densities, occupation numbers, transition strengths (such as dipole strengths in atoms and Gamow-Teller strengths in nuclei) and so on. Statistical spectroscopy is valid only in the $\lambda > \lambda_c$ region as here we can apply the average-fluctuation separation discussed in Sect. 4.3. With fluctuations following GOE, spectral averages over a few spacings will smoothen the fluctuations and they will be close to the actual values as GOE fluctuations are small in size (due to spectral rigidity). Realistic systems such as atoms and nuclei carry orbital angular momentum besides spin. Then, in L–S coupling the many particle (L, S) will be good and similarly in j–j coupling only the total J (note that $\vec{J} = \vec{L} + \vec{S}$) will be good. As EGOE$(1+2)$ results extend in many situations, as discussed in Chap. 6, to EGOE$(1+2)$-**s**, it is plausible to argue that in general EGOE$(1+2)$ results extend to subspaces defined by good quantum numbers. With this, we can assume that the EGOE$(1+2)$ forms [similarly the EGOE$(1+2)$-**s** forms] for various densities apply to EGOEs with LS and/or J symmetry (see Chap. 13 for some detailed discussion on EGOEs with angular momentum J symmetry). Indeed, this has been assumed in nuclear structure studies and verified in many numerical examples using nuclear shell model codes. This approach is called spectral distribution theory in nuclear physics [10] and this subject is being studied from early 70's. Similarly, in the 90's it was also shown that same ideas will apply to atoms [25–28]. Here we will very briefly discuss the EGOE based approach to nuclear level densities and orbit occupancies. Then, as a detailed example, we will discuss the theory for transition strengths with applications to neutrinoless double beta decay (NDBD) nuclear transition matrix elements (NTME).

7.2.1 Level Densities and Occupancies

In spectroscopic studies, one is often interested in the energy region that is not too far from the ground state. Although EGOE(1 + 2) gives Gaussian form for the state densities $I(E)$, this will not be useful as the convergence to Gaussian form is poor for eigenvalues far removed from the centroid. Therefore, in practice it is more appropriate to use strength functions or partial sums of them (called partial densities) and use the results that they will be of Gaussian form for EGOE(1 + 2)'s in the Gaussian domain ($\lambda > \lambda_F$) and intermediate to BW and Gaussian in the $\lambda_c \leq \lambda < \lambda_F$ region. For EGOE(1 + 2), with m fermions in N sp states the basis states k are nothing but the configurations $\widetilde{m} = (m_1, m_2, \ldots, m_N)$ where m_i is number of fermions in the ith sp state; $m_i = 0$ or 1 and $\sum m_i = m$. Similarly for EGOE(1+2)-s, $\widetilde{m} = (m_1, m_2, \ldots, m_\Omega)$ where m_i being the number of fermions in the ith sp level (each double degenerate); $m_i = 0$ or 1 or 2, $\sum m_i = m$ and $N = 2\Omega$. Now, we have

$$\begin{aligned} I^m(E) &= \langle\!\langle \delta(H-E) \rangle\!\rangle^m = \sum_{\widetilde{m}} \langle\!\langle \delta(H-E) \rangle\!\rangle^{\widetilde{m}} \\ &= \sum_{\widetilde{m}} I^{\widetilde{m}}(E) \\ &\xrightarrow{\text{EGOE(1+2)}} \sum_{\widetilde{m}} I_{\mathscr{G}}^{\widetilde{m}}(E) \quad \text{or} \quad \sum_{\widetilde{m}} I_{BW-\mathscr{G}}^{\widetilde{m}}(E). \end{aligned} \tag{7.7}$$

Similarly,

$$\begin{aligned} I^{m,S}(E) &= \langle\!\langle \delta(H-E) \rangle\!\rangle^{m,S} = \sum_{\widetilde{m} \in S} \langle\!\langle \delta(H-E) \rangle\!\rangle^{\widetilde{m},S} \\ &= \sum_{\widetilde{m} \in S} I^{\widetilde{m},S}(E) \\ &\xrightarrow{\text{EGOE(1+2)-s}} \sum_{\widetilde{m} \in S} I_{\mathscr{G}}^{\widetilde{m},S}(E) \quad \text{or} \quad \sum_{\widetilde{m} \in S} I_{BW-\mathscr{G}}^{\widetilde{m},S}(E). \end{aligned} \tag{7.8}$$

Note that the $I^{\widetilde{m}}$ are normalized to their respective dimensions and the corresponding normalized $\rho_{BW-\mathscr{G}}^{\widetilde{m}}$ is given by Eq. (5.27). Extending Eqs. (7.7) and (7.8) to $(\widetilde{m}, L, S)$ or $(\widetilde{m}, J)$ spaces, it is possible to calculate level densities [in level densities the $(2J + 1)$ or $(2L + 1)(2S + 1)$ degeneracy factor is not counted] in nuclei and atoms and apply them to data analysis. For simplicity let us consider only J densities. Then, one can use Eq. (7.8) with J replacing S and evaluating the moments $\langle H^p \rangle^{\widetilde{m},J}$, $p = 1, 2$ using exact methods, though they are cumbersome. This approach has been used successfully recently for Horoi et al. [29–31] for lighter ($A < 80$) nuclei. On the other hand, one can use Eq. (7.7) and then project out J by using the so-called spin-cutoff factors. A variant of this approach was used by French's group for calculating level densities in medium-heavy and heavy nuclei [10, 32, 33]. It is important to point out that the variances $\sigma^2(\widetilde{m}) = \langle H^2 \rangle^{\widetilde{m}}$ involve (we are dropping S or J in this discussion) both $\widetilde{m} \to \widetilde{m}$ [giving internal variance $\sigma^2(\widetilde{m} \to \widetilde{m})$] and $\widetilde{m} \to \widetilde{m}'$ with $\widetilde{m} \neq \widetilde{m}'$ [giving external variances $\sigma^2(\widetilde{m} \to \widetilde{m}')$] matrix elements of H. In general $\sigma^2(\widetilde{m} \to \widetilde{m}')$ being non-zero as $\widetilde{m}$'s are not good symmetry subspaces of H. With $\langle H\rangle^{\widetilde{m}}$ defining the energy of the configuration $\widetilde{m}$ [equivalent to the configuration centroids $E_c(\widetilde{m}) = \langle H\rangle^{\widetilde{m}}$], it is easily seen that the variances will have contributions from both close by and distant configurations. The later will produce large skewness and hence should be ignored are treated differently. For a proper statistical treatment of the $\langle\widetilde{m}'|H|\widetilde{m}\rangle$ matrix elements, it is necessary to extend EGOE(1 + 2) to a partitioned EGOE(1 + 2) by associating an addition quantum number to distinguish close lying from distant configurations. This rather complex EE will be discussed further in Chap. 13.

Occupation numbers determine the single particle/collective structures at low-energies and the thermodynamic behavior at higher energies and hence their importance in spectroscopy. A theory for occupancies follow from the fact that $\widetilde{m}$ are eigenstates of the number operators $\hat{n}_i$ [$i = 1, 2, \ldots, N$ for EGOE(1 + 2) and $i = 1, 2, \ldots, \Omega$ for EGOE(1 + 2)-**s**]. Now, applying Eqs. (7.7) and (7.8)

$$
\begin{aligned}
\langle\hat{n}_i\rangle^E \quad &\equiv \quad \sum_{\widetilde{m}} \frac{I^{\widetilde{m}}(E)}{I^m(E)} m_i(\widetilde{m}) \\
&\xrightarrow{\text{EGOE(1+2): } \lambda_c \le \lambda \le \lambda_{F_k}} \sum_{\widetilde{m}} \frac{I^{\widetilde{m}}_{BW-\mathscr{G}}(E)}{I^m(E)} m_i(\widetilde{m}) \\
&\xrightarrow{\text{EGOE(1+2): } \lambda > \lambda_{F_k}} \sum_{\widetilde{m}} \frac{I^{\widetilde{m}}_{\mathscr{G}}(E)}{I^m(E)} m_i(\widetilde{m})
\end{aligned} \tag{7.9}
$$

and similarly,

$$
\begin{aligned}
\langle n_i\rangle^{E,S} \quad &\equiv \quad \sum_{\widetilde{m}\in S} \frac{I^{\widetilde{m},S}(E)}{I^{m,S}(E)} m_i(\widetilde{m}) \\
&\xrightarrow{\text{EGOE(1+2)-s:} \lambda_c \le \lambda \le \lambda_{F_k}} \sum_{\widetilde{m}\in S} \frac{I^{\widetilde{m},S}_{BW-\mathscr{G}}(E)}{I^{m,S}(E)} m_i(\widetilde{m}) \\
&\xrightarrow{\text{EGOE(1+2)-s: } \lambda > \lambda_{F_k}} \sum_{\widetilde{m}} \frac{I^{\widetilde{m},S}_{\mathscr{G}}(E)}{I^{m,S}(E)} m_i(\widetilde{m}).
\end{aligned} \tag{7.10}
$$

Note that: (i) in [34], the first form of (7.9) was employed with the smoothed $I^{\widetilde{m}}(E)$ constructed using a semi-classical theory for interacting spin systems; (ii) the second form of Eqs. (7.9) and (7.10), assuming that they extend to J spaces, was employed by Flambaum for atoms [25]; (iii) the third form in Eqs. (7.9) and (7.10), i.e. the Gaussian domain result, was used for nuclei by assuming that they extend to good J spaces [10]; (iv) forms in Eq. (7.10) are used in EGOE(1 + 2)-**s** analysis [35] and they should be useful in the studies of mesoscopic systems; (v) it is easy to add the skewness and excess corrections to the Gaussian densities in Eqs. (7.9) and (7.10). More detailed discussion on the forms, as determined by the operation of EGOE(1 + 2)'s, for occupancies and also for transition strength sums was given in [36].

7.2.2 Transition Strengths: Simplified Form for One-Body Operators

For EGOE(2), as shown in Sect. 4.4, transition strength densities take in general bivariate Gaussian form. This result extends to EGOE(1 + 2) in the Gaussian domain while in the BW domain, as discussed in Sect. 5.4, bivariate t-distribution given by Eq. (5.36) is appropriate. However in larger (m,N) spaces, just as with level densities and occupancies, it is necessary to have a theory for strengths with partitioning, i.e. with configurations $\widetilde{m}$. An essential complication here is that not only the Hamiltonian operator but also a general transition operator $\mathcal{O}$ will break the symmetry defining $\widetilde{m}$'s. A plausible way, as given first in [37], is to first construct the transition strength density with $H = h(1)$. Given a transition operator $\mathcal{O}$ the exact form for the transition strength density with $H = h(1)$ is,

$$I_{\mathcal{O}}^{h(1)}(x_i, x_f) = (2k+1)^{-1} \sum_{\widetilde{m}_i, \widetilde{m}_f} \sum_{\gamma_i \in \widetilde{m}_i, \gamma_f \in \widetilde{m}_f, \mu} \left| \langle \widetilde{m}_f, \gamma_f | \mathcal{O}_\mu^k | \widetilde{m}_i, \gamma_i \rangle \right|^2$$
$$\times \delta\big(x_i - \boldsymbol{\varepsilon}[\widetilde{m}_i]\big) \delta\big(x_f - \boldsymbol{\varepsilon}[\widetilde{m}_f]\big); \quad \boldsymbol{\varepsilon}[\widetilde{m}] = \sum_i m_i \varepsilon_i. \tag{7.11}$$

Here ε_i are sp energies. Note that $\widetilde{m}$ are eigenstates of $h(1)$. For operators $\mathcal{O}$ of spherical tensor rank k, we have to apply Eq. (7.11) with $\mathcal{O}$ replaced by $\mathcal{O}_\mu^k$, sum over all μ and divide the result by $(2k+1)$. Also, note that we have dropped S and it can be put back appropriately when needed. Expression for the matrix elements $|\langle \widetilde{m}_f, \gamma_f | \mathcal{O}_\mu^k | \widetilde{m}_i, \gamma_i \rangle|^2$ for one and two-body operators is easy to write down and therefore it is in general straightforward to construct $I_{\mathcal{O}}^{h(1)}$. Examples are discussed ahead. Now we will switch on the interaction $V(2)$ and as in Sect. 5.3.3 we assume that the effective one-body part of $V(2)$ is added to $h(1)$ so that $h(1) \leftrightarrow \boldsymbol{h}$ and $V(2) \leftrightarrow \boldsymbol{V}$. Then, the role of $V(2)$ is to locally spread $I_{\mathcal{O}}^{h(1)}$. The spreading function (normalized) $\rho_{\mathcal{O}}^{\boldsymbol{V}}$ has to be a bivariate distribution. Then the strength density will be a convolution of $I_{\mathcal{O}}^{\boldsymbol{h}}$ and $\rho_{\mathcal{O}}^{\boldsymbol{V}}$,

$$I_{\mathcal{O}}^{H=\boldsymbol{h}+\boldsymbol{V}}(E_i, E_f) = \int I_{\mathcal{O}}^{\boldsymbol{h}}(x, y)\, \rho_{\mathcal{O}}^{\boldsymbol{V}}(E_i - x, E_f - y)\, dx\, dy. \tag{7.12}$$

By putting $\boldsymbol{h} = 0$ in Eq. (7.12), EGOE(1 + 2) → EGOE(2) and this gives, from the results in Sect. 4.4, $\rho_{\mathcal{O}}^{\boldsymbol{V}}$ to be a bivariate Gaussian in the Gaussian domain and a bivariate t-distribution in the BW domain. The integral in Eq. (7.12) will be replaced by a summation when we employ Eq. (7.12) for $\rho_{\mathcal{O}}^{\boldsymbol{h}}$ with partitioning. Then the marginal centroids, variances and the correlation coefficient of $\rho_{\mathcal{O}}^{V}$ have to be defined with respect to $(\widetilde{m}_i, \widetilde{m}_f)$. As $\langle V\rangle^{\widetilde{m}} = 0$, the marginal centroids of will be zero. Similarly, in the binary correlation approximation as discussed ahead, it is plausible that the variances are $\langle V^2\rangle^{\widetilde{m}}$. However, for the correlation coefficient ζ, theory with general partitioning is not available yet and therefore it has be assumed to be a constant and it will be generated by V,

$$\zeta = \frac{\langle \mathcal{O}^\dagger V \mathcal{O} V\rangle}{\langle \mathcal{O}^\dagger \mathcal{O}\rangle\langle V^2\rangle}. \tag{7.13}$$

With all these, we have [37–39],

$$\begin{aligned}
\left|\langle E_f|\mathcal{O}|E_i\rangle\right|^2 &= \sum_{\widetilde{m}_i,\widetilde{m}_f} \frac{I_{\mathcal{G}}^{\widetilde{m}_i}(E_i) I_{\mathcal{G}}^{\widetilde{m}_f}(E_f)}{I^{m_i}(E_i) I^{m_f}(E_f)} \left|\langle \widetilde{m}_f|\mathcal{O}|\widetilde{m}_i\rangle\right|^2 \\
&\quad \times \frac{\rho_{\mathcal{O}:biv-\mathcal{G}}(E_i, E_f, E_c(\widetilde{m}_i), E_c(\widetilde{m}_f), \sigma(\widetilde{m}_i), \sigma(\widetilde{m}_f), \zeta)}{\rho_{\mathcal{G}}^{\widetilde{m}_i}(E_i)\rho_{\mathcal{G}}^{\widetilde{m}_f}(E_f)}; \\
\left|\langle \widetilde{m}_f|\mathcal{O}|\widetilde{m}_i\rangle\right|^2 &= \left[d(\widetilde{m}_i) d(\widetilde{m}_f)\right]^{-1} \sum_{\alpha,\beta} \left|\langle \widetilde{m}_f, \alpha|\mathcal{O}|\widetilde{m}_i, \beta\rangle\right|^2.
\end{aligned} \tag{7.14}$$

Theory for transition strengths given by Eq. (7.14) was applied to a variety of problems in nuclear physics and they are listed in Table 7.1. Here below we will briefly discuss a simplified form of Eq. (7.14) for one-body transition operators.

Given a one-body transition operator $\mathcal{O} = \sum_{\alpha,\beta} \varepsilon_{\alpha\beta} a_\alpha^\dagger a_\beta$, it is possible to simplify Eq. (7.14) to a simple form that involves occupation probabilities and the summations all removed. The steps involved are: (i) evaluating $|\langle \widetilde{m}_f|\mathcal{O}|\widetilde{m}_i\rangle|^2$ gives the $\langle \hat{n}_\beta(1-\hat{n}_\alpha)\rangle^{\widetilde{m}_i}$ term (after ignoring $\delta_{\alpha\beta}$ corrections) with $\hat{n}_\alpha$ giving m_α/N_α and N_α is the degeneracy of the sp orbit α; (ii) replace it by $\langle \hat{n}_\beta(1-\hat{n}_\alpha)\rangle^{E_i}$ and this is valid in the chaotic domain where occupancies vary slowly; (iii) assuming constant spectral widths, i.e. σ^2 in Eq. (7.14) do not depend on $\widetilde{m}$ so that $\sigma^2(\widetilde{m}_i) \to \overline{\sigma_i}^2$ and $\sigma^2(\widetilde{m}_f) \to \overline{\sigma_f}^2$; (iv) converting the sum over $\widetilde{m}_i$ into an integral (note that for a given (α, β) and $\widetilde{m}_i$, there is a unique $\widetilde{m}_f$). These will give the following compact form [40],

$$\begin{aligned}
&\left|\langle E_f|\mathcal{O}|E_i\rangle\right|^2 \\
&\quad = \sum_{\alpha,\beta} |\varepsilon_{\alpha\beta}|^2 \big\langle \hat{n}_\beta(1-\hat{n}_\alpha)\big\rangle^{E_i} \overline{D(E_f)} \mathcal{F}(\Delta = E_f - E_i + \varepsilon_\beta - \varepsilon_\alpha, \overline{\sigma_i}, \overline{\sigma_f}, \zeta_{biv})
\end{aligned} \tag{7.15}$$

where

$$\mathcal{F}(\Delta, \overline{\sigma_i}, \overline{\sigma_f}, \zeta) = \frac{1}{\sqrt{2\pi(\overline{\sigma_i}^2 + \overline{\sigma_f}^2 - 2\zeta\overline{\sigma_i}\,\overline{\sigma_f})}} \exp -\frac{\Delta^2}{2(\overline{\sigma_i}^2 + \overline{\sigma_f}^2 - 2\zeta\overline{\sigma_i}\,\overline{\sigma_f})}. \tag{7.16}$$

Table 7.1 Applications of the theory for transition strengths in nuclear structure

No.	Topic	Authors and references
1	Bound on time reversal non-invariant part of nucleon-nucleon interaction	J.B. French, V.K.B. Kota, A. Pandey and S. Tomsovic *Phys. Rev. Lett.* **58** (1987) 2400 *Ann. Phys. (N.Y.)* **181** (1988) 235
2	Parity breaking matrix elements in compound resonance region	S. Tomsovic, M.B. Johnson, A.C. Hayes and J.D. Bowman *Phys. Rev. C* **62** (2003) 054607
3	Single particle transfer	V. Potbhare and N. Tressler *Nucl. Phys.* **A530** (1991) 171
4	Beta decay half lives and rates for presupernovae stars and r-process	K. Kar, S. Sarkar and A. Ray *APJ* **434** (1994) 662 V.K.B. Kota and D. Majumdar *Z. Phys.* **A351** (1995) 377 K. Kar, S. Chakravarti and V.R. Manfredi *Pramana-J. Phys.* **67** (2006) 363
5	Giant dipole widths	D. Majumdar, K. Kar and A. Ansari *J. Phys.* **G23** (1997) L41; **G24** (1998) 2103

Equation (7.16) extends simply to the situation $\rho_{biv-\mathscr{G}} \to \rho_{biv-t}$ with $\mathscr{F}$ changing to [41],

$$\mathscr{F} = \frac{\Gamma(\frac{\nu+1}{2})}{\sqrt{\pi}\,\Gamma(\frac{\nu}{2})} \frac{1}{\sqrt{\nu(\sigma_1^2+\sigma_2^2-2\zeta\sigma_1\sigma_2)}} \left[1 + \frac{\Delta^2}{\nu(\sigma_1^2+\sigma_2^2-2\zeta\sigma_1\sigma_2)}\right]^{-\frac{\nu+1}{2}}. \tag{7.17}$$

Equation (7.17) reduces to the result derived by Flambaum et al. employing the BW form for the strength functions [25, 42]. Then, with $\nu = 1$ in Eq. (7.17),

$$\mathscr{F}(\Delta, \overline{\Gamma_i}, \overline{\Gamma_f})_{BW} = \frac{1}{2\pi} \frac{\overline{\Gamma_i} + \overline{\Gamma_f}}{\Delta^2 + (\overline{\Gamma_i} + \overline{\Gamma_f})^2/4} \tag{7.18}$$

where $\overline{\Gamma_i}$ and $\overline{\Gamma_f}$ are the average BW spreading widths for the basis states over the initial and final many-particle states respectively. It should be clear that (7.18) underestimates the transition matrix elements since it ignores the effects due to the bivariate correlation coefficient ζ_{biv} that appears in the full theory given by Eq. (7.14). In a significant application in atomic physics, the strongly enhanced low-energy electron recombination observed in Au^{25+} was studied by Flambaum et al. [43] using the theory given by Eqs. (7.15) and (7.18) for transition strengths generated by one-body operators and the operator involved here is the dipole operator. Similarly, recombination of low energy electrons with U^{28+} has been studied in [44]. Equations (7.15) and (7.16) should be useful in the calculation of GT strengths in nuclei.

7.2.3 Neutrinoless Double-β Decay: Binary Correlation Results

Double-β decay (DBD) is an extremely rare weak-interaction process in which two identical nucleons inside the atomic nucleus undergo decay with or without emission of neutrinos. The neutrinoless double-β decay (NDBD or $0\nu\beta^-\beta^-$), where two neutrons change into two protons without emitting any neutrinos, is of fundamental significance as its experimental confirmation will tell us about lepton number violation in nature and that the neutrino is a Majorana particle. In particular, experimental value for NDBD half-life gives a value or a bound on neutrino mass [45] provided the corresponding nuclear transition matrix elements (NTME) are obtained using a reliable nuclear model. Thus, the focus in nuclear physics is to calculate NTME for NDBD candidate nuclei. Half-life for NDBD, for the decay of a initial even-even nucleus from its gs (with $J_i^\pi = 0_i^+$) to the gs of the final even-even nucleus (with $J_f^\pi = 0_f^+$), is given by [46]

$$\left[T_{1/2}^{0\nu}(0_i^+ \to 0_f^+)\right]^{-1} = G^{0\nu}\left|M^{0\nu}(0^+)\right|^2 \left(\frac{\langle m_\nu \rangle}{m_e}\right)^2, \tag{7.19}$$

where $\langle m_\nu \rangle$ is the effective neutrino mass and $G^{0\nu}$ is a phase space integral (kinematical factor) [47, 48]. The $M^{0\nu}$ is the NTME generated by the NDBD transition operator $\mathcal{O}(2:0\nu)$ and it is a sum of a Gamow-Teller like (M_{GT}), Fermi like (M_F) and tensor (M_T) two-body operators. For the discussion in this section, explicit form of the two-body operator $\mathcal{O}(2:0\nu)$ is not essential except the property that this operator changes two neutrons (n) into two protons (p). The NTME $|M^{0\nu}|^2$ can be viewed as a transition strength (matrix element connecting a given initial state to a final state by a transition operator) generated by the two-body transition operator $\mathcal{O}(2:0\nu)$. Therefore the transition strength theory given by Eq. (7.14) can be in principle applied [49] by replacing $\widetilde{m}$ by proton-neutron configurations with fixed J value so that $\widetilde{m}_i \to (\widetilde{m}_p, \widetilde{m}_n)_i J_i = 0$ and $\widetilde{m}_f \to (\widetilde{m}_p, \widetilde{m}_n)_f J_f = 0$. Before proceeding to implement this theory, it is essential to establish that the spreading function $\rho_{\mathcal{O}}^V$ generated by V is of bivariate Gaussian form for $\mathcal{O}(2:0\nu)$ type of operators. In addition, we also need an expression for the bivariate correlation coefficient ζ.

In order to establish that generically $\rho_{\mathcal{O}}^V$ is a bivariate Gaussian, formulas for the first four bivariate moments (and cumulants) are derived (see Sect. 7.2.3.2 ahead) for $\rho_{\mathcal{O}:biv}^{(m_p,m_n)_i,(m_p,m_n)_f:H}(E_i, E_f; E_c^i, E_c^f, \sigma_i, \sigma_f, \zeta)$, the spreading function defined over proton-neutron spaces with H being a two-body Hamiltonian. Note that the third and higher order cumulants are zero for a bivariate Gaussian. As appropriate for heavy nuclei, $H = H_{pp} + H_{nn} + H_{pn}$ and it preserves (m_p, m_n). Similarly, the transition operator $\mathcal{O}$ changes (m_p, m_n) to $(m_p + 2, m_n - 2)$. For these types of H and $\mathcal{O}$ operators, we need averages over the two-orbit configurations (m_p, m_n). For generality, from now onwards in the reminder of this section, these are denoted as (m_1, m_2) with m_1 being number of particles in the first orbit and m_2 in the second orbit. In order to derive formulas for the bivariate moments of $\rho_{\mathcal{O}}^H$, an extended binary correlation theory for two-orbit configurations is needed (formulation given in Chap. 5 and applied in Chaps. 5 and 6 is for a single orbit) and this is as follows.

7.2.3.1 Basic Binary Correlation Results for Two-Orbit Configuration Averages

Let us consider m particles in two orbits with number of sp states being N_1 and N_2 respectively. Now the m-particle space can be divided into configurations (m_1, m_2) with m_1 particles in the #1 orbit and m_2 particles in the #2 orbit such that $m = m_1 + m_2$. Similarly, consider H to be k_H-body operator with fixed body ranks i and j respectively in m_1 and m_2 spaces such that (m_1, m_2) is preserved by H. Then the general form for H is,

$$H(k_H) = \sum_{i+j=k_H;\alpha,\beta,\gamma,\delta} \left[v_H^{\alpha\beta\gamma\delta}(i,j)\right] \alpha_1^\dagger(i)\beta_1(i)\gamma_2^\dagger(j)\delta_2(j). \tag{7.20}$$

To proceed further, we will represent H by a EGOE ensemble such that $v_H^{\alpha\beta\gamma\delta}(i,j)$ are independent $G(0, v_H^2(i,j))$ variables. Thus, the ensemble here is a two-orbit EGOE(k_H) or more precisely a EGOE(k_H)-$[U(N_1) + U(N_2)]$ ensemble with $U(N_1)$ generating m_1 and $U(N_2)$ generating m_2. Now, in the dilute limit

$$\begin{aligned}
&\overline{\left\langle H^2(k_H)\right\rangle^{m_1,m_2}} \\
&\quad = \sum_{i+j=k_H} v_H^2(i,j) \sum_{\alpha,\beta,\gamma,\delta} \left\langle \alpha_1^\dagger(i)\beta_1(i)\gamma_2^\dagger(j)\delta_2(j)\beta_1^\dagger(i)\alpha_1(i)\delta_2^\dagger(j)\gamma_2(j)\right\rangle^{m_1,m_2} \\
&\quad = \sum_{i+j=k_H} v_H^2(i,j) \sum_{\alpha,\beta} \left\langle \alpha_1^\dagger(i)\beta_1(i)\beta_1^\dagger(i)\alpha_1(i)\right\rangle^{m_1} \sum_{\gamma,\delta} \left\langle \gamma_2^\dagger(j)\delta_2(j)\delta_2^\dagger(j)\gamma_2(j)\right\rangle^{m_2} \\
&\quad = \sum_{i+j=k_H} v_H^2(i,j) T(m_1, N_1, i) T(m_2, N_2, j).
\end{aligned} \tag{7.21}$$

Note that the function T is defined by Eqs. (4.12) and (4.13). Just as above, extending the single orbit results for product of four operator given in Sect. 4.2, formula for $\overline{\langle H(k_H)G(k_G)H(k_H)G(k_G)\rangle^{m_1,m_2}}$ is,

$$\begin{aligned}
&\overline{\left\langle H(k_H)G(k_G)H(k_H)G(k_G)\right\rangle^{m_1,m_2}} \\
&\quad = \sum_{i+j=k_H, t+u=k_G} \; \sum_{\alpha_1,\beta_1,\gamma_1,\delta_1,\alpha_2,\beta_2,\gamma_2,\delta_2} v_H^2(i,j) v_G^2(t,u) \\
&\qquad \times \left\langle \alpha_1(i)\beta_1^\dagger(i)\gamma_1(t)\delta_1^\dagger(t)\beta_1(i)\alpha_1^\dagger(i)\delta_1(t)\gamma_1^\dagger(t)\right\rangle^{m_1} \\
&\qquad \times \left\langle \alpha_2(j)\beta_2^\dagger(j)\gamma_2(u)\delta_2^\dagger(u)\beta_2(j)\alpha_2^\dagger(j)\delta_2(u)\gamma_2^\dagger(u)\right\rangle^{m_2}.
\end{aligned} \tag{7.22}$$

Here, G is a two-orbit EGOE(k_G) just as H is a two-orbit EGOE(k_H) and it is assumed that the two EGOEs are independent. Applying Eqs. (4.19) and (4.26) to the two traces in Eq. (7.22) gives to leading order,

$$\begin{aligned}&\overline{\langle H(k_H)G(k_G)H(k_H)G(k_G)\rangle}^{m_1,m_2}\\&\quad=\sum_{i+j=k_H,t+u=k_G} v_H^2(i,j)v_G^2(t,u)F(m_1,N_1,i,t)F(m_2,N_2,j,u). \quad (7.23)\end{aligned}$$

The $F(\cdots)$'s appearing in Eq. (7.23) are given by Eqs. (4.17) and (4.26). Combining Eqs. (7.21) and (7.23), we have

$$\begin{aligned}&\overline{\langle H^4(k_H)\rangle}^{m_1,m_2}\\&\quad=2\Bigg[\sum_{i+j=k_H} v_H^2(i,j)T(m_1,N_1,i)T(m_2,N_2,j)\Bigg]^2\\&\quad+\sum_{i+j=k_H,t+u=k_H} v_H^2(i,j)v_H^2(t,u)F(m_1,N_1,i,t)F(m_2,N_2,j,u). \quad (7.24)\end{aligned}$$

Formulation given here is applied in the next section.

7.2.3.2 Binary Correlation Results for the Bivariate Moments of $\rho_{\mathcal{O}:biv}^{(m_1,m_2),(m_1',m_2'):H}$

With space #1 denoting protons and similarly space #2 neutrons, general form of H for the present purpose is given by Eq. (7.20) and it is represented by a two-orbit EGOE(k_H) defined before. Similarly, transition operator $\mathcal{O}$ for the present purpose is of the form,

$$\mathcal{O}(k_{\mathcal{O}})=\sum_{\gamma,\delta} v_{\mathcal{O}}^{\gamma\delta}(k_{\mathcal{O}})\gamma_1^\dagger(k_{\mathcal{O}})\delta_2(k_{\mathcal{O}}). \quad (7.25)$$

Note that $k_{\mathcal{O}}=2$ for NDBD (similarly, $k_{\mathcal{O}}=1$ for β-decay GT strengths). Again, $\mathcal{O}$ is represented by a EGOE in the sense that $v_{\mathcal{O}}^{\gamma\delta}$ are independent $G(0,v_{\mathcal{O}}^2)$ variables. The operator $H(k_H)$ preserves the two orbit configurations (m_1,m_2) and $\mathcal{O}$ and its hermitian conjugate $\mathcal{O}^\dagger$ do not preserve (m_1,m_2). However, the action of $\mathcal{O}$ and $\mathcal{O}^\dagger$ is simple giving $\mathcal{O}(k_{\mathcal{O}})|m_1,m_2\rangle=|m_1+k_{\mathcal{O}},m_2-k_{\mathcal{O}}\rangle$ and $\mathcal{O}^\dagger(k_{\mathcal{O}})|m_1,m_2\rangle=|m_1-k_{\mathcal{O}},m_2+k_{\mathcal{O}}\rangle$. Thus, given a (m_1,m_2) for an initial state, the (m_1',m_2') for the final state generated by the action of $\mathcal{O}$ is uniquely defined. Now, the transition strength density $I_{\mathcal{O}}(E_i,E_f)$ and the corresponding bivariate moments are

$$\begin{aligned}&I_{\mathcal{O}:biv}^{(m_1,m_2),(m_1',m_2'):H}(E_i,E_f)\\&\quad=I^{(m_1',m_2'):H}(E_f)\big|\big\langle(m_1',m_2')E_f\big|\mathcal{O}\big|(m_1,m_2)E_i\big\rangle\big|^2 I^{(m_1,m_2):H}(E_i), \quad (7.26)\end{aligned}$$

$$\widetilde{M}_{PQ}\big((m_1,m_2)\big)=\overline{\langle\mathcal{O}^\dagger(k_{\mathcal{O}})H^Q(k_H)\mathcal{O}(k_{\mathcal{O}})H^P(k_H)\rangle}^{(m_1,m_2)}. \quad (7.27)$$

Note that, $\widetilde{M}$ are in general non-central and non-normalized moments. Also, the final configuration (m_1',m_2') is not specified in defining $\widetilde{M}$ as it is unique given a

(m_1, m_2). To apply BCA in the derivation of the formulas for the bivariate moments, the EGOEs representing $H(k_H)$ and $\mathcal{O}(k_\mathcal{O})$ are assumed to be independent. We will begin with $\widetilde{M}_{00}(m_1, m_2)$.

Using Eq. (7.25) and applying the basic rules given by Eqs. (4.6) and (4.7), we have

$$\begin{aligned}
\widetilde{M}_{00}(m_1, m_2) &= \overline{\langle \mathcal{O}^\dagger(k_\mathcal{O}) \mathcal{O}(k_\mathcal{O}) \rangle^{m_1, m_2}} \\
&= \sum_{\gamma, \delta} \overline{\left\{ v_\mathcal{O}^{\gamma\delta} \right\}^2} \langle \delta_2^\dagger(k_\mathcal{O}) \gamma_1(k_\mathcal{O}) \gamma_1^\dagger(k_\mathcal{O}) \delta_2(k_\mathcal{O}) \rangle^{m_1, m_2} \\
&= v_\mathcal{O}^2 \binom{\widetilde{m}_1}{k_\mathcal{O}} \binom{m_2}{k_\mathcal{O}}.
\end{aligned} \tag{7.28}$$

Trivially, $\widetilde{M}_{10}(m_1, m_2)$ and $\widetilde{M}_{01}(m_1, m_2)$ will be zero as $H(k_H)$ is represented by a two-orbit EGOE(k_H). Thus, $\widetilde{M}_{PQ}(m_1, m_2)$ are central moments. Moreover, by definition, all the odd-order moments are zero giving $\widetilde{M}_{PQ}(m_1, m_2) = 0$ for $P + Q$ odd. The next non-zero bivariate moment $\widetilde{M}_{11}$ is given by,

$$\begin{aligned}
&\widetilde{M}_{11}(m_1, m_2) \\
&= \overline{\langle \mathcal{O}^\dagger(k_\mathcal{O}) H(k_H) \mathcal{O}(k_\mathcal{O}) H(k_H) \rangle^{m_1, m_2}} \\
&= v_\mathcal{O}^2 \sum_{\alpha_1, \beta_1, \alpha_2, \beta_2, \gamma_1, \delta_2; i+j=k_H} v_H^2(i, j) \langle \gamma_1^\dagger(k_\mathcal{O}) \alpha_1(i) \beta_1^\dagger(i) \gamma_1(k_\mathcal{O}) \beta_1(i) \alpha_1^\dagger(i) \rangle^{m_1} \\
&\quad \times \langle \delta_2(k_\mathcal{O}) \alpha_2(j) \beta_2^\dagger(j) \delta_2^\dagger(k_\mathcal{O}) \beta_2(j) \alpha_2^\dagger(j) \rangle^{m_2}.
\end{aligned} \tag{7.29}$$

Then, contracting over the $\gamma^\dagger \gamma$ and $\delta \delta^\dagger$ operators, respectively in the first and second traces in Eq. (7.28) using Eqs. (4.8) and (4.9) appropriately, we have

$$\begin{aligned}
\widetilde{M}_{11}(m_1, m_2) = v_\mathcal{O}^2 \sum_{i+j=k_H} v_H^2(i, j) \binom{\widetilde{m}_1 - i}{k_\mathcal{O}} \binom{m_2 - j}{k_\mathcal{O}} \\
\times T(m_1, N_1, i) T(m_2, N_2, j).
\end{aligned} \tag{7.30}$$

Note that the formulas for the functions $T(\cdots)$'s appearing in Eq. (7.30) are given by Eqs. (4.12), (4.13) and (4.14). Similarly, the functions $F(\cdots)$'s appearing ahead are given by Eqs. (4.17) and (4.26). For the marginal variances, we have

$$\begin{aligned}
\widetilde{M}_{20}(m_1, m_2) &= \overline{\langle \mathcal{O}^\dagger(k_\mathcal{O}) \mathcal{O}(k_\mathcal{O}) H^2(k_H) \rangle^{m_1, m_2}} \\
&= \widetilde{M}_{00}(m_1, m_2) \overline{\langle H^2(k_H) \rangle^{m_1, m_2}}, \\
\widetilde{M}_{02}(m_1, m_2) &= \overline{\langle \mathcal{O}^\dagger(k_\mathcal{O}) H^2(k_H) \mathcal{O}(k_\mathcal{O}) \rangle^{m_1, m_2}} \\
&= \widetilde{M}_{00}(m_1, m_2) \overline{\langle H^2(k_H) \rangle^{m_1 + k_\mathcal{O}, m_2 - k_\mathcal{O}}}.
\end{aligned} \tag{7.31}$$

In Eq. (7.31), the ensemble averages of $H^2(k_H)$ are given by Eq. (7.21). Now, the bivariate correlation coefficient ζ_{biv} is

$$\zeta_{biv}(m_1,m_2) = \frac{\widetilde{M}_{11}(m_1,m_2)}{\sqrt{\widetilde{M}_{20}(m_1,m_2)\widetilde{M}_{02}(m_1,m_2)}}. \tag{7.32}$$

Clearly, ζ_{biv} will be independent of $v_{\mathcal{O}}^2$.

Proceeding further, derived are formulas for the fourth order moments $\widetilde{M}_{PQ}$, $P+Q=4$. Firstly, for $(PQ)=(40)$ and (04), we have

$$\begin{aligned}
\widetilde{M}_{40}(m_1,m_2) &= \overline{\langle \mathcal{O}^\dagger(k_\mathcal{O})\mathcal{O}(k_\mathcal{O})H^4(k_H)\rangle^{m_1,m_2}} \\
&= \widetilde{M}_{00}(m_1,m_2)\overline{\langle H^4(k_H)\rangle^{m_1,m_2}}, \\
\widetilde{M}_{04}(m_1,m_2) &= \overline{\langle \mathcal{O}^\dagger(k_\mathcal{O})H^4(k_H)\mathcal{O}(k_\mathcal{O})\rangle^{m_1,m_2}} \\
&= \widetilde{M}_{00}(m_1,m_2)\overline{\langle H^4(k_H)\rangle^{m_1+k_\mathcal{O},m_2-k_\mathcal{O}}}.
\end{aligned} \tag{7.33}$$

In Eq. (7.33), the ensemble averages of $H^4(k_H)$ are given by Eq. (7.24). For $(PQ)=(31)$, we have

$$\begin{aligned}
\widetilde{M}_{31}(m_1,m_2) &= \overline{\langle \mathcal{O}^\dagger(k_\mathcal{O})H(k_H)\mathcal{O}(k_\mathcal{O})H^3(k_H)\rangle^{m_1,m_2}} \\
&= \overline{\langle \mathcal{O}^\dagger(k_\mathcal{O})H(k_H)\mathcal{O}(k_\mathcal{O})H(k_H)H(k_H)H(k_H)\rangle^{m_1,m_2}} \\
&\quad + \overline{\langle \mathcal{O}^\dagger(k_\mathcal{O})H(k_H)\mathcal{O}(k_\mathcal{O})H(k_H)H(k_H)H(k_H)\rangle^{m_1,m_2}} \\
&\quad + \overline{\langle \mathcal{O}^\dagger(k_\mathcal{O})H(k_H)\mathcal{O}(k_\mathcal{O})H(k_H)H(k_H)H(k_H)\rangle^{m_1,m_2}}.
\end{aligned} \tag{7.34}$$

First and last terms on RHS of Eq. (7.34) are simple as HH can be taken out of the average leaving with a term similar to $\widetilde{M}_{11}(m_1,m_2)$. For the second term, the $\mathcal{O}^\dagger$ and $\mathcal{O}$ operators are contracted across H operator using Eqs. (4.8) and (4.9) and then one is left with an average of the form $\langle HGHG\rangle$. These will give the final formula,

$$\begin{aligned}
&\widetilde{M}_{31}(m_1,m_2) \\
&\quad = 2\overline{\langle H^2(k_H)\rangle^{m_1,m_2}}\widetilde{M}_{11}(m_1,m_2) \\
&\qquad + \overline{\langle \mathcal{O}^\dagger(k_\mathcal{O})H(k_H)\mathcal{O}(k_\mathcal{O})H(k_H)H(k_H)H(k_H)\rangle^{m_1,m_2}} \\
&\quad = 2\overline{\langle H^2(k_H)\rangle^{m_1,m_2}}\widetilde{M}_{11}(m_1,m_2) + v_\mathcal{O}^2 \sum_{i+j=k_H,\,t+u=k_H} v_H^2(i,j)v_H^2(t,u) \\
&\qquad \times \binom{m_2-j}{k_\mathcal{O}}\binom{\widetilde{m}_1-i}{k_\mathcal{O}} F(m_1,N_1,i,t)F(m_2,N_2,j,u).
\end{aligned} \tag{7.35}$$

Similarly, we have

$$
\begin{aligned}
\widetilde{M}_{13}(m_1,m_2) &= \overline{\langle \mathcal{O}^\dagger(k_\mathcal{O}) H^3(k_H) \mathcal{O}(k_\mathcal{O}) H(k_H)\rangle}^{m_1,m_2} \\
&= \overline{\langle \mathcal{O}^\dagger(k_\mathcal{O}) H(k_H) H(k_H) H(k_H) \mathcal{O}(k_\mathcal{O}) H(k_H)\rangle}^{m_1,m_2} \\
&\quad + \overline{\langle \mathcal{O}^\dagger(k_\mathcal{O}) H(k_H) H(k_H) H(k_H) \mathcal{O}(k_\mathcal{O}) H(k_H)\rangle}^{m_1,m_2} \\
&\quad + \overline{\langle \mathcal{O}^\dagger(k_\mathcal{O}) H(k_H) H(k_H) H(k_H) \mathcal{O}(k_\mathcal{O}) H(k_H)\rangle}^{m_1,m_2} \\
&= 2\overline{\langle H^2(k_H)\rangle}^{m_1+k_\mathcal{O},m_2-k_\mathcal{O}} \widetilde{M}_{11}(m_1,m_2) \\
&\quad + v_\mathcal{O}^2 \sum_{i+j=k_H,t+u=k_H} v_H^2(i,j) v_H^2(t,u) G(t,u) \\
&\quad \times \binom{\widetilde{m}_1 - k_\mathcal{O} - t + i}{i} \binom{m_1 + k_\mathcal{O} - t}{i} \\
&\quad \times \binom{\widetilde{m}_2 - u + k_\mathcal{O} + j}{j} \binom{m_2 - k_\mathcal{O} - u}{j}; \\
G(t,u) &= \binom{\widetilde{m}_1 - t}{k_\mathcal{O}} \binom{m_2 - u}{k_\mathcal{O}} T(m_1,N_1,t) T(m_2,N_2,u).
\end{aligned}
\tag{7.36}
$$

In Eq. (7.36), the first and last terms can be evaluated by first calculating the H^2 average over the intermediate states $|m_1 + k_\mathcal{O}, m_2 - k_\mathcal{O}\rangle$ and then the remaining part is similar to $\widetilde{M}_{11}(m_1,m_2)$. Also, the second average is evaluated by first contracting the two correlated H's that are between $\mathcal{O}^\dagger$ and $\mathcal{O}$ operators and then one is again left with a term similar to $\widetilde{M}_{11}(m_1,m_2)$. Finally, $\widetilde{M}_{22}(m_1,m_2)$ is given by,

$$
\begin{aligned}
\widetilde{M}_{22}(m_1,m_2) &= \overline{\langle \mathcal{O}^\dagger(k_\mathcal{O}) H^2(k_H) \mathcal{O}(k_\mathcal{O}) H^2(k_H)\rangle}^{m_1,m_2} \\
&= \overline{\langle \mathcal{O}^\dagger(k_\mathcal{O}) H(k_H) H(k_H) \mathcal{O}(k_\mathcal{O}) H(k_H) H(k_H)\rangle}^{m_1,m_2} \\
&\quad + \overline{\langle \mathcal{O}^\dagger(k_\mathcal{O}) H(k_H) H(k_H) \mathcal{O}(k_\mathcal{O}) H(k_H) H(k_H)\rangle}^{m_1,m_2} \\
&\quad + \overline{\langle \mathcal{O}^\dagger(k_\mathcal{O}) H(k_H) H(k_H) \mathcal{O}(k_\mathcal{O}) H(k_H) H(k_H)\rangle}^{m_1,m_2} \\
&= \widetilde{M}_{00}(m_1,m_2) \overline{\langle H^2(k_H)\rangle}^{m_1+k_\mathcal{O},m_2-k_\mathcal{O}} \overline{\langle H^2(k_H)\rangle}^{m_1,m_2} \\
&\quad + v_\mathcal{O}^2 \sum_{i+j=k_H,t+u=k_H} v_H^2(i,j) v_H^2(t,u) \binom{\widetilde{m}_1 - i - t}{k_\mathcal{O}} \binom{m_2 - u - j}{k_\mathcal{O}} \\
&\quad \times \big[F(m_1,N_1,i,t) F(m_2,N_2,j,u) \\
&\quad + T(m_1,N_1,i) T(m_1,N_1,t) T(m_2,N_2,j) T(m_2,N_2,u)\big].
\end{aligned}
\tag{7.37}
$$

In Eq. (7.37), the first term is evaluated by first calculating the H^2 average (for the H^2 between $\mathcal{O}^\dagger$ and $\mathcal{O}$ operators) over the intermediate state $|m_1 + k_{\mathcal{O}}, m_2 - k_{\mathcal{O}}\rangle$ and then one is left with product of averages of H^2 and $\mathcal{O}^\dagger\mathcal{O}$ operators. For the third term, first the $\mathcal{O}^\dagger$ and $\mathcal{O}$ operators are contracted across H^2 operator and then we are left with average of the form $\langle H^2\rangle \times \langle H^2\rangle$. Similarly, for the second term, after contracting the $\mathcal{O}^\dagger$ and $\mathcal{O}$ operators across H^2 operator, we are left with an average of the form $\langle HGHG\rangle$. The results for the moments to fourth order given above are reported first in [50].

7.2.3.3 Bivariate Cumulants for NDBD Nuclei

Given the $\widetilde{M}_{PQ}(m_1, m_2)$, the normalized central moments M_{PQ} are $M_{PQ} = \widetilde{M}_{PQ}/\widetilde{M}_{00}$. Then, the reduced moments $\widehat{M}_{PQ}$ are

$$\widehat{M}_{PQ} = \frac{M_{PQ}(m_1, m_2)}{[M_{20}(m_1, m_2)]^{P/2}[M_{02}(m_1, m_2)]^{Q/2}}; \quad P + Q \geq 2. \tag{7.38}$$

In terms of $\widehat{M}_{PQ}$, the fourth order cumulants from Eq. (B.10) are

$$\begin{aligned}
k_{40}(m_1, m_2) &= \widehat{M}_{40}(m_1, m_2) - 3, \qquad k_{04}(m_1, m_2) = \widehat{M}_{04}(m_1, m_2) - 3, \\
k_{31}(m_1, m_2) &= \widehat{M}_{31}(m_1, m_2) - 3\widehat{M}_{11}(m_1, m_2), \\
k_{13}(m_1, m_2) &= \widehat{M}_{13}(m_1, m_2) - 3\widehat{M}_{11}(m_1, m_2), \\
k_{22}(m_1, m_2) &= \widehat{M}_{22}(m_1, m_2) - 2\widehat{M}_{11}^2(m_1, m_2) - 1.
\end{aligned} \tag{7.39}$$

In order to obtain some insight into the values of the fourth order cumulants and the bivariate correlation coefficient for NDBD nuclei, it is assumed that the $v_H^2(i, j) = v_H^2$ independent of (i, j) (note that $k_H = 2$ and $k_{\mathcal{O}} = 2$ in NDBD applications). Then, ζ_{biv} and k_{PQ}, $P + Q = 4$ are functions of only (m_p, m_n, N_p, N_n) and independent of both v_H^2 and $V_{\mathcal{O}}^2$. The ζ_{biv} and k_{PQ}, $P + Q = 4$ are calculated in [50] for several $0\nu\beta^-\beta^-$ decay candidate nuclei using Eqs. (7.28)–(7.37). In these calculations, the function $T(\cdots)$ is evaluated using Eq. (4.12) and the function $F(\cdots)$ using Eq. (4.26). For example, for ^{100}Mo, ^{150}Nd and ^{238}U nuclei, (k_{40}, k_{04}, k_{13}, k_{31}, k_{22}) are (−0.45, −0.42, −0.24, −0.26, −0.20), (−0.27, −0.29, −0.22, −0.20, −0.19) and (−0.18, −0.18, −0.15, −0.15, −0.13) respectively. These results clearly establish that bivariate Gaussian is a good approximation for $0\nu\beta^-\beta^-$ decay transition strength densities (for a good bivariate Gaussian, $|k_{PQ}| \lesssim 0.3$). It is also seen from the numerical calculations that $\zeta_{biv} \sim 0.6$–0.8. It is important to mention that $\zeta_{biv} = 0$ for GOE. Therefore, the transition strength density will be narrow in the (E_i, E_f) plane. All these results show that, one can apply the formulation given by Eq. (7.14) for calculating NTME for NDBD. However, in actual applications it should be recognized that the parent and daughter nuclear states carry angular momentum J as a good quantum number. Therefore, we need $|\langle E_f J_f = 0|\mathcal{O}|E_i J_i = 0\rangle|^2$ (for gs to gs transitions). Then, in order to apply Eq. (7.14), we need $E_c((\widetilde{m}_p, \widetilde{m}_n)_i, J_i = 0)$, $E_c((\widetilde{m}_p, \widetilde{m}_n)_f, J_f = 0)$, $\sigma((\widetilde{m}_p, \widetilde{m}_n)_i, J_i = 0)$, $\sigma((\widetilde{m}_p, \widetilde{m}_n)_f, J_f = 0)$, $|\langle(\widetilde{m}_p, \widetilde{m}_n)_f J_f = 0|\mathscr{O}|(\widetilde{m}_p, \widetilde{m}_n)_i J_i = 0\rangle|^2$ and ζ_{biv}. In this statistical procedure for NDBD NTME calculations, it is possible to assume that ζ_{biv} is a free parameter and its starting value can be taken from the binary correlation theory. Exact calculations of the fixed-J averages is cumbersome (approximations for fixed-J averages are discussed in Chap. 13). However, large scale computer codes are developed recently by Sen'kov et al. [51] and they will give $E_c((\widetilde{m}_p, \widetilde{m}_n), J = 0)$ and $\sigma((\widetilde{m}_p, \widetilde{m}_n), J = 0)$ for medium mass nuclei (these codes may need extensions for heavy nuclei). Also, the methods used by Sen'kov et al. will allow one to derive formulas for $|\langle(\widetilde{m}_p, \widetilde{m}_n)_f J_f = 0|\mathscr{O}|(\widetilde{m}_p, \widetilde{m}_n)_i J_i = 0\rangle|^2$. With these, it is possible in the near future to apply the theory described in this section to NDBD NTME calculations.

References

1. Y. Imry, *Introduction to Mesoscopic Physics* (Oxford University Press, New York, 1997)
2. M. Janssen, *Fluctuations and Localization in Mesoscopic Electron Systems* (World Scientific, Singapore, 2001)
3. T. Guhr, A. Müller-Groeling, H.A. Weidenmüller, Random-matrix theories in quantum physics: common concepts. Phys. Rep. **299**, 189–425 (1998)
4. Y. Alhassid, Statistical theory of quantum dots. Rev. Mod. Phys. **72**, 895–968 (2000)
5. Y. Alhassid, *Mesoscopic Effects in Quantum Dots, Nanoparticles and Nuclei*, AIP Conf. Proc., vol. 777, ed. by V. Zelevinsky (2005), pp. 250–269
6. M. Abramowtiz, I.A. Stegun (eds.), *Handbook of Mathematical Functions*, NBS Applied Mathematics Series, vol. 55 (U.S. Govt. Printing Office, Washington, D.C., 1972)
7. Y. Yoshinaga, A. Arima, Y.M. Zhao, Lowest bound of energies for random interactions and the origin of spin-zero ground state dominance in even-even nuclei. Phys. Rev. C **73**, 017303 (2006)
8. J.J. Shen, Y.M. Zhao, A. Arima, Y. Yoshinaga, Lowest eigenvalue of random Hamiltonians. Phys. Rev. C **77**, 054312 (2008)
9. K.F. Ratcliff, Application of spectral distributions in nuclear spectroscopy. Phys. Rev. C **3**, 117–143 (1971)
10. V.K.B. Kota, R.U. Haq, *Spectral Distributions in Nuclei and Statistical Spectroscopy* (World Scientific, Singapore, 2010)
11. Ph. Jacquod, A.D. Stone, Ground state magnetization for interacting fermions in a disordered potential: kinetic energy, exchange interaction, and off-diagonal fluctuations. Phys. Rev. B **64**, 214416 (2001)
12. M. Vyas, Random interaction matrix ensembles in mesoscopic physics, in *Proceedings of the National Seminar on New Frontiers in Nuclear, Hadron and Mesoscopic Physics*, ed. by V.K.B. Kota, A. Pratap (Allied Publishers, New Delhi, 2010), pp. 23–37
13. Ph. Jacquod, A.D. Stone, Suppression of ground-state magnetization in finite-size systems due to off-diagonal interaction fluctuations. Phys. Rev. Lett. **84**, 3938–3941 (2000)
14. V.K.B. Kota, N.D. Chavda, R. Sahu, One plus two-body random matrix ensemble with spin: analysis using spectral variances. Phys. Lett. A **359**, 381–389 (2006)
15. M. Vyas, Some studies on two-body random matrix ensembles, Ph.D. Thesis, M.S. University of Baroda, India (2012)
16. C.T. Black, D.C. Ralph, M. Tinkham, Spectroscopy of the superconducting gap in individual nanometer-scale aluminum particles. Phys. Rev. Lett. **76**, 688–691 (1996)
17. T. Papenbrock, L. Kaplan, G.F. Bertsch, Odd-even binding effect from random two-body interactions. Phys. Rev. B **65**, 235120 (2002)

18. Y. Alhassid, H.A. Weidenmüller, A. Wobst, Disordered mesoscopic systems with interaction: induced two-body ensembles and the Hartree-Fock approach. Phys. Rev. B **72**, 045318 (2005)
19. S. Schmidt, Y. Alhassid, Mesoscopic competition of superconductivity and ferromagnetism: conductance peak statistics for metallic grains. Phys. Rev. Lett. **101**, 207003 (2008)
20. Y. Alhassid, Ph. Jacquod, A. Wobst, Random matrix model for quantum dots with interactions and the conductance peak spacing distribution. Phys. Rev. B **61**, R13357–R13360 (2000)
21. S.R. Patel, S.M. Cronenwett, D.R. Stewart, A.G. Huibers, C.M. Marcus, C.I. Duruöz, J.S. Harris Jr., K. Campman, A.C. Gossard, Statistics of Coulomb blockade peak spacings. Phys. Rev. Lett. **80**, 4522–4525 (1998)
22. S. Lüscher, T. Heinzel, K. Ensslin, W. Wegscheider, M. Bichler, Signatures of spin pairing in chaotic quantum dots. Phys. Rev. Lett. **86**, 2118–2121 (2001)
23. M. Vyas, V.K.B. Kota, N.D. Chavda, One-plus two-body random matrix ensembles with spin: results for pairing correlations. Phys. Lett. A **373**, 1434–1443 (2009)
24. Y. Alhassid, H.A. Weidenmüller, A. Wobst, Scrambling of Hartree-Fock levels as a universal Brownian-motion process. Phys. Rev. B **76**, 193110 (2007)
25. V.V. Flambaum, A.A. Gribakina, G.F. Gribakin, I.V. Ponomarev, Quantum chaos in many-body systems: what can we learn from the Ce atom. Physica D **131**, 205–220 (1999)
26. J. Karwowski, Statistical theory of spectra. Int. J. Quant. Chem. **51**, 425–437 (1994)
27. J. Karwowski, F. Rajadell, J. Planelles, V. Mas, The first four moments of spectral density distribution of an N-electron Hamiltonian matrix defined in an antisymmetric and spin-adapted model space. At. Data Nucl. Data Tables **61**, 177–232 (1995)
28. J. Planelles, F. Rajadell, J. Karwowski, V. Mas, A diagrammatic approach to statistical spectroscopy of many-fermion Hamiltonians. Phys. Rep. **267**, 161–194 (1996)
29. M. Horoi, J. Kaiser, V. Zelevinsky, Spin- and parity-dependent nuclear level densities and the exponential convergence method. Phys. Rev. C **67**, 054309 (2003)
30. R.A. Sen'kov, M. Horoi, High-performance algorithm to calculate spin- and parity-dependent nuclear level densities. Phys. Rev. C **82**, 024304 (2010)
31. R.A. Sen'kov, M. Horoi, V. Zelevinsky, High-performance algorithm for calculating non-spurious spin- and parity-dependent nuclear level densities. Phys. Lett. B **702**, 413–418 (2011)
32. V.K.B. Kota, D. Majumdar, Application of spectral averaging theory in large shell model spaces: analysis of level density data of fp-shell nuclei. Nucl. Phys. A **604**, 129–162 (1996)
33. J.B. French, S. Rab, J.F. Smith, R.U. Haq, V.K.B. Kota, Nuclear spectroscopy in the chaotic domain: level densities. Can. J. Phys. **84**, 677–706 (2006)
34. F. Borgonovi, G. Celardo, F.M. Izrailev, G. Casati, Semiquantal approach to finite systems of interacting particles. Phys. Rev. Lett. **88**, 054101 (2002)
35. M. Vyas, V.K.B. Kota, N.D. Chavda, Transitions in eigenvalue and wavefunction structure in $(1+2)$-body random matrix ensembles with spin. Phys. Rev. E **81**, 036212 (2010)
36. V.K.B. Kota, Convergence of moment expansions for expectation values with embedded random matrix ensembles and quantum chaos. Ann. Phys. (N.Y.) **306**, 58–77 (2003)
37. J.B. French, V.K.B. Kota, A. Pandey, S. Tomsovic, Statistical properties of many-particle spectra VI. Fluctuation bounds on N-N T-noninvariance. Ann. Phys. (N.Y.) **181**, 235–260 (1988)
38. S. Tomsovic, Bounds on the time-reversal non-invariant nucleon-nucleon interaction derived from transition-strength fluctuations, Ph.D. Thesis, University of Rochester, Rochester, New York (1986)
39. V.K.B. Kota, D. Majumdar, Bivariate distributions in statistical spectroscopy studies: IV. Interacting particle Gamow-Teller strength densities and β-decay rates of fp-shell nuclei for presupernova stars. Z. Phys. A **351**, 377–383 (1995)
40. V.K.B. Kota, R. Sahu, Theory for matrix elements of one-body transition operators in the quantum chaotic domain of interacting particle systems. Phys. Rev. E **62**, 3568–3571 (2000)

41. V.K.B. Kota, N.D. Chavda, R. Sahu, Bivariate t-distribution for transition matrix elements in Breit-Wigner to Gaussian domains of interacting particle systems. Phys. Rev. E **73**, 047203 (2006)
42. V.V. Flambaum, A.A. Gribakina, G.F. Gribakin, M.G. Kozlov, Structure of compound states in the chaotic spectrum of the Ce atom: localization properties, matrix elements, and enhancement of weak perturbations. Phys. Rev. A **50**, 267–296 (1994)
43. V.V. Flambaum, A.A. Gribakina, G.F. Gribakin, C. Harabati, Electron recombination with multicharged ions via chaotic many-electron states. Phys. Rev. A **66**, 012713 (2002)
44. S. Sahoo, G.F. Gribakin, V. Dzuba, Recombination of low energy electrons with U^{28+}, arXiv:physics/0401157v1 [physics.atom-ph]
45. F.T. Avignone III, S.R. Elliott, J. Engel, Double beta decay, Majorana neutrinos, and neutrino mass. Rev. Mod. Phys. **80**, 481–516 (2008)
46. S.R. Elliot, P. Vogel, Double beta decay. Annu. Rev. Nucl. Part. Sci. **52**, 115–151 (2002)
47. F. Boehm, P. Vogel, *Physics of Massive Neutrinos*, 2nd edn. (Cambridge University Press, Cambridge, 1992)
48. J. Kotila, F. Iachello, Phase-space factors for double-β decay. Phys. Rev. C **85**, 034316 (2012)
49. V.K.B. Kota, Nuclear models and statistical spectroscopy for double beta decay, in *Neutrinoless Double Beta Decay*, ed. by V.K.B. Kota, U. Sarkar (Narosa Publishing House, New Delhi, 2008), pp. 63–76
50. M. Vyas, V.K.B. Kota, Spectral distribution method for neutrinoless double-beta decay nuclear transition matrix elements: binary correlation results (2011). arXiv:1106.0395v1 [nucl-th]
51. R.A. Sen'kov, M. Horoi, V. Zelevinsky, A high-performance Fortran code to calculate spin- and parity-dependent nuclear level densities. Comput. Phys. Commun. **184**, 215–221 (2013)

Chapter 8
One Plus Two-Body Random Matrix Ensembles with Parity: EGOE(1 + 2)-π

8.1 EGOE(1 + 2)-π: Definition and Construction

Parity ratios of nuclear level densities, i.e. ratio of number of positive parity states and negative parity states in a atomic nucleus, is an important ingredient in nuclear astrophysical applications [1]. It is possible to understand the general structure of parity ratios by considering embedded random matrix ensembles generated by parity (π) preserving random interactions [2]. With parity, the sp space and the many fermion spaces divide into positive and negative parity spaces. Therefore, we need to start with say N_+ number of positive parity ($\pi = +$) sp states and similarly N_- number of negative parity ($\pi = -1$) sp states. In the first step, one can ignore the internal structure of the sp states in each π space although in nuclei there is a clear separation of the +ve and −ve parity sp levels. With Hamiltonian being a one plus two-body operator preserving parity, the one-body part $\widehat{h}(1)$ is defined by N_+ number of degenerate +ve parity sp states and N_- number of degenerate −ve parity sp states with spacing between them is say Δ. Then we have $N = N_+ + N_-$ sp states. The matrix for the two-body part $\widehat{V}(2)$ will be a 3×3 block matrix in two particle spaces as there are three possible ways to generate two particle states with definite parity: (i) both fermions in +ve parity states; (ii) both fermions in −ve parity states; (iii) one fermion in +ve and other fermion in −ve parity states. Figure 8.1 shows the structure of H operator in the defining space. The matrices A, B and C in the figure correspond to (i), (ii) and (iii) above respectively. For parity preserving interactions only the states (i) and (ii) will be mixed and mixing matrix is D in Fig. 8.1. Note that the matrices A, B and C are symmetric square matrices while D is in general a rectangular mixing matrix. Consider N sp states arranged such that the states 1 to N_+ have +ve parity and states $N_+ + 1$ to N have −ve parity. Then the operator form of H preserving parity is,

V.K.B. Kota, *Embedded Random Matrix Ensembles in Quantum Physics*, Lecture Notes in Physics 884, DOI 10.1007/978-3-319-04567-2_8,

$$
\begin{aligned}
\widehat{H} &= \widehat{h}(1) + \widehat{V}(2);\\
\widehat{h}(1) &= \sum_{i=1}^{N_+} \varepsilon_i^{(+)} \hat{n}_i^{(+)} + \sum_{j=N_++1}^{N} \varepsilon_j^{(-)} \hat{n}_j^{(-)}; \quad \varepsilon_i^{(+)} = 0,\ \varepsilon_j^{(-)} = \Delta,\\
\widehat{V}(2) &= \sum_{\substack{i,j,k,l=1\\(i<j,k<\ell)}}^{N_+} \langle \nu_k \nu_\ell | \widehat{V}(2) | \nu_i \nu_j \rangle a_k^\dagger a_\ell^\dagger a_j a_i\\
&+ \sum_{\substack{i',j',k',\ell'=N_++1\\(i'<j',k'<\ell')}}^{N} \langle \nu_{k'} \nu_{\ell'} | \widehat{V}(2) | \nu_{i'} \nu_{j'} \rangle a_{k'}^\dagger a_{\ell'}^\dagger a_{j'} a_{i'}\\
&+ \sum_{i'',k''=1}^{N_+} \sum_{j'',\ell''=N_++1}^{N} \langle \nu_{k''} \nu_{\ell''} | \widehat{V}(2) | \nu_{i''} \nu_{j''} \rangle a_{k''}^\dagger a_{\ell''}^\dagger a_{j''} a_{i''}\\
&+ \sum_{\substack{P,Q=1\\(P<Q)}}^{N_+} \sum_{\substack{R,S=N_++1\\(R<S)}}^{N} \left[\langle \nu_P \nu_Q | \widehat{V}(2) | \nu_R \nu_S \rangle a_P^\dagger a_Q^\dagger a_S a_R + \text{h.c.} \right].
\end{aligned}
\tag{8.1}
$$

In Eq. (8.1), ν_i's are sp states, $\langle \dots | \widehat{V}(2) | \dots \rangle$ are the two-particle matrix elements and $\hat{n}_i$ are number operators. From now on we will drop the hat over H, h and V when there is no confusion. Note that the four terms in the RHS of the expression for $V(2)$ in Eq. (8.1) correspond respectively to the matrices A, B, C and D shown in Fig. 8.1a. Many particle states for m fermions in the N sp states can be obtained by distributing m_1 fermions in the +ve parity sp states (N_+ in number) and similarly, m_2 fermions in the −ve parity sp states (N_- in number) with $m = m_1 + m_2$. Let us denote each distribution of m_1 fermions in N_+ sp states by $\mathbf{m}_1$ and similarly, $\mathbf{m_2}$ for m_2 fermions in N_- sp states. Many particle basis defined by $(\mathbf{m}_1, \mathbf{m}_2)$ with m_2 even will form the basis for +ve parity states and similarly, with m_2 odd for −ve parity states. In the $(\mathbf{m}_1, \mathbf{m}_2)$ basis with m_2 even (or odd), the H matrix construction reduces to the matrix construction for spinless fermion systems, i.e. Eq. (4.3) will apply. Therefore it is easy to construct the many particle H matrices in +ve and −ve parity spaces. The matrix dimensions d_+ for +ve parity and d_- for −ve parity spaces are given by,

$$
\begin{aligned}
d_+ &= \sum_{m_1,m_2(m_2\ \mathit{even})} d(N_+, N_- : m_1, m_2),\\
d_- &= \sum_{m_1,m_2(m_2\ \mathit{odd})} d(N_+, N_- : m_1, m_2);\\
d(N_+, N_- : m_1, m_2) &= \binom{N_+}{m_1}\binom{N_-}{m_2}.
\end{aligned}
\tag{8.2}
$$

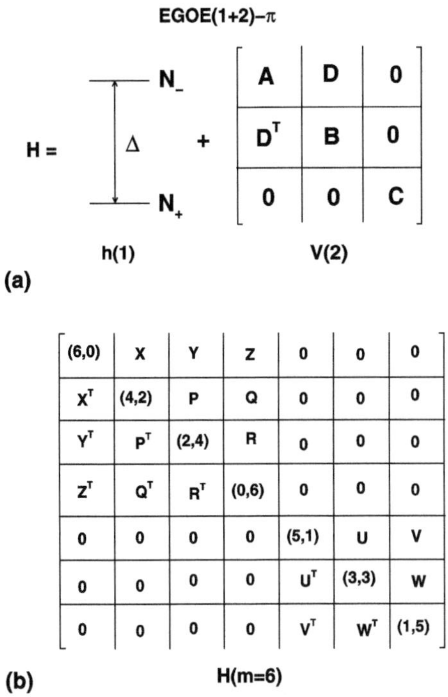

Fig. 8.1 (**a**) Parity preserving one plus two-body H with a sp spectrum defining $h(1)$ along with a schematic form of the $V(2)$ matrix in the two-particle space. Dimension of the matrices A, B and C are $N_+(N_+ - 1)/2$, $N_-(N_- - 1)/2$, and N_+N_-, respectively. Note that D^T is the transpose of the matrix D. (**b**) H matrix for a $m = 6$ system. The upper 4×4 matrix corresponds to +ve parity and the lower 3×3 matrix to negative parity. The (m_1, m_2) values for each of the diagonal block are shown in the figures. Dimensions of the matrices that correspond to the diagonal blocks and the total matrix dimension are given by Eq. (8.2)

The EGOE(1 + 2)-π ensemble is defined by choosing the matrices A, B and C to be independent GOEs with matrix elements variances v_a^2, v_b^2 and v_c^2 respectively [2]. Similarly, matrix elements of the mixing D matrix are chosen to be independent (independent of A, B and C matrix elements) zero centered Gaussian variables with variance v_d^2. Without loss of generality we choose $\Delta = 1$ so that all the v's are in Δ units. This general EGOE(1 + 2)-π model will have too many parameters $(v_a^2, v_b^2, v_c^2, v_d^2, N_+, N_-, m)$ and therefore it is necessary to reduce the number of parameters. A numerically tractable and physically relevant (as discussed ahead) restriction is to choose the matrix elements variances of the diagonal blocks A, B and C to be same and then we have the EGOE(1 + 2)-π model defined by (N_+, N_-, m) and the variance parameters (τ, α) where

$$\frac{v_a^2}{\Delta^2} = \frac{v_b^2}{\Delta^2} = \frac{v_c^2}{\Delta^2} = \tau^2, \qquad \frac{v_d^2}{\Delta^2} = \alpha^2. \tag{8.3}$$

Thus, for EGOE(1 + 2)-π

$$\begin{aligned} & A : \text{GOE}\big(\tau^2\big), \qquad B : \text{GOE}\big(\tau^2\big), \qquad C : \text{GOE}\big(\tau^2\big), \qquad D : \text{GOE}\big(\alpha^2\big); \\ & \quad A, B, C, D \text{ are independent GOE's.} \end{aligned} \tag{8.4}$$

Note that the D matrix is a GOE only in the sense that its matrix elements D_{ij} are independent zero centered Gaussian variables with variance α^2. In the limit $\tau^2 \to \infty$

and $\alpha = \tau$, the model defined by Eqs. (8.1), (8.3) and (8.4) reduces to the simpler model analyzed in [3]. Figure 8.1b shows an example of the matrix structure of EGOE(1 + 2)-π for a $m = 6$ system. Then the H matrix in +ve parity space will be a 4×4 block matrix with the four diagonal blocks mixed by the off-diagonal block matrices X, Y, Z, P, Q and R as shown in the figure. Similarly, the H matrix in −ve parity space will be a 3×3 block matrix with the three diagonal blocks mixed by the off-diagonal block matrices U, V, and W as shown in the figure.

The smoothed +ve and −ve parity densities $\rho^m_\pm(E)$ are a sum of the partial densities $\rho^{m_1,m_2}(E)$,

$$\begin{aligned} \rho^m_\pm(E) &= \big\langle \delta(H-E)\big\rangle^{m,\pm} = \frac{1}{d_\pm} \sum_{m_1,m_2}{}' \, d(m_1,m_2)\rho^{m_1,m_2}(E); \\ \rho^{m_1,m_2}(E) &= \big\langle \delta(H-E)\big\rangle^{m_1,m_2}. \end{aligned} \tag{8.5}$$

Note that the summation in Eq. (8.5) is over m_2 even for +ve parity density and similarly over m_2 odd for −ve parity density. Also, we have dropped N_+ and N_- in $d(N_+, N_-, m_1, m_2)$. In Eq. (8.5), $\rho^m_\pm(E)$ as well as $\rho^{m_1,m_2}(E)$ are normalized to unity. However, in practice, the densities normalized to dimensions are needed and they are denoted by $I^m_\pm(E)$ and $I^{m_1,m_2}(E)$ respectively,

$$I^m_\pm(E) = d_\pm \rho^m_\pm(E) = \sum_{m_1,m_2}{}' \, I^{m_1,m_2}(E); \qquad I^{m_1,m_2}(E) = d(m_1,m_2)\rho^{m_1,m_2}(E). \tag{8.6}$$

In order to understand $I^m_\pm(E)$ and hence the parity ratios, $I^{m_1,m_2}(E)$ are examined [2] via their moments $M_p(m_1,m_2) = \langle H^p\rangle^{m_1,m_2}$ using BCA for the two-orbit averages given in Sect. 7.2.3.

8.2 Binary Correlation Results for Lower Order Moments of Fixed-(m_1, m_2) Partial Densities

For the EGOE(1 + 2)-π Hamiltonian, we have $H = h(1) + V(2) = h(1) + X(2) + D(2)$ with $X(2) = A \oplus B \oplus C$ is the direct sum of the spreading matrices A, B and C and $D(2) = D + \widetilde{D}$ is the off-diagonal mixing matrix. Here, $\widetilde{D} = D^\dagger$ is the transpose of the matrix D. Denoting +ve parity sp space by #1 and −ve parity sp space by #2, operator form of D is

$$D(2) = \sum_{\gamma,\delta} v_D^{\gamma\delta} \gamma_1^\dagger(2)\delta_2(2), \tag{8.7}$$

with $\overline{[v_D^{\gamma,\delta}]^2} = v_D^2 = \alpha^2$. From now on, we will denote $X(2)$ by X and $D(2)$ by D. Note that the operator form of X is given by Eq. (7.20) and $\overline{[v_X^{\alpha\beta\gamma\delta}(i,j)]^2} = v_X^2(i,j) = \tau^2$ with $i + j = 2$. Also, $h(1)$ conserves (m_1, m_2) symmetry while X

preserves (m_1, m_2) symmetry. Using all these, results in Sect. 7.2.3 are applied to derive formulas for $M_r(m_1, m_2)$ with $r \leq 4$. These results will be good in the dilute limit defined by $m_1, N_1, m_2, N_2 \to \infty$, $m/N_1 \to 0$ and $m/N_2 \to 0$ with $m = m_1$ or $m = m_2$. The first moment $M_1(m_1, m_2)$ of the partial densities $\rho^{m_1,m_2}(E)$ is trivially,

$$M_1(m_1, m_2) = \overline{\langle (h+V) \rangle^{m_1,m_2}} = m_2 \tag{8.8}$$

as $\langle h^r \rangle^{m_1,m_2} = (m_2)^r$ and $\overline{\langle V \rangle^{m_1,m_2}} = 0$. The second moment M_2 is,

$$\begin{aligned} M_2(m_1, m_2) &= \overline{\langle (h+V)^2 \rangle^{m_1,m_2}} \\ &= \langle h^2 \rangle^{m_1,m_2} + \overline{\langle V^2 \rangle^{m_1,m_2}} = (m_2)^2 + \overline{\langle V^2 \rangle^{m_1,m_2}}; \end{aligned}$$

$$\begin{aligned} \overline{\langle V^2 \rangle^{m_1,m_2}} &= \overline{\langle X^2 \rangle^{m_1,m_2}} + \overline{\langle D\widetilde{D} \rangle^{m_1,m_2}} + \overline{\langle \widetilde{D}D \rangle^{m_1,m_2}}, \\ \overline{\langle X^2 \rangle^{m_1,m_2}} &= \tau^2 \sum_{i+j=2} T(m_1, N_1, i) T(m_2, N_2, j), \\ \overline{\langle D\widetilde{D} \rangle^{m_1,m_2}} &= \alpha^2 \binom{m_1}{2}\binom{\widetilde{m}_2}{2}, \qquad \overline{\langle \widetilde{D}D \rangle^{m_1,m_2}} = \alpha^2 \binom{\widetilde{m}_1}{2}\binom{m_2}{2}. \end{aligned} \tag{8.9}$$

The second line in Eq. (8.9) follows by using the fact that $X(2)$ and $D(2)$ are independent and $D(2)$ can correlate only with $\widetilde{D}(2)$. In Eq. (8.9), the expression for $\overline{\langle X^2 \rangle^{m_1,m_2}}$ follows directly from Eq. (7.21). The last two equations in Eq. (8.9) can be derived using Eq. (8.7) that gives the definition of the operator $D(2)$ and Eqs. (4.6) and (4.7) appropriately to contract the operators $\gamma^\dagger$ with γ and δ with $\delta^\dagger$. For the $T(\cdots)$'s in Eq. (8.9), Eq. (4.12) is used. Note that, Eq. (8.9) gives the binary correlation formula for $\sigma^2(m_1, m_2)$. Similarly, the third moment M_3 is

$$\begin{aligned} M_3(m_1, m_2) &= \overline{\langle (h+V)^3 \rangle^{m_1,m_2}} \\ &= \langle h^3 \rangle^{m_1,m_2} + 2\langle h \rangle^{m_1,m_2} \overline{\langle V^2 \rangle^{m_1,m_2}} + \overline{\langle XhX \rangle^{m_1,m_2}} \\ &\quad + \overline{\langle Dh\widetilde{D} \rangle^{m_1,m_2}} + \overline{\langle \widetilde{D}hD \rangle^{m_1,m_2}} \\ &= (m_2)^3 + 2m_2 \overline{\langle V^2 \rangle^{m_1,m_2}} + m_2 \overline{\langle X^2 \rangle^{m_1,m_2}} \\ &\quad + (m_2+2) \overline{\langle D\widetilde{D} \rangle^{m_1,m_2}} + (m_2-2) \overline{\langle \widetilde{D}D \rangle^{m_1,m_2}}. \end{aligned} \tag{8.10}$$

In Eq. (8.10), the last three terms on the RHS are evaluated by using the following properties of the operators X, D and $\widetilde{D}$,

$$\begin{aligned} &X(2)|m_1, m_2\rangle \to |m_1, m_2\rangle, \qquad D(2)|m_1, m_2\rangle \to |m_1+2, m_2-2\rangle, \\ &\widetilde{D}(2)|m_1, m_2\rangle \to |m_1-2, m_2+2\rangle. \end{aligned} \tag{8.11}$$

Also, the fixed-(m_1, m_2) averages involving X^2, V^2, $D\widetilde{D}$ and $\widetilde{D}D$ in Eq. (8.10) follow from Eq. (8.9). Now, the formula for the fourth moment M_4 is,

$$\begin{aligned}
M_4(m_1, m_2) &= \overline{\langle (h+V)^4 \rangle^{m_1,m_2}} \\
&= \langle h^4 \rangle^{m_1,m_2} + 3\langle h^2 \rangle^{m_1,m_2} \overline{\langle V^2 \rangle^{m_1,m_2}} + \langle h^2 \rangle^{m_1,m_2} \overline{\langle X^2 \rangle^{m_1,m_2}} \\
&\quad + \overline{\langle D h^2 \widetilde{D} \rangle^{m_1,m_2}} + \overline{\langle \widetilde{D} h^2 D \rangle^{m_1,m_2}} + 2\overline{\langle h X h X \rangle^{m_1,m_2}} \\
&\quad + 2\overline{\langle h D h \widetilde{D} \rangle^{m_1,m_2}} + 2\overline{\langle h \widetilde{D} h D \rangle^{m_1,m_2}} + \overline{\langle V^4 \rangle^{m_1,m_2}} \\
&= (m_2)^4 + 3(m_2)^2 \overline{\langle V^2 \rangle^{m_1,m_2}} + (m_2)^2 \overline{\langle X^2 \rangle^{m_1,m_2}} \\
&\quad + (m_2+2)^2 \overline{\langle D\widetilde{D} \rangle^{m_1,m_2}} + (m_2-2)^2 \overline{\langle \widetilde{D} D \rangle^{m_1,m_2}} \\
&\quad + 2(m_2)^2 \overline{\langle X^2 \rangle^{m_1,m_2}} + 2m_2(m_2+2) \overline{\langle D\widetilde{D} \rangle^{m_1,m_2}} \\
&\quad + 2m_2(m_2-2) \overline{\langle \widetilde{D} D \rangle^{m_1,m_2}} + \overline{\langle V^4 \rangle^{m_1,m_2}}.
\end{aligned} \tag{8.12}$$

The first term in Eq. (8.12) is trivial. The next two terms follow from Eq. (8.9). The terms 4–8 in Eq. (8.12) are also simple and they will follow from Eq. (8.11). The expression for $\overline{\langle V^4 \rangle^{m_1,m_2}}$, which is non-trivial, is,

$$\begin{aligned}
\overline{\langle V^4 \rangle^{m_1,m_2}} &= \overline{\langle X^4 \rangle^{m_1,m_2}} + 3\overline{\langle X^2 \rangle^{m_1,m_2}} \left\{ \overline{\langle D\widetilde{D} \rangle^{m_1,m_2}} + \overline{\langle \widetilde{D} D \rangle^{m_1,m_2}} \right\} \\
&\quad + \overline{\langle D X^2 \widetilde{D} \rangle^{m_1,m_2}} + \overline{\langle \widetilde{D} X^2 D \rangle^{m_1,m_2}} \\
&\quad + 2\overline{\langle X D X \widetilde{D} \rangle^{m_1,m_2}} + 2\overline{\langle X \widetilde{D} X D \rangle^{m_1,m_2}} + \overline{\langle (D+\widetilde{D})^4 \rangle^{m_1,m_2}}.
\end{aligned} \tag{8.13}$$

The formula for the first term in Eq. (8.13) follows from Eq. (7.23),

$$\begin{aligned}
\overline{\langle X^4 \rangle^{m_1,m_2}} &= 2\left\{ \overline{\langle X^2 \rangle^{m_1,m_2}} \right\}^2 + T_1; \\
T_1 &= \tau^4 \sum_{i+j=2, t+u=2} F(m_1, N_1, i, t) F(m_2, N_2, j, u).
\end{aligned} \tag{8.14}$$

Combining Eqs. (8.13) and (8.14), we have,

$$\begin{aligned}
&\overline{\langle V^4 \rangle^{m_1,m_2}} \\
&\quad = 2\left\{ \overline{\langle X^2 \rangle^{m_1,m_2}} \right\}^2 + T_1 + 3\overline{\langle X^2 \rangle^{m_1,m_2}} \left\{ \overline{\langle D\widetilde{D} \rangle^{m_1,m_2}} + \overline{\langle \widetilde{D} D \rangle^{m_1,m_2}} \right\} \\
&\qquad + \left\{ \overline{\langle D X^2 \widetilde{D} \rangle^{m_1,m_2}} + \overline{\langle \widetilde{D} X^2 D \rangle^{m_1,m_2}} \right\} \\
&\qquad + 2\left\{ \overline{\langle X D X \widetilde{D} \rangle^{m_1,m_2}} + \overline{\langle X \widetilde{D} X D \rangle^{m_1,m_2}} \right\} + \overline{\langle (D+\widetilde{D})^4 \rangle^{m_1,m_2}} \\
&\quad = 2\left\{ \overline{\langle X^2 \rangle^{m_1,m_2}} \right\}^2 + 3\overline{\langle X^2 \rangle^{m_1,m_2}} \left\{ \overline{\langle D\widetilde{D} \rangle^{m_1,m_2}} + \overline{\langle \widetilde{D} D \rangle^{m_1,m_2}} \right\} \\
&\qquad + T_1 + T_2 + 2T_3 + T_4.
\end{aligned} \tag{8.15}$$

To simplify the notations, we have introduced T_1, T_2, T_3 and T_4 in Eq. (8.15). The first and second terms in the RHS of the last step in Eq. (8.15) are completely determined by Eq. (8.9). Also, expression for T_1 is given in Eq. (8.14). Now, we will evaluate the terms T_2, T_3 and T_4. Firstly, using Eq. (8.11), we have

$$
\begin{aligned}
T_2 &= \overline{\langle DX^2\widetilde{D}\rangle^{m_1,m_2}} + \overline{\langle \widetilde{D}X^2 D\rangle^{m_1,m_2}} \\
&= \left\{\overline{\langle D\widetilde{D}\rangle^{m_1,m_2}}\right\}\left\{\overline{\langle X^2\rangle^{m_1-2,m_2+2}}\right\} \\
&\quad + \left\{\overline{\langle \widetilde{D}D\rangle^{m_1,m_2}}\right\}\left\{\overline{\langle X^2\rangle^{m_1+2,m_2-2}}\right\}.
\end{aligned}
\tag{8.16}
$$

Formulas for the averages involving X^2, $D\widetilde{D}$ and $\widetilde{D}D$ in Eq. (8.16) are given by Eq. (8.9). Using Eqs. (4.8) and (4.9) appropriately to contract the operators D with $\widetilde{D}$ across the operator X along with the expression for $\overline{\langle X^2\rangle^{m_1,m_2}}$ in Eq. (8.9), we have

$$
\begin{aligned}
T_3 &= \overline{\langle XDX\widetilde{D}\rangle^{m_1,m_2}} + \overline{\langle X\widetilde{D}XD\rangle^{m_1,m_2}} \\
&= \tau^2\alpha^2 \sum_{i+j=2}\left[\binom{m_1-i}{2}\binom{\widetilde{m}_2-j}{2} + \binom{\widetilde{m}_1-i}{2}\binom{m_2-j}{2}\right] \\
&\quad \times T(m_1, N_1, i)T(m_2, N_2, j).
\end{aligned}
\tag{8.17}
$$

Similarly, the expression for T_4 is

$$
\begin{aligned}
T_4 &= \overline{\langle (D+\widetilde{D})^4\rangle^{m_1,m_2}} \\
&= \overline{\langle D^2\widetilde{D}^2\rangle^{m_1,m_2}} + \overline{\langle \widetilde{D}^2 D^2\rangle^{m_1,m_2}} + \overline{\langle D\widetilde{D}D\widetilde{D}\rangle^{m_1,m_2}} \\
&\quad + \overline{\langle \widetilde{D}D\widetilde{D}D\rangle^{m_1,m_2}} + \overline{\langle D\widetilde{D}^2 D\rangle^{m_1,m_2}} + \overline{\langle \widetilde{D}D^2\widetilde{D}\rangle^{m_1,m_2}}.
\end{aligned}
\tag{8.18}
$$

As D can correlate only with $\widetilde{D}$ in leading order, we have

$$
\begin{aligned}
\overline{\langle D^2\widetilde{D}^2\rangle^{m_1,m_2}} &= \overline{\langle DD\widetilde{D}\widetilde{D}\rangle^{m_1,m_2}} + \overline{\langle DD\widetilde{D}\widetilde{D}\rangle^{m_1,m_2}} \\
&= \alpha^4 \sum_{\gamma,\delta,\kappa,\eta} \left\langle \gamma_1^\dagger(2)\delta_2(2)\kappa_1^\dagger(2)\eta_2(2)\delta_2^\dagger(2)\gamma_1(2)\eta_2^\dagger(2)\kappa_1(2)\right\rangle^{m_1,m_2} \\
&\quad + \alpha^4 \sum_{\gamma,\delta,\kappa,\eta} \left\langle \gamma_1^\dagger(2)\delta_2(2)\kappa_1^\dagger(2)\eta_2(2)\eta_2^\dagger(2)\kappa_1(2)\delta_2^\dagger(2)\gamma_1(2)\right\rangle^{m_1,m_2} \\
&= \alpha^4 \sum_{\gamma,\delta,\kappa,\eta} \left\langle \gamma_1^\dagger(2)\kappa_1^\dagger(2)\gamma_1(2)\kappa_1(2)\right\rangle^{m_1}\left\langle \delta_2(2)\eta_2(2)\delta_2^\dagger(2)\eta_2^\dagger(2)\right\rangle^{m_2} \\
&\quad + \alpha^4 \sum_{\gamma,\delta,\kappa,\eta} \left\langle \gamma_1^\dagger(2)\kappa_1^\dagger(2)\kappa_1(2)\gamma_1(2)\right\rangle^{m_1}\left\langle \delta_2(2)\eta_2(2)\eta_2^\dagger(2)\delta_2^\dagger(2)\right\rangle^{m_2}
\end{aligned}
$$

$$
\begin{aligned}
&= 2\alpha^4 \sum_{\gamma,\kappa} \left\langle \gamma_1^\dagger(2)\kappa_1^\dagger(2)\kappa_1(2)\gamma_1(2)\right\rangle^{m_1} \sum_{\delta,\eta} \left\langle \delta_2(2)\eta_2(2)\eta_2^\dagger(2)\delta_2^\dagger(2)\right\rangle^{m_2} \\
&= 2\overline{\langle D\widetilde{D}\rangle^{m_1,m_2}}\;\overline{\langle D\widetilde{D}\rangle^{m_1-2,m_2+2}}.
\end{aligned}
\tag{8.19}
$$

In order to obtain the last step in Eq. (8.19), the operators $\kappa^\dagger\kappa$ and $\gamma^\dagger\gamma$ are contracted using Eq. (4.6) that gives $\binom{m_1-2}{2}$ and $\binom{m_1}{2}$ respectively. Similarly, contracting operators $\eta\eta^\dagger$ and $\delta\delta^\dagger$ using Eq. (4.7) gives $\binom{\widetilde{m}_2-2}{2}$ and $\binom{\widetilde{m}_2}{2}$ respectively. Combining these gives the last step in Eq. (8.19). Also, the third binary pattern $\overline{\langle DD\widetilde{D}\widetilde{D}\rangle^{m_1,m_2}}$ with the two D's correlated (similarly the two $\widetilde{D}$'s) is not considered as it will be $1/N_1$ or $1/N_2$ order smaller compared to the other two binary patterns shown in Eq. (8.19). Similarly, the other terms in Eq. (8.18) are

$$
\begin{aligned}
\overline{\left\langle\widetilde{D}^2D^2\right\rangle^{m_1,m_2}} &= \overline{\langle\widetilde{D}\widetilde{D}DD\rangle^{m_1,m_2}} + \overline{\langle\widetilde{D}\widetilde{D}DD\rangle^{m_1,m_2}} \\
&= 2\overline{\langle\widetilde{D}D\rangle^{m_1,m_2}}\;\overline{\langle\widetilde{D}D\rangle^{m_1+2,m_2-2}}, \\
\overline{\langle D\widetilde{D}D\widetilde{D}\rangle^{m_1,m_2}} &= \overline{\langle D\widetilde{D}D\widetilde{D}\rangle^{m_1,m_2}} + \overline{\langle D\widetilde{D}D\widetilde{D}\rangle^{m_1,m_2}} \\
&= \left\{\overline{\langle D\widetilde{D}\rangle^{m_1,m_2}}\right\}^2 + \overline{\langle D\widetilde{D}\rangle^{m_1,m_2}}\;\overline{\langle\widetilde{D}D\rangle^{m_1-2,m_2+2}}, \\
\overline{\langle D\widetilde{D}\widetilde{D}D\rangle^{m_1,m_2}} &= \overline{\langle D\widetilde{D}\widetilde{D}D\rangle^{m_1,m_2}} + \overline{\langle D\widetilde{D}\widetilde{D}D\rangle^{m_1,m_2}} \\
&= 2\overline{\langle D\widetilde{D}\rangle^{m_1,m_2}}\;\overline{\langle\widetilde{D}D\rangle^{m_1,m_2}}, \\
\overline{\langle\widetilde{D}DD\widetilde{D}\rangle^{m_1,m_2}} &= \overline{\langle\widetilde{D}DD\widetilde{D}\rangle^{m_1,m_2}} + \overline{\langle\widetilde{D}DD\widetilde{D}\rangle^{m_1,m_2}} \\
&= 2\overline{\langle D\widetilde{D}\rangle^{m_1,m_2}}\;\overline{\langle\widetilde{D}D\rangle^{m_1,m_2}}, \\
\overline{\langle\widetilde{D}D\widetilde{D}D\rangle^{m_1,m_2}} &= \overline{\langle\widetilde{D}D\widetilde{D}D\rangle^{m_1,m_2}} + \overline{\langle\widetilde{D}D\widetilde{D}D\rangle^{m_1,m_2}} \\
&= \left\{\overline{\langle\widetilde{D}D\rangle^{m_1,m_2}}\right\}^2 + \overline{\langle\widetilde{D}D\rangle^{m_1,m_2}}\;\overline{\langle D\widetilde{D}\rangle^{m_1+2,m_2-2}}.
\end{aligned}
\tag{8.20}
$$

Combining Eqs. (8.18)–(8.20), we have

$$
\begin{aligned}
T_4 = {} & \left\{\overline{\langle D\widetilde{D}\rangle^{m_1,m_2}}\right\}^2 + \left\{\overline{\langle\widetilde{D}D\rangle^{m_1,m_2}}\right\}^2 \\
& + \overline{\langle D\widetilde{D}\rangle^{m_1,m_2}}\left[2\overline{\langle D\widetilde{D}\rangle^{m_1-2,m_2+2}} + \overline{\langle\widetilde{D}D\rangle^{m_1-2,m_2+2}}\right] \\
& + \overline{\langle\widetilde{D}D\rangle^{m_1,m_2}}\left[2\overline{\langle\widetilde{D}D\rangle^{m_1+2,m_2-2}} + \overline{\langle D\widetilde{D}\rangle^{m_1+2,m_2-2}}\right] \\
& + 4\left\{\overline{\langle D\widetilde{D}\rangle^{m_1,m_2}}\right\}\left\{\overline{\langle\widetilde{D}D\rangle^{m_1,m_2}}\right\}.
\end{aligned}
\tag{8.21}
$$

Finally, combining Eqs. (8.12), (8.14), (8.15), (8.16), (8.17) and (8.21), expression for the fourth moment is,

$$
\begin{aligned}
M_4(m_1,m_2) = {} & (m_2)^4 + 3(m_2)^2\overline{\langle V^2\rangle^{m_1,m_2}} + 3(m_2)^2\overline{\langle X^2\rangle^{m_1,m_2}} \\
& + (m_2+2)^2\overline{\langle D\widetilde{D}\rangle^{m_1,m_2}} + (m_2-2)^2\overline{\langle \widetilde{D}D\rangle^{m_1,m_2}} \\
& + 2m_2(m_2+2)\overline{\langle D\widetilde{D}\rangle^{m_1,m_2}} + 2m_2(m_2-2)\overline{\langle \widetilde{D}D\rangle^{m_1,m_2}} \\
& + 2\left\{\overline{\langle X^2\rangle^{m_1,m_2}}\right\}^2 + 3\overline{\langle X^2\rangle^{m_1,m_2}}\left\{\overline{\langle D\widetilde{D}\rangle^{m_1,m_2}} + \overline{\langle \widetilde{D}D\rangle^{m_1,m_2}}\right\} \\
& + \tau^4 \sum_{i+j=2,t+u=2} F(m_1,N_1,i,t)F(m_2,N_2,j,u) \\
& + \left\{\overline{\langle D\widetilde{D}\rangle^{m_1,m_2}}\right\}\left\{\overline{\langle X^2\rangle^{m_1-2,m_2+2}}\right\} \\
& + \left\{\overline{\langle \widetilde{D}D\rangle^{m_1,m_2}}\right\}\left\{\overline{\langle X^2\rangle^{m_1+2,m_2-2}}\right\} \\
& + 2\tau^2\alpha^2 \sum_{i+j=2}\left[\binom{m_1-i}{2}\binom{\widetilde{m}_2-j}{2} + \binom{\widetilde{m}_1-i}{2}\binom{m_2-j}{2}\right] \\
& \times T(m_1,N_1,i)T(m_2,N_2,j) \\
& + \left\{\overline{\langle D\widetilde{D}\rangle^{m_1,m_2}}\right\}^2 + \left\{\overline{\langle \widetilde{D}D\rangle^{m_1,m_2}}\right\}^2 \\
& + \overline{\langle D\widetilde{D}\rangle^{m_1,m_2}}\left[2\overline{\langle D\widetilde{D}\rangle^{m_1-2,m_2+2}} + \overline{\langle \widetilde{D}D\rangle^{m_1-2,m_2+2}}\right] \\
& + \overline{\langle \widetilde{D}D\rangle^{m_1,m_2}}\left[2\overline{\langle \widetilde{D}D\rangle^{m_1+2,m_2-2}} + \overline{\langle D\widetilde{D}\rangle^{m_1+2,m_2-2}}\right] \\
& + 4\left\{\overline{\langle D\widetilde{D}\rangle^{m_1,m_2}}\right\}\left\{\overline{\langle \widetilde{D}D\rangle^{m_1,m_2}}\right\}. \qquad (8.22)
\end{aligned}
$$

Equations (8.8), (8.9), (8.10), and (8.22), respectively give the first four non-central moments $[M_1(m_1,m_2)$, $M_2(m_1,m_2)$, $M_3(m_1,m_2)$ and $M_4(m_1,m_2)]$. In applying Eq. (8.22), the function $T(\cdots)$ can be calculated using Eq. (4.12) and similarly, the function $F(\cdots)$ using Eq. (4.17) or Eq. (4.26). Formulas for the first four cumulants $[K_1(m_1,m_2)$, $K_2(m_1,m_2)$, $K_3(m_1,m_2)$, $K_4(m_1,m_2)]$ in terms of the non-central moments are [4],

$$
\begin{aligned}
& K_1(m_1,m_2) = M_1(m_1,m_2), \qquad K_2(m_1,m_2) = M_2(m_1,m_2) - M_1^2(m_1,m_2), \\
& K_3(m_1,m_2) = M_3(m_1,m_2) - 3M_2(m_1,m_2)M_1(m_1,m_2) + 2M_1^3(m_1,m_2), \\
& K_4(m_1,m_2) = M_4(m_1,m_2) - 4M_3(m_1,m_2)M_1(m_1,m_2) - 3M_2^2(m_1,m_2) \\
& \qquad + 12M_2(m_1,m_2)M_1^2(m_1,m_2) - 6M_1^4(m_1,m_2). \qquad (8.23)
\end{aligned}
$$

Then, the skewness and excess parameters are

$$
\gamma_1(m_1,m_2) = \frac{K_3(m_1,m_2)}{[K_2(m_1,m_2)]^{3/2}}, \qquad \gamma_2(m_1,m_2) = \frac{K_4(m_1,m_2)}{[K_2(m_1,m_2)]^2}. \qquad (8.24)
$$

After carrying out the simplifications using Eqs. (8.8), (8.9), (8.10), (8.22) and (8.23), it is easily seen that,

$$\gamma_1(m_1, m_2) = \frac{2[\overline{\langle D\widetilde{D}\rangle^{m_1,m_2}} - \overline{\langle \widetilde{D}D\rangle^{m_1,m_2}}]}{\{\overline{\langle D\widetilde{D}\rangle^{m_1,m_2}} + \overline{\langle \widetilde{D}D\rangle^{m_1,m_2}} + \overline{\langle X^2\rangle^{m_1,m_2}}\}^{3/2}}. \tag{8.25}$$

Thus, γ_1 will be non-zero only when $\alpha \neq 0$ and the τ dependence appears only in the denominator. Also, it is seen that for $N_+ = N_-$, $\gamma_1(m_1, m_2) = -\gamma_1(m_2, m_1)$. The expression for γ_2 is more cumbersome. Denoting $\mathcal{D} = \overline{\langle D\widetilde{D}\rangle^{m_1,m_2}}$, $\widetilde{\mathcal{D}} = \overline{\langle \widetilde{D}D\rangle^{m_1,m_2}}$ and $\mathcal{X} = \overline{\langle X^2\rangle^{m_1,m_2}}$ for brevity, we have

$$\gamma_2(m_1, m_2) + 1 = \frac{T_1 + T_2 + 2T_3 + T_4 + (\widetilde{\mathcal{D}} + \mathcal{D})(4 - \mathcal{X}) - 2(\widetilde{\mathcal{D}} + \mathcal{D})^2}{\{\widetilde{\mathcal{D}} + \mathcal{D} + \mathcal{X}\}^2}. \tag{8.26}$$

The formulas for T's, $\mathcal{D}$, $\widetilde{\mathcal{D}}$ and $\mathcal{X}$ given before together with Eq. (8.26) show that, for $N_+ = N_-$, $\gamma_2(m_1, m_2) = \gamma_2(m_2, m_1)$. With, $T_1 \sim \mathcal{X}^2 + C_1$, $T_2 = T_3 \sim \mathcal{X}(\widetilde{\mathcal{D}} + \mathcal{D})$ and $T_4 \sim 3(\widetilde{\mathcal{D}} + \mathcal{D})^2 + C_2$ which are good in the dilute limit ($|C1/T_1|$ and $|C_2/T_4|$ will be close to zero), we have

$$\gamma_2(m_1, m_2) = \frac{C_1 + C_2 + 4(\widetilde{\mathcal{D}} + \mathcal{D})}{\{\widetilde{\mathcal{D}} + \mathcal{D} + \mathcal{X}\}^2}. \tag{8.27}$$

Note that C_1 and $\mathcal{X}$ depend only on τ. Similarly, C_2, $\widetilde{\mathcal{D}}$ and $\mathcal{D}$ depend only on α. The $(\widetilde{\mathcal{D}} + \mathcal{D})$ term in the numerator will contribute to $\gamma_2(m_1, m_2)$ when $\tau = 0$ and α is very small. The approximation $T_2 = T_3 \sim \mathcal{X}(\widetilde{\mathcal{D}} + \mathcal{D})$ is crucial in obtaining the numerator in Eq. (8.27) with no cross-terms involving the α and τ parameters. With this, we have k_4 to be the sum of k_4's coming from $X(2)$ and $D(2)$ matrices [note that, as mentioned before, $X(2) = A \oplus B \oplus C$ and $D(2) = D \oplus \widetilde{D}$]. Equation (8.27) shows that, for $\alpha \ll \tau$, $\gamma_2(m_1, m_2) = C_1/[\overline{\langle X^2\rangle^{m_1,m_2}}]^2$ with $C_1 \sim -9\tau^4 N^4 m^3/16$ for $m_1 = m_2 = m/2$ and $N_1 = N_2 = N$. Evaluating $\overline{\langle X^2\rangle^{m_1,m_2}}$ in the dilute limit then gives $\gamma_2 \sim -4/m$. Similarly, for $\tau \ll \alpha$, we have $\gamma_2(m_1, m_2) = C_2/[\overline{\langle D\widetilde{D}\rangle^{m_1,m_2}} + \overline{\langle \widetilde{D}D\rangle^{m_1,m_2}}]^2$ with $C_2 \sim -\alpha^4 N^4 m^3/16$ and this gives $\gamma_2 \sim -4/m$. Therefore, in the $\tau \ll \alpha$ and $\tau \gg \alpha$ limit, the result for γ_2 is same as the result for spinless fermion EGOE(2) and this shows that for a range of (τ, α) values, $\rho^{m_1,m_2}(E)$ will be close to Gaussian. Moreover, to the extent that Eq. (8.27) applies, the density $\rho^{m_1,m_2}(E)$ is a convolution of the densities generated by $X(2)$ and $D(2)$ operators. Finally, the binary correlation results for $\gamma_1(m, \pm)$ and $\gamma_2(m, \pm)$ of $I_\pm^m(E)$ are found to be close to the exact results obtained using the eigenvalues from EGOE(1 + 2)-π ensembles in a number of numerical examples; see [2] for details.

8.3 State Densities and Parity Ratios

In order to carry out numerical calculations for state densities and parity ratios, it is necessary to have a physically meaningful range of values for the parameters $(\tau, \alpha, m/N_+, N_+/N_-)$ defining EGOE(1 + 2)-π. For $A = 20$–80 nuclei, $\Delta = 3$–5 MeV is reasonable and this along with realistic nuclear effective interactions in $sdfp$ and $fpg_{9/2}$ spaces give $N_+/N_- \sim 0.5$–2.0, $\tau \sim 0.09$–0.24 and $\alpha \sim (0.9$–$1.3) \times \tau$. These deduced values of α and τ clearly point out that one has to consider the more general EGOE(1 + 2)-π defined by Eq. (8.4). Also, $m \lesssim N_+$ or N_-, whichever is lower. Following all these, numerical calculations are carried out in [2] using $\tau = 0.05, 0.1, 0.2, 0.3, \alpha/\tau = 0.5, 1.0, 1.5$ and $N = N_+ + N_- \leq 16$. Also, values of m are chosen such that $m \ll N$ as in the dilute limit it is possible to understand the ensemble results better.

8.3.1 Results for Fixed-Parity State Densities

Figures 8.2 and 8.3 show results for fixed-parity ensemble averaged eigenvalue (state) densities $\overline{I_\pm(E)} = \overline{\langle\langle\delta(H - E)\rangle\rangle^\pm}$. Combining all $\widehat{E} = [E - E_c(m, \pm)]/\sigma(m, \pm)$ values and using a bin-size $\Delta\widehat{E} = 0.2$, histograms for $I_\pm(E)$ are generated. It is seen that the state densities for $\tau \geq 0.1$ are unimodal and close to a Gaussian (multimodal for small τ values). For $V(2) = 0$, the eigenvalue densities will be a sum of spikes at $0, 2\Delta, 4\Delta, \ldots$ for +ve parity densities and similarly at $\Delta, 3\Delta, 5\Delta, \ldots$ for −ve parity densities. As we switch on $V(2)$, the spikes will spread due to the matrices A, B and C in Fig. 8.1 and mix due to the matrix D. The variance $\sigma^2(m_1, m_2)$ can be written as,

$$\sigma^2(m_1, m_2) = \sigma^2(m_1, m_2 \rightarrow m_1, m_2) + \sigma^2(m_1, m_2 \rightarrow m_1 \pm 2, m_2 \mp 2). \quad (8.28)$$

The internal variance $\sigma^2(m_1, m_2 \rightarrow m_1, m_2)$ is due to A, B and C matrices and it receives contribution only from the τ parameter. Similarly, the external variance $\sigma^2(m_1, m_2 \rightarrow m_1 \pm 2, m_2 \mp 2)$ is due to the matrix D and it receives contribution only from the α parameter. When we switch on $V(2)$, as the ensemble averaged centroids generated by $V(2)$ will be zero, the positions of the spikes will be largely unaltered. However, they will start spreading and mixing as τ and α increase. Therefore, the density will be multimodal with the modes well separated for very small (τ, α) values. As τ and α start increasing from zero, the spikes spread and will start overlapping for $\sigma(m_1, m_2) \gtrsim \Delta$. This is the situation with $\tau = 0.05$ shown in Fig. 8.2. However, as τ increases (with $\alpha \sim \tau$), the densities start becoming unimodal as seen from the $\tau = 0.1$ and 0.2 examples in Figs. 8.2 and 8.3. As the particle numbers in the examples shown in Figs. 8.2 and 8.3 are small, the excess parameter $\gamma_2(m, \pi) \sim -0.7$ to -0.8 (skewness parameter $\gamma_1(m, \pi) \sim 0$ in all our examples) and therefore, the densities $I_\pm(E)$ show some deviations from Gaussian form. The smoothed +ve and −ve parity densities are a sum of the partial densities $\rho^{m_1,m_2}(E)$

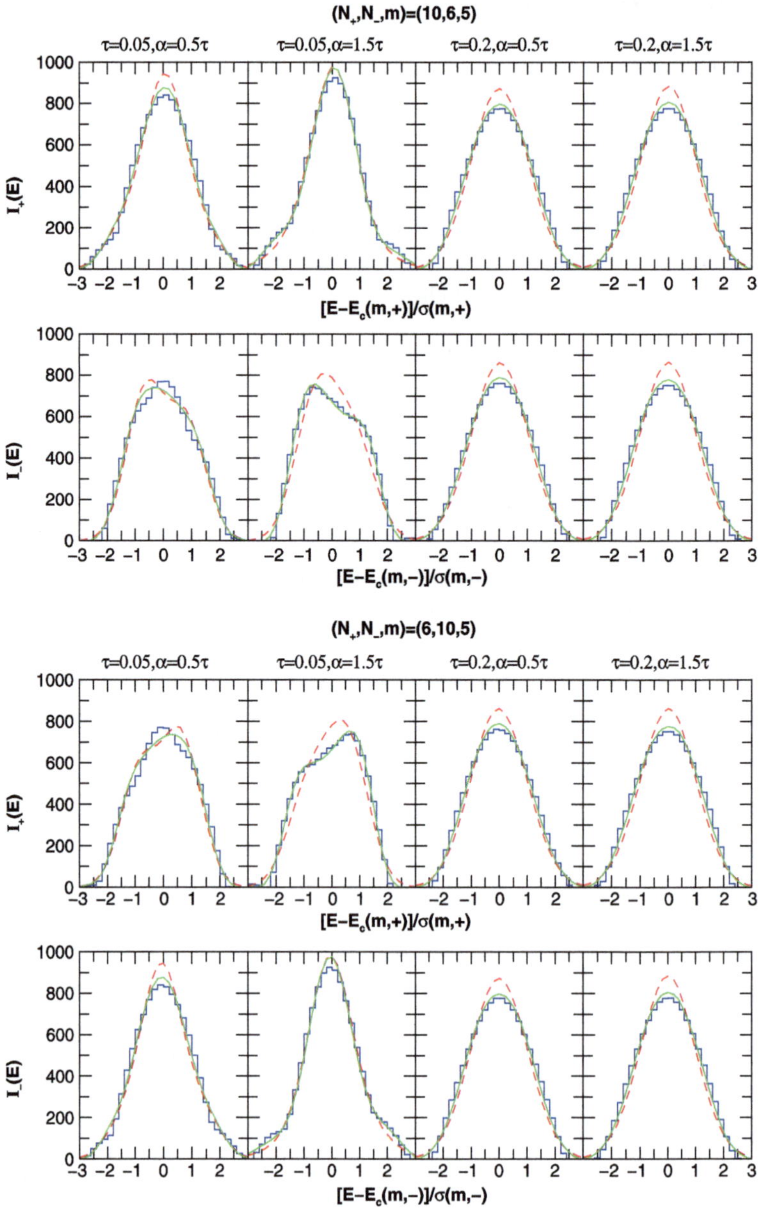

Fig. 8.2 Positive and negative parity state densities for various (τ, α) values for $(N_+, N_-, m) = (10, 6, 5)$ and $(6, 10, 5)$ systems. Histograms are numerical ensemble results. The *dashed (red) curve* corresponds to Gaussian form for $\rho^{m_1,m_2}(E)$ in Eq. (8.6) and similarly, *solid (green) curve* corresponds to Edgeworth corrected Gaussian form with $\gamma_1(m_1, m_2)$ and $\gamma_2(m_1, m_2)$ formulas given in Sect. 8.2. Figures are taken from [2] with permission from American Physical Society

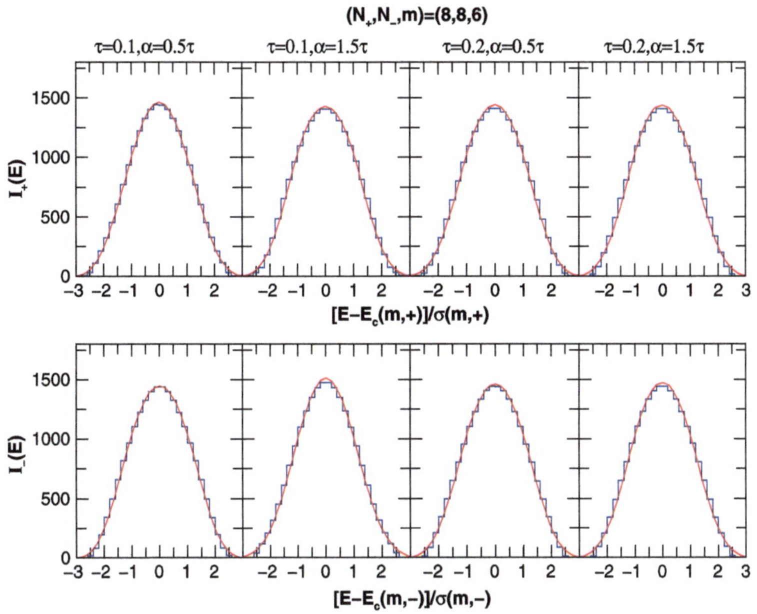

Fig. 8.3 Positive and negative parity state densities for various (τ, α) values for $(N_+, N_-, m) = (8, 8, 6)$. Smoothed curves (*solid red lines*) are obtained using fixed-(m_1, m_2) partial densities and the final state densities are close to Gaussian in form. Figure is taken from [2] with permission from American Physical Society

as given by Eq. (8.6). It is clearly seen from Figs. 8.2 and 8.3 that the sum of partial densities, with the partial densities represented by ED corrected Gaussians, describe extremely well the exact fixed-π densities in these examples. Therefore, for the (τ, α) values in the range determined by nuclear $sdfp$ and $fpg_{9/2}$ interactions, i.e. $\tau \sim 0.1$–0.3 and $\alpha \sim 0.5\tau$–2τ, the partial densities can be well represented by ED corrected Gaussians and total densities are also close to ED corrected Gaussians. On the other hand, for $\tau \ll 0.1$ the sum of ED corrected partial densities still give a good representation of $I_\pm(E)$ as seen from Fig. 8.2. It is possible that the agreements in Figs. 8.2 and 8.3 may become more perfect if we employ, for the partial densities, some non-canonical forms defined by the first four moments as given for example in [5, 6]. However, these forms are not derived by solving EGOE(1+2)-π.

8.3.2 Results for Parity Ratios

Results for parity ratios $I_-^m(E)/I_+^m(E)$ are given in Figs 8.4 and 8.5. As parity ratios need to be calculated at a given value of the excitation energy E, the eigenvalues in both +ve and −ve parity spaces have to be measured with respect to one gs energy E_{gs}. In the results presented in Figs. 8.4 and 8.5, E_{gs} is determined by taking all the +ve and −ve parity eigenvalues of all the members of the ensemble

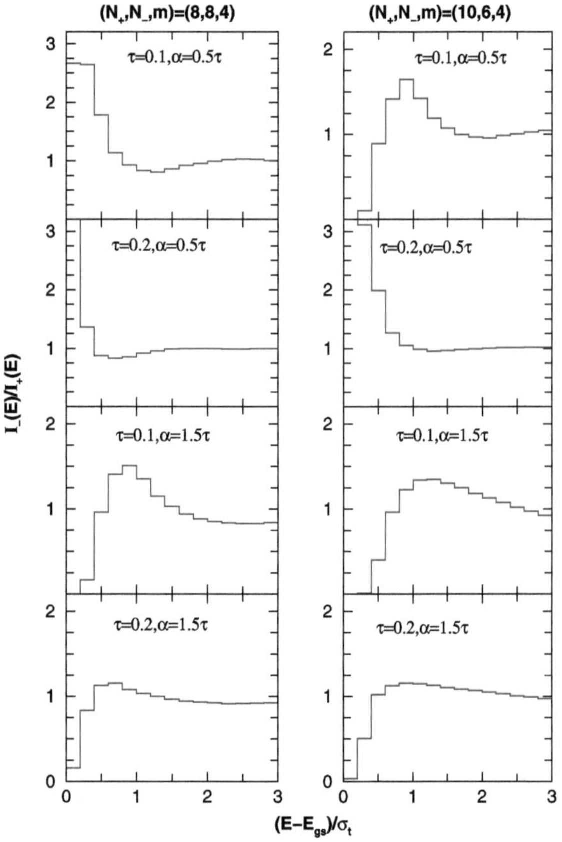

Fig. 8.4 Parity ratios for various (τ, α) values for $(N_+, N_-, m) = (8, 8, 4)$ and $(10, 6, 4)$ systems. Figure is taken from [2] with permission from American Physical Society (Color figure online)

and choosing the lowest of all these as E_{gs}. Similarly, the ensemble averaged total (+ve and −ve parity eigenvalues combined) spectrum width σ_t of the system is used for scaling. Thus, the variable used is $\mathbf{E} = (E - E_{gs})/\sigma_t$. Starting with E_{gs} and using a bin-size of $\Delta\mathbf{E} = 0.2$, the number of states $I_+(\boldsymbol{E})$ with +ve parity and also the number of states $I_-(\mathbf{E})$ with −ve parity in a given bin are calculated and then the ratio $I_-(\mathbf{E})/I_+(\boldsymbol{E})$ is the parity ratio. Note that the results in Figs. 8.4 and 8.5 are shown for $\mathbf{E} = 0 - 3$ as the spectrum span is $\sim 5.5\sigma_t$. To go beyond the middle of the spectrum, for real nuclei, one has to include more sp levels (also a finer splitting of the +ve and −ve parity levels may be needed) and therefore, N_+ and N_- will change. Continuing with this, one obtains the Bethe form for nuclear level densities [7]. General observations from Figs. 8.4 and 8.5 are as follows. (i) The parity ratio $I_-(\mathbf{E})/I_+(\mathbf{E})$ will be zero up to an energy $\mathbf{E}_0$. (ii) Then, it starts increasing and becomes larger than unity at an energy $\mathbf{E}_m$. (iii) From here on, the parity ratio decreases and saturates quickly to unity from an energy $\mathbf{E}_1$. In these examples, $\mathbf{E}_0 \lesssim 0.4$, $\mathbf{E}_m \sim 1$ and $\mathbf{E}_1 \sim 1.5$. It is seen that the curves shift towards left as τ increases. Also the position of the peak shifts to much larger value of $\mathbf{E}_m$ and equilibration gets delayed as α increases for a fixed τ value. Therefore

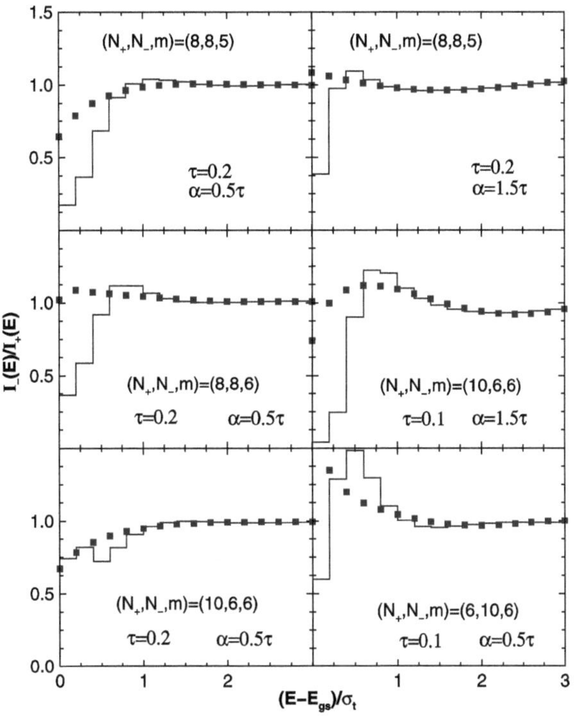

Fig. 8.5 Parity ratios for various (τ, α) values and for various (N_+, N_-, m) systems. *Filled squares (brown)* are obtained using ED form for fixed-(m_1, m_2) partial densities. In calculating γ_2 of the partial densities, function $F(\cdots)$ is needed [see for example Eq. (8.14)] and here the finite-N formula given by (4.26) is used. Figure is taken from [2] with permission from American Physical Society (Color figure online)

for larger τ, the energies $(\mathbf{E}_0, \mathbf{E}_m, \mathbf{E}_1)$ are smaller compared to those for a smaller τ. The three transition energies also depend on (N_+, N_-, m). This general structure of the parity ratios will remain same even when we change $\Delta \to -\Delta$ (i.e. $-$ve parity sp states below the $+$ve parity sp states). General structures (i)–(iii) are also seen in the numerical examples shown in [1] where a method based on the Fermi-gas model has been employed. If $\sigma_t \sim 6$–8 MeV, equilibration in parities is expected around $E \sim 8$–10 MeV and this is clearly seen in the examples in [1]. It is also seen that the equilibration is quite poor for very small values of τ and therefore comparing with the results in [1], it can be argued that very small values of τ are ruled out for atomic nuclei. Hence, it is plausible to conclude that generic results for parity ratios can be derived using EGOE(1 + 2)-π with reasonably large (τ, α) values.

Turning to prediction of parity ratios, it is of interest to compare the numerically calculated parity ratios with those obtained using the ED form for $\rho^{m_1,m_2}(E)$. Some results for this are shown in Fig. 8.5. Here, starting with the absolute ground state energy E_{gs} and using a bin-size (in Fig. 8.5, $\Delta\mathbf{E} = 0.2$), $+$ve and $-$ve parity densities in a given energy bin are be obtained using the smoothed $I^{\pm}(\mathbf{E})$ and their ratio is the parity ratio at a given $\mathbf{E}$. The smoothed $I^{\pm}(\mathbf{E})$ are constructed using the first four moments of $\rho^{m_1,m_2}(E)$ and Eq. (8.6). In the examples shown in Fig. 8.5, I_+ and I_- are close to Gaussians. It is seen that the agreement with exact results is good for $\mathbf{E} \gtrsim 0.5$. However, for smaller $\mathbf{E}$, to obtain a good agreement one should have a better prescription for determining the tail part of the $\rho^{m_1,m_2}(\mathbf{E})$ distributions. Developing a theory for this is an important problem for future.

References

1. D. Mocelj, T. Rauscher, G. Martínez-Pinedo, K. Langanke, L. Pacearescu, A. Faessler, F.-K. Thielemann, Y. Alhassid, Large-scale prediction of the parity distribution in the nuclear level density and application to astrophysical reaction rates. Phys. Rev. C **75**, 045805 (2007)
2. M. Vyas, V.K.B. Kota, P.C. Srivastava, One plus two-body random matrix ensembles with parity: density of states and parity ratios. Phys. Rev. C **83**, 064301 (2011)
3. T. Papenbrock, H.A. Weidenmüller, Abundance of ground states with positive parity. Phys. Rev. C **78**, 054305 (2008)
4. A. Stuart, J.K. Ord, *Kendall's Advanced Theory of Statistics: Distribution Theory* (Oxford University Press, New York, 1987)
5. S.M. Grimes, T.N. Massey, New expansion technique for spectral distribution calculations. Phys. Rev. C **51**, 606–610 (1995)
6. E. Terán, C.W. Johnson, Simple models for shell-model configuration densities. Phys. Rev. C **74**, 067302 (2006)
7. V.K.B. Kota, R.U. Haq, *Spectral Distributions in Nuclei and Statistical Spectroscopy* (World Scientific, Singapore, 2010)

Chapter 9
Embedded GOE Ensembles for Interacting Boson Systems: BEGOE(1 + 2) for Spinless Bosons

In Chaps. 4–8 EE for fermion systems are discussed in detail with analytical and numerical results. In the present chapter and the next chapter, we will consider EE for finite interacting boson systems (called BEE with 'B' for bosons). Unlike for fermion systems, for fixed number (N) of sp states, boson number m can increase beyond N and therefore a dense limit with $m \to \infty$ (complete definition given ahead) is possible and this is one new aspect of boson systems. Also, BEE are important because of the increasing interest in investigating (using experiments and theory) BEC and quantum gases in general. As Asaga et al. state [1]: *In an atomic trap, bosonic atoms occupy partly degenerate single particle states. The interaction will lift the degeneracy. A random matrix approach should reveal the generic features of the resulting system.* In addition, BEE are also important in understanding certain aspects of the Interacting Boson Model (IBM) of atomic nuclei [2–5]. To get started with BEE, we will first consider BEGOE(1 + 2) for spinless boson systems in this chapter.

9.1 Definition and Construction

The BEGOE(2)/BEGUE(2) ensemble for spinless boson systems is generated by defining the two-body Hamiltonian H to be GOE/GUE in two-particle spaces and then propagating it to many-particle spaces by using the geometry of the many-particle spaces [this is in general valid for k-body Hamiltonians, with $k < m$, generating BEGOE(k)/BEGUE(k)]. Consider a system of m spinless bosons occupying N sp states $|\nu_i\rangle$, $i = 1, 2, \ldots, N$; see Fig. 9.1. Then, BEGOE(2) is defined by the Hamiltonian operator,

$$\widehat{H}(2) = \sum_{\nu_i \le \nu_j, \nu_k \le \nu_l} \frac{\langle \nu_k \nu_l | \widehat{H}(2) | \nu_i \nu_j \rangle}{\sqrt{(1+\delta_{ij})(1+\delta_{kl})}} b^\dagger_{\nu_k} b^\dagger_{\nu_l} b_{\nu_i} b_{\nu_j}, \tag{9.1}$$

V.K.B. Kota, *Embedded Random Matrix Ensembles in Quantum Physics*, Lecture Notes in Physics 884, DOI 10.1007/978-3-319-04567-2_9,

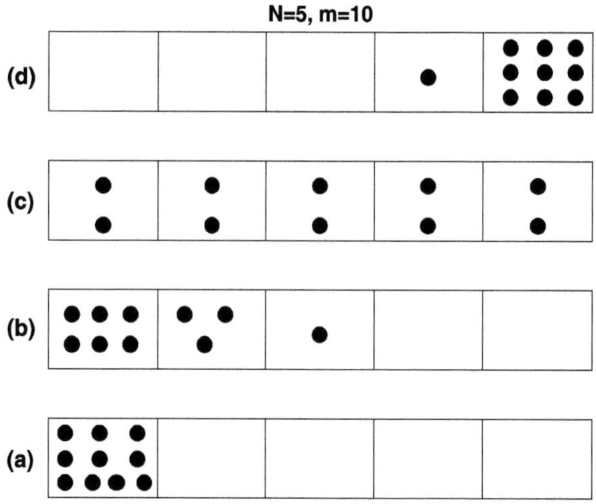

Fig. 9.1 Some m boson configurations or basis states for $m = 10$ spinless bosons in $N = 5$ sp states. Enumeration of the configurations is similar to distributing m particles in N boxes with the conditions that the occupancy of each box lies between zero and m and the maximum number of occupied boxes equals m. In the figure, (**a**) corresponds to the basis state $|(\nu_1)^{10}\rangle$, (**b**) corresponds to the basis state $|(\nu_1)^6(\nu_2)^3\nu_3\rangle$, (**c**) corresponds to the basis state $|(\nu_1)^2(\nu_2)^2(\nu_3)^2(\nu_4)^2(\nu_5)^2\rangle$ and (**d**) corresponds to the basis state $|\nu_4(\nu_5)^9\rangle$

with the symmetries for the symmetrized two-body matrix elements $\langle \nu_k\nu_l|\widehat{H}(2)|\nu_i\nu_j\rangle$ being,

$$\begin{aligned}\langle \nu_k\nu_l \mid \widehat{H}(2) \mid \nu_j\nu_i\rangle &= \langle \nu_k\nu_l \mid \widehat{H}(2) \mid \nu_i\nu_j\rangle,\\ \langle \nu_k\nu_l \mid \widehat{H}(2) \mid \nu_i\nu_j\rangle &= \langle \nu_i\nu_j \mid \widehat{H}(2) \mid \nu_k\nu_l\rangle.\end{aligned} \tag{9.2}$$

Note that $|\nu_i\nu_j\rangle$ denote two-boson symmetric states. The action of the Hamiltonian operator defined by Eq. (9.1) on an the basis states, defined by distributions of bosons in the sp states as shown in Fig. 9.1, generates the H matrix in m-boson spaces. Note that b_{ν_i} and $b^\dagger_{\nu_i}$ in Eq. (9.1) annihilate and create a boson in the sp state $|\nu_i\rangle$, respectively. The Hamiltonian matrix $H(m)$ in m-particle spaces contains three different types of non-zero matrix elements and explicit formulas for these are [6],

$$\left\langle \prod_{r=i,j,\ldots} (\nu_r)^{n_r} \middle| \widehat{H}(2) \middle| \prod_{r=i,j,\ldots} (\nu_r)^{n_r} \right\rangle = \sum_{i\geq j} \frac{n_i(n_j-\delta_{ij})}{(1+\delta_{ij})} \langle \nu_i\nu_j|\widehat{H}(2)|\nu_i\nu_j\rangle,$$

$$\begin{aligned}&\left\langle (\nu_i)^{n_i-1}(\nu_j)^{n_j+1} \prod_{r'=k,l,\ldots} (\nu_{r'})^{n_{r'}} \middle| \widehat{H}(2) \middle| \prod_{r=i,j,\ldots} (\nu_r)^{n_r} \right\rangle\\ &\quad= \sum_{k'} \left[\frac{n_i(n_j+1)(n_{k'}-\delta_{k'i})^2}{(1+\delta_{k'i})(1+\delta_{k'j})}\right]^{1/2} \langle \nu_{k'}\nu_j|\widehat{H}(2)|\nu_{k'}\nu_i\rangle,\end{aligned} \tag{9.3}$$

$$
\left\langle (\nu_i)^{n_i+1}(\nu_j)^{n_j+1}(\nu_k)^{n_k-1}(\nu_l)^{n_l-1} \prod_{r'=m,n,\ldots} (\nu_{r'})^{n_{r'}} \middle| \widehat{H}(2) \middle| \prod_{r=i,j,\ldots} (\nu_r)^{n_r} \right\rangle
$$
$$
= \left[\frac{n_k(n_l - \delta_{kl})(n_i+1)(n_j+1+\delta_{ij})}{(1+\delta_{ij})(1+\delta_{kl})} \right]^{1/2} \langle \nu_i \nu_j | \widehat{H}(2) | \nu_k \nu_l \rangle.
$$

Note that all other m-particle matrix elements are zero due to two-body selection rules. In the second equation in Eq. (9.3), $i \neq j$ and in the third equation, four combinations are possible: (i) $k = l$, $i = j$, $k \neq i$; (ii) $k = l$, $i \neq j$, $k \neq i$, $k \neq j$; (iii) $k \neq l$, $i = j$, $i \neq k$, $i \neq l$; and (iv) $i \neq j \neq k \neq l$. BEGOE(2) for spinless boson systems is defined by Eqs. (9.2) and (9.3) with the H matrix in two-particle spaces represented by GOE(v^2). Now the m-boson BEGOE(2) Hamiltonian matrix ensemble is denoted by $\{H(m)\}$, with $\{H(2)\}$ being a GOE. Note that the $H(m)$ matrix dimension is

$$
d_b(N, m) = \binom{N+m-1}{m} \tag{9.4}
$$

and the number of independent matrix elements is $d_b(N,2)[d_b(N,2)+1]/2$. The subscript 'b' in $d_b(N,m)$ stands for 'bosons'. Using Eqs. (9.2) and (9.3) with GOE representation for H in two-particle spaces, computer codes have been developed for constructing BEGOE(2) ensemble [7].

Extension of BEGOE(2) to BEGOE(1 + 2) incorporating mean-field one-body part is straightforward. The BEGOE(1 + 2) Hamiltonian is,

$$
\{\widehat{H}\}_{\text{BEGOE}(1+2)} = \widehat{h}(1) + \lambda \{\widehat{V}(2)\}; \quad \widehat{h}(1) = \sum_{i=1}^{N} \varepsilon_i \hat{n}_i. \tag{9.5}
$$

The $\widehat{V}(2)$ above is same as $\widehat{H}(2)$ in Eq. (9.1) and the two-particle matrix elements of $\widehat{V}(2)$ are $V_{ijkl} = \langle i, j | \widehat{V}(2) | k, l \rangle$. Similarly, ε_i in Eq. (9.5) are sp energies and they can be fixed or drawn from an appropriate random ensemble as in EGOE(1 + 2). Now on, we will drop the hat over H, h and V when there is no confusion. The m particle matrix for H in Eq. (9.5) follows from Eqs. (9.2) and (9.3) by just adding the $h(1)$ contribution to the diagonal matrix elements,

$$
\left\langle \prod_{r=i,j,\ldots} (\nu_r)^{n_r} \middle| h(1) \middle| \prod_{r=i,j,\ldots} (\nu_r)^{n_r} \right\rangle = \sum_{r=i,j,\ldots} n_r \varepsilon_r. \tag{9.6}
$$

We assume that the sp energies given by $h(1)$ have average spacing Δ. The λ parameter is expressed in units of Δ and we assume without loss of generality $\Delta = 1$. Clearly, it is easy to construct BEGOE(1 + 2) matrices on a computer using the code for BEGOE(2). However, the matrix dimensions makes the calculations prohibitive for larger vales of (m, N). For example $d_b(5,10) = 1001$, $d_b(6,12) = 6188$, $d_b(6,20) = 53130$, $d_b(8,20) = 888030$ and $d_b(10,20) = 10015005$.

It is important to stress that, unlike for fermionic EE, there are only a few BEE investigations in literature [1, 6, 8–11]. Moreover, for interacting spinless boson

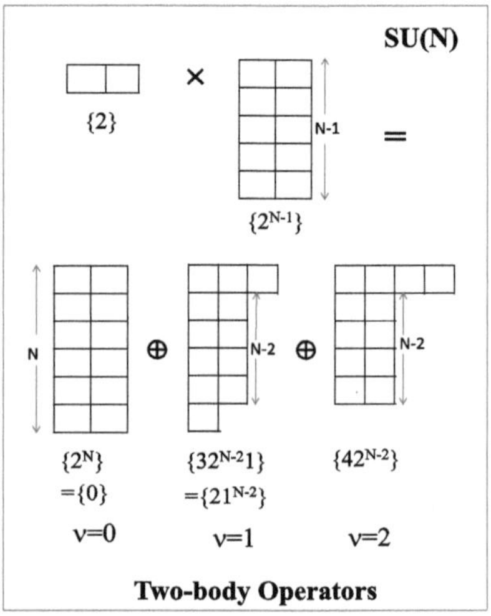

Fig. 9.2 Young tableaux for various tensor parts of two-body operators with respect to $SU(N)$ for spinless boson systems. Figure 5.1a gives the tensor parts for one-body operators (Color figure online)

systems with m bosons in N sp orbitals, dense limit defined by $m \to \infty$, $N \to \infty$ and $m/N \to \infty$ is also possible as m can be greater than N for bosons. Also, many of the results for bosons, as discussed ahead, can be obtained from those for fermions by using $N \to -N$ symmetry and a $m \to N$ symmetry [12–15].

Using BEGOE(1 + 2) codes, in many numerical examples, eigenvalue densities $\rho(E)$ are constructed and they are seen to be close to Gaussian in form. Due to growing matrix dimensions, most of the calculations are restricted to $N = 4, 5$ with $m = 10$–12 giving reasonable examples for the dense limit [6, 9, 16]. See Fig. 5.2 for an example. In order to further confirm that $\rho(E)$ is close to Gaussian for BEGOE(1 + 2), analytical formulas for the first four moments of the eigenvalue density are derived for a given $\widehat{H}(1+2)$. Before turning to them it is useful to mention that a more symmetrized form of V_{ijkl} will be useful and to this end introduced are $\mathcal{V}_{ijkl}$ where

$$\mathcal{V}_{ijkl} = \sqrt{(1+\delta_{ij})(1+\delta_{kl})}\, V_{ijkl}. \tag{9.7}$$

Then, the $V(2)$ operator takes the form

$$V(2) = \frac{1}{4} \sum_{i,j,k,l} \mathcal{V}_{ijkl} b_i^\dagger b_j^\dagger b_k b_l. \tag{9.8}$$

In the next two subsections $\lambda V(2)$ is called $V(2)$.

9.2 Energy Centroids and Spectral Variances: $U(N)$ Algebra

Embedding algebra for BEGOE(1 + 2) is $U(N)$. As one and two boson states transform the $U(N)$ irreps $\{1\}$ and $\{2\}$ in Young tableaux notation, the one and two boson creation operators also transform as {1} and {2}. Then, one and two boson annihilation operators transform as $\{1^{N-1}\}$ and $\{2^{N-1}\}$ respectively. Therefore, $h(1)$ transform as $\{1\} \times \{1^{N-1}\} = [\{1^N\} = \{0\}] \oplus \{21^{N-2}\} = (\nu = 0) + (\nu = 1)$ irreps (or tensors). Note that here we have used $U(N) \leftrightarrow SU(N)$ equivalence. Similarly $V(2)$ transforms as $\{2\} \times \{2^{N-1}\} = [\{2^N\} = \{0\}] \oplus [\{32^{N-2}1\} = \{21^{N-2}\}] \oplus \{42^{N-2}\} = (\nu = 0) + (\nu = 1) + (\nu = 2)$ irreps (or tensors). Figures 5.1a and 9.2 show these decompositions in terms of Young tableaux for one-body and two-body operators respectively. Given $H = h(1) + V(2)$ as defined by Eqs. (9.5)–(9.8), it is possible to write explicitly its $U(N)$ decomposition into various ν parts. Firstly, it is easy to recognize the $\nu = 0$ part as it should be a scalar with respect to $U(N)$, i.e. it should be a polynomial in $\hat{n}$. The result is,

$$
\begin{aligned}
& H^{\nu=0}(1+2) = h^{\nu=0}(1) + V^{\nu=0}(2) = \varepsilon_0 \hat{n} + V_0 \binom{\hat{n}}{2}; \\
& \varepsilon_0 = \frac{1}{N} \sum_i \varepsilon_i, \qquad V_0 = \frac{1}{2N(N+1)} \sum_{i,j} \mathscr{V}_{ijij}.
\end{aligned}
\tag{9.9}
$$

Little thought will give the $\nu = 1$ parts of $h(1)$ and $V(2)$,

$$
\begin{aligned}
H^{\nu=1}(1+2) &= h^{\nu=1}(1) + V^{\nu=1}(2), \\
h^{\nu=1}(1) &= \sum_i \varepsilon_i^{\nu=1} \hat{n}_i; \quad \varepsilon_i^{\nu=1} = \varepsilon_i - \varepsilon_0, \\
V^{\nu=1}(2) &= (\hat{n} - 1) \sum_{i,j} \zeta_{i,j} b_i^\dagger b_j; \\
\zeta_{i,j} &= \frac{1}{N+2} \sum_k \left(\mathscr{V}_{ikjk} - \delta_{ij} \frac{1}{N} \left[\sum_{m,n} \mathscr{V}_{mnmn} \right] \right).
\end{aligned}
\tag{9.10}
$$

Thus, $V^{\nu=1}$ corresponds to an effective (m-dependent) mean-field producing part of $V(2)$ and it is in general off-diagonal in the original mean-field basis, i.e. $\zeta_{ij} \neq 0$ for $i \neq j$. Finally, $V^{\nu=2}(2)$ part is given by

$$
\begin{aligned}
& V^{\nu=2}(2) = V(2) - V^{\nu=0}(2) - V^{\nu=1}(2); \\
& V_{ijij}^{\nu=2} = V_{ijij} - V_0 - \zeta_{i,i} - \zeta_{j,j}, \\
& V_{ikjk}^{\nu=2} = V_{ikjk} - \sqrt{(1+\delta_{ik})(1+\delta_{jk})}\, \zeta(i,j); \quad i \neq j, \\
& V_{ijkl}^{\nu=2} = V_{ijkl}; \quad \text{for all other cases.}
\end{aligned}
\tag{9.11}
$$

Also, we can write the $V^{\nu=2}(2)$ operator as

$$
V^{\nu=2}(2) = \frac{1}{4} \sum_{i,j,k,l} \widetilde{\mathscr{V}}_{ijkl} b_i^\dagger b_j^\dagger b_k b_l; \quad \widetilde{\mathscr{V}}_{ijkl} = \sqrt{(1+\delta_{ij})(1+\delta_{kl})}\, V_{ijkl}^{\nu=2}. \tag{9.12}
$$

Just as for fermion systems, propagation equation for boson systems for the m-particle average of a k-body operator $A(k)$ is simple,

$$\langle A(k)\rangle^m = \binom{m}{k}\langle A(k)\rangle^k. \tag{9.13}$$

Similarly, the various ν pats of $A(k)$ will be orthogonal with respect to averages over the m-boson spaces, i.e. $\langle A^{\nu_1} B^{\nu_2}\rangle^m = \delta_{\nu_1\nu_2}\langle A^{\nu_1} B^{\nu_1}\rangle^m$. As the m particle averages are polynomials in m, using Eq. (9.13) we obtain easily the propagation equations for the energy centroids and spectral variances,

$$\begin{aligned}
E_c(m) &= \langle H(1+2)\rangle^m = m\varepsilon_0 + \binom{m}{2} V_0, \\
\sigma^2(m) &= \langle (H - H^{\nu=0})^2\rangle^m = \langle (H^{\nu=1})^2\rangle^m + \langle (H^{\nu=2})^2\rangle^m; \\
\langle (H^{\nu=1})^2\rangle^m &= \frac{m(m+N)}{N(N+1)} \sum_{i,j} \xi_{ij}(m)\xi_{ji}(m), \\
\xi_{ij}(m) &= \varepsilon_i^{\nu=1}\delta_{ij} + (m-1)\zeta_{ij}, \\
\langle (H^{\nu=2})^2\rangle^m &= \frac{m(m-1)(N+m)(N+m+1)}{N(N+1)(N+2)(N+3)} \frac{1}{4} \sum_{i,j,k,l} \widetilde{\mathscr{V}}_{ijkl}\widetilde{\mathscr{V}}_{klij}.
\end{aligned} \tag{9.14}$$

Using Eq. (9.14) we can calculate ensemble average values for the energy centroids (they come from $h(1)$ only) and spectral variances for any m and with these, Gaussian eigenvalue densities can be constructed. However, to prove that the dense limit gives Gaussian form, formulas for the third and fourth moments are needed as discussed below.

9.3 Third and Fourth Moment Formulas: Gaussian Eigenvalue Density in Dense Limit

For fermion systems, formulas for the third and fourth moments $\langle (H - H^{\nu=0})^i\rangle^m$, $i = 3, 4$ are derived in detail by several authors using diagrammatic methods [17, 18]. They can be extended to boson systems by using $N \to -N$ symmetry [12–15], i.e. by substituting $-N$ for N in the expressions for moments of fermion systems and then taking absolute values of each term, one obtains the expressions for boson systems. The final formulas for 3rd and 4th moments are [12] as follows. Firstly, formula for the 3rd central moment is

$$
\begin{aligned}
\mathcal{M}_3 &= \left\langle \left(H - H^{\nu=0}\right)^3 \right\rangle^m \\
&= \frac{m(N+m)(N+2m)}{N(N+1)(N+2)} X_1 \\
&\quad + \sum_{s=2,3} \binom{m}{s} \binom{N+m+1}{2} \binom{s+2}{2}^{-1} \binom{N+s+1}{s+2}^{-1} X_s + \left\langle (\widetilde{\mathcal{V}})^3 \right\rangle^m ; \\
X_1 &= \sum_{i,j,k} \xi_{ij}(m)\xi_{jk}(m)\xi_{ki}(m), \qquad X_2 = 3D_1 + \frac{3}{2}E_1, \qquad X_3 = 3E_1, \\
D_1 &= \sum_{i,j,k,l} \widetilde{\mathcal{V}}_{ijkl}\xi_{ik}(m)\xi_{jl}(m), \\
E_1 &= \sum_{i,j,k,l,r} \widetilde{\mathcal{V}}_{ijkl}\widetilde{\mathcal{V}}_{krij}\xi_{lr}(m).
\end{aligned}
\tag{9.15}
$$

Formula for $\langle(\widetilde{\mathcal{V}})^3\rangle^m$ is given ahead. Going further, formula for the fourth central moment is,

$$
\begin{aligned}
\mathcal{M}_4 &= \left\langle \left(H - H^{\nu=0}\right)^4 \right\rangle^m \\
&= \frac{m(N+m)}{N(N+1)} M_2 + \sum_{s=2,3} \binom{m}{s} \binom{N+m+1}{2} \binom{s+2}{2}^{-1} \binom{N+s+1}{s+2}^{-1} Y_s \\
&\quad + \sum_{s=2,3} \binom{m}{s+1} \binom{N+m+2}{3} \binom{s+4}{3}^{-1} \binom{N+s+3}{s+4}^{-1} Z_s + \left\langle (\widetilde{\mathcal{V}})^4 \right\rangle^m ; \\
Y_2 &= 12K_1 + 2(G_1 + G_2 + G_3) + F_1 + 3(M_1)^2 + 6M_2, \\
Y_3 &= 24K_1 + 2F_1, \\
Z_2 &= 12G_1 + 6G_2 + 12G_3 + 12G_4 + \frac{3}{2}G_5 + 2F_1 + 12F_2 + 6F_3, \\
Z_3 &= 4F_1 + 24F_2 + 12F_3, \\
M_1 &= \sum_{i,j} \xi_{ij}(m)\xi_{ji}(m), \qquad M_2 = \sum_{i,j,k,l} \xi_{ij}(m)\xi_{jk}(m)\xi_{kl}(m)\xi_{li}(m), \\
K_1 &= \sum_{i,j,k,l,r} \widetilde{\mathcal{V}}_{ijkl}\xi_{ik}(m)\xi_{jr}(m)\xi_{rl}(m), \\
G_1 &= \sum_{i,j,k,l,r,s} \widetilde{\mathcal{V}}_{ijkl}\widetilde{\mathcal{V}}_{ksij}\xi_{ir}(m)\xi_{rs}(m), \qquad G_2 = \sum_{i,j,k,l,r,s} \widetilde{\mathcal{V}}_{ijkl}\widetilde{\mathcal{V}}_{rsij}\xi_{kr}(m)\xi_{ls}(m), \\
G_3 &= \sum_{i,j,k,l,r,s} \widetilde{\mathcal{V}}_{ijkl}\widetilde{\mathcal{V}}_{kris}\xi_{lr}(m)\xi_{sj}(m), \qquad G_4 = \sum_{i,j,k,l,r,s} \widetilde{\mathcal{V}}_{ijkl}\widetilde{\mathcal{V}}_{kris}\xi_{jl}(m)\xi_{rs}(m),
\end{aligned}
$$

$$G_5 = \left[\sum_{i,j,k,l} \widetilde{\mathcal{V}}_{ijkl}\widetilde{\mathcal{V}}_{klij}\right]\left[\sum_{r,s} \xi_{rs}(m)\xi_{sr}(m)\right],$$

$$F_1 = \sum_{i,j,k,l,r,s,p} \widetilde{\mathcal{V}}_{ijkl}\widetilde{\mathcal{V}}_{kprs}\widetilde{\mathcal{V}}_{rsij}\xi_{lp}(m), \qquad F_2 = \sum_{i,j,k,l,r,s,p} \widetilde{\mathcal{V}}_{ijkl}\widetilde{\mathcal{V}}_{lrjs}\widetilde{\mathcal{V}}_{spri}\xi_{kp}(m),$$

$$F_3 = \sum_{i,j,k,l,r,s,p} \widetilde{\mathcal{V}}_{ijkl}\widetilde{\mathcal{V}}_{klrj}\widetilde{\mathcal{V}}_{rpis}\xi_{sp}(m). \tag{9.16}$$

Finally, formula for $\langle \widetilde{\mathcal{V}}^r \rangle^m$, valid for $r = 2, 3$ and 4, is given by

$$\begin{aligned}
\langle (\widetilde{\mathcal{V}})^r \rangle^m &= \sum_{s=2}^{r} \binom{m}{s}\binom{N+s+m-1}{s}\binom{2s}{s}^{-1}\binom{N+2s-1}{2s}^{-1} C_r^s;\\
C_2^2 &= \frac{1}{4}\sum_{i,j,k,l} \widetilde{\mathcal{V}}_{ijkl}\widetilde{\mathcal{V}}_{klij}, \qquad C_3^2 = \frac{1}{8}\sum_{i,j,k,l,r,s} \widetilde{\mathcal{V}}_{ijkl}\widetilde{\mathcal{V}}_{klrs}\widetilde{\mathcal{V}}_{rsij},\\
C_3^3 &= 2C_3^2 + \sum_{i,j,k,l,r,s} \widetilde{\mathcal{V}}_{iljk}\widetilde{\mathcal{V}}_{kslr}\widetilde{\mathcal{V}}_{rjsi}, \qquad C_4^2 = \frac{1}{16}(AA1),\\
C_4^3 &= \frac{1}{4}(AA1) + (CC1) + \frac{1}{2}(BA1) + 2(CA1),\\
C_4^4 &= \frac{3}{8}(AA1) + 6(CC1) + 3(BA1) + 6(CA1) + 3(AB1) + 3\big(C_2^2\big)^2,\\
AA1 &= \sum_{i,j,k,l,r,s,o,p} \widetilde{\mathcal{V}}_{ijkl}\widetilde{\mathcal{V}}_{klrs}\widetilde{\mathcal{V}}_{rsop}\widetilde{\mathcal{V}}_{opij},\\
AB1 &= \sum_{i,j,k,l,r,s,o,p} \widetilde{\mathcal{V}}_{ijkl}\widetilde{\mathcal{V}}_{lrjs}\widetilde{\mathcal{V}}_{sorp}\widetilde{\mathcal{V}}_{pkoi},\\
BA1 &= \sum_{i,j,k,l,r,s,o,p} \widetilde{\mathcal{V}}_{ijkl}\widetilde{\mathcal{V}}_{klis}\widetilde{\mathcal{V}}_{rsop}\widetilde{\mathcal{V}}_{oprj},\\
CA1 &= \sum_{i,j,k,l,r,s,o,p} \widetilde{\mathcal{V}}_{ijkl}\widetilde{\mathcal{V}}_{krso}\widetilde{\mathcal{V}}_{olrp}\widetilde{\mathcal{V}}_{spij},\\
CC1 &= \sum_{i,j,k,l,r,s,o,p} \widetilde{\mathcal{V}}_{ijkl}\widetilde{\mathcal{V}}_{rsjo}\widetilde{\mathcal{V}}_{olsp}\widetilde{\mathcal{V}}_{pkri}.
\end{aligned} \tag{9.17}$$

By numerical construction of various members of BEGOE(1 + 2) with some values for (m, N, λ), formulas given by Eqs. (9.14)–(9.17) have been verified and they in turn provide a good test of the BEGOE(1 + 2) codes. Some examples are as follows [19]. For $m = 8, 12, 20$ and 400 with $N = 5$, $\overline{\gamma_2}$ values are -0.21, -0.11, -0.05 and -0.03 respectively. Similarly, for $m = 12, 20$ and 400 with $N = 12$, the $\overline{\gamma_2}$ values are -0.17, -0.07 and -0.01 respectively. For sufficiently large values for N $(N > 5)$ and $m \gg N$, $|\overline{\gamma_2}| < 0.3$ ($\overline{\gamma_1} \sim 0$ as expected) for BEGOE(2). Analytical formula for $\overline{\gamma_2}$ can be obtained by considering $V^{\nu=2}(2)$. In the strict dense limit, only this part will generate γ_2 for BEGOE(2). Equations (9.17) and (9.14) will give,

$$\begin{aligned}\overline{\gamma_2(m,N)} &\overset{m\to\infty}{\longrightarrow} \frac{(N+2)(N+3)\,P}{(N+5)(N+6)(N+7)} - 3; \\ P &= 6\frac{(N+4)(N+5)^2}{(N+2)(N+3)}\big[\gamma_2(4,N)+3\big] \\ &\quad - 6\frac{(N+4)(N+6)}{(N+2)}\big[\gamma_2(3,N)+3\big] + (N+7)\big[\gamma_2(2,N)+3\big].\end{aligned} \tag{9.18}$$

For sufficiently large N, $\gamma_2(m,N)$ for $m = 2, 3$ and 4 will be given by Eq. (4.32), i.e. $\gamma_2(m) \sim \binom{m}{2}^{-1}\binom{m-2}{2} - 1$. Then, $\gamma_2(2,N) = \gamma_2(3,N) = -1$ and $\gamma_2(4,N) = -5/6$. These and Eq. (9.18) will lead to

$$\gamma_2(\infty, N) = -2(2N+11)/(N+6)(N+7). \tag{9.19}$$

Therefore in the dense limit [6],

$$\gamma_2 \overset{dense\ limit}{\longrightarrow} -\frac{4}{N} \tag{9.20}$$

and this is good for $N \geq 20$. The dense limit result for BEGOE(2) as given by Eq. (9.20) should be compared to the result $\gamma_2(m) \to -4/m$ for EGOE(2) for fermions in the dilute limit. Thus there is a $m \to N$ symmetry between fermions in dilute limit and bosons in the dense limit. Thus, for sufficiently large values of N, BEGOE(2) gives Gaussian eigenvalue densities in the dense limit. However, even for small N as seen from Eq. (9.19), the Gaussian form is valid. For example for $N = 5$, we have $\gamma_2(\infty, N) = -0.32$ and therefore for the dense boson systems $N > 5$ is sufficient for obtaining the Gaussian form.

Simplifications used above are some what complicated for BEGOE(1 + 2). However, it can be shown easily [12] that reasonable $h(1)$ will give Gaussian densities in the dense limit. Combining this with the EGOE(2) Gaussian densities, one can argue that BEGOE(1 + 2) in general, independent of λ value, gives Gaussian eigenvalue densities; see Fig. 5.2. Numerical calculations for sufficiently large value for N ($N > 5$) and $m \gg N$ have indeed shown that $|\overline{\gamma_2}| < 0.3$ (similarly $\overline{\gamma_1}$) for BEGOE(1 + 2). Some examples with $\lambda = 0.025$ and sp energies given by $\varepsilon_i = i + 1/i$ are as follows [19]. With $m = 10$, for $N = 4$, 6 and 8, $(\overline{\gamma_1}, \overline{\gamma_2}) = (0.16, -0.43)$, $(0.13, -0.29)$ and $(0.09, -0.25)$ respectively. Similarly, With $m = 5000$, for $N = 4, 8$ and 12, $(\overline{\gamma_1}, \overline{\gamma_2}) = (0.0, -0.41)$, $(0.0, -0.2)$ and $(0.0, -0.13)$ respectively.

9.4 Average-Fluctuation Separation and Ergodicity in the Spectra of Dense Boson Systems

9.4.1 Average-Fluctuation Separation

For boson systems we will consider BEGOE(2) and the dense limit defined by $m \to \infty$, $N \to \infty$ and $m/N \to \infty$. As discussed in Sect. 4.3.1, level motion in BEGOE(2) is given by Eqs. (4.44) and (4.50) as the eigenvalue density in the dense limit is close to a Gaussian. To apply Eq. (4.50), we need $\tilde{\Sigma}_{rr}$. A formula for this is obtained as follows.

In two particle space, the H matrix is GOE and therefore the two particle matrix elements $H_{\alpha\beta}$ are independent Gaussian variables with $\overline{H_{\alpha\beta}} = 0$, $\overline{H^2_{\alpha\alpha}} = 2v^2$ and $\overline{H^2_{\alpha\beta}} = v^2$ for $\alpha \neq \beta$. Now the two particle variance is,

$$\begin{aligned}\overline{\langle H^2\rangle^{m=2}} &= \binom{N+1}{2}^{-1} \sum_{\alpha,\beta} \overline{H^2_{\alpha\beta}} \\ &= \binom{N+1}{2}^{-1} \left\{\binom{N+1}{2}\left\{\binom{N+1}{2} - 1\right\} v^2 + \binom{N+1}{2} 2v^2\right\}. \end{aligned} \tag{9.21}$$

For large N, the above equation simplifies to $\overline{\langle H^2\rangle^{m=2}} = N^2v^2/2$. Therefore the m-particle variance $\sigma^2(m)$, from Eq. (9.14), is

$$\begin{aligned}\sigma^2(m) &= \langle H^2\rangle^m \to \langle (H^{\nu=2})^2\rangle^m \\ &= \frac{m(m-1)(N+m)(N+m+1)}{N(N+1)(N+2)(N+3)} \langle\langle H^2\rangle\rangle^{m=2}. \end{aligned} \tag{9.22}$$

Then in the dense limit, using the normalization $\langle H^2\rangle^{m=2} = \sigma^2(2) = N^2v^2/2 = 1$, we have

$$\sigma^2(m) = \binom{m}{2}^2 \binom{N}{2}^{-1}. \tag{9.23}$$

Now, the variance Σ_{11} of the centroid fluctuations is given by

$$\begin{aligned}\Sigma_{11} &= \overline{\langle H\rangle^m \langle H\rangle^m} - \overline{\langle H\rangle^m}\;\overline{\langle H\rangle^m} = \overline{\frac{m(m-1)}{N(N+1)} \sum_{\alpha} H_{\alpha\alpha} \frac{m(m-1)}{N(N+1)} \sum_{\beta} H_{\beta\beta}} \\ &= \frac{m^4}{N^4} \sum_{\alpha} \overline{H^2_{\alpha\alpha}} = \frac{m^4}{N^4} \binom{N+1}{2} 2v^2 = 2\frac{m^4}{N^4} \langle H^2(2)\rangle^{m=2} \\ \Rightarrow \quad \tilde{\Sigma}_{11} &= 2\frac{m^4}{N^4}. \end{aligned} \tag{9.24}$$

In the last step in Eq. (9.24) we have used the normalization that $\sigma^2(2) = 1$ and also $\overline{H_{\alpha\alpha} H_{\beta\beta}} = 0$ for $\alpha \neq \beta$. Similarly the expression for the variance of the variance fluctuations $\Sigma_{22} = \overline{\langle H^2\rangle^m \langle H^2\rangle^m} - \overline{\langle H^2\rangle^m}\ \overline{\langle H^2\rangle^m}$ is derived as follows. First we use

$$\begin{aligned}\left\langle H^2\right\rangle^m &= \frac{m(m-1)(N+m)(N+m+1)}{N(N+1)(N+2)(N+3)}\left\langle H^2\right\rangle^2 \\ &\overset{\substack{m\to\infty,\\ N\to\infty\\ m/N\to\infty}}{=} \frac{m^4}{N^4} 2 \sum_{\alpha\geq\beta} H^2_{\alpha\beta}.\end{aligned} \tag{9.25}$$

Then,

$$\begin{aligned}\Sigma_{22} &= 4\frac{m^8}{N^8} \sum_{\substack{\alpha\geqslant\beta\\ \gamma\geqslant\delta}} \overline{H^2_{\alpha\beta} H^2_{\gamma\delta}} - 4\frac{m^8}{N^8} \sum_{\alpha\geqslant\beta} \overline{H^2_{\alpha\beta}} \sum_{\gamma\geqslant\delta} \overline{H^2_{\gamma\delta}} \\ &= 4\frac{m^8}{N^8}\left\{\sum_{\alpha\geqslant\beta} \overline{H^4_{\alpha\beta}} + \sum_{\substack{\alpha\beta\neq\gamma\delta\\ \alpha\geqslant\beta\\ \gamma\geqslant\delta}} \overline{H^2_{\alpha\beta}}\ \overline{H^2_{\gamma\delta}} - \sum_{\alpha\geqslant\beta} \overline{H^2_{\alpha\beta}} \sum_{\gamma\geqslant\delta} \overline{H^2_{\gamma\delta}}\right\} \\ &= 4\frac{m^8}{N^8}\left\{\sum_{\alpha\geqslant\beta} \overline{H^4_{\alpha\beta}} - \sum_{\alpha\geqslant\beta} \left(\overline{H^2_{\alpha\beta}}\right)^2 + \sum_{\substack{\alpha\geqslant\beta\\ \gamma\geqslant\delta}} \overline{H^2_{\alpha\beta}}\ \overline{H^2_{\gamma\delta}} - \sum_{\substack{\alpha\geqslant\beta\\ \gamma\geqslant\delta}} \overline{H^2_{\alpha\beta}}\ \overline{H^2_{\gamma\delta}}\right\} \\ &= 4\frac{m^8}{N^8}\left\{3\sum_{\alpha\geqslant\beta} \upsilon^4 - \sum_{\alpha\geqslant\beta} \upsilon^4\right\} \\ &= 4\frac{m^8}{N^8} 2\left\{\frac{1}{2}\binom{N+1}{2}\left(\binom{N+1}{2}+1\right)\right\}\upsilon^4 \\ &\simeq 4\frac{m^8}{N^8}\left(\left\langle H^2(2)\right\rangle^{m=2}\right)^2.\end{aligned} \tag{9.26}$$

Now using the normalization that $\sigma^2(2) = 1$ gives

$$\tilde{\Sigma}_{22} = 4\frac{m^8}{N^8}. \tag{9.27}$$

Following Eqs. (9.24) and (9.27), it is conjectured [6] that in general $\tilde{\Sigma}_{\zeta\zeta}$ is,

$$\tilde{\Sigma}_{\zeta\zeta} = \overline{\left\langle H^\zeta\right\rangle^m \left\langle H^\zeta\right\rangle^m} = 2\zeta\frac{m^{4\zeta}}{N^{4\zeta}} = 2\zeta\binom{m}{2}^{2\zeta}\binom{N}{2}^{-2\zeta}. \tag{9.28}$$

Note that for a k-body Hamiltonian, it is plausible that $\tilde{\Sigma}_{\zeta\zeta} = 2\zeta \binom{m}{k}^{2\zeta} \binom{N}{k}^{-2\zeta}$. Combining (9.23), (9.28) and (4.50) will give,

$$\overline{\langle S_\zeta^2 \rangle} = 2\zeta \binom{N}{2}^{-\zeta}. \tag{9.29}$$

Substituting this in Eq. (4.44) gives for the level motion in dense limit for BEGOE(2),

$$\begin{aligned}
\frac{\overline{(\delta E)^2}}{\overline{D(E)}^2} &\overset{\text{BEGOE(2)}}{=} \binom{N+m-1}{m}^2 \binom{m}{2}^2 \binom{N}{2}^{-2} [\rho_{\mathscr{G}}(E)]^2 \\
&\quad \times \left\{ \sum_{\zeta \geq 1} (\zeta!)^{-2} 2\zeta \binom{N}{2}^{-\zeta} [He_{\zeta-1}(\hat{E})]^2 \right\} \\
&\overset{\hat{E}=0}{\rightarrow} \frac{1}{\pi} \frac{\binom{N+m-1}{m}^2}{\binom{N}{2}} \left\{ 1 + \frac{1}{12} \binom{N}{2}^{-2} + \frac{1}{320} \binom{N}{2}^{-4} + \cdots \right\}.
\end{aligned} \tag{9.30}$$

Thus, just as for fermions (see Chap. 4), as ζ increases, deviations in $\overline{(\delta E)^2}$ from the leading term rapidly go to zero due to the $\binom{N}{2}^{-2r}$, $r = 1, 2, \ldots$ terms in Eq. (9.30). There will be no change until $\zeta \sim N/2$, thereby defining separation. Beyond this, as pointed out first for bosons by Patel et al. [6] using numerical calculations, for $\zeta \gg N/2$ the deviations grow, i.e. fluctuations set in and they will tend to that of GOE. This is further tested using more numerical calculations in [16]. Comparing Eq. (9.30) with Eq. (4.54), one sees again $m \leftrightarrow N$ symmetry between dilute fermion and dense boson systems.

9.4.2 Ergodicity in BEGOE(2)

An important question raised by Asaga et al. [8], investigating BEGUE(k) is that the bosonic ensembles are not ergodic. This was inferred from the study of level fluctuations for large number of bosons in two and three sp states. Turning to boson systems it is seen from Eqs. (9.24) and (9.27), in the dense limit, scaled Σ_{11} and Σ_{22} are

$$\begin{aligned}
\widehat{\Sigma}_{11} &= \frac{\Sigma_{11}(m)}{\overline{\langle H^2 \rangle^m}} \rightarrow \frac{4}{N^2}, \\
\widehat{\Sigma}_{22} &= \frac{\Sigma_{22}(m)}{\{\overline{\langle H^2 \rangle^m}\}^2} \rightarrow \frac{16}{N^4}
\end{aligned} \tag{9.31}$$

for BEGOE(2) and they remain valid even for BEGOE(k). Secondly, as $m \rightarrow \infty$ and N finite, still the BEGOE(k) matrix dimension is infinity. Thus, we have a situation where the matrix dimension is infinite and the centroid and variance fluctuations are

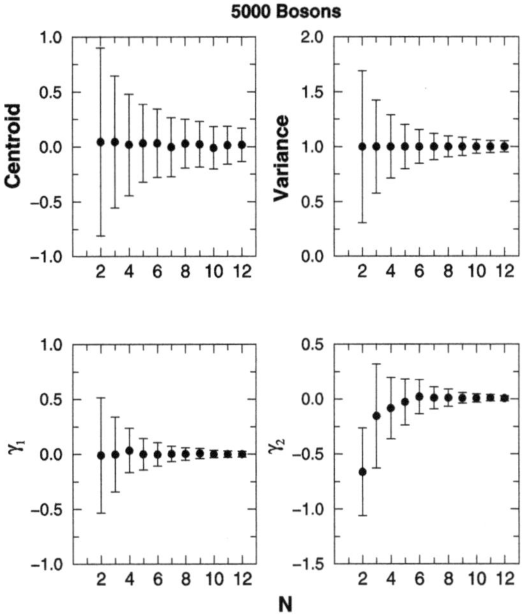

Fig. 9.3 Centroid (E_c), variance (σ^2), skewness (γ_1) and excess (γ_2) parameters for the eigenvalue density of a 200 member BEOE(2) systems with 5000 bosons in N sp states and $N = 2$–12. Figure shows average values (*filled circle*) and widths of the fluctuations (*vertical bars*) of (**a**) E_c normalized by $\{\overline{\sigma^2}\}^{1/2}$, (**b**) σ^2 normalized by $\overline{\sigma^2}$, (**c**) γ_1 and (**d**) γ_2. Figure is taken from [9] with permission from Elsevier

not zero. Therefore, BEGOE(k) [similarly BEGUE(k)] is not ergodic if the dense limit is defined by $m \to \infty$ and N finite [1, 8]. However if we follow the definition used in the beginning of this section, then in the dense limit with sufficiently large N value fluctuations in centroids and variances will tend to zero; see also Fig. 9.3. Going beyond this, fluctuations in γ_1 and γ_2 have been studied numerically for sufficiently large N values and very large m values using the formulas given in Sect. 9.3. As seen from Fig. 9.3, numerical results clearly establish that the variances in γ_1 and γ_2 rapidly go to zero in the dense limit as N increases. Thus in the dense limit defined by $m \to \infty$, $N \to \infty$ and $m/N \to \infty$, BEGOE(k) [also BEGUE(k) discussed in Chap. 11] will be ergodic [9].

9.5 Poisson to GOE Transition in Level Fluctuations: λ_c Marker

In Chaps. 5 and 6 it is seen that fermion systems exhibit three transition markers and these play an important role in statistical nuclear spectroscopy and in mesoscopic physics as discussed in Chap. 7. Further applications will be discussed in Chap. 15 ahead. Then, an important questions is: does BEGOE$(1+2)$ also exhibit three similar transition markers. This is answered in the affirmative in the present and the next two subsections.

Numerical calculations for $N = 4, 5$ systems with $m = 10$–12 have been carried out in [9] and they clearly showed that, as the interaction strength λ in Eq. (9.5) varies, BEGOE$(1+2)$ exhibits Poisson to GOE transition in level fluctuations and there is a λ_c marker for this transition just as for EGOE$(1+2)$. Figure 9.4 shows

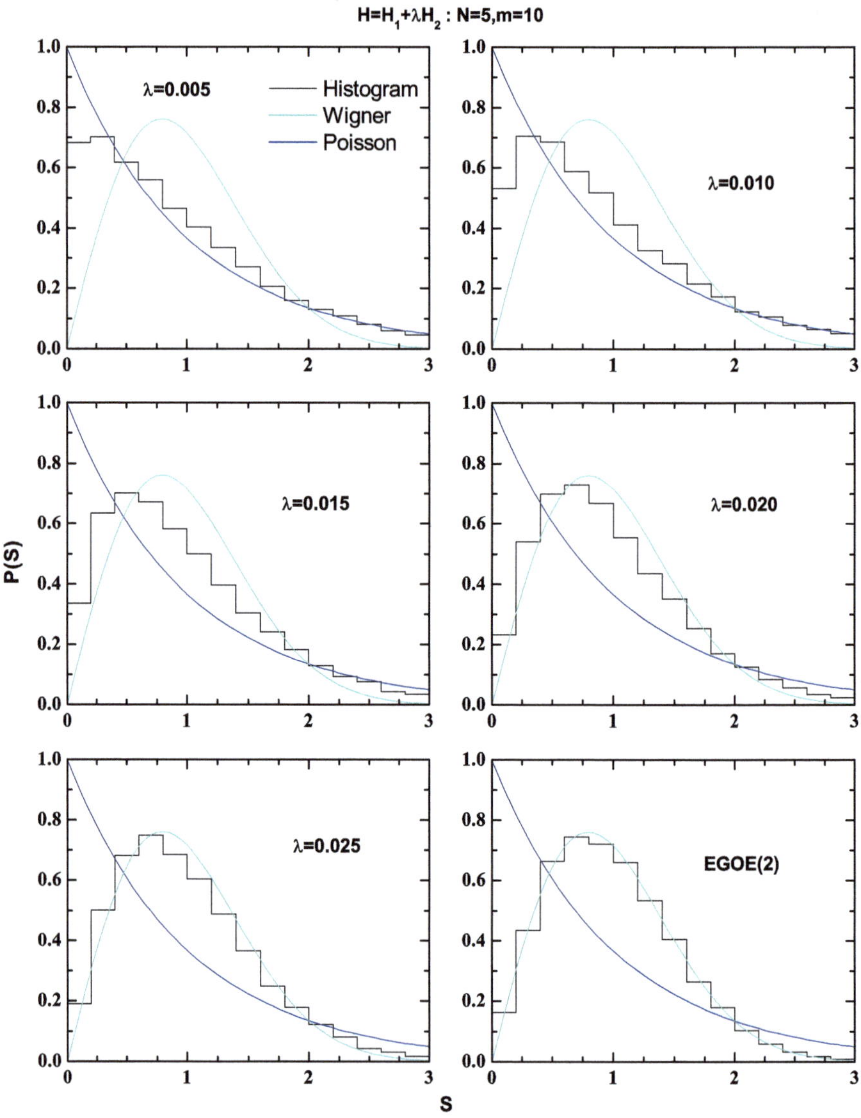

Fig. 9.4 Ensemble averaged NNSD histogram with $H = h(1) + \lambda V(2)$, for various λ values for a BEGOE(1+2) system with $N = 5$ and $m = 10$. Note that in the figure, $H_1 = h(1)$ and $H_2 = V(2)$. BEGOE results are compared with Poisson and Wigner (GOE) forms. It is seen clearly that the system exhibits Poisson to GOE transition in NNSD

an example. For $\lambda = 0$ there are deviations from Poisson form as the sp energies chosen are $\varepsilon_i = i + 1/i$ (see also Chaps. 5 and 6). The transition marker λ_c can be determined for example by using Eq. (5.18). This gives for example, $\lambda_c = 0.025$, 0.018 and 0.015 for $m = 8$, 12 and 16 (with $N = 4$) respectively. Similarly, $\lambda_c =$

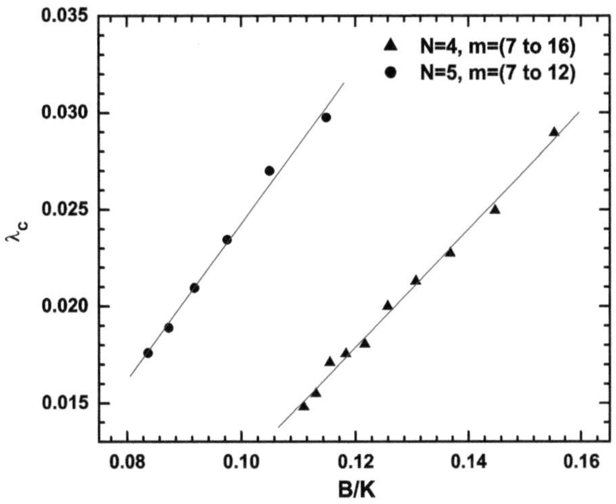

Fig. 9.5 Calculated critical interaction strength λ_c vs B/K. *Filled circles* are for $N = 5$ (with $m = 7 - 16$) and *filed triangles* are for $N = 5$ (with $m =$ 7–12). Figure is taken from [9] with permission from Elsevier (Color figure online)

0.027, 0.021 and 0.018 for $m = 8$, 10 and 12 (with $N = 5$) respectively. In order to verify if λ_c values for BEGOE(1 + 2) follow AJS criterion, an attempt has been made in [9]. According to AJS, λ_c is proportional to the spacing between states directly coupled by $V(2)$. With B giving the span of the directly coupled states ($B \propto N\Delta$) and K the number of directly coupled states, $\lambda_c \propto B/K$. However till now there is no success in deriving a formula for K for boson systems. In [9], K is determined by explicit counting in many numerical examples with $N = 4, 5$. Plot of λ_c vs B/K, constructed using this, as shown in Fig. 9.5, verifies that AJS is indeed applicable to BEGOE(1 + 2). It should be noted that though λ_c is proportional to B/K, the slope is seen to be N dependent.

9.6 BW to Gaussian Transition in Strength Functions: λ_F Marker

Strength functions $F_k(E)$ defined with respect to the basis states $|k\rangle$, which are the eigenstates of $h(1)$ with energy $E_k = \langle k|H|k\rangle$, as discussed before in Chap. 5, give information about localization (or delocalization) of the wavefunctions. Just as for fermionic systems (see Chaps. 5 and 6), increasing λ beyond λ_c, it is seen that the strength functions $F_k(E)$ generated by BEGOE(1 + 2) exhibit BW to Gaussian transition [10] giving a transition marker $\lambda_F > \lambda_c$. Figure 9.6 shows an example. In the calculations, strength functions $F_\xi(E)$ with $\xi - \delta \le E_k \le \xi + \delta$ are averaged and plotted as $F_\xi(E)$ in Fig. 9.6; $\delta = 0.025$ for $\lambda \le 0.035$ and 0.1 for $\lambda > 0.035$. The calculated $F_\xi(E)$ histograms are fitted to a simple function $F_\xi(E : \mu)$ interpolating

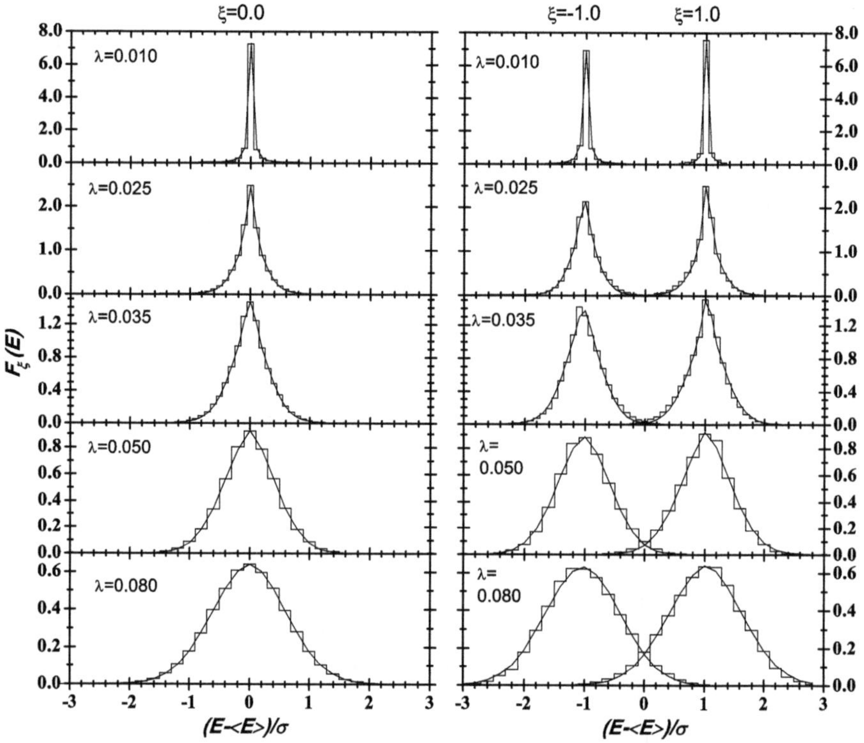

Fig. 9.6 Ensemble averaged $F_\xi(E)$ histograms for a 20 member BEGOE(1 + 2) with $N = 5$, $m = 10$. Results are shown for $\xi = 0, \pm 1$ and for various λ values. Best fit curves obtained using Eq. (9.32) are also shown for each ξ and λ. All energies are scaled using σ, the spectral width. It is seen clearly that the system exhibits BW (for very small λ, it is close to a delta-function) to Gaussian transition in strength functions. Figure is taken from [10] with permission from Elsevier

BW and Gaussian forms,

$$
\begin{aligned}
F_\xi(E:\mu) &= \mu F_{BW:\xi}(E) + (1-\mu) F_{\mathscr{G}:\xi}(E); \\
F_{BW:\xi}(E) &= \frac{1}{2\pi} \frac{\Gamma}{(E-\xi)^2 + \Gamma^2/4}, \\
F_{\mathscr{G}:\xi}(E) &= \frac{1}{\sqrt{2\pi}\,\sigma} \exp -(E-\xi)^2/2\sigma^2
\end{aligned}
\tag{9.32}
$$

with (μ, Γ, σ) being the free parameters. As seen from Fig. 9.6, the fits are quite good. As μ defines the shape of $F_\xi(E)$, this is the most important parameter in Eq. (9.32). Weighted average of μ as a function of λ is shown in Fig. 9.7 and average is calculated as $\mu = [\sum \mu(\xi) \exp -\xi^2/2]/[\sum \exp -\xi^2/2]$; $\mu(\xi)$ represents μ-value that corresponds to $F_\xi(E)$ for a given λ. Using Fig. 9.7 and a criterion for onset of

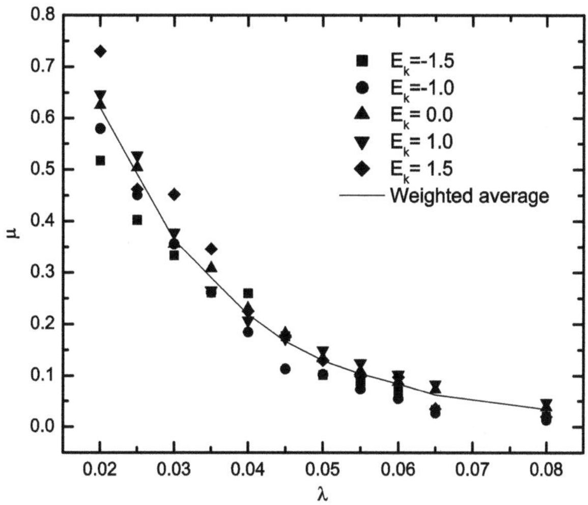

Fig. 9.7 Parameter μ vs λ for various ξ values (they are called E_k in the figure). *Continuous curve* gives weighted average of μ. Figure is taken from [10] with permission from Elsevier

Gaussian behavior, one can deduce the λ_F value. In [10, 20], the criterion used is

$$R(\lambda_F) = 0.7;$$
$$R(\lambda) = \frac{\sum_i \{F_\xi^\lambda(E_i) - F_{BW:\xi}(E)\}^2}{\sum_i \{F_{\mathscr{G}:\xi}(E_i) - F_{BW:\xi}(E)\}^2}. \tag{9.33}$$

The interpolating function $F_\xi(E:\mu)$ gives $R(\lambda_F) = (1 - \mu^2) = 0.7 \Rightarrow \mu = 0.163$. Thus, there will be Gaussian behavior for $\mu \le 0.163$ with onset at 0.163. This together with the results in Fig. 9.7 give $\lambda_F \sim 0.05$ for the $N = 5$, $m = 10$ system considered in Fig. 9.6. Although we have clear demonstration that as λ going beyond λ_c, strength functions make a transition from BW to Gaussian form in the dense limit of BEGOE(1 + 2), just as with λ_c, there is no formula yet for the λ_F marker in terms of (N, m).

9.7 Thermalization Region: λ_t Marker

9.7.1 NPC, S^{info} and S^{occ}

As we increase λ beyond λ_F, BEGOE(1 + 2) generates a region of thermalization. Before discussing this, we consider NPC, S^{info} and S^{occ} in the dense limit. Firstly, for $\lambda > \lambda_F$, it has been well verified that Eq. (5.23) describes NPC in $h(1)$ basis and similarly Eq. (5.25) for $\exp(S^{info})$. Some examples are shown in Figs. 9.8 and 9.9 and given in these figures are also the values of the correlation coefficient ζ. In Fig. 9.9, S^{info} in both $h(1)$ and $V(2)$ basis is shown and the importance of this will be discussed ahead. As there is no restriction on number of bosons in a given

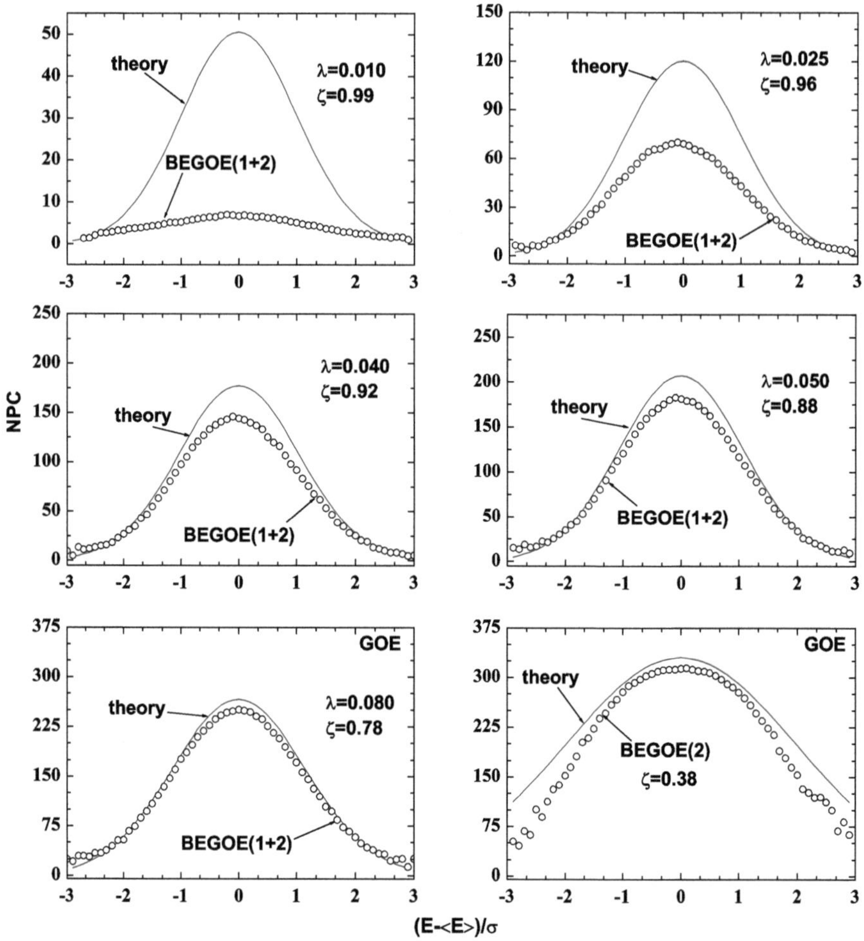

Fig. 9.8 NPC vs E for different λ values for a 20 member BEGOE(1 + 2) with $N = 5$, $m = 10$. In the figures 'theory' corresponds to Eq. (5.23). Values of ζ, the correlation coefficient, are also shown in the figures (Color figure online)

sp level, definition of $S^{occ}(E)$ will be different for bosons, i.e. Eq. (5.32) will not apply. The definition is,

$$S^{occ}(E) = -\sum_i \langle E|\hat{n}_i|E\rangle \big\{\ln\langle E|\hat{n}_i|E\rangle\big\}. \tag{9.34}$$

Here, $\langle E|\hat{n}_i|E\rangle$ is the occupancy of the i-th sp state at energy E. Applying Eq. (5.31) and carrying out simplifications by treating ε_i as a continuous variable, formula for

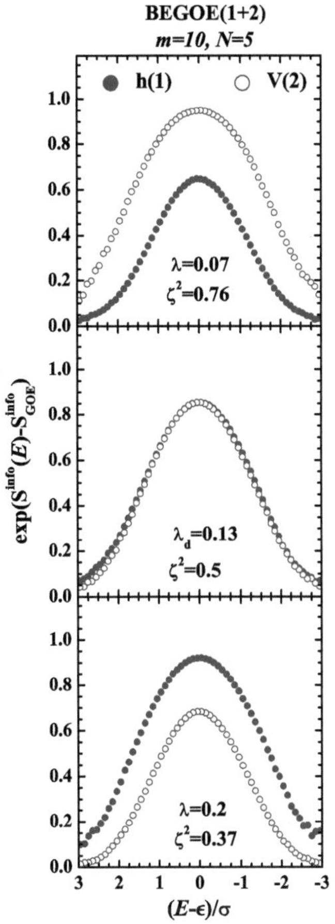

Fig. 9.9 Information entropy vs E in the $h(1)$ and $V(2)$ basis for a 100 member BEGOE$(1+2)$ ensemble with $N=5$, $m=10$ for different λ values. Results averaged over bin size 0.1 are shown as *circles*; *filled circles* correspond to $h(1)$ basis and *open circles* correspond to $V(2)$ basis. Ensemble averaged ζ^2 values are also given in the figure. Note that at the duality point $\lambda=\lambda_d$, the results in $h(1)$ and $V(2)$ basis coincide. Although not shown in the figures, the BEGOE$(1+2)$ results follow Eq. (5.25). See Sect. 9.7.2 for details. Figure is constructed using the results in [21] (Color figure online)

$S^{occ}(E)$, valid in the $\lambda > \lambda_F$ has been derived in [9] giving

$$\exp\{S^{occ}(E) - \exp S^{occ:max}\} = \exp - \left[\left(\frac{N+m}{N} \right) \frac{\zeta^2 \hat{E}^2}{2} \right]. \tag{9.35}$$

Result of Eq. (9.35) is compared with numerical examples in Fig. 9.10.

In order to apply the formulas for NPC, S^{info} and S^{occ}, we need the correlation coefficient ζ and it is defined by Eq. (5.21). Formula for this follows from the results in Sect. 9.2 and the fact that number of off-diagonal and diagonal two-particle matrix elements are $N(N+1)(N+2)(N-1)/4$ and $N(N+1)/2$ respectively. Secondly, for the $V(2)$ matrix, variance of the these off-diagonal elements is λ^2 while that of the diagonal elements is $2\lambda^2$. Then we have,

$$\zeta^2(m,N) = \frac{\frac{m(N+m)}{N(N+1)} \sum_i \tilde{\varepsilon}_i^2 + \lambda^2 \{\frac{m(m-1)(N+m)(N+m+1)}{(N+2)(N+3)}\}}{\frac{m(N+m)}{N(N+1)} \sum_i \tilde{\varepsilon}_i^2 + \lambda^2 \{\frac{m(m-1)(N+m)(N+m+1)(N^2+N+2)}{4(N+2)(N+3)}\}}. \tag{9.36}$$

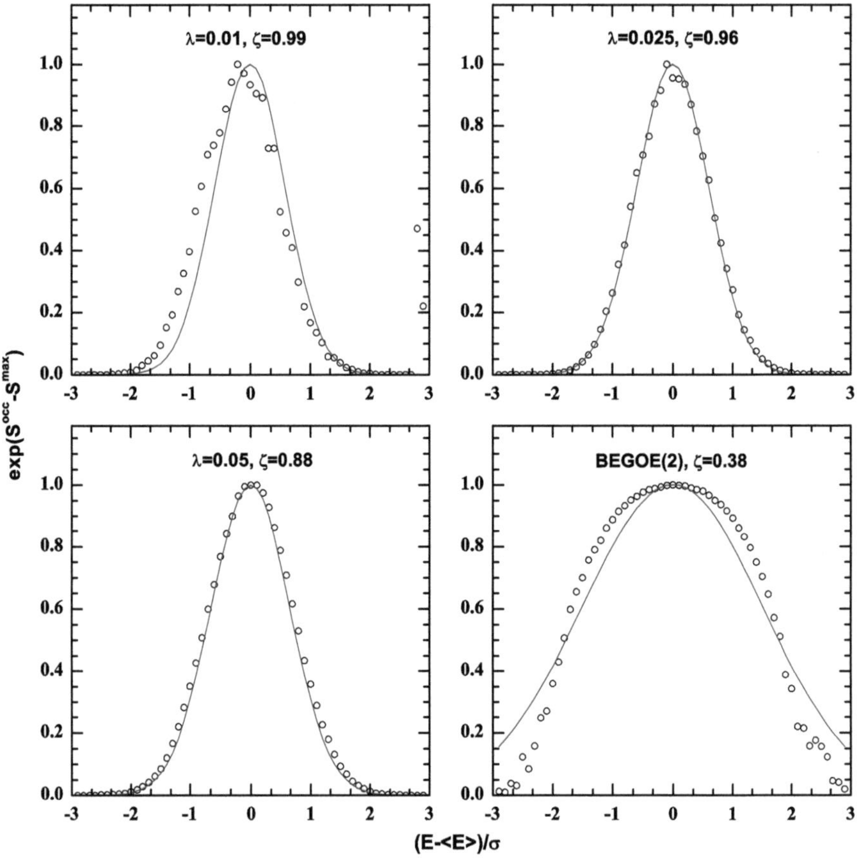

Fig. 9.10 S^{occ} vs E for the same system used in Fig. 9.8. In the figure, *open circles* correspond to the results from the ensemble calculations and the *continuous (red) curves* correspond to Eq. (9.35). Ensemble averaged ζ values are also given in the figure (Color figure online)

Numerical calculations in [10] showed that Eq. (9.36) is good for any λ.

9.7.2 *Thermalization in* BEGOE(1 + 2)

Thermalization in interacting boson systems was investigated by Borgonovi et al. [22], using a simple symmetrized coupled two-rotor model. They explored different definitions of temperature and compared the occupancy number distribution with the Bose-Einstein distribution. Their conclusion is: "For chaotic eigenstates, the distribution of occupation numbers can be approximately describe by the Bose-Einstein distribution, although the system is isolated and consists of two particles only. In this case a strong enough interaction plays the role of a heat bath, thus leading to thermalization". In order to establish that this is a generic property of interacting boson systems, thermalization in BEGOE(1 + 2) was investigated in [21] using different definitions of temperature and entropy and the results are as follows.

Temperature can be defined in a number of different ways in the standard thermodynamical treatment. These definitions of temperature are known to give same result in the thermodynamical limit i.e. near a region where thermalization occurs [23]. Four definitions of temperature ($T = \beta^{-1}$) are:

- β_c: defined using the canonical expression between energy and temperature,

$$\langle E \rangle_{\beta_c} = \frac{\sum_i E_i \exp[-\beta_c E_i]}{\sum_i \exp[-\beta_c E_i]} \tag{9.37}$$

  where E_i are the eigenvalues of the Hamiltonian.
- β_{fit}: defined using occupation numbers obtained by making use of the standard canonical distribution,

$$\langle n_k \rangle^E = \frac{\sum_i \langle n_k \rangle^{E_i} \exp[-\beta_{fit} E_i]}{\sum_i \exp[-\beta_{fit} E_i]}. \tag{9.38}$$

  Here k is single particle state index and E_i are eigenvalues. In applying Eq. (9.38), the constraint $\sum_k \langle n_k \rangle^E = m$ should be taken into account.
- β_{BE}: defined using Bose-Einstein distribution for the occupation numbers,

$$\langle n_k \rangle^E = 1 / \{\exp[\beta_{BE}(E)(\varepsilon_k - \mu(E))] - 1\}. \tag{9.39}$$

  Here μ is the chemical potential. Although, this expression is derived for a system with large number of non-interacting particles in contact with a thermostat, it can be used even in isolated systems with relatively few particles [24, 25].
- β_T: defined using state density $\rho(E)$ generated by H. Note that

$$\beta_T = \frac{d \ln[\rho(E)]}{dE}. \tag{9.40}$$

Figure 9.11 shows ensemble averaged values of β, computed via various definitions described above, for a 100 member BEGOE(1+2) ensemble with $m = 10$ and $N = 5$ as a function of $\hat{E} = (E - \varepsilon)/\sigma$, for various λ values. The β values are calculated from $\hat{E} = -1.5$ to the center of the spectrum, where temperature is infinity. The edges of the spectrum have been avoided as (i) density of states is small near the edges of the spectrum and (ii) eigenstates near the edges are not fully chaotic. Since the state density for BEGOE(1 + 2) is Gaussian irrespective of λ values, β_T as a function of energy gives straight line. Dotted lines shown in the plots represents β_T results in Fig. 9.11. It is clearly seen that for $\lambda < \lambda_c$ [for $(m, N) = (10, 5)$, $\lambda_c \sim 0.02$ and $\lambda_F \sim 0.05$], all the definitions give different values of β. Whereas in the region $\lambda \le \lambda_F$, temperature found from BE distribution, β_{BE}, turns out to be completely different from other temperatures. As in this region, the structure of eigenstates is not chaotic enough leading to strong variation in the distribution of the occupation

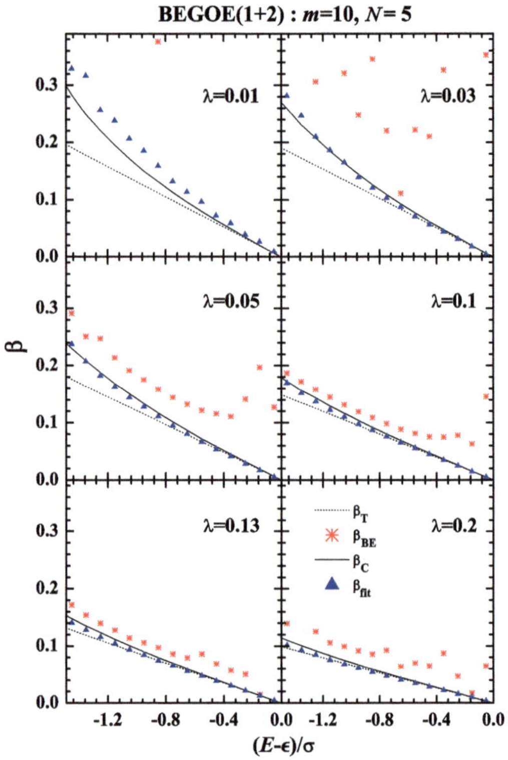

Fig. 9.11 Ensemble averaged inverse of temperature β as a function energy E, for different values of two body interaction strength λ for the BEGOE$(1+2)$ system considered in Fig. 9.9. Results for different definitions of inverse of temperature β are given. In the calculations, the sp energies are chosen to be independent Gaussian random variables. With some modification, figure is taken from [21] with permission from Elsevier

numbers and therefore strong fluctuations in β_{BE}. Moreover, near the center of the spectrum (i.e. as $T \to \infty$), value of the denominator in Eq. (9.39) becomes very small, which leads to large variation in β_{BE} values form member to member. Further increase in $\lambda > \lambda_F$, in the chaotic region, all definitions give essentially same value for the temperature for $\lambda \sim \lambda_t$. It is seen from Fig. 9.11 that the matching between different values of β is good near $\lambda = \lambda_t = 0.13$ for the $N = 5$, $m = 10$ example.

For further establishing that $\lambda \sim \lambda_t$ defines thermodynamic region, used are three different definitions of entropy and these are [as in EGOE$(1+2)$ and EGOE$(1+2)$-**s** studies] thermodynamic entropy S^{ther}, information entropy S^{info} and occupancy entropy S^{occ}. The following measure, introduced in [26] (see also Chap. 15) has been used to obtain λ_t:

$$\Delta_s(\lambda) = \left\{ \int_{-\infty}^{\infty} \left[\left(R_E^{info} - R_E^{ther}\right)^2 + \left(R_E^{sp} - R_E^{ther}\right)^2 \right] dE \right\}^{1/2} \Big/ \left\{ \int_{-\infty}^{\infty} R_E^{ther} dE \right\}, \tag{9.41}$$

where $R_E^{\alpha} = \exp[S^{\alpha}(E) - S^{\alpha}_{max}]$. In the thermodynamic region the values of the different entropies should be very close to each other, hence the minimum of Δ_s gives the value of λ_t. In Fig. 9.12, results shown for ensemble averages $\overline{\Delta_s(\lambda)}$ (blue stars) obtained for a 100 member BEGOE$(1+2)$ ensemble with 10 bosons in 5 single particle states as a function of the two-body interaction strength λ. The second

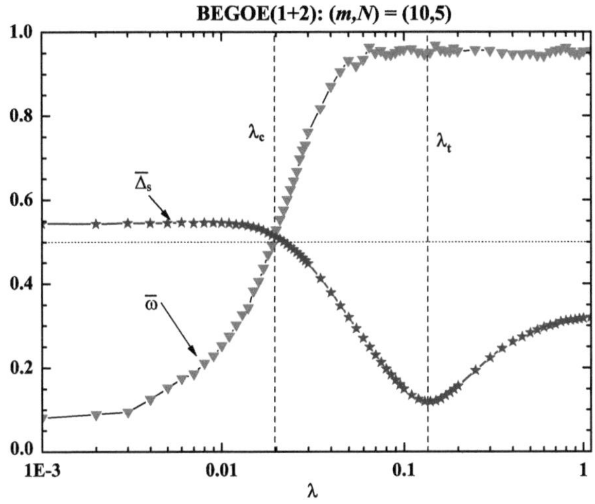

Fig. 9.12 Ensemble averaged values $\overline{\omega}$ and $\overline{\Delta_s}$ as a function of two body interaction strength λ for the BEGOE(1 + 2) system considered in Fig. 9.9. The *vertical dashed lines* represent the position of λ_c and λ_t. In the calculation single particle energies are taken as independent real Gaussian random variables. Here $\lambda_c \simeq 0.02$ and $\lambda_t \simeq 0.13$. Figure is taken from [21] with permission from Elsevier (Color figure online)

vertical dashed line indicates the position of λ_t where ensemble averaged $\overline{\Delta_s(\lambda)}$ is minimum. For the present example, we obtained $\lambda_t \simeq 0.13$. This value of λ_t is same as obtained using different definitions of temperature. In order to show that $\lambda_c \ll \lambda_t$, the NNSD as a function of λ are fitted to Brody distribution and extracted the Brody parameter ω. Then the chaos marker λ_c is determined by the condition $\omega(\lambda) = 1/2$. Ensemble averaged values of $\overline{\omega(\lambda)}$ are shown in Fig. 9.12 and the λ_c value is shown by a vertical dash line in the figure.

To derive a formula for λ_t, considered is duality in BEGOE(1 + 2). The duality region (see Chaps. 5 and 6) $\lambda = \lambda_d$ is the region (in λ space) where all wave functions look alike and it is expected to correspond to the thermodynamic region defined by $\lambda = \lambda_t$. To examine duality, $S^{info}(E)$ in $h(1)$ basis and in $V(2)$ basis are compared. Figure 9.9 shows some numerical results and it is seen that the values of $S^{info}(E)$ in these two basis coincide at $\lambda = 0.13$ giving value for the duality marker $\lambda_d \simeq 0.13$ for the $N = 5, m = 10$ example. This value is very close to the value of marker λ_t and therefore, λ_d region can be interpreted as the thermodynamic region in the sense that all different definitions of temperature and entropy coincide in this region. As discussed in Sect. 5.3.5, λ_t is given by $\zeta^2(\lambda_t) = 0.5$. In addition, Eq. (9.36) gives the (m, N) dependence of the marker,

$$\lambda_t = 2\sqrt{\frac{(N+2)X}{N(N+1)(N-2)(m-1)(N+m-1)}}, \quad X = \sum_{i=1}^{N} \tilde{\varepsilon_i}^2. \tag{9.42}$$

For uniform sp spectrum with $\varepsilon_i = i$, $X = N(N+1)(N-1)/12$ and then,

$$\lambda_t = \sqrt{\frac{(N-1)(N+2)}{3(N-2)(m-1)(N+m+1)}}.$$

For $m = 10$ and $N = 5$, this gives $\lambda_t \approx 0.15$. For the sp energies that are used in the calculations in Figs. 9.11 and 9.12, $X = N(N^2 + 5)/12$ and then

$$\lambda_t = \sqrt{\frac{(N+2)(N^2+5)}{3(N+1)(N-2)(m-1)(N+m+1)}}.$$

For $m = 10$ and $N = 5$, this gives $\lambda_t \approx 0.16$ as compared to the numerically found value $\lambda_t = 0.13$. In the dense limit, Eq. (9.42) gives $\lambda_t \sim \frac{1}{m}\sqrt{\frac{N}{3}}$. Similarly, in the dilute limit, it gives $\lambda_t \sim \frac{1}{\sqrt{3m}}$ in agreement with the EGOE(1 + 2) result given in Chap. 5.

References

1. T. Asaga, L. Benet, T. Rupp, H.A. Weidenmüller, Spectral properties of the k-body embedded Gaussian ensembles of random matrices for bosons. Ann. Phys. (N.Y.) **298**, 229–247 (2002)
2. F. Iachello, A. Arima, *The Interacting Boson Model* (Cambridge University Press, Cambridge, 1987)
3. F. Iachello, R.D. Levine, *Algebraic Theory of Molecules* (Oxford University Press, New York, 1995)
4. A. Frank, P. Van Isacker, *Algebraic Methods in Molecular and Nuclear Physics* (Wiley, New York, 1994)
5. V.K.B. Kota, Group theoretical and statistical properties of interacting boson models of atomic nuclei: recent developments, in *Focus on Boson Research*, ed. by A.V. Ling (Nova Science Publishers Inc., New York, 2006), pp. 57–105
6. K. Patel, M.S. Desai, V. Potbhare, V.K.B. Kota, Average-fluctuations separation in energy levels in dense interacting boson systems. Phys. Lett. A **275**, 329–337 (2000)
7. M. Vyas, Some studies on two-body random matrix ensembles, Ph.D. Thesis, M.S. University of Baroda, India (2012)
8. T. Asaga, L. Benet, T. Rupp, H.A. Weidenmüller, Non-ergodic behaviour of the k-body embedded Gaussian random ensembles for bosons. Europhys. Lett. **56**, 340–346 (2001)
9. N.D. Chavda, V. Potbhare, V.K.B. Kota, Statistical properties of dense interacting Boson systems with one plus two-body random matrix ensembles. Phys. Lett. A **311**, 331–339 (2003)
10. N.D. Chavda, V. Potbhare, V.K.B. Kota, Strength functions for interacting bosons in a mean-field with random two-body interactions. Phys. Lett. A **326**, 47–54 (2004)
11. M. Vyas, N.D. Chavda, V.K.B. Kota, V. Potbhare, One plus two-body random matrix ensembles for boson systems with F-spin: analysis using spectral variances. J. Phys. A, Math. Theor. **45**, 265203 (2012)
12. V.K.B. Kota, V. Potbhare, Shape of the eigenvalue distribution for bosons in scalar space. Phys. Rev. C **21**, 2637–2642 (1980)
13. V.K.B. Kota, A symmetry for the widths of the eigenvalue spectra of boson and fermion systems. J. Phys. Lett. **40**, L579–L582 (1979)
14. V.K.B. Kota, Studies on the goodness of IBA group symmetries: centroids, widths and partial widths for irreducible representations of IBA group symmetries. Ann. Phys. (N.Y.) **134**, 221–258 (1981)
15. P. Cvitanovic, A.D. Kennedy, Spinors in negative dimensions. Phys. Scr. **26**, 5–14 (1982)
16. R.J. Leclair, R.U. Haq, V.K.B. Kota, N.D. Chavda, Power spectrum analysis of the average-fluctuation density separation in interacting particle systems. Phys. Lett. A **372**, 4373–4378 (2008)

17. B.J. Dalton, S.M. Grimes, J.P. Vary, S.A. Williams (eds.), *Moment Methods in Many Fermion Systems* (Plenum, New York, 1980)
18. S.S.M. Wong, *Nuclear Statistical Spectroscopy* (Oxford University Press, New York, 1986)
19. N.D. Chavda, Study of random matrix ensembles for bosonic systems, Ph.D. Thesis, M.S. University of Baroda, Vadodara, India (2004)
20. V.K.B. Kota, R. Sahu, Breit-Wigner to Gaussian transition in strength functions, arXiv:nucl-th/0006079
21. N.D. Chavda, V.K.B. Kota, V. Potbhare, Thermalization in one- plus two-body ensembles for dense interacting boson systems. Phys. Lett. A **376**, 2972–2976 (2012)
22. F. Borgonovi, I. Guarneri, F.M. Izrailev, G. Casati, Chaos and thermalization in a dynamical model of two interacting particles. Phys. Lett. A **247**, 140–144 (1998)
23. M. Rigol, V. Dunjko, M. Olshanii, Thermalization and its mechanism for generic isolated quantum systems. Nature (London) **452**, 854–858 (2008)
24. V.V. Flambaum, F.M. Izrailev, Statistical theory of finite Fermi systems based on the structure of chaotic eigenstates. Phys. Rev. E **56**, 5144–5159 (1997)
25. V.V. Flambaum, G.F. Gribakin, F.M. Izrailev, Correlations within eigenvectors and transition amplitudes in the two-body random interaction model. Phys. Rev. E **53**, 5729–5741 (1996)
26. V.K.B. Kota, A. Relaño, J. Retamosa, M. Vyas, Thermalization in the two-body random ensemble, J. Stat. Mech. P10028 (2011)

Chapter 10
Embedded GOE Ensembles for Interacting Boson Systems: BEGOE(1 + 2)-F and BEGOE(1 + 2)-$S1$ for Bosons with Spin

Going beyond the embedded ensembles for spinless boson systems, it is possible to analyze BEGOE for two species boson systems in terms of bosons carrying a fictitious ($F = \frac{1}{2}$) spin such that the two projections of the boson correspond to the two species. With GOE embedding, this gives BEGOE(1 + 2)-F ensemble [1]. Similarly, because of the interest in spinor BEC and also in the IBM-3 model of atomic nuclei, it is useful to study BEE with bosons carrying spin $S = 1$ degree of freedom. With GOE embedding, this gives BEGOE(1+2)-$S1$ ensemble [2]. Results for these two ensembles are presented in this chapter.

10.1 BEGOE(1 + 2)-F for Two Species Boson Systems

For a two species boson system, it is possible to introduce a fictitious spin, called F-spin for the bosons, such that the two projections of F represent the two species. Then, for m bosons the total fictitious spin F takes values $\frac{m}{2}, \frac{m}{2} - 1, \ldots, 0$ or $\frac{1}{2}$. For such a system with m number of bosons in Ω number of single particle levels, each doubly degenerate, it is possible to define an embedded Gaussian orthogonal ensemble of random matrices generated by random two-body interactions that conserve the total F-spin and this random matrix ensemble is denoted by BEGOE(1 + 2)-F. With degenerate single particle orbitals we have BEGOE(2)-F. The embedding in BEGOE(1 + 2)-F is generated by the Lie algebra $U(2\Omega) \supset U(\Omega) \otimes SU(2)$ with $SU(2)$ generating F-spin. Some applications of BEGOE(1 + 2)-F are: (i) the ensemble is directly applicable to the proton-neutron interacting boson model (pn-IBM) of atomic nuclei [3] and gives, as discussed ahead, some generic structures generated by F-spin used in this model; (ii) it is possible to use the ensemble as a generic model for interacting boson systems, with internal degrees of freedom, in the study of various issues related to thermalization in finite quantum systems [4–6]; (iii) this ensemble will allow us to obtain deeper understanding of the similarities and differences in statistical properties of interacting fermion and boson systems (in addition to using embedded ensembles with spinless fermions/bosons, it is possible

V.K.B. Kota, *Embedded Random Matrix Ensembles in Quantum Physics*, Lecture Notes in Physics 884, DOI 10.1007/978-3-319-04567-2_10,

to use ensembles with spin degree of freedom for fermions/ bosons); (iv) it is possible to apply this ensemble to two component boson systems such as those discussed for example in [7].

10.1.1 Definition and Construction of BEGOE(1 + 2)-F

Let us consider a system of m ($m > 2$) bosons with F-spin degree of freedom and occupying Ω number of sp levels. For convenience, in the remaining part of this paper, we use the notation $\mathscr{F}$ for the F-spin quantum number of a single boson, f for the F-spin carried by a two boson system and for $m > 2$ boson systems F for the F-spin. Therefore, $\mathscr{F} = \frac{1}{2}$, $f = 0$ or 1 and $F = \frac{m}{2}, \frac{m}{2} - 1, \ldots, 0$ or $\frac{1}{2}$. Similarly, the space generated by the sp levels $i = 1, 2, \ldots, \Omega$ is referred as orbital space. Then the sp states of a boson are denoted by $|i; \frac{1}{2}, m_{\mathscr{F}}\rangle$ with $i = 1, 2, \ldots, \Omega$ and $m_{\mathscr{F}} = \frac{1}{2}$ (spin up) or $-\frac{1}{2}$ (spin down). Note that $m_{\mathscr{F}}$ are the eigenvalues of the z-component $\hat{F}_z$ of the F-spin operator $\hat{F}$ for a single boson. With Ω number of orbital degrees of freedom and two spin ($m_{\mathscr{F}}$) degrees of freedom, total number of sp states is $N = 2\Omega$. Going further, two boson states that are symmetric in the total orbital × spin space are denoted by $|(ij); f, m_f\rangle$ with $f = \frac{1}{2} \times \frac{1}{2} = 0$ or 1. Then, $m_f = 0$ for $f = 0$ and $m_f = 1, 0, -1$ for $f = 1$. Similarly, for $f = 1$ we have $i \geq j$ (or equivalently $i \leq j$) and for $f = 0$ we have $i < j$ (or equivalently $i > j$). This gives, without counting the m_f quantum number, for $f = 0$ and $f = 1$, number of sates to be $\Omega(\Omega - 1)/2$ and $\Omega(\Omega + 1)/2$ respectively. For further discussion, we need boson creation ($b^\dagger_{---}$) and annihilation (b_{---}) operators. In terms of them, the sp states are $|i; \frac{1}{2}, m_{\mathscr{F}}\rangle = b^\dagger_{i;\frac{1}{2},m_{\mathscr{F}}}|0\rangle$. Similarly, the two boson states are $|(ij); f, m_f\rangle = \frac{1}{\sqrt{(1+\delta_{ij})}}(b^\dagger_{i;\frac{1}{2}} b^\dagger_{j;\frac{1}{2}})^f_{m_f}|0\rangle$. Note that here we are using spin (angular momentum) coupled representation.

For one plus two-body Hamiltonians preserving m particle F-spin, the one-body Hamiltonian $\widehat{h}(1)$ is

$$\widehat{h}(1) = \sum_{i=1}^{\Omega} \varepsilon_i n_i \tag{10.1}$$

where the orbitals i are doubly degenerate, n_i are number operators and ε_i are sp energies (it is in principle possible to consider $\widehat{h}(1)$ with off-diagonal energies ε_{ij}). A two-body Hamiltonian operator $\widehat{V}(2)$ preserving F-spin is given by,

$$\widehat{V}(2) = \sum_{i,j,k,\ell;f,m_f}{}^{'} \frac{V^f_{ijk\ell}}{\sqrt{(1+\delta_{ij})(1+\delta_{k\ell})}} (b^\dagger_{i:\frac{1}{2}} b^\dagger_{j:\frac{1}{2}})^f_{m_f} \left[(b^\dagger_{k:\frac{1}{2}} b^\dagger_{\ell:\frac{1}{2}})^f_{m_f}\right]^\dagger. \tag{10.2}$$

The ‘prime’ over the summation symbol in (10.2) indicates that the summation over i, j, k and ℓ is restricted to $i \geq j$ and $k \geq \ell$ for $f = 1$ and $i > j$ and $k > \ell$ for $f = 0$.

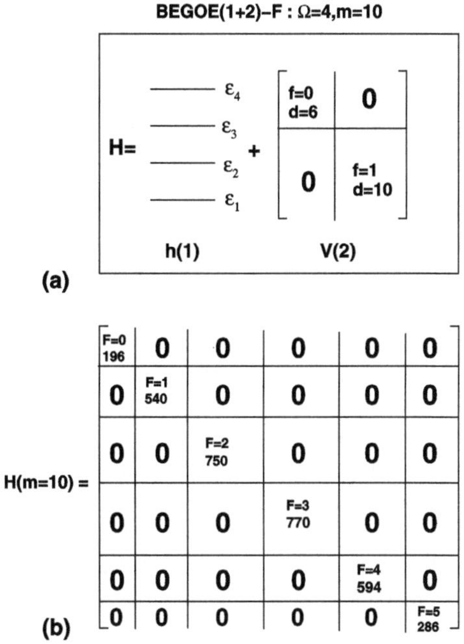

Fig. 10.1 Figure illustrating the block diagonal structure of $V(2)$ and $H(m)$ matrices for a $\Omega = 4$ and $m = 10$ boson system. (**a**) sp levels generated by $h(1)$ operator and the matrix of the $V(2)$ operator in two boson space. Note that the sp levels are doubly degenerate. (**b**) Decomposition of the H matrix in m particle space into direct sum of matrices with fixed F-spin value. There is a BEGOE(1 + 2)-F ensemble in each (m, F) space corresponding to each of the diagonal block in (**b**). Note that the matrix elements in the off-diagonal blocks in (**a**) and (**b**) are all zero

The symmetrized (with respect to the total orbital × spin space) two-body matrix elements $V^f_{ijk\ell} = \langle (ij)f, m_f | \widehat{V}(2) | (k\ell)f, m_f \rangle$ are independent of the m_f quantum number and this ensures that $\widehat{V}(2)$ preserves F-spin. It is seen from (10.2) that $\widehat{V}(2) = \widehat{V}^{f=0}(2) + \widehat{V}^{f=1}(2)$. Then the matrix of $\widehat{V}(2)$ in two boson spaces will be a 2×2 block matrix and the two diagonal blocks correspond to $f = 0$ and 1 respectively and the off-diagonal block is zero, i.e. the matrix is a direct sum of $f = 0$ and $f = 1$ matrices. See Fig. 10.1a for an example of $\widehat{h}(1)$ spectrum in one boson space and $\widehat{V}(2)$ in two boson spaces.

The BEGOE(2)-F ensemble for a given (m, F) system is generated by first defining the two parts of the two-body Hamiltonian to be independent GOE(1)s in the two-particle spaces [one for $\widehat{V}^{f=0}(2)$ and other for $\widehat{V}^{f=1}(2)$]. Now, the $V(2)$ ensemble defined by $\{\widehat{V}(2)\} = \{\widehat{V}^{f=0}(2)\} + \{\widehat{V}^{f=1}(2)\}$ is propagated to the (m, F)-spaces by using the geometry (direct product structure) of the m-particle spaces. By adding the $\widehat{h}(1)$ part, the BEGOE(1 + 2)-F is defined by the operator

$$\{\widehat{H}\}_{\text{BEGOE(1+2)-}F} = \widehat{h}(1) + \lambda_0 \{\widehat{V}^{f=0}(2)\} + \lambda_1 \{\widehat{V}^{f=1}(2)\}. \tag{10.3}$$

Here λ_0 and λ_1 are the strengths of the $f = 0$ and $f = 1$ parts of $\widehat{V}(2)$ respectively. The mean-field one-body Hamiltonian $\widehat{h}(1)$ is defined by sp energies ε_i with average spacing Δ. Without loss of generality, we put $\Delta = 1$ so that λ_0 and λ_1 are in the units of Δ. In principle, many other choices for the sp energies are possible. Thus BEGOE(1+2)-F is defined by the five parameters $(\Omega, m, F, \lambda_0, \lambda_1)$. The H matrix dimension $d_b(\Omega, m, F)$ for a given (m, F) is

$$d_b(\Omega, m, F) = \frac{(2F+1)}{(\Omega-1)} \binom{\Omega + m/2 + F - 1}{m/2 + F + 1} \binom{\Omega + m/2 - F - 2}{m/2 - F}, \tag{10.4}$$

and they satisfy the sum rule $\sum_F (2F+1) d_b(\Omega, m, F) = \binom{N+m-1}{m}$. For example: (i) $d_b(4, 10, F) = 196$, 540, 750, 770, 594 and 286 for $F = 0$–5; (ii) $d_b(4, 11, F) = 504$, 900, 1100, 1056, 780 and 364 for $F = 1/2$–$11/2$; (iii) $d_b(5, 10, F) = 1176$, 3150, 4125, 3850, 2574 and 1001 for $F = 0$–5; (iv) $d_b(6, 12, F) = 13860$, 37422, 50050, 49049, 36855, 20020 and 6188 for $F = 0$–6; and (v) $d_b(6, 16, F) = 70785$, 198198, 286650, 321048, 299880, 235620, 151164, 72675 and 20349 for $F = 0$–8.

Given ε_i and V^f_{ijkl}, the many particle Hamiltonian matrix for a given (m, F) is obtained by first constructing H matrix in M_F representation (M_F is the F_z quantum number). This is easy to carry out using Eqs. (9.3) and (6.4). From the (m, M_F) matrix, (m, F) matrices can be obtained by projecting spin F using the $\hat{F}^2$ operator just as it was done for fermion systems with spin degree of freedom in Chap. 6. Alternatively, it is possible to construct the H matrix directly in a good F basis using angular-momentum algebra. So far in literature for BEGOE(1 + 2)-F only the M_F representation is used for constructing the H matrices [1]. Note that, states with $M_F = M_F^{min} = 0$ for even m and $M_F = M_F^{min} = \frac{1}{2}$ for odd m will contain all F values. The dimension of this basis space then is $\mathcal{D}(\Omega, m, M_F^{min}) = \sum_F d_b(\Omega, m, F)$. For example, $\mathcal{D}(4, 10, 0) = 3136$, $\mathcal{D}(4, 11, \frac{1}{2}) = 4704$, $\mathcal{D}(5, 10, 0) = 15876$, $\mathcal{D}(6, 12, 0) = 213444$ and $\mathcal{D}(6, 16, 0) = 1656369$. Now we will present some results valid in the dense limit defined by $m \to \infty$, $\Omega \to \infty$, $m/\Omega \to \infty$ and F is fixed. After spin projection, the H matrix constructed in M_F basis for a given m will be block diagonal with one block for each F value with matrix dimensions given by Eq. (10.4). Figure 10.1b shows an example for the block diagonal form and each diagonal block in $H(m)$ represents a BEGOE(1 + 2)-F in (m, F) spaces.

10.1.2 Gaussian Eigenvalue Density and Poisson to GOE Transition in Level Fluctuations

Gaussian behavior for the fixed-(m, F) eigenvalue densities $\rho^{m,F}(E)$ has been verified in many examples for BEGOE(1 + 2)-F. Figure 5.2 shows an example. The Gaussian form is essentially independent of λ (also F). As discussed in Chaps. 4, 5 and 9, the Gaussian form for the eigenvalue density is generic for embedded ensembles of spinless fermion and boson systems. In addition, in Chap. 6 it was shown that the ensemble averaged fixed-(m, S) eigenvalue densities for the fermionic EGOE(1 + 2)-**s** also take Gaussian form. Hence, from the results shown in Fig. 5.2, it is plausible to conclude that the Gaussian form is generic for EE (both bosonic and fermionic) with good quantum numbers. With the eigenvalue density being close to Gaussian, it is useful to derive formulas for the eigenvalue centroids and ensemble

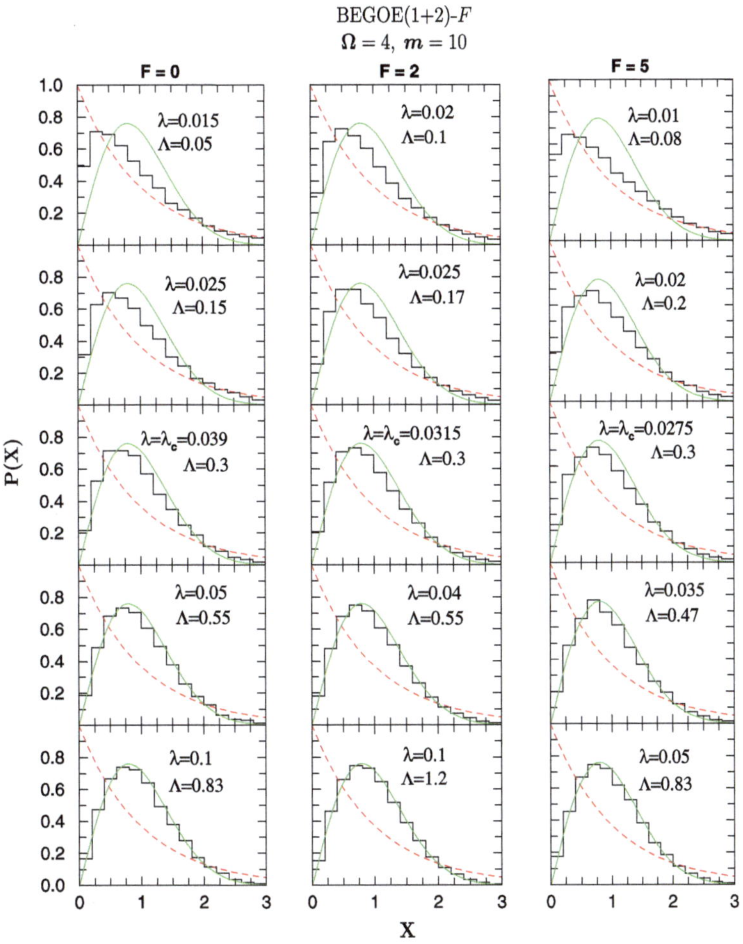

Fig. 10.2 NNSD for a 100 member BEGOE(1 + 2)-F ensemble with $\Omega = 4$, $m = 10$ and $F = 0$, 2 and 5. Calculated NNSD are compared to the Poisson (*red dashed*) and Wigner (GOE) (*green solid*) forms. Values of the interaction strength λ and the transition parameter Λ are given in the figure. The values of Λ are deduced as discussed in Chaps. 5 and 6. The chaos marker λ_c corresponds to $\Lambda = 0.3$ and its values are shown in the figure. Bin-size is 0.2 for the histograms. Figure is taken from [1] with permission from IOP publishing

averaged spectral variances. These in turn, as discussed ahead, will also allow us to study the lowest two moments of the two-point function. From now on, we will drop the 'hat' over the operators H, $h(1)$ and $V(2)$. Before turning to the propagation equations, let us mention that BEGOE(1 + 2)-F also generates in level fluctuations Poisson to Gaussian transition in NNSD. Results for a 100 member BEGOE(1 + 2)-F ensemble with $\Omega = 4$, $m = 10$ and total spins $F = 0, 2$ and 5, for λ varying from 0.01 to 0.1 are shown in Fig. 10.2. As λ increases from zero, there is generically Poisson to GOE transition (as we use sp energies to be $\varepsilon_i = i + 1/i$, the $\lambda = 0$

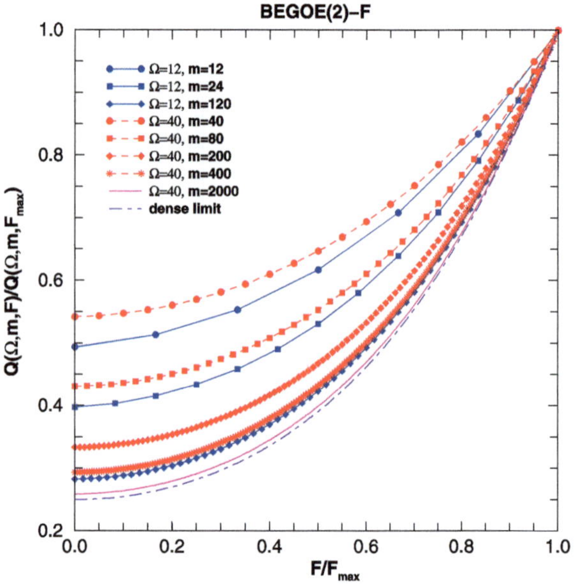

Fig. 10.3 BEGOE(2)-F variance propagator $Q(\Omega, m, F)/Q(\Omega, m, F_{max})$ vs F/F_{max} for various values of Ω and m. Formula for $Q(\Omega, m, F)$ follows from Eqs. (10.6) and (10.7). Note that the results in the figure are for $\lambda_0 = \lambda_1 = \lambda$ and therefore independent of λ; here $h(1) = 0$. Dense limit *(dot-dashed) curve* corresponds to the asymptotic formula given in Chap. 12 with $m = 2000$. Figure is taken from [1] with permission from IOP publishing

limit will not give strictly a Poisson). As seen from the figure, the transition marker $\lambda_c = 0.039, 0.0315, 0.0275$ for $F = 0$, 2 and 5 respectively. Thus λ_c decreases with increasing F-spin and this is opposite to the situation for fermion systems. For a fixed Ω value, as discussed in Chaps. 5 and 6, the λ_c is inversely proportional to K, where K is the number of many-particle states [defined by $h(1)$] that are directly coupled by the two-body interaction. For fermion systems, K is proportional to the variance propagator but not for boson systems as discussed earlier in Sect. 9.5 and at present, also for BEGOE(1 + 2)-F we don't have a formula for K. However, if we use the variance propagator $Q(\Omega, m, F)$ [see Eqs. (10.6), (10.7) and Fig. 10.3 ahead] for K (as it is used for fermion systems), then qualitatively we understand the decrease in λ_c with increasing F-spin.

10.1.3 Propagation Formulas for Energy Centroids and Spectral Variances

Given a general (1 + 2)-body Hamiltonian $H = h(1) + V(2)$, which is a typical member of BEGOE(1 + 2)-F, the eigenvalue centroids will be polynomials in the number operator and the $\hat{F}^2$ operator. As H is of maximum body rank 2, the polynomial form for the eigenvalue centroids is $\langle H \rangle^{m,F} = E_c(m, F) = a_0 + a_1 m + a_2 m^2 + a_3 F(F + 1)$. Solving for the a's in terms of the centroids in one and two particle spaces, the propagation formula for the eigenvalue (or energy) centroids is,

$$\begin{aligned}
\langle H\rangle^{m,F} = E_c(m,F) &= \big[\langle h(1)\rangle^{1,\frac{1}{2}}\big]m + \lambda_0\langle\langle V^{f=0}(2)\rangle\rangle^{2,0}\frac{P^0(m,F)}{4\Omega(\Omega-1)} \\
&\quad + \lambda_1\langle\langle V^{f=1}(2)\rangle\rangle^{2,1}\frac{P^1(m,F)}{4\Omega(\Omega+1)}; \\
P^0(m,F) &= \big[m(m+2) - 4F(F+1)\big], \\
P^1(m,F) &= \big[3m(m-2) + 4F(F+1)\big], \\
\langle h(1)\rangle^{1,\frac{1}{2}} &= \overline{\varepsilon} = \Omega^{-1}\sum_{i=1}^{\Omega}\varepsilon_i, \\
\langle\langle V^{f=0}(2)\rangle\rangle^{2,0} &= \sum_{i<j} V_{ijij}^{f=0}, \qquad \langle\langle V^{f=1}(2)\rangle\rangle^{2,1} = \sum_{i\le j} V_{ijij}^{f=1}.
\end{aligned} \tag{10.5}$$

Just as for the eigenvalue centroids, polynomial form for the spectral variances

$$\sigma^2_{H=h(1)+V(2)}(m,F) = \langle H^2\rangle^{m,F} - \big[E_c(m,F)\big]^2$$

is $\sum_{p=0}^{4} a_p m^p + \sum_{q=0}^{2} b_q m^q F(F+1) + c_0[F(F+1)]^2$. Applying $\Omega \to -\Omega$ transformation to the propagation equation for the spectral variances for fermion systems with spin given by Eqs. (6.9)–(6.12), propagation equation for $\sigma^2_{H=h(1)+V(2)}(m,F)$ in terms of inputs that contain the single particle energies ε_i defining $h(1)$ and the two particle matrix elements V^f_{ijkl} has been derived in [1]. Using this equation it is easy to obtain the formula for ensemble averaged spectral variances (also ensemble averaged covariances in eigenvalue centroids as discussed in Chap. 12). For the choice $\lambda_0 = \lambda_1 = \lambda$, the $\overline{\sigma^2_H(m,F)}$ for BEGOE(2)-F takes the simple form

$$\begin{aligned}
&\overline{\sigma^2_H(m,F)} \xrightarrow{\lambda_0=\lambda_1=\lambda} \lambda^2 Q(\Omega,m,F); \\
&Q(\Omega,m,F) = \sum_{f=0,1}(\Omega-1)\big(\Omega-2(-1)^f\big)(\Omega+2)P^{\nu=1,f}(m,F) \\
&\qquad + \frac{(\Omega-3)(\Omega^2+\Omega+2)}{2(\Omega-1)}P^{\nu=2,f=0}(m,F) \\
&\qquad + \frac{(\Omega-1)(\Omega+2)}{2}P^{\nu=2,f=1}(m,F)
\end{aligned} \tag{10.6}$$

where

$$P^{\nu=1,f=0}(m,F) = \frac{[(m+2)m^\star/2 - \langle F^2\rangle]P^0(m,F)}{8(\Omega-2)(\Omega-1)\Omega(\Omega+1)},$$

$$P^{\nu=1,f=1}(m,F) = \frac{8\Omega(m-1)(\Omega+2m-4)\langle F^2\rangle + (\Omega-2)P^2(m,F)P^1(m,F)}{8(\Omega-1)\Omega(\Omega+1)(\Omega+2)^2},$$

$$
\begin{aligned}
P^{\nu=2,f=0}(m,F) &= \left[m^\star(m^\star-1)-\langle F^2\rangle\right]P^0(m,F)/\left[8\Omega(\Omega+1)\right],\\
P^{\nu=2,f=1}(m,F) &= \left\{\left[\langle F^2\rangle\right]^2\left(3\Omega^2+7\Omega+6\right)/2+3m(m-2)m^\star\left(m^\star+1\right)\right.\\
&\quad \times(\Omega-1)(\Omega-2)/8+\left[\langle F^2\rangle/2\right]\left[(5\Omega+3)(\Omega-2)mm^\star\right.\\
&\quad \left.\left.+\,\Omega(\Omega-1)(\Omega+1)(\Omega-6)\right]\right\}\\
&\quad /\left[(\Omega-1)\Omega(\Omega+2)(\Omega+3)\right];\\
P^2(m,F) &= 3(m-2)m^\star/2+\langle F^2\rangle,\quad m^\star=\Omega+m/2,\ \langle F^2\rangle=F(F+1).
\end{aligned}
\tag{10.7}
$$

Note that P^0 and P^1 are defined by Eq. (10.5). A plot of $Q(\Omega,m,F)/Q(\Omega,m,F_{max})$ vs F/F_{max} for various Ω and m values is shown in Fig. 10.3. It is clearly seen that the propagator value increases as F-spin increases and this is just opposite to the result for fermion systems discussed in Chap. 6. An important consequence of this is BEGOE(2)-F gives ground states with $F = F_{max}$ [for the fermionic EGOE(2)-**s**, random interactions give $S = 0$ ground states]. We will consider this now.

10.1.4 *Preponderance of* $F_{max} = m/2$ *Ground States and Natural Spin Order*

Effect of random interactions in the pn-IBM model with F-spin quantum number has been studied by Yoshida et al. [8]. They found that random interactions conserving F-spin generate predominance of maximum F-spin (F_{max}) ground states. It should be noted that the low-lying states generated by $pn - sd$IBM correspond to those of sdIBM and all sdIBM states will have $F = F_{max}$. Thus random interactions preserve the property that the low-lying states generated by $pn - sd$IBM are those of sdIBM. Similarly, using nuclear shell model with isospin conserving interactions (here protons and neutrons correspond to the two projections of isospin $\mathbf{t} = \frac{1}{2}$), Kirson and Mizrahi [9] showed that random interactions generate natural isospin ordering. Denoting the lowest eigenvalue state (les) for a given many nucleon isospin T by $E_{les}(T)$, the natural isospin ordering corresponds to $E_{les}(T_{min}) \le E_{les}(T_{min}+1) \le \cdots$; for even-even $N = Z$ nuclei, $T_{min} = 0$. Therefore, one can ask if BEGOE(1 + 2)-F generates $F = F_{max}$ ground states and also a spin ordering [for boson systems, natural spin ordering (NSO) corresponds to $E_{les}(F_{max}) \le E_{les}(F_{max}-1)\cdots$], i.e. are the results in [8] are generic to interacting boson systems with F-spin and so also NSO. In this analysis, Majorana force or the space exchange operator has to be considered.

10.1.4.1 $U(\Omega)$ Algebra and Space Exchange Operator

In terms of boson creation and annihilation operators $b^\dagger_{i,\frac{1}{2},m_{\mathscr{F}}}$ and $b_{i,\frac{1}{2},m_{\mathscr{F}}}$ with $i = 1, 2, \ldots, \Omega$, it easy to identify that the $4\Omega^2$ number of one-body operators $A^r_{ij;\mu}$,

$$A^r_{ij;\mu} = \big(b^\dagger_{i,\frac{1}{2}} \tilde{b}_{j,\frac{1}{2}}\big)^r_\mu; \quad r = 0, 1, \tag{10.8}$$

generate $U(2\Omega)$ algebra. In (10.8), $\tilde{b}_{i,\frac{1}{2},m_{\mathscr{F}}} = (-1)^{\frac{1}{2}+m_{\mathscr{F}}} b_{i,\frac{1}{2},-m_{\mathscr{F}}}$ and $r = \frac{1}{2} \times \frac{1}{2}$. The $U(2\Omega)$ irreps are denoted trivially by the particle number m as they must be symmetric irreps $\{m\}$. The Ω^2 number of operators A^0_{ij} generate $U(\Omega)$ algebra and similarly there is a $U(2)$ algebra generated by the number operator $\hat{n}$ and the F-spin generators F^1_μ,

$$\hat{n} = \sqrt{2}\sum_i A^0_{ii}; \qquad F^1_\mu = \frac{1}{\sqrt{2}}\sum_i A^1_{ii;\mu}. \tag{10.9}$$

Then, we have the group-subgroup algebra $U(2\Omega) \supset U(\Omega) \otimes SU(2)$ with $SU(2)$ generated by F^1_μ. As the $U(2)$ irreps are two-rowed, the $U(\Omega)$ irreps have to be two-rowed and they are labeled by $\{m_1, m_2\}$ with $m = m_1 + m_2$ and $F = (m_1 - m_2)/2$; $m_1 \geq m_2 \geq 0$. Thus, with respect to $U(\Omega) \otimes SU(2)$ algebra, many boson states are labeled by $|\{m_1, m_2\}, \xi\rangle$ or equivalently by $|(m, F), \xi\rangle$, where ξ are extra labels required for a complete specification of the states. The quadratic Casimir operator of the $U(\Omega)$ algebra is,

$$C_2\big[U(\Omega)\big] = 2\sum_{i,j} A^0_{ij} \cdot A^0_{ji} \tag{10.10}$$

and its eigenvalues are $\langle C_2[U(\Omega)]\rangle^{\{m_1,m_2\}} = m_1(m_1 + \Omega - 1) + m_2(m_2 + \Omega - 3)$ or equivalently,

$$\big\langle C_2\big[U(\Omega)\big]\big\rangle^{(m,F)} = \frac{m}{2}(2\Omega + m - 4) + 2F(F + 1). \tag{10.11}$$

Note that the Casimir invariant of $SU(2)$ is $\hat{F}^2$ with eigenvalues $F(F + 1)$.

Majorana operator $\widetilde{M}$ acting on a two-particle state exchanges the spatial coordinates of the particles (index i) and leaves the F-spin quantum numbers ($m_{\mathscr{F}}$) unchanged. The operator form of $\widetilde{M}$ is

$$\widetilde{M} = \frac{\kappa}{2} \sum_{i,j,m_{\mathscr{F}},m'_{\mathscr{F}}} \big(b^\dagger_{j,m_{\mathscr{F}}} b^\dagger_{i,m'_{\mathscr{F}}}\big)\big(b^\dagger_{i,m_{\mathscr{F}}} b^\dagger_{j,m'_{\mathscr{F}}}\big)^\dagger. \tag{10.12}$$

Equation (10.12) gives, with κ a constant, $\widetilde{M} = \frac{\kappa}{2}\{C_2[U(\Omega)] - \Omega\hat{n}\}$. Then, we have

$$\widetilde{M} = \kappa\left\{\hat{n}\left(\frac{\hat{n}}{4} - 1\right) + \hat{F}^2\right\}. \tag{10.13}$$

As seen from (10.13), exchange interaction with $\kappa > 0$ generates gs with $F = F_{min} = 0(\frac{1}{2})$ for even(odd) m (this is opposite to the result for fermion systems where the exchange interaction generates gs with $S = S_{max} = m/2$). Now we will study the interplay between random interactions and the Majorana force in generating gs spin structure in boson systems. Note that for states with boson number fixed, $\widetilde{M} \propto \hat{F}^2$ and therefore $\hat{F}^2$ can be treated as the exchange interaction.

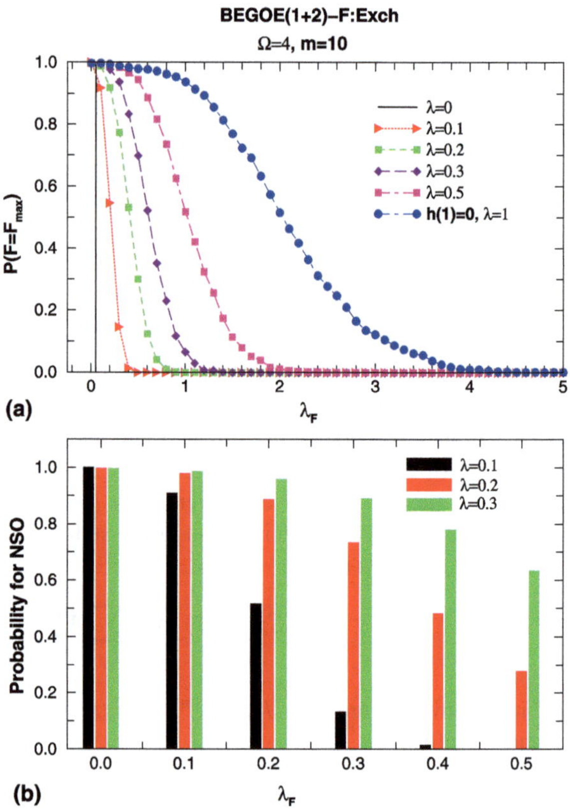

Fig. 10.4 (**a**) Probability for ground states to have spin $F = F_{max}$ as a function of the exchange interaction strength $\lambda_F \geq 0$. (**b**) Probability for natural spin order (NSO) as a function of λ_F. Results are shown for a 500 member BEGOE(1 + 2)-F : Exch ensemble generated by H defined by Eq. (10.14) for a system with $\Omega = 4$ and $m = 10$. Values of the interaction strength λ are shown in the figure. Figure is taken from [1] with permission from IOP publishing

10.1.4.2 Numerical Results for $F_{max} = m/2$ Ground States

In order to understand the gs structure in BEGOE(1 + 2)-F, the probability $P(F = F_{max})$ for the gs to be with F-spin $F_{max} = m/2$ has been studied in [1] by adding the exchange term $\lambda_F \hat{F}^2$ with $\lambda_F > 0$ to the BEGOE(1 + 2)-F Hamiltonian,

$$\{\widehat{H}\}_{\text{BEGOE(1+2)-}F\text{:Exch}} = \widehat{h}(1) + \lambda\big[\big\{\widehat{V}^{f=0}(2)\big\} + \big\{\widehat{V}^{f=1}(2)\big\}\big] + \lambda_F \hat{F}^2. \quad (10.14)$$

Note that the operator $\hat{F}^2$ is simple in the (m, F) basis. Figure 10.4a gives the probability $P(F = F_{max})$ for the ground states to have $F = F_{max}$ as a function of exchange interaction strength λ_F and for various $\lambda = \lambda_0 = \lambda_1$ values. Similarly, Fig. 10.4b shows the results for NSO. Calculations are carried out for ($\Omega = 4$, $m = 10$) system using a 500 member ensemble and sp energies $\varepsilon_i = i + 1/i$. Let us begin with pure random two-body interactions. Then $h(1) = 0$ in (10.14). Now, in the absence of the exchange interaction ($\lambda_F = 0$), as seen from Fig. 10.4a, ground states will have $F = F_{max}$, i.e. the probability $P(F = F_{max}) = 1$. The variance propagator (see Fig. 10.3) derived earlier gives a simple explanation for this by applying the JS prescription. Thus, pure random interactions generate preponderance of $F = F_{max}$ ground states. On the other hand, the exchange interaction acts in opposite direction by generating $F = F_{min}$ ground states. Therefore, by adding the exchange interaction to the $\{V(2)\}$ ensemble, $P(F = F_{max})$ starts decreasing as the strength λ_F ($\lambda_F > 0$) starts increasing. For the example considered in Fig. 10.4, for $\lambda_F > 4$, we have $P(F = F_{max}) \sim 0$. The complete variation with λ_F is shown in Fig. 10.4a marked $h(1) = 0$ and $\lambda = 1$. Similarly, on the other end, for $\lambda = 0$ in Eq. (10.14), we have $H = h(1)$ in the absence of the exchange interaction. In this situation, as all the bosons can occupy the lowest sp state and therefore gs spin $F = F_{max}$ giving $P(F = F_{max}) = 1$. When the exchange interaction is turned on, $P(F = F_{max})$ remains unity until λ_F equals the spacing between the lowest two sp states divided by m and then $P(F = F_{max})$ drops to zero. Variation of $P(F = F_{max})$ with λ_F for several values of λ between 0.1 and 0.5 show that there is a critical value (λ_F^c) of λ_F after which $P(F = F_{max}) = 0$ and its value increases with λ. Also, the variation of $P(F = F_{max})$ with λ_F becomes slower as λ increases. In summary, results in Fig. 10.4a clearly show that with random interactions there is preponderance of $F = F_{max} = m/2$ ground states. This is unlike for fermions where there is preponderance of $S = S_{min} = 0(\frac{1}{2})$ ground states for m even (odd). With the addition of the exchange interaction, $P(F = F_{max})$ decreases and finally goes to zero for $\lambda_F \geq \lambda_F^c$ and the value of λ_F^c increases with λ.

10.1.4.3 Natural Spin Ordering

For the system considered in Fig. 10.4a, for each member of the ensemble, eigenvalue of the lowest state for each F-spin is calculated and using these, obtained is the total number of members N_λ having NSO as a function of λ_F for $\lambda = 0.1, 0.2$ and 0.3 using the Hamiltonian given in (10.14). As stated before, the NSO here corresponds to (as $F = F_{max}$ is the F-spin of the gs of the system) $E_{les}(F_{max}) \leq E_{les}(F_{max} - 1) \leq E_{les}(F_{max} - 2) \leq \cdots$. Results for the probability for NSO are shown in Fig. 10.4b. In the absence of the exchange interaction, as seen from the figure, NSO is found in all the members independent of λ. Thus random interactions strongly favor NSO. The presence of exchange interaction reduces the probability for NSO. Comparing Figs. 10.4a and b, it is clearly seen that with increasing exchange interaction strength, probability for gs state spin to be $F = F_{max}$ is preserved for much larger values of λ_F (with a fixed λ) compared to the NSO. Therefore for preserving both $F = F_{max}$ gs and the NSO with high probability, the λ_F value has to be small. It is plausible to argue that the results in Figs. 10.4a and b obtained using BEGOE(1 + 2)-F are generic for boson systems with F-spin.

10.1.5 BEGOE(1 + 2)-M_F

Consider a system of m bosons occupying Ω number of sp orbitals each with spin $\mathscr{F} = \frac{1}{2}$ so that the number of sp states $N = 2\Omega$. The sp states are denoted by $|\nu_i, m_{\mathscr{F}}\rangle$, $i = 1, 2, \ldots, \Omega$ and $m_{\mathscr{F}} = \pm\frac{1}{2}$. The average spacing between the ν_i states is assumed to be Δ and between two $m_{\mathscr{F}}$ states for a given ν_i to be $\Delta_{m_{\mathscr{F}}}$. For constructing the H matrix in good M_F representation, we arrange the sp states $|i, m_{\mathscr{F}} = \pm\frac{1}{2}\rangle$ in such a way that the first Ω states have $m_{\mathscr{F}} = \frac{1}{2}$ and the remaining Ω states have $m_{\mathscr{F}} = -\frac{1}{2}$. Many-particle states for m bosons in the 2Ω sp states, arranged as explained above, can be obtained by distributing m_1 bosons in the $m_{\mathscr{F}} = \frac{1}{2}$ sp states (Ω in number) and similarly, m_2 fermions in the $m_{\mathscr{F}} = -\frac{1}{2}$ sp states (Ω in number) with $m = m_1 + m_2$. Thus, $M_F = (m_1 - m_2)/2$. Let us denote each distribution of m_1 fermions in $m_{\mathscr{F}} = \frac{1}{2}$ sp states by $\mathbf{m_1}$ and similarly, $\mathbf{m}_2$ for m_2 fermions in $m_{\mathscr{F}} = -\frac{1}{2}$ sp states. Many-particle basis defined by $(\mathbf{m}_1, \mathbf{m}_2)$ with $m_1 - m_2 = 2M_F$ will form the basis for BEGOE(1 + 2)-M_F. As the two-particle m_f can take values ± 1 and 0, the two-body part of the Hamiltonian preserving M_F will be $\widehat{V}(2) = \lambda_0 \widehat{V}^{m_f=0}(2) + \lambda_1 \widehat{V}^{m_f=1}(2) + \lambda_{-1} \widehat{V}^{m_f=-1}(2)$ with the corresponding two-particle matrix being a direct sum matrix generated by $\widehat{V}^{m_f}(2)$. Therefore, the BEGOE(1 + 2)-M_F Hamiltonian is

$$\widehat{H} = \widehat{h}(1) + \lambda_0\big\{\widehat{V}^{m_f=0}(2)\big\} + \lambda_1\big\{\widehat{V}^{m_f=1}(2)\big\} + \lambda_{-1}\big\{\widehat{V}^{m_f=-1}(2)\big\}. \qquad (10.15)$$

In Eq. (10.15), the $\{\widehat{V}^{m_f}(2)\}$ ensembles in two-particle spaces are represented by independent GOE(1)'s and λ_{m_f}'s are their corresponding strengths. The action of the Hamiltonian operator defined by Eq. (10.15) on the $(\mathbf{m}_1, \mathbf{m}_2)$ basis states with a given M_F generates the BEGOE(1 + 2)-M_F ensemble in m-particle spaces. Therefore, BEGOE(1 + 2)-M_F is defined by six parameters $(\Omega, m, \Delta_{m_{\mathscr{F}}}, \lambda_0, \lambda_1, \lambda_{-1})$ [we put $\Delta = 1$ so that $\Delta_{m_{\mathscr{F}}}$ and λ_{m_f}'s are in the units of Δ]. In the $(\mathbf{m}_1, \mathbf{m}_2)$ basis with a given M_F, the H matrix construction reduces to the matrix construction for spinless boson systems. The H matrix dimension for a given M_F is $\sum_{F \geq M_F} d_b(\Omega, m, F)$. Finally, pairing can also be introduced in this ensemble using the algebra $U(2\Omega) \supset SO(2\Omega) \supset SO(\Omega) \otimes SO(2)$ with $SO(2)$ generating M_F; see [10]. Analysis of BEGOE(1 + 2)-M_F will be useful in two component BEC studies [7].

10.2 BEGOE(1 + 2)-$S1$ Ensemble for Spin One Boson Systems

Another interesting extension of BEGOE is to a system of bosons carrying spin $S = 1$ degree of freedom. With random two-body interactions preserving many boson spin S then generates the ensemble called BEGOE(2)-$S1$ [2]. In the presence of a mean-field, the corresponding ensemble is BEGOE(1 + 2)-$S1$. Some basic properties of this ensemble are discussed in this section. BEGOE(1 + 2)-$S1$ ensembles will be useful for spinor BEC discussed in [11, 12] and in the analysis of IBM-3 model of atomic nuclei (here spin S is isospin T of the bosons in IBM-3) [13, 14].

10.2.1 Definition and Construction

Let us consider a system of m ($m > 2$) bosons with spin 1 ($S = 1$) degree of freedom and occupying Ω number of sp levels. For convenience, in the remaining part of this section, we will use the notation $\mathbf{s}$ for the spin quantum number of a single boson, s for the spin carried by a two boson system and for $m > 2$ boson systems S for the spin. Therefore, $\mathbf{s} = 1$; $s = 0$, 1 and 2; $S = m, m-1, \ldots, 0$. Similarly, the $\hat{S}_z$ eigenvalue is denoted by $m_\mathbf{s}$, m_s and M_S respectively. Now on, the space generated by the sp levels $i = 1, 2, \ldots, \Omega$ is referred as orbital space. Then, the sp states of a boson are denoted by $|i; \mathbf{s} = 1, m_\mathbf{s}\rangle$ with $i = 1, 2, \ldots, \Omega$ and $m_\mathbf{s} = +1$, 0 and -1. With Ω number of orbital degrees of freedom and three spin ($m_\mathbf{s}$) degrees of freedom, total number of sp states is $N = 3\Omega$. Going further, two boson (normalized) states that are symmetric in the total orbital $\times$ spin space are denoted by $|(ij); s, m_s\rangle$ with $s = 1 \times 1 = 0$, 1 and 2; however, for $i = j$ only $s = 0, 2$ are allowed.

For one plus two-body Hamiltonians preserving m-particle spin S, the one-body Hamiltonian $h(1)$ is defined by the sp energies ε_i; $i = 1, 2, \ldots, \Omega$. Its operator form is,

$$\widehat{h}(1) = \sum_{i=1}^{\Omega} \varepsilon_i \hat{n}_i \tag{10.16}$$

where $\hat{n}_i = \sum_{m_\mathbf{s}} \hat{n}_{i:m_\mathbf{s}} = \sum_{m_\mathbf{s}} b^\dagger_{i,m_\mathbf{s}} b_{i,m_\mathbf{s}}$. Similarly, the two-body Hamiltonian $V(2)$ is defined by the two-body matrix elements $V^s_{ijkl}(2) = \langle (kl)s, m_s | \widehat{V}(2) | (ij)s, m_s\rangle$ with the two-particle spin s taking values 0, 1 and 2. These matrix elements are independent of the m_s quantum number. The $V(2)$ matrix in two-particle spaces will be a direct sum three matrices generated by the three $\widehat{V}^s(2)$ operators respectively. Then the BEGOE(1 + 2)-$S1$ Hamiltonian is

$$\{\widehat{H}(1+2)\} = \widehat{h}(1) + \lambda_0\{\widehat{V}^{s=0}(2)\} + \lambda_1\{\widehat{V}^{s=1}(2)\} + \lambda_2\{\widehat{V}^{s=2}(2)\} \tag{10.17}$$

with three parameters $(\lambda_0, \lambda_1, \lambda_2)$. Now, BEGOE(1 + 2)-$S1$ ensemble for a given (m, S) system is generated by defining the three parts of $\widehat{V}(2)$ in two-particle spaces to be independent GOE(1)'s and then propagating each member of the $\{\widehat{H}(1+2)\}$ to the m-particle spaces with a given spin S by using the geometry (direct product structure) of the m-particle spaces. A method for carrying out the propagation is discussed ahead. With $\widehat{h}(1)$ given by Eq. (10.16), the sp levels will be triply degenerate with average spacing Δ. Without loss of generality we put $\Delta = 1$ so that the λ's in Eq. (10.17) will be in units of Δ. Note that BEGOE(1 + 2)-$S1$ reduces to BEGOE(2)-$S1$ for $\widehat{h}(1) = 0$ or in the limit $\lambda_i \to \infty$ for $i = 1, 2$ and 3 (equivalently, for sufficiently large values of λ_i).

For generating a many-particle basis, firstly, the sp states are arranged such that the first Ω number of sp states have $m_\mathbf{s} = 1$, next Ω number of sp states have

$m_s = 0$ and the remaining Ω sp states have $m_s = -1$. Now, the many-particle states for m bosons can be obtained by distributing m_1 bosons in the $m_s = 1$ sp states, m_2 bosons in the $m_s = 0$ sp states and similarly, m_3 bosons in the $m_s = -1$ sp states with $m = m_1 + m_2 + m_3$. Thus, $M_S = (m_1 - m_3)$. Let us denote each distribution of m_1 bosons in $m_s = 1$ sp states by $\mathbf{m}_1$, m_2 bosons in $m_s = 0$ sp states by $\mathbf{m}_2$ and similarly, $\mathbf{m}_3$ for m_3 bosons in $m_s = -1$ sp states. Configurations defined by $(\mathbf{m}_1, \mathbf{m}_2, \mathbf{m}_3)$ will form a basis for constructing H matrix in m boson spaces. Action of the Hamiltonian operator defined by Eq. (10.17) on $(\mathbf{m}_1, \mathbf{m}_2, \mathbf{m}_3)$ basis states with fixed-$(m, M_S = 0)$ generates the ensemble in (m, M_S) spaces. It is important to note that the construction of the m-particle H matrix in fixed-$(m, M_S = 0)$ space reduces to the problem of BEGOE(1 + 2) for spinless boson systems and hence Eq. (9.3) will apply. For this, we need to convert the H operator into M_S representation. Two boson states in M_S representation can be written as $|i, m_s; j, m_s'\rangle$; $m_s = m_s + m_s'$. Then, the two particle matrix elements are $V'_{i,m_s^{f1};j,m_s^{f2};k,m_s^{i1};\ell,m_s^{i2}}(2) = \langle i, m_s^{f1}; j, m_s^{f2} \mid \widehat{V}(2) \mid k, m_s^{i1}; \ell, m_s^{i2}\rangle$. It is easy to apply angular momentum algebra and derive formulas for these in terms of $V^s_{ijkl}(2)$. The final formulas are,

$$
\begin{aligned}
V'_{i,1;j,1;k,1;\ell,1}(2) &= V^{s=2}_{ijkl}(2), \\
V'_{i,1;j,0;k,1;\ell,0}(2) &= \frac{\sqrt{(1+\delta_{ij})(1+\delta_{k\ell})}}{2}\big[V^{s=1}_{ijkl}(2) + V^{s=2}_{ijkl}(2)\big], \\
V'_{i,1;j,-1;k,1;\ell,-1}(2) &= \frac{\sqrt{(1+\delta_{ij})(1+\delta_{k\ell})}}{6}\big[2V^{s=0}_{ijkl}(2) + 3V^{s=1}_{ijkl}(2) + V^{s=2}_{ijkl}(2)\big], \\
V'_{i,0;j,0;k,0;\ell,0}(2) &= \left[\frac{1}{3}V^{s=0}_{ijkl}(2) + \frac{2}{3}V^{s=2}_{ijkl}(2)\right], \\
V'_{i,1;j,-1;k,0;\ell,0}(2) &= \frac{\sqrt{(1+\delta_{ij})}}{3}\big[V^{s=2}_{ijkl}(2) - V^{s=0}_{ijkl}(2)\big].
\end{aligned}
\tag{10.18}
$$

All other V' matrix elements follow by symmetries. The fact that the sp energies ε are independent of m_s, Eq. (10.18) above and Eq. (9.3) will allow one to construct the H-matrix in $(\mathbf{m}_1, \mathbf{m}_2, \mathbf{m}_3)$ basis for a given value of m and $M_S = 0$. Then, $\hat{S}^2$ operator is used for projecting states with good S, i.e. to covert the H-matrix into direct sum of matrices with block matrices for each allowed S value. Eigenvalues of the two-body part of $\hat{S}^2$ in the two-particle $s = 0$, 1 and 2 spaces are -4, -2 and 2 respectively. This procedure has been implemented and computer programmes are developed. Some numerical results obtained using these programmes will be discussed in the next subsections. Let us add that the BEGOE(1 + 2)-$S1$ ensemble is defined by five parameters $(\Omega, m, \lambda_0, \lambda_1, \lambda_2)$ with λ_s in units of Δ.

10.2.2 $U(\Omega) \otimes [SU(3) \supset SO(3)]$ Embedding Algebra

Embedding algebra for BEGOE(1 + 2)-$S1$ is not unique and following the earlier results for the IBM-3 model of atomic nuclei [13, 14], it is possible to identify two algebras. They are: (i) $U(3\Omega) \supset U(\Omega) \otimes [U(3) \supset SO(3)]$; (ii) $U(3\Omega) \supset SO(3\Omega) \supset SO(\Omega) \otimes SO(3)$. Here we will consider (i) and in Appendix F (ii) is discussed.

Firstly, the spectrum generating algebra $U(3\Omega)$ is generated by the $(3\Omega)^2$ number of operators $u_q^k(i,j)$ where

$$u_q^k(i,j) = \left(b_{i;\mathbf{s}=1}^\dagger \tilde{b}_{j;\mathbf{s}=1}\right)_q^k; \quad k = 0, 1, 2 \text{ and } i, j = 1, 2, \ldots, \Omega. \tag{10.19}$$

Note that u^k are given in angular momentum coupled representation with $k = \mathbf{s} \times \mathbf{s} = 0, 1, 2$. Also, $\tilde{b}_{i;1,m_\mathbf{s}} = (-1)^{1+m_\mathbf{s}} b_{i;1,-m_\mathbf{s}}$. The quadratic Casimir invariant of $U(3\Omega)$ is

$$\hat{C}_2\big(U(3\Omega)\big) = \sum_{i,j,k} u^k(i,j) \cdot u^k(j,i). \tag{10.20}$$

Note that $T^k \cdot U^k = (-1)^k \sqrt{(2k+1)}(T^k U^k)^0$. In terms of the number operator $\hat{n}$,

$$\hat{n} = \sum_{i,m_\mathbf{s}} b_{i;1,m_\mathbf{s}}^\dagger b_{i;1,m_\mathbf{s}}, \tag{10.21}$$

we have

$$\hat{C}_2\big(U(3\Omega)\big) = \hat{n}(\hat{n} + 3\Omega - 1). \tag{10.22}$$

All m-boson states transform as the symmetric irrep $\{m\}$ w.r.t. $U(3\Omega)$ algebra and

$$\big\langle \hat{C}_2\big(U(3\Omega)\big)\big\rangle^{\{m\}} = m(m + 3\Omega - 1). \tag{10.23}$$

Using the results given in [15] it is easy to write the generators of the algebras $U(\Omega)$ and $SU(3)$ in $U(3\Omega) \supset U(\Omega) \otimes SU(3)$. The $U(\Omega)$ generators are $g(i,j)$ where,

$$g(i,j) = \sqrt{3}\left(b_{i;\mathbf{s}=1}^\dagger \tilde{b}_{j;\mathbf{s}=1}\right)^0; \quad i, j = 1, 2, \ldots, \Omega \tag{10.24}$$

and they are Ω^2 in number. Similarly, $SU(3)$ algebra is generated by the eight operators $h_q^{k=1,2}$ where,

$$h_q^k = \sum_i \left(b_{i;\mathbf{s}=1}^\dagger \tilde{b}_{i;\mathbf{s}=1}\right)_q^k. \tag{10.25}$$

It is useful to mention that $(h^0, h^1_q, h^2_{q'})$ generate $U(3)$ algebra and $U(3) \supset SU(3)$. The quadratic Casimir invariants of $U(\Omega)$ and $SU(3)$ algebras are,

$$\begin{aligned} \hat{C}_2\big(U(\Omega)\big) &= \sum_{i,j} g(i,j) \cdot g(j,i), \\ \hat{C}_2\big(SU(3)\big) &= \frac{3}{2} \sum_{k=1,2} k^k \cdot h^k. \end{aligned} \tag{10.26}$$

The irreps of $U(\Omega)$ can be represented by Young tableaux $\{f\} = \{f_1, f_2, \dots, f_\Omega\}$, $\sum_i f_i = m$. However, as we are dealing with boson systems (i.e. the only allowed $U(3\Omega)$ irrep being $\{m\}$), the irreps of $U(\Omega)$ and $U(3)$ should be represented by the same $\{f\}$. Therefore, $\{f\}$ will be maximum of three rows. The $U(\Omega)$ and $SU(3)$ equivalence gives a relationship between their quadratic Casimir invariants,

$$\begin{aligned} \hat{C}_2\big(U(\Omega)\big) &= \hat{C}_2\big(U(3)\big) + (\Omega - 3)\hat{n}, \\ \hat{C}_2\big(U(3)\big) &= \sum_{k=0,1,2} h^k \cdot h^k = \frac{2}{3}\hat{C}_2\big(SU(3)\big) + \frac{\hat{n}^2}{3}. \end{aligned} \tag{10.27}$$

These relations are easy to prove using Eqs. (10.24)–(10.26). Given the $U(\Omega)$ irrep $\{f_1 f_2 f_3\}$, the corresponding $SU(3)$ irrep in Elliott's notation [16] is given by $(\lambda = f_1 - f_2, \mu = f_2 - f_3)$. Thus,

$$\begin{aligned} &\{m\}_{U(3\Omega)} \rightarrow \big[\{f_1 f_2 f_3\}_{U(\Omega)}\big]\big[(\lambda\mu)_{SU(3)}\big]; \\ &f_1 + f_2 + f_3 = m, \qquad f_1 \geq f_2 \geq f_3 \geq 0, \\ &\lambda = f_1 - f_2, \qquad \mu = f_2 - f_3. \end{aligned} \tag{10.28}$$

Using Eq. (10.28) it is easy to write, for a given m, all the allowed $SU(3)$ and equivalently $U(\Omega)$ irreps. Eigenvalues of $\hat{C}_2(SU(3))$ are given by

$$\big\langle \hat{C}_2\big(SU(3)\big)\big\rangle^{(\lambda\mu)} = C_2(\lambda\mu) = \big[\lambda^2 + \mu^2 + \lambda\mu + 3(\lambda + \mu)\big]. \tag{10.29}$$

Let us add that $SU(3)$ algebra also has a cubic invariant $\hat{C}_3(SU(3))$ and its matrix elements are [17],

$$\big\langle \hat{C}_3\big(SU(3)\big)\big\rangle^{(\lambda\mu)} = C_3(\lambda\mu) = \frac{2}{9}(\lambda - \mu)(2\lambda + \mu + 3)(\lambda + 2\mu + 3). \tag{10.30}$$

The $SO(3)$ subalgebra of $SU(3)$ generates spin S. The spin generators are

$$S^1_q = \sqrt{2} h^1_q, \qquad \hat{S}^2 = C_2\big(SO(3)\big) = S^1 \cdot S^1, \qquad \big\langle \hat{S}^2 \big\rangle^S = S(S+1). \tag{10.31}$$

Given a $(\lambda\mu)$, the allowed S values follow from Elliott's rules [14, 16] and this introduces a 'K' quantum number,

$$\begin{aligned} K &= \min(\lambda, \mu), \min(\lambda, \mu) - 2, \ldots, 0 \text{ or } 1, \\ S &= \max(\lambda, \mu), \max(\lambda, \mu) - 2, \ldots, 0 \text{ or } 1 \quad \text{for } K = 0, \\ &= K, K + 1, K + 2, \ldots, K + \max(\lambda, \mu) \quad \text{for } K \neq 0. \end{aligned} \tag{10.32}$$

Equation (10.32) gives $d_{(\lambda\mu)}(S)$, the number of times a given S appears in a $(\lambda\mu)$ irrep. Similarly the number of sub-states that belong to a $U(\Omega)$ irrep $\{f_1 f_2 f_3\}$ are given by $d_\Omega(f_1 f_2 f_3)$ where [10],

$$d_\Omega(f_1 f_2 f_3) = \begin{vmatrix} d_\Omega(f_1) & d_\Omega(f_1 + 1) & d_\Omega(f_1 + 2) \\ d_\Omega(f_2 - 1) & d_\Omega(f_2) & d_\Omega(f_2 + 1) \\ d_\Omega(f_3 - 2) & d_\Omega(f_3 - 1) & d_\Omega(f_3) \end{vmatrix}. \tag{10.33}$$

Here, $d_\Omega(\{g\}) = \binom{\Omega+g-1}{g}$ and $d_\Omega(\{g\}) = 0$ or $g < 0$. Note that the determinant in Eq. (10.33) involves only symmetric $U(\Omega)$ irreps. Using the $U(3\Omega) \supset U(\Omega) \otimes [U(3) \supset SO(3)]$ algebra, m bosons states can be written as

$$|m; \{f_1 f_2 f_3\}\alpha; (\lambda\mu) K S M_S\rangle.$$

Here, the number of α values is $d_\Omega(f_1 f_2 f_3)$, the K values follow from Eq. (10.32) and $-S \le M_S \le S$. Similarly, m and $(\lambda\mu)$ give a unique $\{f_1 f_2 f_3\}$. Therefore H-matrix dimension in fixed-(m, S) space is given by

$$d_b(m, S) = \sum_{\{f_1 f_2 f_3\} \in m} d_\Omega(f_1 f_2 f_3) d_{(\lambda\mu)}(S), \tag{10.34}$$

and they will satisfy the sum rule $\sum_S (2S + 1) d_b(m, S) = \binom{3\Omega+m-1}{m}$. Also, the dimension $D(m, M_S = 0)$ of the H-matrix in the basis discussed earlier is $D(m, M_S = 0) = \sum_{S \in m} d_b(m, S)$. For example, for $(\Omega = 4, m = 8)$, the dimensions for $S = 0 - 8$ are 714, 1260, 2100, 1855, 1841, 1144, 840, 315 and 165 respectively. Similarly, for $(\Omega = 6, m = 10)$, the dimensions for $S = 0 - 10$ are 51309, 123585, 183771, 189630, 178290, 133497, 94347, 51645, 27027, 9009 and 3003 respectively.

10.2.3 Results for Spectral Properties: Propagation of Energy Centroids and Spectral Variances

Using the method described in the previous subsection, in some examples BEGOE(2)-$S1$ and BEGOE(1 + 2)-$S1$ ensembles are constructed and numerical analysis of the eigenvalue density and spectral fluctuations are carried out. Results from a 100-member BEGOE(2)-$S1$ ensemble with $m = 8$ and $\Omega = 4$ are shown in Fig. 10.5. In the calculations, the strengths of the two-body interaction in the three channels are chosen to be equal, i.e. $\lambda_0 = \lambda_1 = \lambda_2$ and the spectra of each member is first zero centered and scaled to unit width. It is seen from the figure

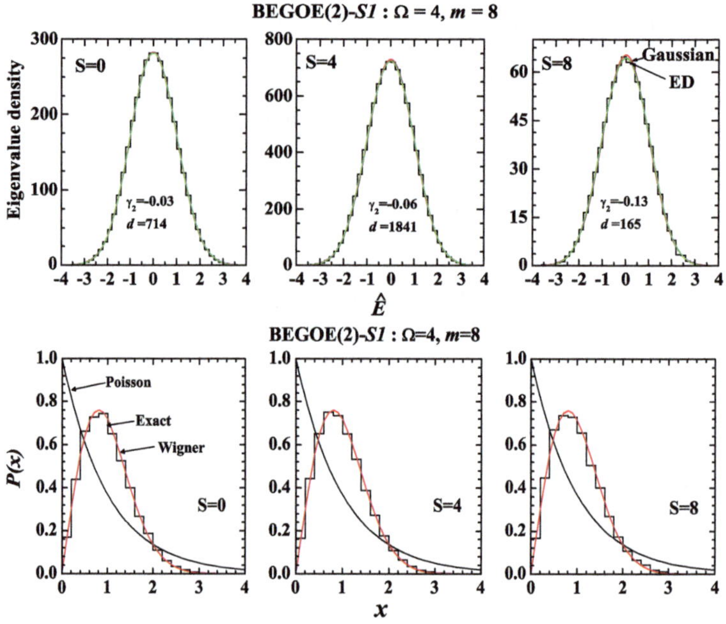

Fig. 10.5 Ensemble averaged eigenvalue density $\rho^{m,S}(\widehat{E})$ vs $\widehat{E} = E - E_c(m, S)/\sigma$ and ensemble averaged Nearest Neighbor Spacing Distribution (NNSD). Results are for a 100 member BEGOE(2)-$S1$ with $\Omega = 4$, $m = 8$ and spin $S = 0$, 4 and 8. Eigenvalue densities are compared with Gaussian and Edgeworth corrected Gaussians (ED) forms. Values of (γ_2) parameters are shown in the figures and $\gamma_1 \sim 0$ in all cases. In the plots, the bin size is 0.2 and the eigenvalue densities are normalized to dimension $d_b(m, S)$. In the NNSD figures, the spacing x is in the units of local mean spacing and the results are compared with Poisson and GOE (Wigner) forms

that the ensemble averaged eigenvalue densities are close to Gaussian. Similarly the NNSD is close to Wigner form. Combining these results with those in Chap. 9 and Sect. 10.1 we can conclude that for finite isolated interacting boson systems the eigenvalue density will be generically of Gaussian form and fluctuations, in absence of the mean-field, follow GOE. With a mean-field [i.e. for BEGOE(1 + 2)-$S1$], as seen from Sect. 10.1, the interaction strength has to be larger than a critical value for the fluctuations to change from Poisson like to GOE. Numerical examples verifying this for BEGOE(1 + 2)-$S1$ are given in [2].

As the eigenvalue density is close to Gaussian, it is of interest to derive formulas for energy centroids and spectral variances in terms of sp energies ε_i and the two-particle $V(2)$ matrix elements V^S_{ijkl}. They will also allow us to study, numerically, fluctuations in energy centroids and spectral variances. Simple propagation equation for the fixed-(m, S) energy centroids $\langle H \rangle^{m,S}$ in terms of the scalars $\hat{n}$ and S^2 operators [their eigenvalues are m and $S(S+1)$] is not possible. This is easily seen from the fact that up to 2 bosons, we have 5 states ($m = 0, S = 0$; $m = 1, S = 1$; $m = 2, S = 0, 1, 2$) but only 4 scalar operators (1, $\hat{n}$, $\hat{n}^2$, $\hat{S}^2$). For the missing operator we can use $\hat{C}_2(SU(3))$ but then only fixed-$(m, (\lambda\mu)S)$ averages will propagate [18]. The propagation equation is,

$$\begin{aligned}
\langle \widehat{H}(1+2)\rangle^{m,(\lambda\mu),S} &= \langle \widehat{h}(1)+\widehat{V}(2)\rangle^{m,(\lambda,\mu),S} = m\langle \widehat{h}(1)\rangle^{1,(10),1} \\
&+ \left[-\frac{m}{6}+\frac{m^2}{18}+\frac{C_2(\lambda\mu)}{9}-\frac{S(S+1)}{6}\right]\langle \widehat{V}(2)\rangle^{2,(20),0} \\
&+ \left[-\frac{5m}{6}+\frac{5m^2}{18}+\frac{C_2(\lambda\mu)}{18}+\frac{S(S+1)}{6}\right]\langle \widehat{V}(2)\rangle^{2,(20),2} \\
&+ \left[\frac{m}{2}+\frac{m^2}{6}-\frac{C_2(\lambda\mu)}{6}\right]\langle \widehat{V}(2)\rangle^{2,(01),1}.
\end{aligned} \tag{10.35}$$

Now, summing over all $(\lambda\mu)$ irreps that contain a given S will give $\langle \widehat{H}(1+2)\rangle^{m,S}$. This is useful in verifying the codes developed for constructing BEGOE(1 + 2)-$S1$ members. Propagation equation for spectral variances $\langle [\widehat{H}(1+2)]^2\rangle^{m,S}$ is more complicated. Just as with energy centroids, it is possible to propagate the variances $\langle [\widehat{H}(1+2)]^2\rangle^{m,(\lambda\mu),S}$. Towards this, first it should be noted that up to $m=4$, there are 19 states as shown in Table 10.1. Therefore, for propagation we need 19 $SO(3)$ scalars that are of maximum body rank 4. For this the invariants $\hat{n}$, $\hat{S}^2$, $\hat{C}_2(SU(3))$ and $\hat{C}_3(SU(3))$ will not suffice as they will give only 15 scalar operators. The missing three operators can be constructed using the $SU(3) \supset SO(3)$ integrity basis operators that are 3- and 4-body in nature [17, 18]. One definition of these operators is given in [17] and let us call them X_3^{DR} and $X_4^{DR}(k)$. In terms of these, it is possible to define the operators $\hat{X}_3$ and $\hat{X}_4$ such that their averages over $(\lambda\mu)S$ spaces are integers giving

$$\hat{X}_3 = -\frac{5}{\sqrt{10}} X_3^{DR}, \qquad \hat{X}_4 = 5X_4^{DR}(1). \tag{10.36}$$

Formulas for the averages $X_i((\lambda\mu), S) = \langle \hat{X}_i\rangle^{(\lambda\mu),S}$ can be written in terms of $SU(3) \supset SO(3)$ reduced Wigner coefficients and programmes for these are given in [19]. Averages for $\hat{X}_3$ and $\hat{X}_4$ over the 19 states with $m \le 4$ are given in Table 10.1 and they will not depend on Ω. Note that Eqs. (10.29) and (10.30) respectively will give $C_2(\lambda\mu)$ and $C_3(\lambda\mu)$. Propagation equation for the spectral variances over fixed-$(\lambda\mu)$, S spaces can be written as,

$$\begin{aligned}
&\langle \widehat{H}^2\rangle^{m,(\lambda\mu),S} = \sum_{i=1}^{19} a_i \mathscr{C}_i; \\
&\mathscr{C}_1 = 1, \quad \mathscr{C}_2 = m, \quad \mathscr{C}_3 = m^2, \quad \mathscr{C}_4 = m^3, \quad \mathscr{C}_5 = m^4, \\
&\mathscr{C}_6 = C_2(\lambda\mu), \quad \mathscr{C}_7 = mC_2(\lambda\mu), \quad \mathscr{C}_8 = m^2C_2(\lambda\mu), \\
&\mathscr{C}_9 = S(S+1), \quad \mathscr{C}_{10} = mS(S+1), \\
&\mathscr{C}_{11} = m^2S(S+1), \quad \mathscr{C}_{12} = S(S+1)C_2(\lambda\mu), \quad \mathscr{C}_{13} = [S(S+1)]^2,
\end{aligned} \tag{10.37}$$

Table 10.1 For boson numbers $m \leq 4$, listed are $\{f\}$, $(\lambda\mu)$, S, $\langle \hat{X}_3 \rangle^{(\lambda,\mu)S}$ and $\langle \hat{X}_4 \rangle^{(\lambda,\mu)S}$

m	$\{f\}$	$(\lambda\mu)$	S	$\langle \hat{X}_3 \rangle$	$\langle \hat{X}_4 \rangle$
0	{0}	(00)	0	0	0
1	{1}	(10)	1	5	−25
2	{2}	(20)	0	0	0
			2	21	−147
	{11}	(01)	1	−5	−25
3	{3}	(30)	1	9	−81
			3	54	−486
	{21}	(11)	1	0	−135
			2	0	−81
	{111}	(00)	0	0	0
4	{4}	(40)	0	0	0
			2	33	−363
			4	110	−1210
	{31}	(21)	1	−7	−121
			2	21	−459
			3	18	−246
	{22}	(02)	0	0	0
			2	−21	−147
	{211}	(10)	1	5	−25

$$\mathscr{C}_{14} = \left[C_2(\lambda\mu)\right]^2, \qquad \mathscr{C}_{15} = C_3(\lambda\mu), \qquad \mathscr{C}_{16} = mC_3(\lambda\mu),$$
$$\mathscr{C}_{17} = X_3\big((\lambda\mu), S\big), \qquad \mathscr{C}_{18} = mX_3\big((\lambda\mu), S\big), \qquad \mathscr{C}_{19} = X_4\big((\lambda\mu), S\big).$$

Using $\langle \widehat{H}^2 \rangle^{m,(\lambda\mu),S}$ for $m \leq 4$ as inputs, one can solve Eq. (10.37) to obtain the a_i's. Then, Eq. (10.37) can be used to calculate $\langle \widehat{H}^2 \rangle^{m,(\lambda\mu),S}$ for any m, $(\lambda\mu)$ and S. However we need numerical values for $X_3((\lambda\mu), S)$ and $X_4((\lambda\mu), S)$. As an example $X_3((\lambda\mu), S)$ and $X_4((\lambda\mu), S)$ values are shown for $m = 6$ in Table 10.2. Spectral variances $\langle \widehat{H}^2 \rangle^{m,S}$ over fixed-S spaces can be obtained easily using $\langle \widehat{H}^2 \rangle^{m,(\lambda\mu),S}$.

Let us add that there are other methods [20] based on the (m_1, m_2, m_3) configurations introduced before. Note that m_1 is number of bosons with $m_s = +1$, m_2 is number of bosons with $m_s = 0$ and m_3 is number of bosons with $m_s = -1$ so that $m = m_1 + m_2 + m_3$ and $M_S = m_1 - m_3$. Also, (m_1, m_2, m_3) can be thought of as a three orbit configuration with degeneracy for each orbit being Ω. Using the results in [21] and Eq. (10.18), it is possible to write the propagation equation for $\langle \widehat{H}^2 \rangle^{m_1,m_2,m_3}$. These will give directly $\langle\langle \widehat{H}^2 \rangle\rangle^{m,M_S}$ by summing the traces over all (m_1, m_2, m_3) that give the same M_S value. Now, the simple subtraction law $\langle\langle \widehat{H}^2 \rangle\rangle^{m,S} = \langle\langle \widehat{H}^2 \rangle\rangle^{m,M_S=S} - \langle\langle \widehat{H}^2 \rangle\rangle^{m,M_S=S+1}$ will give $\langle \widehat{H}^2 \rangle^{m,S}$. The propagation equations are explicitly given in [2].

Using the propagation equations for $\langle \widehat{H}^p \rangle^{m,S}$, $p = 1, 2$, it is possible to calculate spectral variances for each member of the ensemble. This will allow us to examine

Table 10.2 For boson number $m = 6$, listed are $(\lambda\mu)$, S, $\langle\hat{X}_3\rangle^{(\lambda,\mu)S}$ and $\langle\hat{X}_4\rangle^{(\lambda,\mu)S}$

m	$(\lambda\mu)$	S	$\langle\hat{X}_3\rangle$	$\langle\hat{X}_4\rangle$
6	(60)	0	0	0
	(60)	2	45	−675
	(60)	4	150	−2250
	(60)	6	315	−4725
	(41)	1	−9	−297
	(41)	2	27	−891
	(41)	3	36	−702
	(41)	4	132	−2466
	(41)	5	117	−1431
	(30)	1	9	−81
	(30)	3	54	−486
	(03)	1	−9	−81
	(03)	3	−54	−486
	(11)	1	0	−135
	(11)	2	0	−81
	(00)	0	0	0
	(22)	0	0	0
	(22)	2	0	−603
	(22)	3	0	−990
	(22)	4	0	−450

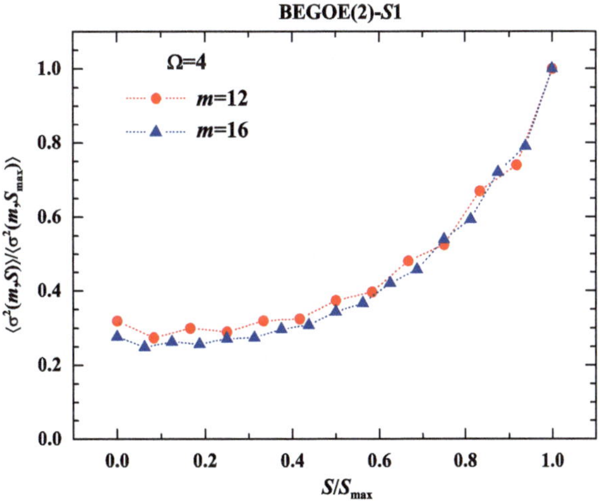

Fig. 10.6 Ensemble averaged fixed-S variances scaled by that of the maximum spin as a function of S/S_{max}. Results are for a 200 member BEGOE(2)-$S1$ ensembles with ($\Omega = 4, m = 12$) and ($\Omega = 4, m = 16$)

numerically, the variation of ensemble averaged spectral variances with spin S even for large (Ω, m) values. Figure 10.6 shows results for the variation of the average of spectral variances with S for $\Omega = 4$ and $m = 12$ and 16. It is clearly seen from the figure that the variances are almost constant for lower spins and increases for S close to S_{max}; a similar result is known for fermion systems [22]. Also, the width of the fluctuations in spectral widths is much smaller (see Sect. 12.6 for numerical examples). Let us add that near constancy of spectral widths is a feature of many-body chaos as discussed in Chaps. 5, 6 and 14.

10.2.4 Summary and Comments on Ground State Spin Structure

In summary, for BEGOE(1 + 2)-$S1$ we have: (i) the form of the fixed-(m, S) eigenvalue density is close to a Gaussian; (ii) for strong enough interaction, level fluctuations follow GOE; (iii) fluctuations in energy centroids are large as shown, with numerical examples, in Chap. 12 ahead; (iv) spectral widths are almost constant for lower spins ($S < S_{max}/2$) and increase with S close to the S_{max}. In addition, in BEGOE(1 + 2)-$S1$ spaces [also in BEGOE(1 + 2)-F spaces] it is possible to introduce pairing algebras and analyze pairing effects in systems modeled by these ensembles. Appendix F gives some details of these pairing algebras. Finally, it is possible to investigate the ground state spin structure in BEGOE(1 + 2)-$S1$. Firstly, the exchange or the Majorana operator ($\widehat{H}_{exch}$) that changes the space labels (i, j) in a two-particle space without changing the spin labels m_s is related in a simple manner to $\hat{C}_2(SU(3))$,

$$\widehat{H}_{exch} = \frac{2}{3}\hat{C}_2\big(U(3)\big) + \frac{1}{3}\hat{n}^2 - 3\hat{n}. \tag{10.38}$$

Now, the simple model Hamiltonian $\widehat{H}_M = \alpha \widehat{H}_{exch} + \beta \hat{S}^2$ generates the basic spin structure of the ground states as $\widehat{H}_{exch} \propto \hat{C}_2(SU(3))$. For $\beta = 0$ and $\alpha < 0$, the ground state for a m boson system is labeled by the $SU(3)$ irrep $(m, 0)$. As this contains all the spins $S = m, m - 2, \ldots, 0$ or 1 and they are all degenerate, we have just $SU(3)$ ground state, labeled by $(m, 0)$ irrep, with no specific choice for spin. On the other hand, if $\alpha < 0$ and $\beta > 0$, the ground state spin is $S = 0$ for m even and $S = 1$ for odd spin. Similarly, if $\alpha < 0$ and $\beta < 0$, the ground state spin is $S = m$. The three basic structures, (i) $SU(3)$ ground state labeled by $(m, 0)$ irrep, (ii) $S = 0$ (for even m) or $S = 1$ (for odd m) ground state; (iii) $S = m$ ground state for spin-one boson systems, depending on (α, β) values, were also discussed recently for spin-one Bose-Hubbard model [23]. Going beyond the simple $\widehat{H}_M$, it is possible to add random two-body interaction and also pairing and mean-field parts and investigate to what extent the three basic structures (i)–(iii) survive as we change the strengths of the added three parts. Numerical calculations for this are challenging.

References

1. M. Vyas, N.D. Chavda, V.K.B. Kota, V. Potbhare, One plus two-body random matrix ensembles for boson systems with F-spin: analysis using spectral variances. J. Phys. A, Math. Theor. **45**, 265203 (2012)

2. H. Deota, N.D. Chavda, V.K.B. Kota, V. Potbhare, M. Vyas, Random matrix ensemble with random two-body interactions in the presence of a mean-field for spin one boson systems. Phys. Rev. E **88**, 022130 (2013)
3. F. Iachello, A. Arima, *The Interacting Boson Model* (Cambridge University Press, Cambridge, 1987)
4. L.F. Santos, F. Borgonovi, F.M. Izrailev, Chaos and statistical relaxation in quantum systems of interacting particles. Phys. Rev. Lett. **108**, 094102 (2012)
5. L.F. Santos, F. Borgonovi, F.M. Izrailev, Onset of chaos and relaxation in isolated systems of interacting spins: energy shell approach. Phys. Rev. E **85**, 036209 (2012)
6. V.K.B. Kota, A. Relaño, J. Retamosa, M. Vyas, Thermalization in the two-body random ensemble, J. Stat. Mech. P10028 (2011)
7. E. Altman, W. Hofstetter, E. Demler, M.D. Lukin, Phase diagram of two-component bosons on an optical lattice. New J. Phys. **5**, 113 (2003)
8. N. Yoshida, Y.M. Zhao, A. Arima, Proton-neutron interacting boson model under random two-body interactions. Phys. Rev. C **80**, 064324 (2009)
9. M.W. Kirson, J.A. Mizrahi, Random interactions with isospin. Phys. Rev. C **76**, 064305 (2007)
10. V.K.B. Kota, Group theoretical and statistical properties of interacting boson models of atomic nuclei: recent developments, in *Focus on Boson Research*, ed. by A.V. Ling (Nova Science Publishers Inc., New York, 2006), pp. 57–105
11. G. Pelka, K. Byczuk, J. Tworzydlo, Paired phases and Bose-Einstein condensation of spin-one bosons with attractive interactions. Phys. Rev. A **83**, 033612 (2011)
12. J. Guzman, G.-B. Jo, A.N. Wenz, K.W. Murch, C.K. Thomas, D.M. Stamper-Kurn, Long-time-scale dynamics of spin textures in a degenerate $F = 1$ ^{87}Rb spinor Bose gas. Phys. Rev. A **84**, 063625 (2011)
13. J.E. García-Ramos, P. Van Isacker, The interacting boson model with $SU(3)$ charge symmetry and its applications to even-even $N \approx Z$ nuclei. Ann. Phys. (N.Y.) **274**, 45–75 (1999)
14. V.K.B. Kota, Spectra and E2 transition strengths for $N = Z$ even-even nuclei in IBM-3 dynamical symmetry limits with good s and d boson isospins. Ann. Phys. (N.Y.) **265**, 101–133 (1998)
15. V.K.B. Kota, $O(36)$ symmetry limit of IBM-4 with good s, d and sd boson spin-isospin Wigner's $SU(4) \sim O(6)$ symmetries for $N \approx Z$ odd-odd nuclei. Ann. Phys. (N.Y.) **280**, 1–34 (2000)
16. J.P. Elliott, Collective motion in the nuclear shell model. I. Classification schemes for states of mixed configurations. Proc. R. Soc. Lond. Ser. A **245**, 128–145 (1958)
17. J.P. Draayer, G. Rosensteel, $U(3) \rightarrow R(3)$ integrity-basis spectroscopy. Nucl. Phys. A **439**, 61–85 (1985)
18. V.K.B. Kota, Two-body ensembles with group symmetries for chaos and regular structures. Int. J. Mod. Phys. E **15**, 1869–1883 (2006)
19. Y. Akiyama, J.P. Draayer, A users guide to Fortran programs for Wigner and Racah coefficients of SU_3. Comput. Phys. Commun. **5**, 405–415 (1973)
20. S.S.M. Wong, *Nuclear Statistical Spectroscopy* (Oxford University Press, New York, 1986)
21. F.S. Chang, J.B. French, T.H. Thio, Distribution methods for nuclear energies, level densities and excitation strengths. Ann. Phys. (N.Y.) **66**, 137–188 (1971)
22. V.K.B. Kota, R.U. Haq, *Spectral Distributions in Nuclei and Statistical Spectroscopy* (World Scientific, Singapore, 2010)
23. H. Katsura, H. Tasaki, Ground states of the spin-1 Bose-Hubbard model. Phys. Rev. Lett. **110**, 130405 (2013)

Chapter 11
Embedded Gaussian Unitary Ensembles: Results from Wigner-Racah Algebra

A long standing question for the embedded ensembles is about their analytical tractability. Amenability to mathematical treatment is one of the four conditions laid down by Dyson [1] for the validity of a random matrix ensemble. To address this issue, in this chapter we will consider embedded unitary ensembles. It is important to recall that out of the three classical ensembles, GUE is mathematically much easier. Simplest embedded unitary ensemble is the embedded Gaussian unitary ensemble of two-body interactions [EGUE(2)] for spinless fermion systems. For m fermions in N sp states, the embedding is generated by the $SU(N)$ algebra. Although EE are known for many years, only recently [2], after the first indications implicit in [3, 4], it is established that the $SU(N)$ Wigner-Racah algebra solves EGUE(2) and also the more general EGUE(k) [as well as EGOE(k)]. These results, with $U(N)$ algebra, extend to BEGUE(k) for spinless bosons in N sp states (see Sects. 11.2 and 11.3 and [5]). For EGUE(2)-**s** for fermions with spin and EGUE(2)-$SU(4)$ for fermions with Wigner's spin-isospin $SU(4)$ symmetry, the embedding algebras, with Ω number of spatial degrees of freedom for a single fermion, are $U(\Omega) \otimes SU(2)$ and $U(\Omega) \otimes SU(4)$ respectively [6, 7]. Similarly, the embedding algebras for BEGUE(2)-F for two-species boson systems with F-spin and BEGUE(2)-$SU(3)$ for spin one bosons are $U(\Omega) \otimes SU(2)$ and $U(\Omega) \otimes SU(3)$ respectively [8, 9]. Again, the Wigner-Racah algebra of these algebras solve the corresponding embedded unitary ensembles. As discussed in Sect. 11.3, all these ensembles can be unified into EGUE(2)-$[U(\Omega) \otimes SU(r)]$. All these results, discussed in some detail in the next seven sections, obtained after more than 30 years of the introduction of embedded ensembles, conclusively establish that two-body random matrix ensembles are amenable to mathematical treatment and thus satisfy Dyson's criterion. Here, Wigner-Racah algebra of the embedding Lie algebras plays the central role.

V.K.B. Kota, *Embedded Random Matrix Ensembles in Quantum Physics*, Lecture Notes in Physics 884, DOI 10.1007/978-3-319-04567-2_11,

11.1 Embedded Gaussian Unitary Ensemble for Spinless Fermions with k-Body Interactions: EGUE(k)

In this section we deal with EGUE(k), i.e. fermions with a general k-body Hamiltonian although for nuclei, atoms and mesoscopic systems $k=2$ is most important. For a system of m spinless fermions in N sp states, one has the unitary groups $SU(N)$, $U(N_k)$ and $U(N_m)$, $N_r = \binom{N}{r}$, with EGUE(k) invariant under $U(N_k)$ and the embedding in m-particle spaces is defined by $SU(N)$; note that a GUE in m particle spaces is invariant under $U(N_m)$ but not the EGUE(k), $k<m$. Analytical results for EGUE(k) follow from the tensorial decomposition of H with respect to $SU(N)$ and the $SU(N)$ Wigner-Racah algebra; in the end Wigner coefficients disappear as expected [note that the Wigner coefficients involve the sub-algebras of $SU(N)$] and all the expressions for the moments involve only $SU(N)$ Racah coefficients. Firstly, sp creation operator $a_i^\dagger$ for any i-th sp state transforms as the irrep $\{1\}$ of $U(N)$ and similarly a product of r creation operators transform, as we have fermions, as the irrep $\{1^r\}$ in Young tableaux notation. Let us add that a $U(N)$ irrep $\{\lambda_1, \lambda_2, \ldots, \lambda_N\}$ defines the corresponding $SU(N)$ irrep as $\{\lambda_1 - \lambda_N, \lambda_2 - \lambda_N, \ldots, \lambda_{N-1} - \lambda_N\}$ with $N-1$ rows. The $U(\Omega) \leftrightarrow SU(\Omega)$ correspondence is used throughout and therefore we use $U(\Omega)$ and $SU(\Omega)$ interchangeably. A normalized r-particle creation operator $A^\dagger(f_r\alpha_r)$ behaves as the $SU(N)$ irrep (tensor) $\{1^r\}$. Similarly a r-particle annihilation operator behaves as $\overline{\{1^r\}} = \{1^{N-r}\}$. Tensorial multiplication gives, $\{1^r\} \otimes \overline{\{1^r\}} \to \sum g_\nu \oplus = \sum \{2^\nu 1^{N-2\nu}\} \oplus$, $\nu = 0, 1, \ldots, r$. Note that $g_0 = \{0\}$ for $SU(N)$ and $g_\nu = \overline{g_\nu}$. Also, the ν here is same as the tensorial rank ν used in Chaps. 5 and 6. $SU(N)$ irreducible tensors $B_k(g_\nu\omega_\nu)$ are defined by,

$$B_k(g_\nu\omega_\nu) = \sum_{\alpha_k, \alpha'_k} A^\dagger\big(\{1^k\}\alpha_k\big) A\big(\{1^k\}\alpha'_k\big) \big\langle \{1^k\}\alpha_k \overline{\{1^k\}\alpha'_k} \big| g_\nu\omega_\nu \big\rangle, \tag{11.1}$$

where $\langle -- \mid -- \rangle$'s are $SU(N)$ Wigner coefficients and α's are the other labels for completely specifying the k particle states [they can be specified by any subgroup chain contained in $SU(N)$]. An important property of $B_k(g_\nu\omega_\nu)$ is that they are orthogonal with respect to the traces over k particle spaces. Given a k-body Hamiltonian

$$H(k) = \sum_{v_a, v_b} V_{v_a v_b}(k) A^\dagger\big(\{1^k\}v_a\big) A\big(\{1^k\}v_b\big), \tag{11.2}$$

where $V_{v_a v_b}(k)$ are matrix elements of $H(k)$ in k-particle space, the $V(k)$ matrix is chosen to be GUE, i.e. $V_{v_a v_b}(k)$ are independent Gaussian variables with zero center and variance given by (with bar denoting ensemble average),

$$\overline{V_{v_a v_b}(k) V_{v_c v_d}(k)} = \big(\lambda^2/N_k\big) \delta_{v_a v_d} \delta_{v_b v_c}. \tag{11.3}$$

Action of $H(k)$ on a given complete set of m-particle basis states will generate EGUE(k) in m-particle spaces. The m-particle matrix elements of $H(k)$ are, with

$s=m-k$,

$$\begin{aligned}
&H_{v_m^1 v_m^2}(k)\\
&\quad=\left\langle\{1^m\}v_m^1\left|H(k)\right|\{1^m\}v_m^2\right\rangle\\
&\quad=\binom{m}{k}\sum_{v_a,v_b,v_s}\left\langle\{1^k\}v_a\{1^s\}v_s\middle|\{1^m\}v_m^1\right\rangle^*\left\langle\{1^k\}v_b\{1^s\}v_s\middle|\{1^m\}v_m^2\right\rangle V_{v_a v_b}(k).
\end{aligned} \tag{11.4}$$

Unitary decomposition of $H(k)$ in terms of the $SU(N)$ tensors $B_k(g_\nu\omega_\nu)$ is,

$$H(k)=\sum_{g_\nu,\omega_\nu}W_{g_\nu\omega_\nu}(k)B_k(g_\nu\omega_\nu) \tag{11.5}$$

and the W's will be independent Gaussian variables with

$$\overline{W_{g_\nu\omega_\nu}(k)W_{g_\mu\omega_\mu}(k)}=\frac{\lambda^2}{N_k}\delta_{g_\nu g_\mu}\delta_{\omega_\nu\omega_\mu}. \tag{11.6}$$

Using Eqs. (11.1)–(11.5) and the sum-rules for $SU(N)$ Wigner coefficients, the result given by Eq. (11.6) can be proved.

Correlations generated by EGUE(k) in m particle spaces follow from the matrix A of the second moments, i.e.

$$A_{\alpha_m^1\alpha_m^4;\alpha_m^3\alpha_m^2}=\overline{\left\langle\{1^m\}\alpha_m^1\left|H(k)\right|\{1^m\}\alpha_m^2\right\rangle\left\langle\{1^m\}\alpha_m^3\left|H(k)\right|\{1^m\}\alpha_m^4\right\rangle}. \tag{11.7}$$

First substituting the $H(k)$ in terms of B_k's as given by Eq. (11.5), then using the Wigner-Eckart theorem for $SU(N)$ and finally applying Eq. (11.6) for carrying out the ensemble average will give

$$\begin{aligned}
&\overline{\left\langle\{1^m\}\alpha_m^1\left|H(k)\right|\{1^m\}\alpha_m^2\right\rangle\left\langle\{1^m\}\alpha_m^3\left|H(k)\right|\{1^m\}\alpha_m^4\right\rangle}\\
&\quad=\frac{\lambda^2}{N_k}\sum_{g_\nu\omega_\nu,\nu=0,1,\dots,k}\left|\left\langle\{1^m\}\right\|B_k(g_\nu)\left\|\{1^m\}\right\rangle\right|^2\\
&\qquad\times\left\langle\{1^m\}\alpha_m^1\overline{\{1^m\}\alpha_m^2}\middle|g_\nu\omega_\nu\right\rangle\left\langle\{1^m\}\alpha_m^3\overline{\{1^m\}\alpha_m^4}\middle|g_\nu\omega_\nu\right\rangle;\\
&\left|\left\langle\{1^m\}\right\|B_k(g_\nu)\left\|\{1^m\}\right\rangle\right|^2\\
&\quad=\frac{\binom{N}{m}^2\binom{m}{k}^2}{d(g_\nu)\binom{N}{m-k}}\left[U\left(\{1^m\}\{1^{N-k}\}\{1^m\}\{1^k\};\{1^{m-k}\}\{2^\nu 1^{N-2\nu}\}\right)\right]^2\\
&\quad=\Lambda^\nu(N,m,m-k),\\
&\Lambda^\nu(N,m,r)=\binom{m-\nu}{r}\binom{N-m+r-\nu}{r}.
\end{aligned} \tag{11.8}$$

In Eq. (11.8), $U(---)$ are $SU(N)$ Racah coefficients, $\langle--\|--\|--\rangle$ are $SU(N)$ reduced matrix elements and $d(g_\nu)=d(\nu)=\binom{N}{\nu}^2-\binom{N}{\nu-1}^2$. In the final step used is the formula given in [10] for $SU(N)$ U-coefficients. An alternative expression for the covariance in Eq. (11.7) follows from the Biedenharn-Elliott sum rule for $SU(N)$ [2, 11, 12],

$$\begin{aligned}
&\overline{\langle\{1^m\}\alpha_m^1|H(k)|\{1^m\}\alpha_m^2\rangle\langle\{1^m\}\alpha_m^3|H(k)|\{1^m\}\alpha_m^4\rangle} \\
&\quad= \sum_{g_\mu\omega_\mu,\mu=0,1,\ldots,m-k} \frac{\lambda^2}{N_k}\Lambda^\mu(N,m,k) \\
&\quad\quad\times \langle\{1^m\}\alpha_m^1\overline{\{1^m\}\alpha_m^4}|g_\mu\omega_\mu\rangle\langle\{1^m\}\alpha_m^3\overline{\{1^m\}\alpha_m^2}|g_\mu\omega_\mu\rangle. \qquad (11.9)
\end{aligned}$$

To derive Eq. (11.9), the two $SU(N)$ Wigner coefficients in Eq. (11.8) are first transformed into the two Wigner coefficients appearing in Eq. (11.9) multiplied by a $SU(N)$ Racah coefficient by a Racah transform. This new Racah coefficient multiplied by the two Racah coefficients in Eq. (11.8) is then reduced to the square of a Racah coefficient using Biedenharn-Elliott sum rule. Then the final Racah coefficient [see Eq. (11.10) below] is simplified using the formulas in [10]. Equation (11.9) gives the eigenvalue decomposition of the matrix of second moments with the first part in the sum giving eigenvalues E_μ and the product of the two Wigner coefficients giving eigenvectors. The eigenvalues E_μ are given by,

$$E_\mu = \frac{\lambda^2}{N_k}\Lambda^\mu(N,m,k) = \frac{\lambda^2}{N_k}\frac{(N_m)^2\binom{m}{k}^2}{d(g_\mu)(N_k)}\big[U(f_m f_{N-m+k} f_m f_{m-k}; f_k g_\mu)\big]^2. \qquad (11.10)$$

Equations (11.8) and (11.9) lead to remarkably simple expressions for the variance and the excess parameter for the eigenvalue density. Obviously, ensemble averaged centroid is zero and the variance is

$$\overline{\langle H^2\rangle^m} = \frac{1}{N_m}\sum_{v_m^i,v_m^j}\overline{H_{v_m^i v_m^j}H_{v_m^j v_m^i}} = \frac{\lambda^2}{N_k}\Lambda^0(N,m,k). \qquad (11.11)$$

This result follows easily from (11.9) and the sum rule $\sum_{v_m^i}\langle\{1^m\}v_m^i\overline{\{1^m\}v_m^i} \mid g_\mu\omega_\mu\rangle = \sqrt{N_m}\,\delta_{\mu,0}$. Now the fourth moment, dropping λ^2/N_k factor, is

$$\begin{aligned}
&\overline{\langle H^4\rangle^m} \\
&\quad= \frac{1}{N_m}\sum_{v_m^i,v_m^j,v_m^{k'},v_m^l}\overline{H_{v_m^i v_m^j}H_{v_m^j v_m^{k'}}H_{v_m^{k'} v_m^l}H_{v_m^l v_m^i}} \\
&\quad= \frac{1}{N_m}\sum_{v_m^i,v_m^j,v_m^{k'},v_m^l}\Bigg\{2\Bigg[\sum_{g_\nu,\omega_\nu}\langle f_m v_m^i|B_k(g_\nu\omega_\nu)|f_m v_m^j\rangle\langle f_m v_m^j|B_k(g_\nu\omega_\nu)|f_m v_m^{k'}\rangle\Bigg] \\
&\quad\quad\times\Bigg[\sum_{g_\mu,\omega_\mu}\langle f_m v_m^{k'}|B_k(g_\mu\omega_\mu)|f_m v_m^l\rangle\langle f_m v_m^l|B_k(g_\mu\omega_\mu)|f_m v_m^i\rangle\Bigg]
\end{aligned}$$

$$+\left[\sum_{g_\nu,\omega_\nu}\left\langle f_m v_m^i\middle|B_k(g_\nu\omega_\nu)\middle|f_m v_m^j\right\rangle\left\langle f_m v_m^{k'}\middle|B_k(g_\nu\omega_\nu)\middle|f_m v_m^l\right\rangle\right]$$

$$\times\left[\sum_{g_\mu,\omega_\mu}\left\langle f_m v_m^j\middle|B_k(g_\mu\omega_\mu)\middle|f_m v_m^{k'}\right\rangle\left\langle f_m v_m^l\middle|B_k(g_\mu\omega_\mu)\middle|f_m v_m^i\right\rangle\right]\Bigg\}. \tag{11.12}$$

Here we have used Eqs. (11.5) and (11.6) and applied Wigner Eckart theorem. Now, formula for the excess parameter follows easily by using both Eqs. (11.8) and (11.9) together with the orthonormal properties of $SU(N)$ Wigner coefficients. The final formula is [2],

$$\begin{aligned}\gamma_2(N,m,k) &= \frac{\overline{\langle H^4\rangle^m}}{[\overline{\langle H^2\rangle^m}]^2}-3\\ &= \left[(N_m)^{-1}\sum_{\nu=0}^{min\{k,m-k\}}\frac{\Lambda^\nu(N,m,m-k)\Lambda^\nu(N,m,k)d(g_\nu)}{[\Lambda^0(N,m,k)]^2}\right]-1.\end{aligned} \tag{11.13}$$

In the dilute limit Eq. (11.13) reduces to the binary correlation result given by Eq. (4.32). Thus EGUE(k) generates Gaussian densities. For a complete proof, higher order cumulants should be studied. In principle, the formalism given above applies to k_6 but the exact formula is not yet derived. At this stage it is useful to remark that for EGOE(k),

$$\overline{V_{v_a v_b}(k)V_{v_c v_d}(k)}=\left(\lambda^2/N_k\right)\{\delta_{v_a v_d}\delta_{v_b v_c}+\delta_{v_a v_c}\delta_{v_b v_d}\}, \tag{11.14}$$

and in the dilute limit EGUE(k) result for γ_2 reduces to that of EGOE(k); see [3] for details.

Going beyond the lower order moments of the state density, it is also possible to derive formulas for the lower order moments

$$\Sigma_{rr}\left(m,m'\right)=\overline{\left\langle H^r\right\rangle^m\left\langle H^r\right\rangle^{m'}}-\overline{\left\langle H^r\right\rangle^m}\,\overline{\left\langle H^r\right\rangle^{m'}} \tag{11.15}$$

with $r=1$ and 2, of the two-point correlation function,

$$S^{m,m'}\left(E,E'\right)=\overline{\rho^m(E)\rho^{m'}\left(E'\right)}-\overline{\rho^m(E)}\,\overline{\rho^{m'}\left(E'\right)}. \tag{11.16}$$

The final formulas are [13],

$$\hat{\Sigma}_{11}\left(m,m'\right)=\frac{\Sigma_{11}(m,m')}{\sqrt{\overline{\langle H^2\rangle^m}\;\overline{\langle H^2\rangle^{m'}}}}=\sqrt{\frac{\Lambda^0(N,m,m-k)}{N_m\Lambda^0(N,m,k)}\frac{\Lambda^0(N,m',m'-k)}{N_{m'}\Lambda^0(N,m',k)}}, \tag{11.17}$$

and

$$\hat{\Sigma}_{22}(m,m') = \frac{\Sigma_{22}(m,m')}{\overline{\langle H^2\rangle^m}\,\overline{\langle H^2\rangle^{m'}}}$$
$$= \frac{2}{N_m N_{m'}} \sum_{\nu=0}^{k} \frac{\Lambda^\nu(N,m,m-k)\Lambda^\nu(N,m',m'-k)}{\Lambda^0(N,m,k)\Lambda^0(N,m',k)} d(\nu). \quad (11.18)$$

The result for $\overline{\langle H\rangle^m \langle H\rangle^{m'}}$ and hence for $\hat{\Sigma}_{11}$, follows easily from the simple trace formula $\langle H(k)\rangle^m = \binom{m}{k}\langle H(k)\rangle^k$ or alternatively by applying Eq. (11.8) and using the fact that only $\nu = 0$ terms will contribute to $\langle H\rangle^m$. Similarly, Σ_{22} formula has been derived using

$$\overline{\langle H^2\rangle^m \langle H^2\rangle^{m'}} = [N_m N_{m'}]^{-1} \sum_{a,b,c,d} \overline{|H_{a,b}(m)|^2 |H_{c,d}(m')|^2}$$
$$= \overline{\langle H^2\rangle^m}\,\overline{\langle H^2\rangle^{m'}} + 2[N_m N_{m'}]^{-1} \sum_{a,b,c,d} \left\{\overline{H_{a,b}(m) H_{c,d}(m')}\right\}^2 \quad (11.19)$$

where $H_{a,b}(m) = \langle m,a|H|m,b\rangle$ is a m-particle matrix element. Note that we have used $\overline{x^2y^2} = \overline{x^2}\,\overline{y^2} + 2(\overline{xy})^2$. Applying Eq. (11.8) to the second term in the second equality and using orthonormal properties of $SU(N)$ Wigner coefficients will give finally the formula for $\hat{\Sigma}_{22}(m,m')$. The formulas for $\hat{\Sigma}_{rr}(m,m)$, $r = 1, 2$ were derived first in [2, 3]. It is important to remind that Σ_{rr} is the (rr)-th bivariate moment of the two point function. Before turning to EGUE/EGOE with spin degree of freedom, it is important to mention that in the standard applications of GUE/GOE, correlations between levels with different m will be zero [i.e. $\hat{\Sigma}_{11}(m,m') = 0$ and $\hat{\Sigma}_{22}(m,m') = 0$] as independent GUE/GOE description for levels with different m has to be used. Therefore results given by Eqs. (11.17)–(11.19) provide useful signatures for EGUE/EGOE and in Chap. 12 this will be discussed in more detail.

11.2 Embedded Gaussian Unitary Ensemble for Spinless Boson Systems: BEGUE(k)

For spinless bosons in N sp states with a general k-body Hamiltonian, we have BEGUE(k). As pointed out in [2], it is striking that all the EGUE(k) results of Sect. 11.1 translate directly to those of BEGUE(k) by applying the well known $N \to -N$ symmetry [14, 15], i.e. in the fermion results replace N by $-N$ and then take the absolute value of the final result. For example, the m boson space dimension N_m^B is

$$N_m^B = \left|\binom{-N}{m}\right| = \binom{N+m-1}{m}. \quad (11.20)$$

More importantly the eigenvalues E_μ of the matrix of the second moments follow from Eq. (11.10) by using $N \to -N$ symmetry,

$$\Lambda_B^\nu(N,m,k) \to \left| \binom{m-\nu}{k} \binom{-N-m+k-\nu}{k} \right| = \binom{m-\nu}{k} \binom{N+m+\nu-1}{k}. \tag{11.21}$$

This result was explicitly derived in [5]. Moreover, for bosons $\{k\} \otimes \{k^{N-1}\} \to g_\nu = \{2\nu, \nu^{N-2}\}$, $\nu = 0, 1, \ldots, k$. Also, the $N \to -N$ symmetry and Eq. (11.20) will give $d^B(g_\nu) = \{(N+\nu-1)_\nu\}^2 - \{(N+\nu-2)_{\nu-1}\}^2$ and this is same as Eq. (15) of [5]. Similarly Eqs. (11.11), (11.13), (11.17) and (11.18) for $\langle H^2 \rangle$, $\gamma_2(N,m,k)$, Σ_{11} and Σ_{22} respectively extend directly to BEGUE(k) with $\Lambda^\nu(N,m,k)$ replaced by $\Lambda_B^\nu(N,m,k)$ defined in Eq. (11.21) and similarly replacing N_m by N_m^B and $d(g_\nu)$ by $d^B(g_\nu)$. Detailed derivations given in [5] are in agreement with these. In addition, for fermions to bosons there is also a $m \leftrightarrow N$ symmetry and this connects fermion results (say for M_p and Σ_{pq}) in dilute limit to boson results in dense limit as discussed in Sect. 9.4 and [14].

11.3 EGUE(2)-$SU(r)$ Ensembles: General Formulation

Consider a system of m fermion or bosons in Ω number of sp levels each r-fold degenerate. Then the SGA is $U(r\Omega)$ and it is possible to consider $U(r\Omega) \supset U(\Omega) \otimes SU(r)$ algebra. Now, for random two-body Hamiltonians preserving $SU(r)$ symmetry, one can introduce embedded GUE with $U(\Omega) \otimes SU(r)$ embedding and this ensemble is called EGUE(2)-$SU(r)$. Ensembles with $r = 2$ and 4 for fermions correspond to fermions with spin (or isospin [16]) and spin-isospin $SU(4)$ symmetry [17–19] respectively. Similarly, for bosons $r = 2, 3$ are of interest. Also $r = 1$ gives back EGUE(2) and BEGUE(2) both. It is important to note that the distinction between fermions and bosons is in the $U(\Omega)$ irreps that need to be considered. Now, we will give a formulation in terms of $SU(\Omega)$ Wigner-Racah algebra that is valid for any $r \geq 1$ [20].

Let us begin with the normalized two-particle states $|f_2 F_2; v_2 \beta_2\rangle$ where the $U(r)$ irreps $F_2 = \{1^2\}$ and $\{2\}$ and the corresponding $U(\Omega)$ irreps f_2 are $\{2\}$ (symmetric) and $\{1^2\}$ (antisymmetric) respectively for fermions and $\{1^2\}$ (antisymmetric) and $\{2\}$ (symmetric) respectively for bosons. Similarly v_2 are additional quantum numbers that belong to f_2 and β_2 belong to F_2. As f_2 uniquely defines F_2, from now on we will drop F_2 unless it is explicitly needed and also we will use the $f_2 \leftrightarrow F_2$ equivalence whenever needed. With $A^\dagger(f_2 v_2 \beta_2)$ and $A(f_2 v_2 \beta_2)$ denoting creation and annihilation operators for the normalized two particle states, a general two-body Hamiltonian operator $\widehat{H}$ preserving $SU(r)$ symmetry can be written as

$$\widehat{H} = \widehat{H}_{\{2\}} + \widehat{H}_{\{1^2\}} = \sum_{f_2, v_2^i, v_2^f, \beta_2;\, f_2 = \{2\},\{1^2\}} H_{f_2 v_2^i v_2^f}(2) A^\dagger\big(f_2 v_2^f \beta_2\big) A\big(f_2 v_2^i \beta_2\big). \tag{11.22}$$

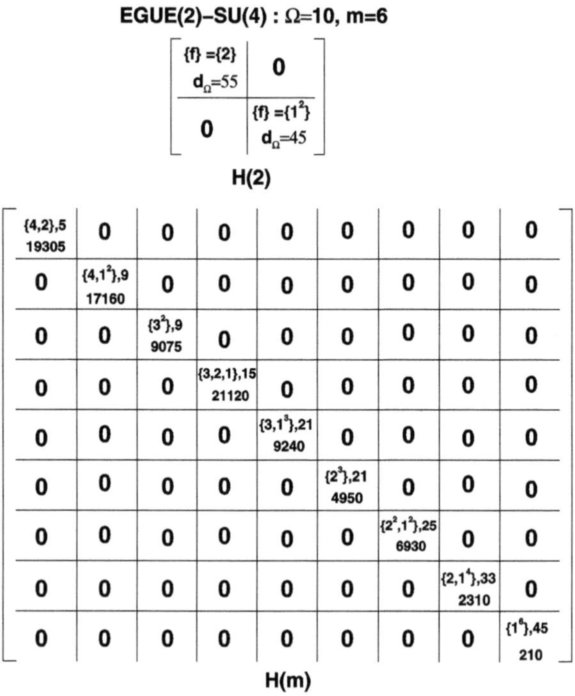

Fig. 11.1 (**a**) EGUE(2)-$SU(4)$ ensemble for fermions in the defining space. (**b**) Decomposition of the H matrix in ($\Omega = 10$, $m = 6$) space into direct sum of matrices with fixed $SU(\Omega)$ irrep f_m. There is a EGUE(2)-$SU(4)$ ensemble in each f_m space corresponding to each diagonal block in the figure. Shown also next to each f_m in the figure, is the eigenvalue $\langle \hat{C}_2(SU(4)) \rangle^{f_m}$ of the quadratic Casimir invariant of $SU(4)$. Similarly, below each f_m shown is the matrix dimension

In Eq. (11.22), $H_{f_2 v_2^i v_2^f}(2) = \langle f_2 v_2^f \beta_2 \mid H \mid f_2 v_2^i \beta_2 \rangle$ independent of the β_2's. The uniform summation over β_2 in Eq. (11.22) ensures that $\widehat{H}$ is $SU(r)$ scalar and therefore it will not connect states with different f_2's. However, $\widehat{H}$ is not a $SU(r)$ invariant operator. Just as the two particle states, we can denote the m particle states by $|f_m v_m^f \beta_m^F\rangle$; $F_m = \widetilde{f_m}$ for fermions and $F_m = f_m$ for bosons. Action of $\widehat{H}$ on these states generates states that are degenerate with respect to β_m^F but not v_m^f. Therefore for a given f_m, there will be $d_\Omega(f_m)$ number of levels each with $d_r(\widetilde{f_m})$ number of degenerate states. Formula for the dimension $d_\Omega(f_m)$ is [21],

$$d_\Omega(f_m) = \prod_{i<j=1}^{\Omega} \frac{f_i - f_j + j - i}{j - i}, \tag{11.23}$$

where $f_m = \{f_1, f_2, \ldots\}$. Equation (11.23) also gives $d_r(F_m)$ with the product ranging from $i = 1$ to r and replacing f_i by F_i. As $\widehat{H}$ is a $SU(r)$ scalar, the m particle H matrix will be a direct sum of matrices with each of them labeled by the f_m's with dimension $d_\Omega(f_m)$. Thus

$$H(m) = \sum_{f_m} H_{f_m}(m) \oplus . \tag{11.24}$$

Figure 11.1 shows an example for Eq. (11.24) with $r = 4$ for fermions. As seen from Eq. (11.22), the H matrix in two particle spaces is a direct sum of the two matrices

$H_{f_2}(2)$, one in the $f_2 = \{2\}$ space and the other in $\{1^2\}$ space. Similarly, for the 6 particle example shown in Fig. 11.1 there are 9 f_m's and therefore the H matrix is a direct sum of 9 matrices. It should be noted that the matrix elements of $H_{f_m}(m)$ matrices receive contributions from both $H_{\{2\}}(2)$ and $H_{\{1^2\}}(2)$.

Embedded random matrix ensemble EGUE(2)-$SU(r)$ for a m fermion or boson systems with a fixed f_m, i.e. $\{H_{f_m}(m)\}$, is generated by the ensemble of H operators given in Eq. (11.22) with $H_{\{2\}}(2)$ and $H_{\{1^2\}}(2)$ matrices replaced by independent GUE ensembles of random matrices,

$$\{H(2)\} = \{H_{\{2\}}(2)\}_{\text{GUE}} \oplus \{H_{\{1^2\}}(2)\}_{\text{GUE}}. \tag{11.25}$$

In Eq. (11.25), $\{--\}$ denotes ensemble. Random variables defining the real and imaginary parts of the matrix elements of $H_{f_2}(2)$ are independent Gaussian variables with zero center and variance given by (with bar representing ensemble average),

$$\overline{H_{f_2 v_2^1 v_2^2}(2) H_{f_2' v_2^3 v_2^4}(2)} = \delta_{f_2 f_2'} \delta_{v_2^1 v_2^4} \delta_{v_2^2 v_2^3} (\lambda_{f_2})^2. \tag{11.26}$$

Also, the independence of the $\{H_{\{2\}}(2)\}$ and $\{H_{\{1^2\}}(2)\}$ GUE ensembles imply,

$$\begin{aligned}
&\overline{\left[H_{\{2\}v_2^1 v_2^2}(2)\right]^P \left[H_{\{1^2\}v_2^3 v_2^4}(2)\right]^Q} \\
&\quad = \left\{\overline{\left[H_{\{2\}v_2^1 v_2^2}(2)\right]^P}\right\}\left\{\overline{\left[H_{\{1^2\}v_2^3 v_2^4}(2)\right]^Q}\right\} \quad \text{for } P \text{ and } Q \text{ even}, \\
&\quad = 0 \quad \text{for } P \text{ or } Q \text{ odd}.
\end{aligned} \tag{11.27}$$

Action of $\widehat{H}$ defined by Eq. (11.22) on m particle basis states with a fixed f_m, along with Eqs. (11.26)–(11.27) generates EGUE(2)-$SU(r)$ ensemble $\{H_{f_m}(m)\}$; it is labeled by the $U(\Omega)$ irrep f_m with matrix dimension $d_\Omega(f_m)$.

As discussed before for EGUE(k) for fermions in Sect. 11.1 and similarly for bosons in Sect. 11.2, tensorial decomposition of $\widehat{H}$ with respect to the embedding algebra $U(\Omega) \otimes SU(r)$ plays a crucial role in generating analytical results; as before $U(\Omega)$ and $SU(\Omega)$ are used interchangeably. As $\widehat{H}$ preserves $SU(r)$, it transforms as the irrep $\{0\}$ with respect to the $SU(r)$ algebra. However with respect to $SU(\Omega)$, the tensorial characters, in Young tableaux notation, for $f_2 = \{2\}$ are $\mathbf{F}_\nu = \{0\}$, $\{21^{\Omega-2}\}$ and $\{42^{\Omega-2}\}$ with $\nu = 0, 1$ and 2 respectively. Similarly for $f_2 = \{1^2\}$ they are $\mathbf{F}_\nu = \{0\}$, $\{21^{\Omega-2}\}$ and $\{2^2 1^{\Omega-4}\}$ with $\nu = 0, 1, 2$ respectively. Note that $\mathbf{F}_\nu = f_2 \times \overline{f_2}$ where $\overline{f_2}$ is the irrep conjugate to f_2 and the $\times$ denotes Kronecker product. Young tableaux for the $\mathbf{F}_\nu$ are same as those in Figs. 9.2 and 5.1b for $f_2 = \{2\}$ and $\{1^2\}$ respectively with N replaced by Ω in the figures. Now, we can define unitary tensors B's that are scalars in $SU(r)$ space,

$$\begin{aligned}
B(f_2 \mathbf{F}_\nu \omega_\nu) = &\sum_{v_2^i, v_2^f, \beta_2} A^\dagger(f_2 v_2^f \beta_2) A(f_2 v_2^i \beta_2) \langle f_2 v_2^f \overline{f_2}\, \overline{v_2^i} | \mathbf{F}_\nu \omega_\nu \rangle \\
&\times \langle F_2 \beta_2 \overline{F_2}\, \overline{\beta_2} \mid 00 \rangle.
\end{aligned} \tag{11.28}$$

In Eq. (11.28), $\langle f_2 - - - \rangle$ are $SU(\Omega)$ Wigner coefficients and $\langle F_2 - - - \rangle$ are $SU(r)$ Wigner coefficients. The expansion of $\widehat{H}$ in terms of B's is,

$$\widehat{H} = \sum_{f_2, \mathbf{F}_\nu, \omega_\nu} W(f_2 \mathbf{F}_\nu \omega_\nu) B(f_2 \mathbf{F}_\nu \omega_\nu). \tag{11.29}$$

The expansion coefficients W's follow from the orthogonality of the tensors B's with respect to the traces over fixed f_2 spaces. Then we have the most important relation needed for all the results given ahead,

$$\overline{W(f_2 \mathbf{F}_\nu \omega_\nu) W\big(f_2' \mathbf{F}_\nu' \omega_\nu'\big)} = \delta_{f_2 f_2'} \delta_{\mathbf{F}_\nu \mathbf{F}_\nu'} \delta_{\omega_\nu \omega_\nu'} (\lambda_{f_2})^2 d_r(F_2). \tag{11.30}$$

This is derived starting with Eq. (11.29) and using Eqs. (11.25)–(11.28). Also used are the sum rules for Wigner coefficients appearing in Eq. (11.28).

Turning to m particle H matrix elements, first we denote the $U(\Omega)$ and $U(r)$ irreps by f_m and F_m respectively. Correlations generated by EGUE(2)-$SU(r)$ between states with (m, f_m) and $(m', f_{m'})$ follow from the covariance between the m-particle matrix elements of H. Now using Eqs. (11.29) and (11.30) along with the Wigner-Eckart theorem applied using $SU(\Omega) \otimes SU(r)$ Wigner-Racah algebra (see for example [22]) will give

$$\begin{aligned}
&\overline{H_{f_m v_m^i v_m^f} H_{f_{m'} v_{m'}^i v_{m'}^f}} \\
&= \overline{\big\langle f_m F_m v_m^f \beta \big| H \big| f_m F_m v_m^i \beta \big\rangle \big\langle f_{m'} F_{m'} v_{m'}^f \beta' \big| H \big| f_{m'} F_{m'} v_{m'}^i \beta' \big\rangle} \\
&= \sum_{f_2, \mathbf{F}_\nu, \omega_\nu} \frac{(\lambda_{f_2})^2}{d_\Omega(f_2)} \sum_{\rho, \rho'} \langle f_m | \| B(f_2 \mathbf{F}_\nu) \| | f_m \rangle_\rho \langle f_{m'} | \| B(f_2 \mathbf{F}_\nu) \| | f_{m'} \rangle_{\rho'} \\
&\quad \times \big\langle f_m v_m^i \mathbf{F}_\nu \omega_\nu \mid f_m v_m^f \big\rangle_\rho \big\langle f_{m'} v_{m'}^i \mathbf{F}_\nu \omega_\nu \mid f_{m'} v_{m'}^f \big\rangle_{\rho'}; \\
&\langle f_m | \| B(f_2 \mathbf{F}_\nu) \| | f_m \rangle_\rho = \sum_{f_{m-2}} F(m) \frac{\mathscr{N}_{f_{m-2}}}{\mathscr{N}_{f_m}} \frac{U(f_m \overline{f_2} f_m f_2; f_{m-2} \mathbf{F}_\nu)_\rho}{U(f_m \overline{f_2} f_m f_2; f_{m-2} \{0\})}.
\end{aligned} \tag{11.31}$$

Here the summation in the last equality is over the multiplicity index ρ and this arises as $f_m \times \mathbf{F}_\nu$ gives in general more than once the irrep f_m. In Eq. (11.31), $F(m) = -m(m-1)/2$, $d_\Omega(f_m)$ is dimension with respect to $U(\Omega)$ as given by Eq. (11.23) and $\langle \ldots \mid \ldots \rangle$ and $U(\ldots)$ are $SU(\Omega)$ Wigner and Racah coefficients respectively. Similarly, $\mathscr{N}_{f_m}$ is dimension with respect to the S_m group,

$$\mathscr{N}_{f_m} = \frac{m! \prod_{i<k=1}^{r} (\ell_i - \ell_k)}{\ell_1! \ell_2! \cdots \ell_r!}; \quad \ell_i = f_i + r - i. \tag{11.32}$$

Note that r denotes total number of rows in the Young tableaux for f_m.

Lower order cross correlations between states with different (m, f_m) are given by the normalized bivariate moments $\hat{\Sigma}_{rr}(m, f_m : m', f_{m'})$, $r = 1, 2$ of the two-point function S^ρ where, with $\rho^{m,f_m}(E)$ defining fixed-(m, f_m) density of states,

$$\begin{aligned}
S^{mf_m:m'f_{m'}}\left(E, E'\right) &= \overline{\rho^{m,f_m}(E)\rho^{m',f_{m'}}\left(E'\right)} - \overline{\rho^{m,f_m}(E)}\,\overline{\rho^{m',f_{m'}}\left(E'\right)}; \\
\hat{\Sigma}_{11}\left(m, f_m : m', f_{m'}\right) &= \overline{\langle H\rangle^{m,f_m}\langle H\rangle^{m',f_{m'}}} \Big/ \sqrt{\overline{\langle H^2\rangle^{m,f_m}}\,\overline{\langle H^2\rangle^{m',f_{m'}}}}, \\
\hat{\Sigma}_{22}\left(m, f_m : m', f_{m'}\right) &= \overline{\langle H^2\rangle^{m,f_m}\langle H^2\rangle^{m',f_{m'}}} \Big/ \left[\overline{\langle H^2\rangle^{m,f_m}}\,\overline{\langle H^2\rangle^{m',f_{m'}}}\right] - 1.
\end{aligned} \tag{11.33}$$

In Eq. (11.33), $\overline{\langle H^2\rangle^{m,f_m}}$ is the second moment (or variance) of the eigenvalue density $\overline{\rho^{m,f_m}(E)}$ and its centroid $\overline{\langle H\rangle^{m,f_m}} = 0$ by definition. We begin with $\overline{\langle H\rangle^{m,f_m}\langle H\rangle^{m',f_{m'}}}$. As $\langle H\rangle^{m,f_m}$ is the trace of H (divided by dimensionality) in (m, f_m) space, only $\mathbf{F}_\nu = \{0\}$ will generate this. Then trivially,

$$\begin{aligned}
\overline{\langle H\rangle^{m,f_m}\langle H\rangle^{m',f_{m'}}} &= \sum_{f_2} \frac{(\lambda_{f_2})^2}{d_\Omega(f_2)} P^{f_2}(m, f_m) P^{f_2}(m', f_{m'}); \\
P^{f_2}(m, f_m) &= F(m) \sum_{f_{m-2}} \left[\mathcal{N}_{f_{m-2}}/\mathcal{N}_{f_m}\right].
\end{aligned} \tag{11.34}$$

In terms of m particle H matrix elements, $\overline{\langle H^2\rangle^{m,f_m}}$ is

$$\overline{\langle H^2\rangle^{m,f_m}} = \left[d(f_m)\right]^{-1} \sum_{v_m^1, v_m^2} \overline{H_{f_m v_m^1 v_m^2} H_{f_m v_m^2 v_m^1}}.$$

Applying Eq. (11.31) and the orthonormal properties of the $SU(\Omega)$ Wigner coefficients lead to

$$\overline{\langle H^2\rangle^{m,f_m}} = \sum_{f_2} \frac{(\lambda_{f_2})^2}{d_\Omega(f_2)} \sum_{\nu=0,1,2} \mathcal{Q}^\nu(f_2 : m, f_m) \tag{11.35}$$

where

$$\mathcal{Q}^\nu(f_2 : m, f_m) = \left[F(m)\right]^2 \sum_{f_{m-2}, f'_{m-2}} \frac{\mathcal{N}_{f_{m-2}}}{\mathcal{N}_{f_m}} \frac{\mathcal{N}_{f'_{m-2}}}{\mathcal{N}_{f_m}} X_{UU}\left(f_2; f_{m-2}, f'_{m-2}; \mathbf{F}_\nu\right). \tag{11.36}$$

The X_{UU} function involves $SU(\Omega)$ Racah coefficients,

$$\begin{aligned}
&X_{UU}\left(f_2; f_{m-2}, f'_{m-2}; \mathbf{F}_\nu\right) \\
&\quad = \sum_\rho \frac{U(f_m, \overline{f_2}, f_m, f_2; f_{m-2}, \mathbf{F}_\nu)_\rho U(f_m, \overline{f_2}, f_m, f_2; f'_{m-2}, \mathbf{F}_\nu)_\rho}{U(f_m, \overline{f_2}, f_m, f_2; f_{m-2}, \{0\}) U(f_m, \overline{f_2}, f_m, f_2; f'_{m-2}, \{0\})}.
\end{aligned} \tag{11.37}$$

Summation over the multiplicity index ρ in Eq. (11.37) arises naturally in applications to physical problems as all the physically relevant results should be independent of ρ which is a label for equivalent $SU(\Omega)$ irreps. Let us add that,

$$\mathscr{Q}^{\nu=0}(f_2:m,f_m)=\left[P^{f_2}(m,f_m)\right]^2. \tag{11.38}$$

Equations (11.34)–(11.36) and Table 4 of [7] will allow one to calculate covariances $\hat{\Sigma}_{11}$ in energy centroids. For the covariances $\hat{\Sigma}_{22}$ in spectral variances, the formula is [7]

$$\begin{aligned}
\hat{\Sigma}_{22}(m,f_m;m',f_{m'}) &= \frac{X_{\{2\}}+X_{\{1^2\}}+4X_{\{1^2\}\{2\}}}{\overline{\langle H^2\rangle^{m,f_m}}\;\overline{\langle H^2\rangle^{m',f_{m'}}}};\\
X_{f_2} &= \frac{2(\lambda_{f_2})^4}{[d_\Omega(f_2)]^2}\sum_{\nu=0,1,2}\left[d(\mathbf{F}_\nu)\right]^{-1}\mathscr{Q}^\nu(f_2:m,f_m)\mathscr{Q}^\nu\left(f_2:m',f_{m'}\right),\\
X_{\{1^2\}\{2\}} &= \frac{\lambda^2_{\{2\}}\lambda^2_{\{1^2\}}}{d_\Omega(\{2\})d_\Omega(\{1^2\})}\sum_{\nu=0,1}\left[d(\mathbf{F}_\nu)\right]^{-1}\mathscr{R}^\nu(m,f_m)\mathscr{R}^\nu\left(m',f_{m'}\right).
\end{aligned} \tag{11.39}$$

Here $d(\mathbf{F}_\nu)$ are dimension of the irrep $\mathbf{F}_\nu$, and we have $d(\{0\})=1$, $d(\{2,1^{\Omega-2}\})=\Omega^2-1$, $d(\{4,2^{\Omega-2}\})=\Omega^2(\Omega+3)(\Omega-1)/4$, and $d(\{2^2,1^{\Omega-4}\})=\Omega^2(\Omega-3)(\Omega+1)/4$. Note that $\mathscr{Q}^\nu(f_2:m,f_m)$ are defined by Eq. (11.36). The function $\mathscr{R}^\nu(m,f_m)$ also involve $SU(\Omega)$ U-coefficients,

$$\begin{aligned}
\mathscr{R}^\nu(m,f_m) &= \left[F(m)\right]^2\sum_{f_{m-2},f'_{m-2}}\frac{\mathscr{N}_{f_{m-2}}}{\mathscr{N}_{f_m}}\frac{\mathscr{N}_{f'_{m-2}}}{\mathscr{N}_{f_m}}Y_{UU}\left(f_{m-2},f'_{m-2};\mathbf{F}_\nu\right);\\
&Y_{UU}\left(f_{m-2},f'_{m-2};\mathbf{F}_\nu\right)\\
&= \sum_\rho\frac{U(f_m,\{1^{\Omega-2}\},f_m,\{1^2\};f_{m-2},\mathbf{F}_\nu)_\rho U(f_m,\{2^{\Omega-1}\},f_m,\{2\};f'_{m-2},\mathbf{F}_\nu)_\rho}{U(f_m,\{1^{\Omega-2}\},f_m,\{1^2\};f_{m-2},\{0\})U(f_m,\{2^{\Omega-1}\},f_m,\{2\};f'_{m-2},\{0\})}.
\end{aligned} \tag{11.40}$$

In $Y_{UU}(f_{m-2},f'_{m-2};\mathbf{F}_\nu)$, f_{m-2} comes from $f_m\otimes\{1^{\Omega-2}\}$ and f'_{m-2} comes from $f_m\otimes\{2^{\Omega-1}\}$. Similarly, the summation is over $\nu=0$ and 1 only as $\nu=2$ parts for $f_2=\{2\}$ and $\{1^2\}$ are different. It is useful to note that,

$$\mathscr{R}^{\nu=0}(m,f_m)=P^{\{2\}}(m,f_m)P^{\{1^2\}}(m,f_m). \tag{11.41}$$

Formulas for X_{UU} and Y_{UU} are given in [7] and they are simplified version of the formulas given in [23]. For illustration, some of these results are collected in Table 11.1. These and Eqs. (11.33)–(11.41) will allow one to derive analytical/numerical results for spectral variances and covariances in energy centroids and spectral variances for any EGUE(2)-$SU(r)$ for fermion or boson systems.

Table 11.1 Formulas for $X_{UU}(f_2; f_{m-2}, f'_{m-2}; \mathbf{F}_\nu)$ and $Y_{UU}(f_{m-2}, f'_{m-2}; \mathbf{F}_\nu)$ with $\nu = 1, 2$

$\{f_{m-2}\}\{f'_{m-2}\}$	$X_{UU}(\{1^2\}; f_{m-2}, f'_{m-2}; \{2^\nu, 1^{\Omega-2\nu}\})$
$\{f(ab)\}\{f(ab)\}$	$\frac{\Omega}{(\Omega-2)}\{\delta_{\nu,2} + \frac{(\Omega-1)(\Omega-2)}{2\Pi_a^{(b)}\Pi_b^{(a)}}\delta_{\nu,2} + (3-2\nu)\frac{(\Omega-1)}{2} \times [(1+\frac{1}{\tau_{ab}})\frac{1}{\Pi_b^{(a)}} + (1-\frac{1}{\tau_{ab}})\frac{1}{\Pi_a^{(b)}} - \frac{4}{\Omega}\delta_{\nu,1}]\}$
$\{f(ab)\}\{f(ac)\}$	$\frac{\Omega(\Omega-1)}{2(\Omega-2)}\{\frac{2}{(\Omega-1)}\delta_{\nu,2} - \frac{4}{\Omega}\delta_{\nu,1} + (3-2\nu)\frac{1}{\Pi_a^{(bc)}}\}$
$\{f_{m-2}\}\{f'_{m-2}\}$	$X_{UU}(\{2\}; f_{m-2}, f'_{m-2}; \{2\nu, \nu^{\Omega-2}\})$
$\{f(ab)\}\{f(ab)\}$	$\frac{\Omega(\Omega+1)}{2}\{\frac{1}{\Pi_a^{(b)}\Pi_b^{(a)}}\delta_{\nu,2} + \frac{2}{(\Omega+1)(\Omega+2)}\delta_{\nu,2} + (3-2\nu)\frac{1}{(\Omega+2)}[\frac{(\tau_{ab}-1)^2}{\tau_{ab}(\tau_{ab}+1)}\frac{1}{\Pi_b^{(a)}} + \frac{(\tau_{ab}+1)^2}{\tau_{ab}(\tau_{ab}-1)}\frac{1}{\Pi_a^{(b)}} - \frac{4}{\Omega}\delta_{\nu,1}]\}$
$\{f(aa)\}\{f(aa)\}$	$\frac{\Omega}{(\Omega+2)}\{\delta_{\nu,2} + (3-2\nu)\frac{2(\Omega+1)}{\Pi'_a} + \frac{(\Omega+1)(\Omega+2)}{2\Pi''_a}\delta_{\nu,2} - \frac{2(\Omega+1)}{\Omega}\delta_{\nu,1}\}$
$\{f(aa)\}\{f(bb)\}$	$-\frac{2(\Omega+1)}{(\Omega+1)}\delta_{\nu,1} + \frac{\Omega}{(\Omega+2)}\delta_{\nu,2}$
$\{f(aa)\}\{f(ab)\}$	$\frac{\Omega}{(\Omega+2)}\{\delta_{\nu,2} + (3-2\nu)\frac{(\Omega+1)(\tau_{ab}+1)}{(\tau_{ab}-1)\Pi_a^{(b)}} - \frac{2(\Omega+1)}{\Omega}\delta_{\nu,1}\}$
$\{f_{m-2}\}\{f'_{m-2}\}$	$Y_{UU}(f_{m-2}, f'_{m-2}; \{2, 1^{\Omega-2}\})$
$\{f(ab)\}\{f(ab)\}$	$-\frac{\Omega}{2}[\frac{(\Omega^2-1)}{(\Omega^2-4)}]^{1/2}\{(1+\frac{1}{\tau_{ab}})\frac{1}{\Pi_a^{(b)}} + (1-\frac{1}{\tau_{ab}})\frac{1}{\Pi_b^{(a)}} - \frac{4}{\Omega}\}$
$\{f(ab)\}\{f(ac)\}$	$-\frac{\Omega}{2}[\frac{(\Omega^2-1)}{(\Omega^2-4)}]^{1/2}\{(1+\frac{1}{\tau_{ac}})\frac{1}{\Pi_a^{(b)}} - \frac{4}{\Omega}\}$
$\{f(ab)\}\{f(aa)\}$	$-\Omega[\frac{(\Omega^2-1)}{(\Omega^2-4)}]^{1/2}\{\frac{1}{\Pi_a^{(b)}} - \frac{2}{\Omega}\}$

11.3.1 Results for BEGUE(2): $r = 1$

Simplest of the EGUE(2)-$SU(r)$ are the EGUEs with $r = 1$ and they corresponds to EGUE(2) and BEGUE(2) depending on totally antisymmetric or symmetric f_m one considers. Also they correspond to $k = 2$ in Sects. 11.1 and 11.2 respectively. For illustration we consider BEGUE(2) in some detail. For this ensemble, in order to apply the formulas for $\langle H^2 \rangle$, $\hat{\Sigma}_{11}$ and $\hat{\Sigma}_{22}$, first we need the formulas for X_{UU} and Y_{UU}. Some of these, taken from Tables 4 and 7 of [7], are given in Table 11.1. For applying these formulas, we need the 'axial distances' τ_{ij} for the boxes i and j in a given Young tableaux. Given a $f_m = \{f_1, f_2, \ldots, f_\Omega\}$ we have,

$$\tau_{ij} = f_i - f_j + j - i. \tag{11.42}$$

In terms of τ_{ij} the functions $\Pi_a^{(b)}$, $\Pi_b^{(a)}$, $\Pi_a^{(bc)}$, Π'_a and Π''_a are defined as,

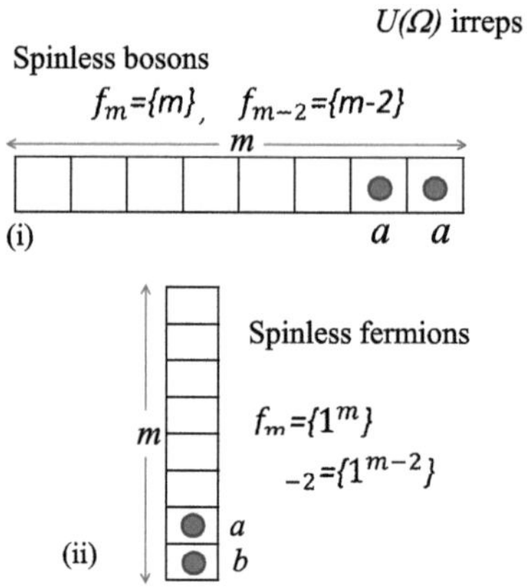

Fig. 11.2 Young tableaux denoting the $SU(\Omega)$ irreps $f_m = \{m\}$ and $\{1^m\}$ as appropriate for **(i)** spinless boson and **(ii)** spinless fermion systems. Removal of two boxes generating $m-2$ particle irreps f_{m-2} for these systems are also shown in the figure. For **(i)** only the irrep $f_2 = \{2\}$ will apply and similarly for **(ii)** only $\{1^2\}$ will apply. Figure is taken from [20] with permission from American Institute of Physics (Color figure online)

$$\begin{aligned}
\Pi_a^{(b)} &= \prod_{i=1,2,\ldots,\Omega;i\neq a,i\neq b} (1-1/\tau_{ai}) \\
\Pi_b^{(a)} &= \prod_{i=1,2,\ldots,\Omega;i\neq a,i\neq b} (1-1/\tau_{bi}) \\
\Pi_a^{(bc)} &= \prod_{i=1,2,\ldots,\Omega;i\neq a,i\neq b,i\neq c} (1-1/\tau_{ai}); \quad a\neq b\neq c, \\
\Pi_a' &= \prod_{i=1,2,\ldots,\Omega;i\neq a} (1-1/\tau_{ai}) \\
\Pi_a'' &= \prod_{i=1,2,\ldots,\Omega;i\neq a} (1-2/\tau_{ai}).
\end{aligned} \tag{11.43}$$

With these we can calculate X_{UU} and Y_{UU}; see [7] for full discussion. For BE-GUE(2), the algebra $U(\Omega)\otimes SU(r)$ with $r=1$ reduces to just $U(\Omega)$ or $SU(\Omega)$. Similarly, f_m is the totally symmetric irrep $\{m\}$ and $f_{m-2}=\{m-2\}$. Therefore to generate f_{m-2} only the action of removal of $\{2\}$ from f_m is allowed. Denoting the last two boxes of f_m by a and a (note that we can remove only boxes from the right end to get proper Young tableaux and also boxes in a given row must have the same symbol to apply the results in Table 11.1) as shown in Fig. 11.2, we have

$$\begin{aligned}
\tau_{ai} &= m+i-1, \\
\Pi_a' &= \frac{m}{m+\Omega-1}, \\
\Pi_a'' &= \frac{m(m-1)}{(m+\Omega-1)(m+\Omega-2)}.
\end{aligned} \tag{11.44}$$

Similarly $\mathscr{N}_{f_m} = 1$ and $\mathscr{N}_{f_{m-2}} = 1$ as both are symmetric irreps. Now the formulas in Table 11.1 will give X_{UU} and there by $\mathscr{Q}^\nu$ in Eq. (11.36),

$$\begin{aligned}
\mathscr{Q}^{\nu=0}\big(\{2\}; m, \{m\}\big) &= \frac{m^2(m-1)^2}{4}, \\
\mathscr{Q}^{\nu=1}\big(\{2\}; m, \{m\}\big) &= \frac{m^2(m-1)^2}{4} \frac{2(\Omega+m)(\Omega^2-1)}{m(\Omega+2)}, \\
\mathscr{Q}^{\nu=2}\big(\{2\}; m, \{m\}\big) &= \frac{m^2(m-1)^2}{4} \frac{\Omega^2(\Omega-1)(\Omega+m)(\Omega+m+1)}{2(\Omega+2)m(m-1)}.
\end{aligned} \tag{11.45}$$

These and Eq. (11.35) will give,

$$\langle H^2\rangle^{\{m\}} = \lambda_{\{2\}}^2 \binom{m}{2} \binom{\Omega+m-1}{2} = \lambda_{\{2\}}^2 \Lambda_B^0(\Omega, m, 2). \tag{11.46}$$

This agrees with the result stated in Sect. 11.2. As $P^{\{2\}}(m, \{m\}) = -m(m-1)/2$, we have easily,

$$\begin{aligned}
&\hat{\Sigma}_{11}\big(\{m\}, \{m'\}\big) \\
&= \frac{2\sqrt{m(m-1)(m')(m'-1)}}{\Omega(\Omega+1)\sqrt{(\Omega+m-1)(\Omega+m-2)(\Omega+m'-1)(\Omega+m'-2)}}.
\end{aligned} \tag{11.47}$$

Again, this agrees with the result stated in Sect. 11.2. Further, $\hat{\Sigma}_{22}$ is determined only by $X_{\{2\}}$ defined in Eq. (11.39) and then, using Eq. (11.45), we have

$$\begin{aligned}
&\hat{\Sigma}_{22}\big(\{m\}, \{m'\}\big) \\
&= \frac{2}{36\binom{\Omega+2}{3}^2(\Omega+3)\binom{\Omega+m-1}{2}\binom{\Omega+m'-1}{2}} \\
&\quad \times \bigg[4\Omega^2(\Omega-1)\binom{\Omega+m+1}{2}\binom{\Omega+m'+1}{2} \\
&\quad + 4(\Omega+2)^2(\Omega+3)\binom{m}{2}\binom{m'}{2} \\
&\quad + 4\big(\Omega^2-1\big)(\Omega+3)(m-1)(\Omega+m) \\
&\quad \times \big(m'-1\big)\big(\Omega+m'\big)\bigg].
\end{aligned} \tag{11.48}$$

For $m = m'$, it can be verified that Eq. (11.48) reduces to

$$\hat{\Sigma}_{22}\big(\{m\},\{m\}\big) = \frac{2}{(\Omega_m^B)^2}\sum_{\nu=0}^{2}\frac{[\Lambda_B^\nu(\Omega,m,m-2)]^2 d^B(g_\nu)}{[\Lambda_B^0(\Omega,m,2)]^2} \tag{11.49}$$

as expected from Sect. 11.2; Eq. (11.49) agrees with the result given for BEGUE(k) in [5]. Finally, it is useful to mention that in the $m \longrightarrow \infty$ and N finite limit we have,

$$\begin{aligned}\hat{\Sigma}_{11}\big(\{m\},\{m\}\big) &= \frac{2}{\Omega(\Omega+1)},\\ \hat{\Sigma}_{22}\big(\{m\},\{m\}\big) &= 8\frac{\Omega^2(\Omega-1)+(\Omega+2)^2(\Omega+3)+4(\Omega^2-1)(\Omega+3)}{\Omega^2(\Omega+1)^2(\Omega+2)^2(\Omega+3)}.\end{aligned} \tag{11.50}$$

Non-vanishing of $\hat{\Sigma}_{11}$ and $\hat{\Sigma}_{22}$ for finite N in the $m \longrightarrow \infty$ is interpreted in [5, 24] as non-ergodicity of BEGUE ensembles. See the discussion in Chap. 9 for the resolution of this problem.

In the next four sections we will consider specific $SU(r)$'s and present results that are appropriate for some physical systems.

11.4 Embedded Gaussian Unitary Ensemble for Fermions with Spin: EGUE(2)-$SU(2)$ with $r = 2$

Embedded Gaussian Unitary Ensemble for fermions with spin $\mathbf{s} = \frac{1}{2}$ degree of freedom corresponds to $r = 2$ in Sect. 11.3 and this ensemble, applicable to mesoscopic systems with mobile electrons carrying spin degree of freedom, is denoted by EGUE(2)-$SU(2)$ or EGUE(2)-**s**. For this ensemble, the $U(\Omega)$ irreps for m fermion systems with spin S are $f_m = \{2^p 1^q\}$ where $m = 2p + q$ and $S = q/2$. Formulas for $\langle H^2\rangle^{m,S}$ and the normalized bivariate moments $\hat{\Sigma}_{rr}(m,S:m',S')$, $r = 1,2$ of the two-point correlation function $S^{mS:m'S'}(E,E')$ follow from the formulation given in Sect. 11.3. It is easily seen that with $\langle S^2\rangle = S(S+1)$,

$$\begin{aligned}\overline{\langle H\rangle^{m,S}\langle H\rangle^{m',S'}} &= \sum_{f_2(s_2)}\frac{(\lambda_{f_2})^2}{d_\Omega(f_2)}P^{s_2}(m,S)P^{s_2}(m',S');\\ P^{s_2}(m,S) &= \big[(2s_2+1)m(m-4s_2+2)+4(2s_2-1)\langle S^2\rangle\big]/8, \quad s_2 = 0,1.\end{aligned} \tag{11.51}$$

To proceed further we need X_{UU} and Y_{UU}. The f_{m-2} irreps obtained by removing $\{2\}$ or $\{1^2\}$ from f_m follow from Fig. 11.3. Note that all three choices (i)–(iii) shown in the figure will apply for $\{1^2\}$ and only (i) will apply to $\{2\}$. Using the formulas in Table 11.1, the final formula for $\overline{\langle H^2\rangle^{(m,S)}}$, in terms of $m^x = (\Omega - \frac{m}{2})$ is

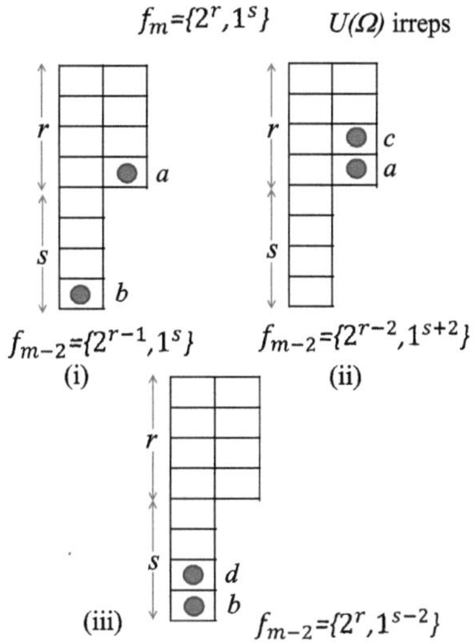

Fig. 11.3 Young tableaux denoting the two-column $SU(\Omega)$ irreps $f_m = \{2^r 1^s\}$ appropriate for EGUE(2)-$SU(2)$. Removal of two boxes generating $m-2$ particle irreps f_{m-2} are also shown in the figure. For **(i)** both the irreps $f_2 = \{2\}$ and $\{1^2\}$ will apply while for **(ii)** and **(iii)** only $\{1^2\}$ will apply (Color figure online)

$$
\begin{aligned}
\overline{\langle H^2\rangle^{m,S}} &= \sum_{f_2} \frac{(\lambda_{f_2})^2}{d(f_2)} \sum_{\nu=0,1,2} \mathscr{Q}^\nu(f_2 : m, S); \\
\mathscr{Q}^0\big(\{2\} : m, S\big) &= \big[P^0(m, S)\big]^2, \\
\mathscr{Q}^1\big(\{2\} : m, S\big) &= \big[(\Omega+1)P^0(m, S)/2\big]\big[m^x(m+2)/2 + \langle S^2\rangle\big], \\
\mathscr{Q}^2\big(\{2\} : m, S\big) &= \big[\Omega(\Omega+3)P^0(m, S)/4\big]\big[m^x(m^x+1) - \langle S^2\rangle\big], \\
\mathscr{Q}^0\big(\{1^2\} : m, S\big) &= \big[P^1(m, S)\big]^2, \\
\mathscr{Q}^1\big(\{1^2\} : m, S\big) &= \frac{(\Omega-1)}{16(\Omega-2)}\big[8(\Omega+2)P^1(m, S)P^2(m, S) \\
&\quad + 8\Omega(m-1)(\Omega-2m+4)\langle S^2\rangle\big], \\
\mathscr{Q}^2\big(\{1^2\} : m, S\big) &= \frac{\Omega}{8(\Omega-2)}\big[(3\Omega^2-7\Omega+6)\big(\langle S^2\rangle\big)^2 \\
&\quad + 3m(m-2)m^x\big(m^x-1\big)(\Omega+1)(\Omega+2)/4 \\
&\quad + \langle S^2\rangle\big\{-mm^x(5\Omega-3)(\Omega+2) \\
&\quad + \Omega(\Omega-1)(\Omega+1)(\Omega+6)\big\}\big]; \\
P^2(m, S) &= 3m^x(m-2)/2 - \langle S^2\rangle.
\end{aligned} \tag{11.52}
$$

Further, Eqs. (11.51) and (11.52) will give $\hat{\Sigma}_{11}$ for any (m, S, m', S', Ω). For $\hat{\Sigma}_{22}$ the only unknowns are $\mathscr{R}^\nu$ and they are given by

$$
\begin{aligned}
\mathscr{R}^0\big(\{2\}\{1^2\} : mS\big) &= P^0(m, S)P^1(m, S), \\
\mathscr{R}^1\big(\{2\}\{1^2\} : mS\big) &= -\frac{1}{2}\sqrt{\frac{(\Omega^2-1)(\Omega+2)}{(\Omega-2)}}\,P^0(m, S)P^2(m, S).
\end{aligned} \tag{11.53}
$$

Finally, let us consider the excess parameter $\overline{\gamma_2(m,S)} = \{\overline{\langle H^4\rangle^{m,S}}/[\overline{\langle H^2\rangle^{m,S}}]^2\} - 3$ and this is the most important (as $\overline{\langle H^3\rangle^{m,S}} = 0$) lower order shape parameter for fixed-(m,S) density of states $\overline{\rho^{m,S}(E)}$. General expression, derived using $SU(\Omega)$ algebra given in [11], for the fourth moment $\overline{\langle H^4\rangle^{m,S}}$ in terms of U-coefficients involves the multiplicity labels ρ's. However, for the physically interesting situation with $S=0$ (i.e. $f_m = \{2^r\}, r = m/2$), all the multiplicity labels will be unity and then $\overline{\gamma_2(m,S=0)}$ is given by [6],

$$
\begin{aligned}
\left[\overline{\gamma_2(m,S=0)}+1\right] &= \left[\overline{\langle H^2\rangle^{m,S=0}}\right]^{-2} \sum_{f_2^a,f_2^b} \frac{(\lambda_{f_2^a})^2(\lambda_{f_2^b})^2}{d_\Omega(f_2^a)d_\Omega(f_2^b)} \\
&\quad \times \sum_{\nu_1,\nu_2} \frac{d_\Omega(f_m)}{\sqrt{d_\Omega(F_{\nu_1})d_\Omega(F_{\nu_2})}} \left|\langle f_m \,|||\, B\left(f_2^a F_{\nu_1}\right) |||\, f_m\rangle\right|^2 \\
&\quad \times \left|\langle f_m \,|||\, B\left(f_2^b F_{\nu_2}\right) |||\, f_m\rangle\right|^2 U(f_m \overline{f_m} f_m f_m; F_{\nu_1} F_{\nu_2}).
\end{aligned}
\tag{11.54}
$$

In Eq. (11.54), $f_2^a = \{2\}, \{1^2\}$ and similarly f_2^b. This expression is pleasing and it is possible to obtain the triple barred coefficients using the tables in [23] and Eq. (11.31). But still we need $U(f_m \overline{f_m} f_m f_m; F_{\nu_1} F_{\nu_2})$ coefficient and deriving a formula for this needs further advances in $SU(N)$ Racah algebra. Thus, our present knowledge of $SU(N)$ Wigner-Racah algebra will not allow us to go too far in analytically solving EGUE(2)-**s** and even the simpler EGUE(2).

11.5 Embedded Gaussian Unitary Ensemble for Fermions with Wigner's Spin-Isospin $SU(4)$ Symmetry: EGUE(2)-$SU(4)$ with $r = 4$

Wigner introduced in 1937 [17] the spin-isospin $SU(4)$ supermultiplet scheme for atomic nuclei. There is good evidence for the goodness of this symmetry in some parts of the periodic table [25] and also more recently there is new interest in $SU(4)$ symmetry for heavy $N \sim Z$ nuclei [18, 19]. Therefore it is clearly of importance to study embedded Gaussian unitary ensemble of random matrices generated by random two-body interactions with $SU(4)$ symmetry and this corresponds to EGUE(2)-$SU(4)$ with $r = 4$ in Sect. 11.3. Before giving some analytical results for EGUE(2)-$SU(4)$, we will first turn to a brief discussion of the $SU(4)$ algebra.

Let us consider a system with m nucleons distributed in Ω number of orbits each with spin ($\mathbf{s} = \frac{1}{2}$) and isospin ($\mathbf{t} = \frac{1}{2}$) degrees of freedom. Then the total number of sp states is $N = 4\Omega$ and the spectrum generating algebra is $U(4\Omega)$. The sp states in uncoupled representation are $a^\dagger_{i,\alpha}|0\rangle = |i,\alpha\rangle$ with $i = 1, 2, \ldots, \Omega$ denoting the spatial orbits and $\alpha = 1, 2, 3, 4$ are the four spin-isospin states $|m_{\mathbf{s}}, m_{\mathbf{t}}\rangle = |\frac{1}{2}, \frac{1}{2}\rangle$, $|\frac{1}{2}, -\frac{1}{2}\rangle$, $|-\frac{1}{2}, \frac{1}{2}\rangle$ and $|-\frac{1}{2}, -\frac{1}{2}\rangle$ respectively. The $(4\Omega)^2$ number of operators $C_{i\alpha;j\beta}$

generate $U(4\Omega)$ algebra. For m fermions, all states belong to the $U(4\Omega)$ irrep $\{1^m\}$. In uncoupled notation, $C_{i\alpha;j\beta} = a^\dagger_{i,\alpha} a_{j,\beta}$. Similarly $U(\Omega)$ and $U(4)$ algebras are generated by A_{ij} and $B_{\alpha\beta}$ respectively, where $A_{ij} = \sum_{\alpha=1}^{4} C_{i\alpha;j\alpha}$ and $B_{\alpha\beta} = \sum_{i=1}^{\Omega} C_{i\alpha;i\beta}$. The number operator $\hat{n}$, the spin operator $\hat{S} = S^1_\mu$, the isospin operator $\hat{T} = T^1_\mu$ and the Gamow-Teller operator $\sigma\tau = (\sigma\tau)^{1,1}_{\mu,\mu'}$ of $U(4)$ in spin-isospin coupled notation are [26],

$$
\begin{aligned}
\hat{n} &= 2\sum_i \mathscr{A}^{0,0}_{ii;0,0}, \qquad S^1_\mu = \sum_i \mathscr{A}^{1,0}_{ii;\mu,0}, \qquad T^1_\mu = \sum_i \mathscr{A}^{0,1}_{ii;0,\mu}, \\
(\sigma\tau)^{1,1}_{\mu,\mu'} &= \sum_i \mathscr{A}^{1,1}_{ii;\mu,\mu'}; \qquad \mathscr{A}^{s,t}_{ij;\mu_s,\mu_t} = (a^\dagger_i \tilde{a}_j)^{s,t}_{\mu_s,\mu_t}.
\end{aligned}
\tag{11.55}
$$

Note that $\tilde{a}_{j;\mu_s,\mu_t} = (-1)^{1+\mu_s+\mu_t} a_{j;-\mu_s,-\mu_t}$. These 16 operators form $U(4)$ algebra. Dropping the number operator, we have $SU(4)$ algebra. For the $U(4)$ algebra, the irreps are characterized by the partitions $\{F\} = \{F_1, F_2, F_3, F_4\}$ with $F_1 \geq F_2 \geq F_3 \geq F_4 \geq 0$ and $m = \sum_{i=1}^{4} F_i$. Note that F_α are the eigenvalues of $B_{\alpha\alpha}$. Due to the antisymmetry constraint on the total wavefunction, the $U(\Omega)$ irrep $\{f\} = \{\widetilde{F}\}$ which is obtained by changing rows to columns in $\{F\}$; note that $F_i \leq \Omega$ and $f_i \leq 4$. Before proceeding further, let us examine the quadratic Casimir invariants of $U(\Omega)$, $U(4)$ and $SU(4)$ algebras. For example,

$$
\begin{aligned}
C_2\big[U(\Omega)\big] &= \sum_{i,j} A_{ij}A_{ji} = \hat{n}\Omega - \sum_{i,j,\alpha,\beta} a^\dagger_{i,\alpha} a^\dagger_{j,\beta} a_{j,\alpha} a_{i,\beta}, \\
C_2\big[U(4)\big] &= \sum_{\alpha,\beta} B_{\alpha,\beta}B_{\beta,\alpha} \quad \Rightarrow \quad C_2\big[U(\Omega)\big] + C_2\big[U(4)\big] = \hat{n}(\Omega + 4).
\end{aligned}
\tag{11.56}
$$

Also, in terms of spin, isospin and Gamow-Teller operators, $C_2[SU(4)] = S^2 + T^2 + (\sigma\tau)\cdot(\sigma\tau)$ and

$$
\big\langle C_2\big[U(4)\big]\big\rangle^{\{F\}} = \sum_{i=1}^{4} F_i(F_i + 5 - 2i) = \left\langle C_2\big[SU(4)\big] + \frac{\hat{n}^2}{4}\right\rangle^{\{F\}}. \tag{11.57}
$$

The space exchange or Majorana operator $\widetilde{M}$ that exchanges the spatial coordinates of the particles (the index i) and leaves the spin-isospin quantum numbers unchanged allow us to understand the significance of $SU(4)$ symmetry,

$$
\widetilde{M}\big|i,\alpha,\alpha'; j,\beta,\beta'\big\rangle = \big|j,\alpha,\alpha'; i,\beta,\beta'\big\rangle, \tag{11.58}
$$

where α, β are labels for spin and α', β' are labels for isospin. As $|i,\alpha,\alpha'; j,\beta,\beta'\rangle = a^\dagger_{i,\alpha,\alpha'} a^\dagger_{j,\beta,\beta'}|0\rangle$, Eqs. (11.58), (11.56) and (11.57) in that order will give,

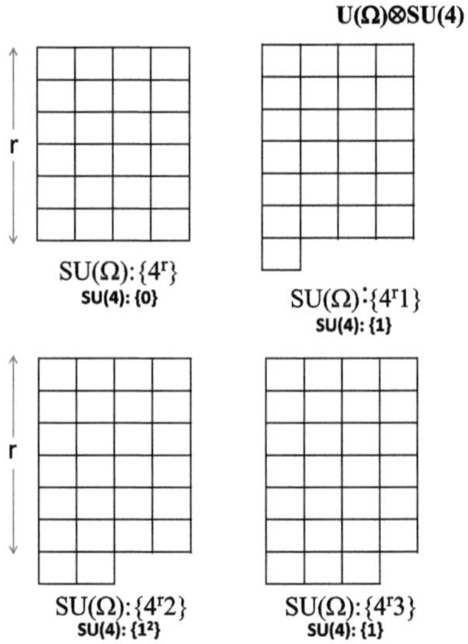

Fig. 11.4 Young tableaux denoting the special $SU(\Omega)$ irreps $f_m^{(p)} = \{4^r, p\}$, $p = 0, 1, 2, 3$ considered in EGUE(2)-$SU(4)$ analysis with $U(\Omega) \otimes SU(4)$ embedding algebra. The corresponding $SU(4)$ irreps are also given in the figure (Color figure online)

Table 11.2 $P^{f_2}(m, f_m)$ for $f_m = \{4^r, p\}$; $p = 0, 1, 2$ and 3 and $\{f_2\} = \{2\}, \{1^2\}$

f_m	$P^{f_2}(m, f_m)$	
	$f_2 = \{2\}$	$f_2 = \{1^2\}$
$\{4^r\}$	$-3r(r+1)$	$-5r(r-1)$
$\{4^r, 1\}$	$-\frac{3r}{2}(2r+3)$	$-\frac{5r}{2}(2r-1)$
$\{4^r, 2\}$	$-(3r^2+6r+1)$	$-5r^2$
$\{4^r, 3\}$	$-\frac{3}{2}(r+2)(2r+1)$	$-\frac{5r}{2}(2r+1)$

$$
\begin{aligned}
2\kappa \widetilde{M} &= 2\kappa \sum_{i,j,\alpha,\beta,\alpha',\beta'} \left(a^\dagger_{j,\alpha,\alpha'} a^\dagger_{i,\beta,\beta'}\right)\left(a^\dagger_{i,\alpha,\alpha'} a^\dagger_{j,\beta,\beta'}\right)^\dagger \\
&= \kappa\left\{C_2\left[U(\Omega)\right] - \Omega\hat{n} = 4\hat{n} - C_2\left[U(4)\right]\right\} \\
&= 2\kappa\left\{2\hat{n}\left(1 - \frac{\hat{n}}{16}\right) - \frac{1}{2}C_2\left[SU(4)\right]\right\}.
\end{aligned}
\tag{11.59}
$$

The preferred $U(\Omega)$ irrep for the ground state of a m nucleon system is the most symmetric one. Therefore $\langle C_2[U(\Omega)]\rangle$ should be maximum for the ground state irrep. This implies, as seen from Eq. (11.59), the strength κ of $\widetilde{M}$ must be negative. As a consequence, as follows from the last equality in Eq. (11.59), the ground states are labeled by $SU(4)$ irreps with smallest eigenvalue for the quadratic Casimir invariant consistent with a given (m, T_z), $T = |T_z|$. Therefore, for $N = Z$ even-even, $N = Z$ odd-odd and $N = Z \pm 1$ odd-A nuclei the $U(\Omega)$ irreps for the gs are $\{4^r\}$, $\{4^r, 2\}$, $\{4^r, 1\}$ and $\{4^r, 3\}$ with spin-isospin structure being $(0, 0)$, $(1, 0) \oplus (0, 1)$, $(\frac{1}{2}, \frac{1}{2})$, and $(\frac{1}{2}, \frac{1}{2})$ respectively. For convenience, the gs $U(\Omega)$ irreps are denoted by

Table 11.3 $\overline{\langle H^2\rangle^{m,f_m}}$, $\mathscr{Q}^{\nu=1,2}(f_2:m,f_m)$ and $\mathscr{R}^{\nu=1}(m,f_m)$ for some examples

f_m	$\overline{\langle H^2\rangle^{m,f_m}}$
$\{4^r\}$	$\frac{r(\Omega-r+4)}{2}[\lambda^2_{\{2\}}3(r+1)(\Omega-r+3)+\lambda^2_{\{1^2\}}5(r-1)(\Omega-r+5)]$
$\{4^r,1\}$	$\frac{r(\Omega-r+4)}{4}[\lambda^2_{\{2\}}\{6r(\Omega-r+1)+9\Omega+15\}$ $+\lambda^2_{\{1^2\}}5\{2r(\Omega-r+5)-\Omega-9\}]$
$\{4^r,2\}$	$\lambda^2_{\{2\}}\frac{1}{2}[3r^4-6(\Omega+2)r^3+(3\Omega^2+6\Omega-5)r^2$ $+(\Omega+2)(6\Omega+17)r+\Omega(\Omega+1)]$ $+\lambda^2_{\{1^2\}}\frac{5r}{2}(\Omega-r+4)\{(\Omega+4)r-r^2-3\}$
$\{4^r,3\}$	$\frac{1}{4}[\lambda^2_{\{2\}}3(r+2)(\Omega-r+2)(2r\Omega-2r^2+6r+\Omega+1)$ $+\lambda^2_{\{1^2\}}5r(\Omega-r+4)(2r\Omega-2r^2+6r+\Omega-1)]$

f_m	f_2	ν	$\mathscr{Q}^{\nu}(f_2:m,f_m)$
$\{4^r\}$	$\{2\}$	1	$\frac{9r(r+1)^2(\Omega-r)(\Omega+1)(\Omega+4)}{2(\Omega+2)}$
		2	$\frac{3r\Omega(r+1)(\Omega-r+1)(\Omega-r)(\Omega+4)(\Omega+5)}{4(\Omega+2)}$
	$\{1^2\}$	1	$\frac{25r(r-1)^2(\Omega-r)(\Omega-1)(\Omega+4)}{2(\Omega-2)}$
		2	$\frac{5r\Omega(r-1)(\Omega+3)(\Omega+4)(\Omega-r)(\Omega-r-1)}{4(\Omega-2)}$

f_m	$\mathscr{R}^{\nu=1}(m,f_m)$
$\{4^r\}$	$-\frac{15r}{2}\sqrt{\frac{\Omega^2-1}{\Omega^2-4}}(r^2-1)(\Omega-r)(\Omega+4)$

$f_m^{(p)}$ where

$$f_m^{(p)}=\{4^r,p\};\quad m=4r+p \text{ and } p=\operatorname{mod}(m,4). \tag{11.60}$$

For the special $SU(\Omega)$ irreps in Eq. (11.60), and shown in Fig. 11.4, analytical formulas are much simpler than for a general $SU(\Omega)$ irrep [7].

The formalism given in Sect. 11.3 was applied in detail in [7]. For example, formulas for $P^{f_2}(m,f_m)$ are given in Table 11.2 for $\{f_m^{(p)}\}$ irreps. Evaluating all the $\mathscr{Q}$'s as given in detail in [7], analytical formulas for $\mathscr{Q}^{\nu}(f_2:m,f_m)$ and also for $\overline{\langle H^2\rangle^{m,f_m}}$ are obtained for $\{f_m^{(p)}\}$ irreps. Some of these results are given in Table 11.3. Equations (11.34)–(11.36) and Tables 4 and 7 of [7] will allow us to calculate covariances $\hat{\Sigma}_{11}$ in energy centroids for any irrep. On the other hand, the results in Tables 11.2 and 11.3 will give formulas for $\hat{\Sigma}_{11}$ for $\{f_m^{(p)}\}$ irreps. Similarly, the $\mathscr{R}$ formula given in Table 11.3 will us to calculate $\hat{\Sigma}_{22}$ for the irrep $\{4^r\}$. Note that, $\mathscr{Q}^{\nu=0}(f_2:m,f_m)=[P^{f_2}(m,f_m)]^2$ and $\mathscr{R}^{\nu=0}(m,f_m)=P^{\{2\}}(m,f_m)P^{\{1^2\}}(m,f_m)$.

11.6 Embedded Gaussian Unitary Ensemble for Bosons with F-Spin: BEGUE(2)-$SU(2)$ with $r = 2$

For two species boson systems with F-spin, following the discussion in Chap. 10, we have BEGUE(2)-$SU(2)$ or BEGUE(2)-F. For this ensemble, results in Sect. 11.3 with $r = 2$ will be applicable. For such a m boson system, the $SU(\Omega)$ irreps will be two rowed denoted by $f_m = \{m - r, r\}$ with $F = \frac{m}{2} - r$. With this, there are three allowed f_{m-2} irreps as shown in Fig. 11.5. The irreps in (i) and (iii) in the figure can be obtained by removing $f_2 = \{2\}$ from f_m. However for (ii) in the figure both $\{2\}$ and $\{1^2\}$ will apply. For $f_{m-2} = \{m - r - 2, r\}$ irrep [this corresponds to (i) in Fig. 11.5] we have

$$\begin{aligned}
\tau_{a2} &= m - 2r + 1, \\
\tau_{ai} &= m - r + i - 1; \quad i = 3, 4, \ldots, \Omega, \\
\Pi'_a &= \frac{(m - 2r)(m - r + 1)}{(m - 2r + 1)(m - r + \Omega - 1)}, \\
\Pi''_a &= \frac{(m - 2r - 1)(m - r)(m - r + 1)}{(m - 2r + 1)(m - r + \Omega - 1)(m - r + \Omega - 2)}.
\end{aligned} \tag{11.61}$$

Similarly for $f_{m-2} = \{m - r, r - 2\}$ irrep [this corresponds to (iii) in Fig. 11.5] we have

$$\begin{aligned}
\tau_{b1} &= 2r - m - 1, \\
\tau_{bi} &= r + i - 2, \quad i = 3, 4, \ldots, \Omega \\
\Pi'_a &= \frac{(r)(2r - m - 2)}{(2r - m - 1)(r + \Omega - 2)}, \\
\Pi''_a &= \frac{(2r - m - 3)(r)(r - 1)}{(2r - m - 1)(r + \Omega - 2)(r + \Omega - 3)}.
\end{aligned} \tag{11.62}$$

Finally, for $f_{m-2} = \{m - r - 1, r - 1\}$ irrep [this corresponds to (ii) in Fig. 11.5] we have

$$\begin{aligned}
\tau_{ab} &= m - 2r + 1 = 2F + 1, \\
\tau_{ai} &= m - r + i - 1, \qquad \tau_{bi} = r + i - 2; \quad i = 3, 4, \ldots, \Omega, \\
\Pi_a^{(b)} &= \frac{(m - r + 1)}{(m - r + \Omega - 1)}, \\
\Pi_b^{(a)} &= \frac{(r)}{(r + \Omega - 2)}.
\end{aligned} \tag{11.63}$$

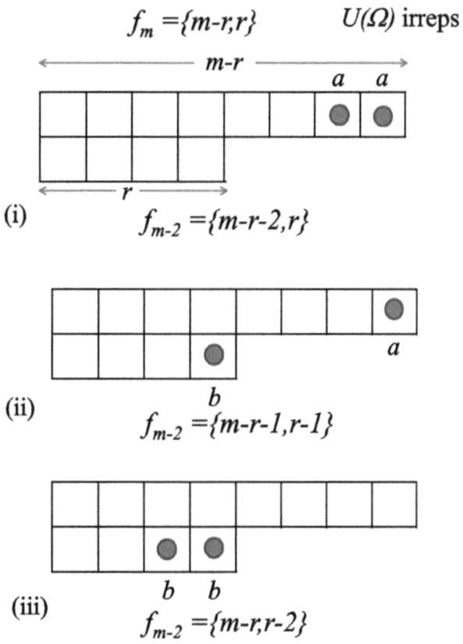

Fig. 11.5 Young tableaux denoting the two-rowed $SU(\Omega)$ irreps $f_m = \{m-r, r\}$ appropriate for BEGUE(2)-$SU(2)$. Removal of two boxes generating $m-2$ particle irreps f_{m-2} are also shown in the figure. For **(ii)** both the irreps $f_2 = \{2\}$ and $\{1^2\}$ will apply and for **(i)** and **(iii)** only $\{2\}$ will apply. Figure is taken from [20] with permission from American Institute of Physics (Color figure online)

These and $\mathscr{N}_{f_{m-2}}/\mathscr{N}_{f_m}$ will give the formulas for the lower order moments of one and two point functions as described in Sect. 11.3. The dimension ratios are,

$$\begin{aligned}\frac{\mathscr{N}_{\{m-r-2,r\}}}{\mathscr{N}_{\{m-r,r\}}} &= \frac{(m-r)(m-r+1)(m-2r-1)}{m(m-1)(m-2r+1)},\\ \frac{\mathscr{N}_{\{m-r-1,r-1\}}}{\mathscr{N}_{\{m-r,r\}}} &= \frac{r(m-r+1)}{m(m-1)},\\ \frac{\mathscr{N}_{\{m-r,r-2\}}}{\mathscr{N}_{\{m-r,r\}}} &= \frac{r(r-1)(m-2r+3)}{m(m-1)(m-2r+1)}.\end{aligned} \tag{11.64}$$

Using Eqs. (11.61)–(11.64) and the expressions in Table 11.1, it is possible to derive analytical formulas for the P's, $\mathscr{Q}$'s and $\mathscr{R}$'s that define $\langle H^2\rangle$, $\hat{\Sigma}_{11}$ and $\hat{\Sigma}_{22}$. The final formulas (obtained in [20] using MATHEMATICA) are, with (m, F) defining f_m,

$$\begin{aligned}P^{\{2\}}(m,F) &= \frac{1}{8}\big[3m(m-2)+4F(F+1)\big],\\ P^{\{1^2\}}(m,F) &= \frac{1}{8}\big[m(m+2)-4F(F+1)\big],\\ \mathscr{Q}^{\nu=0}\big(\{2\}:m,F\big) &= \big[P^{\{2\}}(m,F)\big]^2,\\ \mathscr{Q}^{\nu=0}\big(\big\{1^2\big\}:m,F\big) &= \big[P^{\{1^2\}}(m,F)\big]^2,\\ \mathscr{Q}^{\nu=1}\big(\{2\}:m,F\big) &= \frac{(\Omega+1)}{16(\Omega+2)}\\ &\quad\times\big[2(\Omega-2)P^{\{2\}}(m,F)\big\{3(2\Omega+m)(m-2)+4F(F+1)\big\}\\ &\quad+8\Omega(m-1)(\Omega+2m-4)F(F+1)\big],\end{aligned}$$

$$
\begin{aligned}
\mathcal{Q}^{\nu=1}\big(\{1^2\}:m,F\big) &= \frac{(\Omega-1)P^{\{1^2\}}(m,F)}{8}\big[(2\Omega+m)(m+2)-4F(F+1)\big],\\
\mathcal{Q}^{\nu=2}\big(\{2\}:m,F\big) &= \frac{(\Omega)}{8(\Omega+2)}\Big[(3\Omega^2+7\Omega+6)\big[F(F+1)\big]^2\\
&\quad +\frac{3}{16}m(m-2)(2\Omega+m)(2\Omega+m+2)(\Omega-1)(\Omega-2)\\
&\quad +\frac{F(F+1)}{2}\big\{m(2\Omega+m)(5\Omega+3)(\Omega-2)\\
&\quad +2\Omega\big(\Omega^2-1\big)(\Omega-6)\big\}\Big],\\
\mathcal{Q}^{\nu=2}\big(\{1^2\}:m,F\big) &= \frac{\Omega(\Omega-3)P^{\{1^2\}}(m,F)}{16}\\
&\quad\times\big[(2\Omega+m)(2\Omega+m-2)-4F(F+1)\big],\\
\mathcal{R}^{\nu=0}(m,F) &= P^{\{2\}}(m,F)P^{\{1^2\}}(m,F),\\
\mathcal{R}^{\nu=1}(m,F) &= \sqrt{\frac{\Omega^2-1}{\Omega^2-4}}\frac{(2-\Omega)P^{\{1^2\}}(m,F)}{8}\big\{4\big[F(F+1)-3\Omega\big]\\
&\quad +3m(2\Omega+m-2)\big\}.
\end{aligned}
\tag{11.65}
$$

Note that Eq. (11.65) is closely related to the BEGOE(2)-F results given by Eq. (10.7). More importantly, they are related to the EGUE(2)-$SU(2)$ results by $\Omega \to -\Omega$ transformation.

11.7 Embedded Gaussian Unitary Ensemble for Spin One Bosons: BEGUE(2)-*SU*(3) with $r = 3$

Spin one boson systems, as discussed in Chap. 10, posses $U(3\Omega) \supset U(\Omega) \otimes [SU(3) \supset SO(3)]$ symmetry. For these systems, it is possible to consider interactions preserving the $SU(3)$ symmetry. This gives, for the GUE version, BEGUE(2)-$SU(3)$ that corresponds to $r = 3$ in Sect. 11.3. As $U(3)$ irreps will have, in Young tableaux representation, maximum 3 rows, the $U(\Omega)$ irrep also will have maximum three rows. Given m bosons in Ω number of sp levels, the allowed $U(\Omega)$ irreps are $\{f_1, f_2, f_3, f_4, \ldots, f_\Omega\} = \{f_1, f_2, f_3\}$ with $f_1 + f_2 + f_3 = m$, $f_1 \ge f_2 \ge f_3 \ge 0$ and $f_i = 0$ for $i = 4, 5, \ldots, \Omega$. For $f_2 = 0$ and $f_3 = 0$, we have totally symmetric irreps with $\{f_1\} = \{m\}$ and for these irreps all the results derived in Sect. 11.3.1 will apply directly. Similarly, for $f_2 \ne 0$ and $f_3 = 0$, all the results of Sect. 11.6 will apply. Thus, the non-trivial irreps for BEGUE(2)-$SU(3)$ are the m-boson irreps $f_m = \{f_1, f_2, f_3\}$ with $f_3 \ne 0$. Given a f_m, in general there will be six f_{m-2}

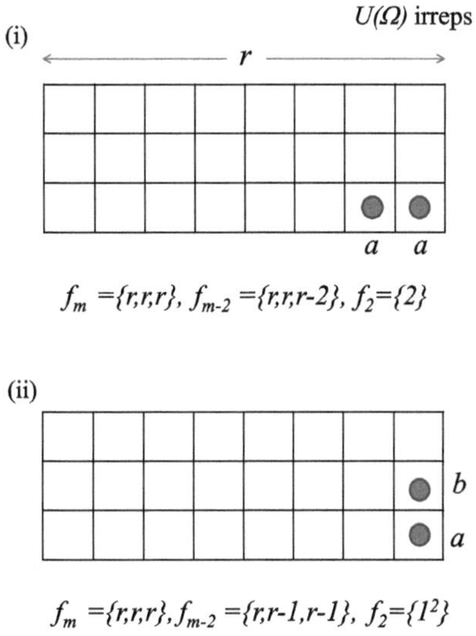

Fig. 11.6 Young tableaux denoting the three-rowed $SU(\Omega)$ irreps $f_m = \{r, r, r\}$, $m = 3r$ appropriate for BEGUE(2)-$SU(3)$. Removal of two boxes generating $m-2$ particle irreps f_{m-2} are also shown in the figure. For **(i)** only the irrep $f_2 = \{2\}$ will apply while for **(ii)** only $\{1^2\}$ will apply. Figure is taken from [20] with permission from American Institute of Physics (Color figure online)

and they are $\{f_1 - 2, f_2, f_3\}$, $\{f_1, f_2 - 2, f_3\}$, $\{f_1, f_2, f_3 - 2\}$, $\{f_1 - 1, f_2 - 1, f_3\}$, $\{f_1 - 1, f_2, f_3 - 1\}$, $\{f_1, f_2 - 1, f_3 - 1\}$. Therefore, as seen from Sect. 11.3, deriving analytical formulas for P's, $\mathscr{Q}$'s and $\mathscr{R}$'s that determine $\langle H^2\rangle$, $\hat{\Sigma}_{11}$ and $\hat{\Sigma}_{22}$ will be cumbersome. One situation that is amenable to analytical treatment is for the irreps $\{n + p, n, n\}$ where $m = 3n + p$ with $p = 0$, 1 and 2 [these are similar to the $\{4^r, p\}$ irreps considered for EGUE(2)-$SU(4)$]. Here we will present the results for $p = 0$ and for others see [20]. For this class of irreps, the f_{m-2} are simple as shown in Fig. 11.6. For $f_{m-2} = \{n, n, n-2\}$, Π'_a and Π''_a are needed and they are given by,

$$\Pi'_a = \frac{3n}{\Omega + n - 3}, \qquad \Pi''_a = \frac{6n(n-1)}{(\Omega + n - 3)(\Omega + n - 4)}. \tag{11.66}$$

Similarly, for $f_{m-2} = f_{n,n-1,n-1}$ we need τ_{ab}, $\Pi_a^{(b)}$ and $\Pi_b^{(a)}$ and they are,

$$\tau_{ab} = -1, \qquad \Pi_a^{(b)} = \frac{3n}{2(\Omega + n - 3)}, \qquad \Pi_b^{(a)} = \frac{2(n+1)}{(\Omega + n - 2)}. \tag{11.67}$$

In addition, ratio of the S_Ω dimensions needed are,

$$\frac{\mathscr{N}_{n,n,n-2}}{\mathscr{N}_{n,n,n}} = \frac{2(n-1)}{(3n-1)}, \qquad \frac{\mathscr{N}_{n,n-1,n-1}}{\mathscr{N}_{n,n,n}} = \frac{n+1}{(3n-1)}. \tag{11.68}$$

With these, carrying out simplification of the formulas given in Table 11.1 will give the following formulas (with $\pi = 1$ for $\{2\}$ and -1 for $\{1^2\}$),

$$
\begin{aligned}
&P^{f_2}\big(m,\{n,n,n\}\big) = -\frac{6}{3-\pi}n(n-\pi),\\
&\mathscr{Q}^{\nu=0}\big(f_2:m,\{n,n,n\}\big) = \big[P^{f_2}\big(m,\{n,n,n\}\big)\big]^2,\\
&\mathscr{Q}^{\nu=1}\big(f_2:m,\{n,n,n\}\big) = \frac{3(3+\pi)^2(\Omega+\pi)(\Omega-3)n(n-\pi)^2(\Omega+n)}{8(\Omega+2\pi)},\\
&\mathscr{Q}^{\nu=2}\big(f_2:m,\{n,n,n\}\big)\\
&\quad= \frac{3(3+\pi)\Omega(\Omega-3+\pi)(\Omega-3)n(n-\pi)(\Omega+n)(\Omega+n+\pi)}{16(\Omega+2\pi)},\\
&\mathscr{R}^{\nu=0}\big(m,\{n,n,n\}\big) = P^{\{2\}}\big(m,\{n,n,n\}\big)P^{\{1^2\}}\big(m,\{n,n,n\}\big),\\
&\mathscr{R}^{\nu=1}\big(m,\{n,n,n\}\big) = -\sqrt{\frac{\Omega^2-1}{\Omega^2-4}}\,3(\Omega-3)n\big(n^2-1\big)(\Omega+n).
\end{aligned}
\tag{11.69}
$$

Using these equations one can calculate the variances $\langle H^2\rangle$ and the covariances $\hat{\Sigma}_{11}$ and $\hat{\Sigma}_{22}$ for irreps of the type $\{n,n,n\}$. For example, Eq. (11.35) can be simplified to give a compact formula for spectral variances,

$$
\begin{aligned}
\big\langle H^2\big\rangle^{m,\{n,n,n\}} &= \lambda_{\{2\}}^2\left[\frac{3}{2}n(n-1)(\Omega+n-3)(\Omega+n-4)\right]\\
&\quad+\lambda_{\{1^2\}}^2\left[\frac{3}{4}n(n+1)(\Omega+n-2)(\Omega+n-3)\right].
\end{aligned}
\tag{11.70}
$$

Using the tables in [7] and the results in Sect. 11.3, one can calculate numerically $\hat{\Sigma}_{11}$ and $\hat{\Sigma}_{22}$ for any f_m. Applications of this will be discussed in Chap. 12.

References

1. F.J. Dyson, A class of matrix ensembles. J. Math. Phys. **13**, 90–97 (1972)
2. V.K.B. Kota, SU(N) Wigner-Racah algebra for the matrix of second moments of embedded Gaussian unitary ensemble of random matrices. J. Math. Phys. **46**, 033514 (2005)
3. L. Benet, T. Rupp, H.A. Weidenmüller, Spectral properties of the k-body embedded Gaussian ensembles of random matrices. Ann. Phys. **292**, 67–94 (2001)
4. Z. Pluhař, H.A. Weidenmüller, Symmetry properties of the k-body embedded unitary Gaussian ensemble of random matrices. Ann. Phys. (N.Y) **297**, 344–362 (2002)
5. T. Asaga, L. Benet, T. Rupp, H.A. Weidenmüller, Spectral properties of the k-body embedded Gaussian ensembles of random matrices for bosons. Ann. Phys. (N.Y.) **298**, 229–247 (2002)
6. V.K.B. Kota, $U(2\Omega)\supset U(\Omega)\otimes SU(2)$ Wigner-Racah algebra for embedded Gaussian unitary ensemble of random matrices with spin. J. Math. Phys. **48**, 053304 (2007)
7. M. Vyas, V.K.B. Kota, Spectral properties of embedded Gaussian unitary ensemble of random matrices with Wigner's $SU(4)$ symmetry. Ann. Phys. (N.Y.) **325**, 2451–2485 (2010)

8. M. Vyas, N.D. Chavda, V.K.B. Kota, V. Potbhare, One plus two-body random matrix ensembles for boson systems with F-spin: analysis using spectral variances. J. Phys. A, Math. Theor. **45**, 265203 (2012)
9. H. Deota, N.D. Chavda, V.K.B. Kota, V. Potbhare, M. Vyas, Random matrix ensemble with random two-body interactions in the presence of a mean-field for spin one boson systems. Phys. Rev. E **88**, 022130 (2013)
10. K.T. Hecht, A simple class of $U(N)$ Racah coefficients and their applications. Commun. Math. Phys. **41**, 135–156 (1975)
11. P.H. Butler, Coupling coefficients and tensor operators for chains of groups. Philos. Trans. R. Soc. Lond. **277**, 545–585 (1975)
12. P.H. Butler, *Point Group Symmetry Applications: Methods and Tables* (Plenum, New York, 1981)
13. V.K.B. Kota, Two-body ensembles with group symmetries for chaos and regular structures. Int. J. Mod. Phys. E **15**, 1869–1883 (2006)
14. V.K.B. Kota, V. Potbhare, Shape of the eigenvalue distribution for bosons in scalar space. Phys. Rev. C **21**, 2637–2642 (1980)
15. P. Cvitanovic, A.D. Kennedy, Spinors in negative dimensions. Phys. Scr. **26**, 5–14 (1982)
16. J.B. French, Isospin distributions in nuclei, in *Isospin in Nuclear Physics*, ed. by D.H. Wilkinson (North-Holland, Amsterdam, 1969), pp. 259–295
17. E.P. Wigner, On the consequences of the symmetry of the nuclear Hamiltonian on the spectroscopy of nuclei. Phys. Rev. **51**, 106–119 (1937)
18. P. Van Isacker, D.D. Warner, D.S. Brenner, Test of Wigner's spin-isospin symmetry from double binding energy differences. Phys. Rev. Lett. **74**, 4607–4610 (1995)
19. R.C. Nayak, V.K.B. Kota, SU(4) symmetry and Wigner energy in the infinite nuclear matter mass model. Phys. Rev. C **64**, 057303 (2001)
20. M. Vyas, V.K.B. Kota, Embedded Gaussian unitary ensembles with $U(\Omega) \otimes SU(r)$ embedding generated by random two-body interactions with $SU(r)$ symmetry. J. Math. Phys. **53**, 123303 (2012)
21. B.G. Wybourne, *Symmetry Principles and Atomic Spectroscopy* (Wiley, New York, 1970)
22. K.T. Hecht, J.P. Draayer, Spectral distributions and the breaking of isospin and supermultiplet symmetries in nuclei. Nucl. Phys. A **223**, 285–319 (1974)
23. K.T. Hecht, Summation relation for $U(N)$ Racah coefficients. J. Math. Phys. **15**, 2148–2156 (1974)
24. T. Asaga, L. Benet, T. Rupp, H.A. Weidenmüller, Non-ergodic behaviour of the k-body embedded Gaussian random ensembles for bosons. Europhys. Lett. **56**, 340–346 (2001)
25. J.C. Parikh, *Group Symmetries in Nuclear Structure* (Plenum, New York, 1978)
26. V.K.B. Kota, J.A. Castilho Alcarás, Classification of states in $SO(8)$ proton-neutron pairing model. Nucl. Phys. A **764**, 181–204 (2006)

Chapter 12
Symmetries, Self Correlations and Cross Correlations in Embedded Ensembles

Correlations between levels with different quantum numbers generated by EEs are very important as these *cross correlations* are absent in the description of levels of interacting particle systems if we use classical GOE or GUE or GSE ensembles. In the description using classical ensembles, one assumes independent GOE or GUE or GSE description for levels with different quantum numbers. As discussed already in Chaps. 4 and 9, self correlations, i.e. correlations between levels with same quantum numbers, are also important for EEs. In Sects. 12.1–12.3 results are presented for the correlations between matrix structure, symmetries and self and cross correlations in embedded ensembles using fermionic EGUE(2), EGUE(2)-**s** and EGUE(2)-$SU(4)$ ensembles. Similarly, Sect. 12.4 deals with bosonic BEGUE(2). Finally results for EGOE(2)-**s** and BEGOE(2)-F ensembles for self and cross correlations are presented in Sects. 12.5 and 12.6 respectively. It is important to emphasize that "cross correlations" is one of the very important new aspect of EE.

12.1 Matrix Structure for Fermionic EGUE(2)-$SU(r)$, $r = 1, 2, 4$

In order to understand the structure of EEs, here first we will consider the matrix structure of fermionic EGUE(2), EGUE(2)-$SU(2)$ [same as EGUE(2)-**s**] and EGUE(2)-$SU(4)$ matrices. Let us consider the example of 8 fermions in $N = 24$ sp states. Then one finds three distinct features and they are as follows:

(i) For spinless fermion systems, we have EGUE(2) with a two particle GUE of dimension 276 and the number of independent variables [denoted by $i_2(0)$] is 76, 176. These generate the m fermion EGUE(2) ensemble with H matrices of dimension $d_f(24, 8) = 7, 35, 471$. For fermions with spin symmetry, we have EGUE(2)-**s** with $\Omega = 12$. This ensemble is generated by independent GUEs in two particle spin $s = 0$ and $s = 1$ spaces with dimensions 78 and 66 respectively. Then the number of independent variables [denoted by $i_2(2)$] for this system is 10, 440. The H matrix dimensions for EGUE(2)-**s** ensembles for the 8 particle system with spins $S = 0, 1, 2, 3$ and 4 are $d_f(12, 8, S) = 70785$,

V.K.B. Kota, *Embedded Random Matrix Ensembles in Quantum Physics*, Lecture Notes in Physics 884, DOI 10.1007/978-3-319-04567-2_12,

113256, 51480, 9009 and 495 respectively. Going further, with $SU(4)$ symmetry we have EGUE(2)-$SU(4)$ ensembles with $\Omega = 6$. These ensembles are generated by two independent GUE's in $f_2 = \{2\}$ and $\{1^2\}$ spaces with dimensions 21 and 15 respectively. Then the number of independent variables [denoted by $i_2(4)$] for this system is 666. The H matrix dimensions for EGUE(2)-$SU(4)$ ensembles for the 8 particle system with $f_8 = \{2^2, 1^4\}$, $\{2^3, 1^2\}$, $\{2^4\}$, $\{3, 1^5\}$, $\{3^2, 1^3\}$, $\{3, 2^2, 1\}$, $\{3^2, 1^2\}$, $\{3^2, 2\}$, $\{4, 1^4\}$, $\{4^2, 1^2\}$, $\{4, 2^2\}$, $\{4, 3, 1\}$ and $\{4^2\}$ are 15, 105, 105, 21, 384, 1050, 1176, 1470, 315, 2430, 2520, 4410 and 1764 respectively. Thus i_2 will be considerably reduced as the symmetry increases (with fixed N), i.e. $i_2(4) \ll i_2(2) \ll i_2(0)$. Similarly the H matrix dimensions decrease as we go from EGUE(2) to EGUE(2)-**s** to EGUE(2)-$SU(4)$.

(ii) For further insight, let us consider the fraction of independent matrix elements $\mathscr{I}(m, f_m)$, for $m \gg 2$ for the EGUE(2)-$SU(4)$ ensemble, defined as the ratio of $i_2(4)$ to the total number (without counting the hermitian conjugates) of matrix elements,

$$\mathscr{I}(m, f_m) = \frac{i_2(4)}{[d_\Omega(f_m)]^2}. \tag{12.1}$$

Similarly, for EGUE(2) and EGUE(2)-**s** ensembles, we can define the fraction of independent matrix elements as $\mathscr{I}(m) = i_2(0)/[d_f(N, m)]^2$ and $\mathscr{I}(m, S) = i_2(2)/[d_f(\Omega, m, S)]^2$ respectively. In the above example, for EGUE(2), EGUE(2)-**s** with $S = 0$ and EGUE(2)-$SU(4)$ with $f_8 = \{4^2\}$, we have $\mathscr{I} = 1.4 \times 10^{-7}$, 2×10^{-6} and 2×10^{-4} respectively. Therefore the H matrices with more symmetry are characterized by relatively large fraction of independent matrix elements and thus they go more towards GOE.

(iii) Due to the two-body selection rules, many of the m particle matrix elements of the EGUE(2) ensembles will be zero. In order to understand the sparse nature of the EGUE matrices, one can introduce a sparsity index $\mathbf{S}$ with $\mathbf{S}^{-1}$ defined as the ratio of number of m-particle states that are directly coupled by the two-body interaction to the m-particle matrix dimension. Note that $\mathbf{S}^{-1}(m) = K(m)/d_f(N, m)$ for EGUE(2) and $K(m)$ is K defined by Eq. (5.16). Similarly, $\mathbf{S}^{-1}(m, S) = K(m, S)/d_f(\Omega, m, S)$ for EGUE(2)-**s** and as argued in Chap. 6, $K(m, S)$ can be equated to the variance propagator $P(\Omega, m, S)$ and formula for this is given by Eq. (6.19). For EGUE(2)-$SU(4)$, given the two-particle variances to be $\lambda_{f_2}^2 = \lambda^2$, the variances $\overline{\langle \widehat{H}^2 \rangle^{m, f_m}}$ can be written as $\sigma^2(m, f_m) = \lambda^2 P^{SU(4)}(m, f_m)$ with $P^{SU(4)}(m, f_m)$ given by Eq. (11.35). Though not well verified, the connectivity factor for EGUE(2)-$SU(4)$ can be taken as $K(m, f_m) \sim P^{SU(4)}(m, f_m)$. Therefore, for the EGUE(2)-$SU(4)$ ensemble, $\mathbf{S}^{-1}(m, f_m) = K(m, f_m)/d_\Omega(f_m)$. For example, from Table 11.3 we have $K(m = 4r, f_m = \{4^r\}) = r(\Omega - r + 4)\{2r(2\Omega - 2r + 9) - \Omega - 8\}$ and $K(m = 4r + 1, f_m = \{4^r, 1\}) = r(\Omega - r + 4)\{4r(2\Omega - 2r + 7) + 2\Omega - 15\}/2$. For the 8 particle example (with $N = 24$) considered before, the connectivity factors K are 4284, 1440 and 864 respectively for EGUE(2), EGUE(2)-**s** with $S = 0$ and EGUE(2)-$SU(4)$ with $f_8 = \{4^2\}$. These give $\mathbf{S}^{-1} = 5.8 \times 10^{-3}$, 0.02

and 0.49 respectively for these ensembles. Therefore as symmetry increases, in general, the many particle EGUE matrices will become more dense.

Consequences of (i)–(iii) will be discussed in the next section.

12.2 Self Correlations in EGUE(2)-$SU(r)$ for Fermions: Role of Symmetries

Self correlations in energy centroids and spectral variances for EGUE(2), EGUE(2)-**s** and EGUE(2)-$SU(4)$ correspond to $\hat{\Sigma}_{rr}(m,m)$, $\hat{\Sigma}_{rr}(m,S:m,S)$ and $\hat{\Sigma}_{rr}(m,f_m:m,f_m)$ respectively. Higher order self correlations are not studied yet in literature. Significance of self correlations is that they will affect level motion in the ensembles as already discussed in Sects. 4.3 and 9.4. Further significance of the magnitude of the self correlations follows by comparing the results with the corresponding ones for EGUE(2), EGUE(2)-**s** and EGUE(2)-$SU(4)$ for fixed number of sp states. Table 12.1 gives the results for $N=24$ and 40. Then, $\Omega=12$ and 20 for EGUE(2)-**s** and $\Omega=6$ and 10 for EGUE(2)-$SU(4)$. Analytical formulas for $\hat{\Sigma}_{11}^{1/2}$ and $\hat{\Sigma}_{22}^{1/2}$ are given in Sect. 11.1 for EGUE(2), in Sect. 11.4 for EGUE(2)-**s** and in Sect. 11.5 for EGUE(2)-$SU(4)$. It is seen from Table 12.1 that the magnitude of the covariances in energy centroids and spectral variances increases by a factor of 3 when we go from EGUE(2) $\rightarrow$ EGUE(2)-**s** $\rightarrow$ EGUE(2)-$SU(4)$. As discussed before, the fraction of independent matrix elements $\mathscr{I}$ increases with symmetry and also the sparsity (**S**) decreases and therefore the EGUE(2)-$SU(4)$ matrices will be dense leading to a more complete mixing of the basis states compared to EGUE(2) and EGUE(2)-**s**. Thus, there is a correlation between (i) increase in fluctuations defined by $\hat{\Sigma}_{11}$ and $\hat{\Sigma}_{22}$ and (ii) the matrices $H_{f_m}(m)$ becoming more dense as we go from EGUE(2) $\rightarrow$ EGUE(2)-**s** $\rightarrow$ EGUE(2)-$SU(4)$. As fluctuations (in energy centroids and spectral variances) are growing with increasing symmetry, it is plausible to conclude that symmetries play a significant role in generating chaos. Analyzing nuclear shell model matrices with J symmetry [they correspond to EGOE(2)-J ensemble described in Chap. 13], a similar conclusion was reached in [1] by Papenbrock and Weidenmüller and as they state: "While the number of independent random variables decreases drastically as we follow this sequence, the complexity of the (fixed) matrices which support the random variables, increases even more. In that sense, we can say that in the TBRE, chaos is largely due to the existence of (an incomplete set of) symmetries."

12.3 Cross Correlations in EGUE(2)-$SU(r)$: A New Signature

One of the most significant aspect of embedded ensembles is that they generate cross correlations in spectra [2]. For example, cross correlations in energy centroids and spectral variances for EGUE(2), EGUE(2)-**s** and EGUE(2)-$SU(4)$ correspond

Table 12.1 Variation in the self correlations in energy centroids ($\hat{\Sigma}_{11}$) and spectral variances ($\hat{\Sigma}_{22}$) with symmetry. For EGUE(2)-**s**, $\Omega = N/2$ and the results are for $S = 0$ for even m and $S = 1/2$ for odd m. Similarly for EGUE(2)-$SU(4)$, $\Omega = N/4$ and the results are for the $f_m^{(p)}$ irreps

N	m	$[\hat{\Sigma}_{11}]^{1/2}$			$[\hat{\Sigma}_{22}]^{1/2}$		
		EGUE(2)	EGUE(2)-**s**	EGUE(2)-$SU(4)$	EGUE(2)	EGUE(2)-**s**	EGUE(2)-$SU(4)$
24	6	0.017	0.043	0.125	0.0056	0.017	0.069
	7	0.021	0.055	0.144	0.0059	0.019	0.076
	8	0.026	0.066	0.160	0.0064	0.021	0.083
	9	0.031	0.081	0.196	0.0069	0.025	0.099
	10	0.037	0.094	0.229	0.0077	0.028	0.117
	11	0.044	0.112	0.256	0.0087	0.034	0.134
	12	0.051	0.128	0.276	0.0099	0.039	0.148
40	12	0.0139	0.038	0.105	0.00222	0.0079	0.035
	13	0.0157	0.044	0.120	0.00234	0.0086	0.039
	14	0.0176	0.048	0.134	0.00247	0.0093	0.044
	15	0.0196	0.054	0.146	0.00262	0.0103	0.049
	16	0.0218	0.06	0.156	0.0028	0.0112	0.053
	17	0.0241	0.067	0.174	0.003	0.0125	0.061
	18	0.0267	0.073	0.192	0.00324	0.0138	0.069
	19	0.0294	0.081	0.206	0.00352	0.0156	0.078
	20	0.0325	0.088	0.218	0.00385	0.0174	0.085

to $\hat{\Sigma}_{rr}(m,m')$ with $m \neq m'$, $\hat{\Sigma}_{rr}(m,S:m',S=S')$ with $m \neq m'$ and/or $S \neq S'$ and $\hat{\Sigma}_{rr}(m,f_m:m',f_{m'})$ with $m \neq m'$ and /or $f_m \neq f_{m'}$ respectively. For EGUE(2), simple formulas for Σ_{rr} are given in Sect. 11.1 and for other ensembles they are more complicated. However, results in Sects. 11.4 and 11.5 will allow one to obtain easily numerical results for cross correlations in EGUE(2)-**s** and EGUE(2)-$SU(4)$. For these two ensembles results for some examples are shown in Figs. 12.1 and 12.2 respectively. The results in these figures can be understood using asymptotic formulas for $[\hat{\Sigma}_{11}]^{1/2}$ and $[\hat{\Sigma}_{22}]^{1/2}$. For EGUE(2) in the dilute limit defined by $N, m \to \infty$, $m/N \to 0$, Eqs. (11.17) and (11.18) will give,

$$\hat{\Sigma}_{11} \stackrel{dilute-limit}{\longrightarrow} \frac{2mm'}{N^4}, \qquad \hat{\Sigma}_{22} \to \frac{4}{N^4}. \tag{12.2}$$

The results in Eqs. (11.17) and (11.18) extend to EGOE and for example, for large N, there will be an extra factor 2 for $\hat{\Sigma}_{11}$ and $\hat{\Sigma}_{22}$ in (12.2); see [3] and Eq. (4.53). Turning to EGUE(2)-**s**, in the dilute limit defined by $\Omega \to \infty$, $m \to \infty$, $m/\Omega \to 0$ and $S \ll m$, simplifying Eqs. (11.51) and (11.52) for $\hat{\Sigma}_{11}$ and similarly for $\hat{\Sigma}_{22}$,

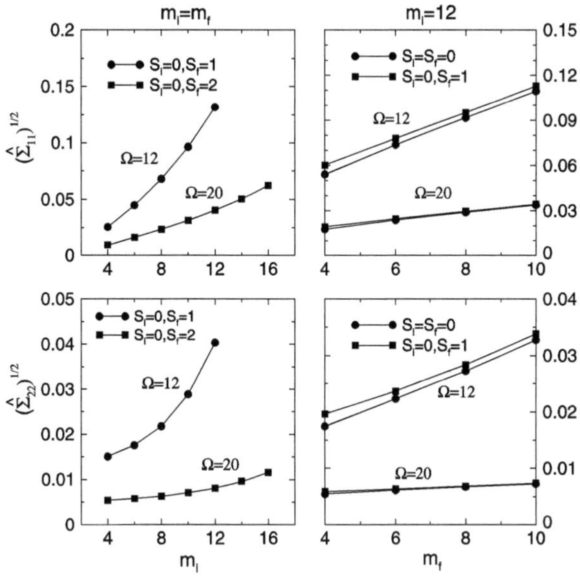

Fig. 12.1 Cross correlations in energy centroids ($[\hat{\Sigma}_{11}]^{1/2}$) and spectral variances ($[\hat{\Sigma}_{22}]^{1/2}$) for several examples for EGUE(2)-**s**. Results in the figures are obtained using the formulas in Sect. 11.4. Similar results are obtained for EGOE(2)-**s** as shown in Fig. 12.4 ahead

will give

$$\begin{aligned}
&\hat{\Sigma}_{rr}\big(m, S : m', S'\big) \\
&\quad \xrightarrow{\Omega\to\infty, m \text{ fixed}} \frac{r^4}{2\Omega^4}\left\{\sum_{f_2(s_2)} (\lambda_{f_2})^{2r} P^{s_2}(m, S) P^{s_2}\big(m', S'\big)\right\} \\
&\qquad \times \left\{\sum_{f_2(s_2)} (\lambda_{f_2})^2 P^{s_2}(m, S) \sum_{f_2'(s_2')} (\lambda_{f_2'})^2 P^{s_2'}\big(m', S'\big)\right\}^{-\frac{r}{2}}; \quad r = 1, 2.
\end{aligned} \tag{12.3}$$

Thus, for finite $\lambda_{\{2\}}/\lambda_{\{1^2\}}$, the $\hat{\Sigma}_{rr} \to 0$ as Ω approaches ∞ and there will be no cross correlations. However, in realistic situations, where Ω will be finite, there will be correlations between states with different or same (m, S) as shown in Fig. 12.1. As seen from the figures, the correlations for centroids are $\sim$10 % and in variances they are much smaller $\sim$3 %. It is interesting and important to look for these correlations in data for fermion systems with spectra for different spins. Finally, for EGUE(2)-$SU(4)$ for $\hat{\Sigma}_{qq}(m, f_m; m', f_{m'})$ with $q = 1, 2$, for all $f_m^{(p)} = \{4^r, p\}$ irreps, in the dilute limit defined by $\Omega \to \infty$, $r \gg 1$ and $r/\Omega \to 0$, we have

$$\begin{aligned}
&\hat{\Sigma}_{qq}\big(m, f_m^{(p)}; m', f_{m'}^{(p)}\big) \\
&\quad \xrightarrow{\Omega\to\infty, r\gg 1} \frac{4q}{\Omega^4} \frac{\sum_{f_2} \lambda_{f_2}^{2q} P^{f_2}(m, f_m^{(p)}) P^{f_2}(m', f_{m'}^{(p)})}{[\{\sum_{f_2} \lambda_{f_2}^2 P^{f_2}(m, f_m^{(p)})\}\{\sum_{f_2} \lambda_{f_2}^2 P^{f_2}(m', f_{m'}^{(p)})\}]^{q/2}}; \\
&\quad q = 1, 2.
\end{aligned} \tag{12.4}$$

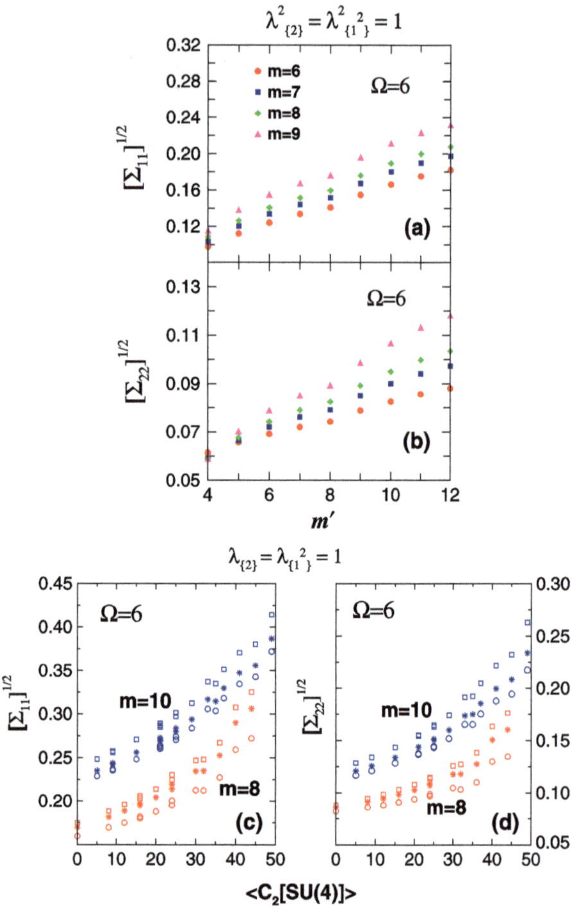

Fig. 12.2 Self and cross correlations in energy centroids and spectral variances for $\Omega = 6$ examples for EGUE(2)-$SU(4)$. Shown in (**a**) and (**b**) are results as a function of m and m' (with fixed f_m and $f_{m'}$): (**a**) $[\hat{\Sigma}_{11}(m, f_m; m', f_{m'})]^{1/2}$ for $\Omega = 6$; (**b**) $[\hat{\Sigma}_{22}(m, f_m; m', f_{m'})]^{1/2}$ for $\Omega = 6$. Note that $\hat{\Sigma}$ is written as Σ in the figures. Results in the figure are for $f_m = f_m^{(p)}$ and $f_{m'} = f_{m'}^{(p)}$. Shown in (**c**) and (**d**) are results as a function of f_m and $f_{m'}$ (with fixed $m = m'$). Results are shown for the first (*circle*), second (*star*) and fourth (*square*) lowest $U(\Omega)$ irreps (ordered according to $\langle C_2[SU(4)]\rangle^{\tilde{f}_m}$) with all other irreps for $m = m' = 8$ (*red*) and 10 (*blue*) as a function of $\langle C_2[SU(4)]\rangle^{\tilde{f}_m}$. Note that for a given value of the eigenvalue of $C_2[SU(4)]$ in some cases there are more than one f_m with the same eigenvalue. Figures (**a**)–(**d**) are taken from [4] with permission from Elsevier

As $\Omega \to \infty$, $\hat{\Sigma}_{qq}(m, f_m^{(p)}; m', f_{m'}^{(p)}) \to 0$ for $q = 1, 2$ and there will be no correlations. For finite Ω, there will be cross correlations between states with different or same (m, f_m) and examples for these are shown in Fig. 12.2. The increase in the cross correlations with m' for fixed f_m and similar increase with $\langle C_2[SU(4)]\rangle^{\tilde{f}_m}$ with fixed m seen from Fig. 12.2 could possibly be exploited in deriving experimental signatures for cross correlations. Comparing the results in Fig. 12.1 with those in Fig. 12.2, we see that just as in Table 12.1, correlations are larger for EGUE(2)-$SU(4)$ compared to EGUE(2)-**s**.

12.4 Self and Cross Correlations in BEGUE(2)-$SU(3)$

Turning to BEGUE ensembles, the formulation in Chap. 11 can be applied to boson systems with $U(r\Omega) \supset U(\Omega) \otimes SU(r)$ by employing, for m bosons, the symmetric irrep $\{m\}$ for $U(r\Omega)$. Analytical formulas for self ans cross correlations in energy

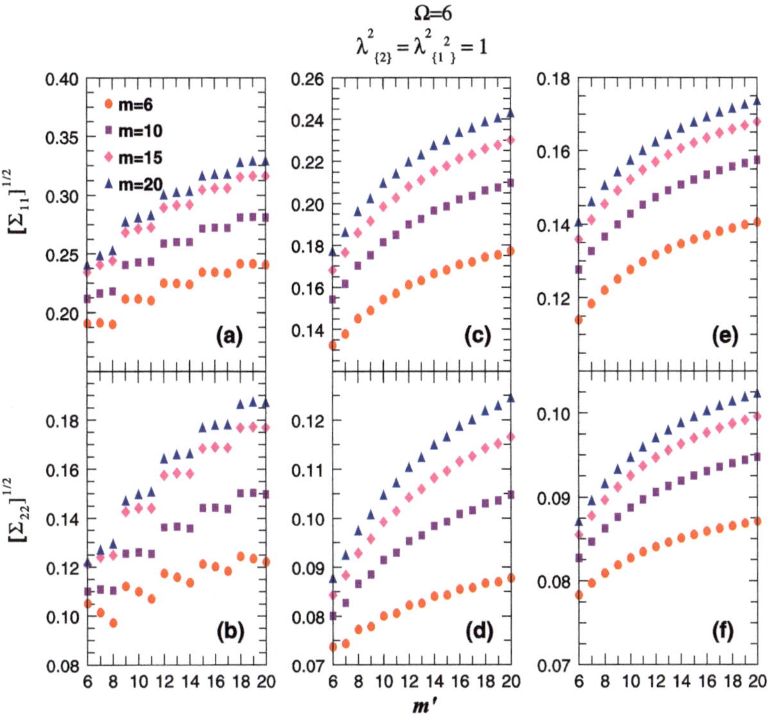

Fig. 12.3 Centroid correlations $\hat{\Sigma}_{11}$ and variance correlations $\hat{\Sigma}_{22}$ for BEGUE(2)-$SU(3)$ ensemble. Note that $\hat{\Sigma}$ is written as Σ in the figure. Figure is taken from [5] with permission from American Institute of Physics

centroids and spectral variances for $r = 1$ are given in Sect. 11.3.1 and those for $r = 2$ will follow from Sect. 11.6. Going further, as an example, results for $r = 3$ for self ($m = m'$) and cross correlations ($m \neq m'$) in energy centroids and spectral variances as a function of m and m' for $\Omega = 6$ (with fixed f_m and $f_{m'}$) are shown in Fig. 12.3. In the figure, results are for: (a) $[\Sigma_{11}(m, f_m; m', f_{m'})]^{1/2}$ with $\{f_m\} = \{m/3, m/3, m/3\}$ for $m \mod 3 = 0$, $\{f_m\} = \{(m+2)/3, (m-1)/3, (m-1)/3\}$ for $m \mod 3 = 1$ and $\{f_m\} = \{(m+4)/3, (m-2)/3, (m-2)/3\}$ for $m \mod 3 = 2$ and similarly $f_{m'}$ is defined; (b) $[\Sigma_{22}(m, f_m; m', f_{m'})]^{1/2}$ with $\{f_m\} = \{m/3, m/3, m/3\}$ for $m \mod 3 = 0$, $\{f_m\} = \{(m+2)/3, (m-1)/3, (m-1)/3\}$ for $m \mod 3 = 1$ and $\{f_m\} = \{(m+4)/3, (m-2)/3, (m-2)/3\}$ for $m \mod 3 = 2$ and similarly $f_{m'}$ is defined; (c) $[\Sigma_{11}(m, f_m; m', f_{m'})]^{1/2}$ with $\{f_m\} = \{m/2, m/2\}$ for $m \mod 2 = 0$ and $\{f_m\} = \{(m+1)/2, (m-1)/2\}$ for $m \mod 2 = 1$ and similarly $f_{m'}$ is defined; (d) $[\Sigma_{22}(m, f_m; m', f_{m'})]^{1/2}$ with $\{f_m\} = \{m/2, m/2\}$ for $m \mod 2 = 0$ and $\{f_m\} = \{(m+1)/2, (m-1)/2\}$ for $m \mod 2 = 1$ and similarly $f_{m'}$ is defined; (e) $[\Sigma_{11}(m, f_m; m', f_{m'})]^{1/2}$ with $\{f_m\} = \{m\}$ and $\{f_{m'}\} = \{m'\}$; (f) $[\Sigma_{22}(m, f_m; m', f_{m'})]^{1/2}$ with $\{f_m\} = \{m\}$ and $\{f_{m'}\} = \{m'\}$. Results in the figure show that [5]: (i) the centroid and variance fluctuations increase with m' for fixed m and vice-verse; (ii) they are larger for three rowed irreps compared to those for one

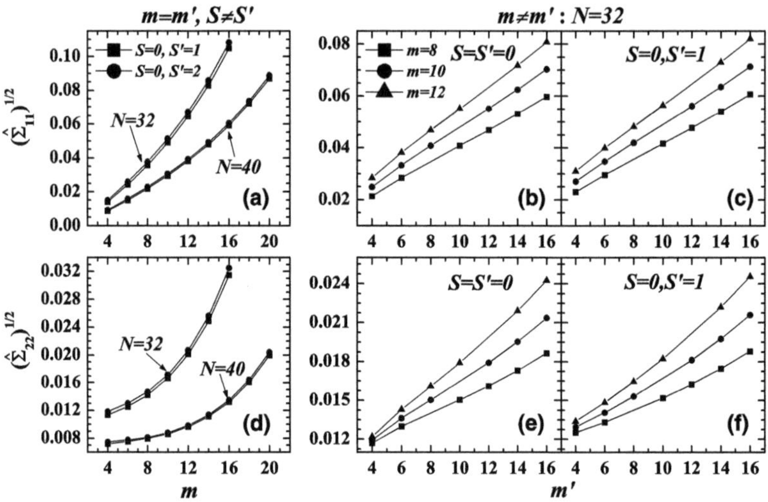

Fig. 12.4 Centroid correlations $\hat{\Sigma}_{11}$ and variance correlations $\hat{\Sigma}_{22}$ for various EGOE(2)-s systems with $(m, S) \neq (m', S')$. In all the calculations $\lambda_0 = \lambda_1 = 1$ in the EGOE(2) Hamiltonian and used is a 200 member ensemble along with the trace propagation formulas given in Sect. 6.3. Note that $N = 2\Omega$ and the values of N are shown in the figures. Figure is taken from [7] with permission from World Scientific

rowed irreps; (iii) centroid fluctuations are much larger than variance fluctuations as seen also for EGUE(2)-s and EGUE(2)-$SU(4)$ ensembles. Similar trends are also seen for $m = m'$ but varying $f_{m'}$ with fixed f_m. More importantly, the centroid and variance fluctuations are smallest for the ground state i.e., the most symmetric irrep for bosons. It is also seen from the figure that the covariances in energy centroids are ~15–25 % and the covariances in spectral variances are ~8–15 %.

12.5 Self and Cross Correlations in EGOE(2)-s

Centroid and variance formulas given in Chap. 6 for (1 + 2)-body Hamiltonians preserving spin S, will allow one to numerically calculate $\overline{\langle H^p\rangle^{m,s}\langle H^p\rangle^{m',s'}}$, $p = 1, 2$ for each member of EGOE(2)-s and therefore one can calculate $\hat{\Sigma}_{rr}(m, S : m', S')$, $r = 1, 2$. Some results, obtained using this approach, are shown in Fig. 12.4. Asymptotic formulas for the covariances will help us in understanding the results in these figures.

In the dilute limit defined by $\Omega \to \infty$, $m \to \infty$, $m/\Omega \to 0$ and $S \ll m$, centroid and variance trace propagation formulas will give,

$$\hat{\Sigma}_{11}(m, S : m', S') \xrightarrow{\text{dilute limit}} \frac{5(m/\Omega)(m'/\Omega)}{4\Omega^2} + \frac{1}{\Omega^4}\left\{\frac{(m/\Omega)}{(m'/\Omega)} S'(S'+1) + \frac{(m'/\Omega)}{(m/\Omega)} S(S+1)\right\}. \tag{12.5}$$

Equation (12.5) shows that for $m = m'$: (i) for low (S, S') values the spin dependence is week in $\hat{\Sigma}_{11}$; (ii) for a given m, as Ω increases $\hat{\Sigma}_{11}$ decreases, (iii) for a fixed Ω, $\hat{\Sigma}_{11}$ increases with m. All these results are seen clearly in Fig. 12.4. Similarly Eq. (12.5) gives: (i) $(\hat{\Sigma}_{11})^{1/2} \simeq 0.035$ for $m/\Omega = 1/2$, $\Omega = 16$ and (S, S') small and this is verified in Figs. 12.4a, b, c; (ii) as seen in Figs. 12.4b, c, for fixed m as m' increases $(\hat{\Sigma}_{11})^{1/2}$ increases and the spin dependence is weak for low spin values. Also, as expected, the first term in Eq. (12.5) has the same structure as in the dilute limit for EGOE(2) (see Sect. 4.3). Figures 12.4d, e, f give the results for $\hat{\Sigma}_{22}(m, S : m', S')$. It is seen that the variance correlations, given by $(\hat{\Sigma}_{22})^{1/2}$, for low spin members are $\lesssim 3$ %. Finally, in the dilute limit

$$\hat{\Sigma}_{22}\big(m, S : m', S'\big) \to \frac{10}{\Omega^4} + \frac{8}{\Omega^4}\left\{\frac{S(S+1)}{m^2} + \frac{S'(S'+1)}{(m')^2}\right\}. \tag{12.6}$$

The first term here has the same structure as in EGOE(2).

12.6 Self and Cross Correlations in BEGOE(2)-F and BEGOE(2)-$S1$

Calculating energy centroids and spectral variances for each member of the BEGOE(2)-F ensemble using the formulation described in Sect. 10.1.3, covariances $\hat{\Sigma}_{rr}(m, F : m', F')$, $r = 1, 2$ have been studied in [8]. Some numerical results obtained in this work are shown in Fig. 12.5. All the calculations are carried out using $\lambda_0 = \lambda_1 = \lambda$ so that $\hat{\Sigma}_{11}$ and $\hat{\Sigma}_{22}$ are independent of λ. From the numerical results for $\hat{\Sigma}_{11}$ it is seen that: (i) for $m \gg \Omega$, the ΔE_c (width of the fluctuations in energy centroids) is ~20 % for $F = 0$ and it goes down to ~15 % for $F = F_{max} = m/2$ for $\Omega = 12$; (ii) going from $\Omega = 12$ to 40, ΔE_c decreases to ~2–7 % ; (iii) for fixed (m, Ω), there is decrease in ΔE_c with increasing F value; (iv) for fixed (m, F) and very large m value, there is a sharp decrease in ΔE_c with increasing Ω up to $\Omega \sim 20$ and then it slowly converges to zero. Similarly, from the results shown in Figs. 12.5d, e for both self correlations giving the width $\Delta\langle H^2\rangle^{m,F}$ of variances and cross correlations $[\Sigma_{22}]^{1/2}$ with $(m, F) \neq (m', F')$, it is seen that $[\Sigma_{22}]^{1/2}$ are always much smaller than $[\Sigma_{11}]^{1/2}$ just as for EGOE(2) for spinless fermion systems and EGOE(2)-**s**. It is also seen from Figs. 12.5d, e that for $\Omega = 12$, widths of the fluctuations in the variances $\langle H^2\rangle^{m,F}$ are ~3–5 %. Similarly for large m, with Ω very small, the widths are quite large but they decrease fast with increasing Ω. Finally, for $\Omega = 12$, the cross correlations are ~4 %.

It is possible to understand the results for self and cross correlations, in energy centroids, i.e. $[\hat{\Sigma}_{11}(m, F : m', F')]^{1/2}$, with $(m, F) = (m', F')$ and $(m, F) \neq (m', F')$ respectively, using the asymptotic structure of the variance propagator $Q(\Omega, m, F)$ defined in Sect. 10.1.3. In the dense limit defined by $m \to \infty$, $\Omega \to \infty$,

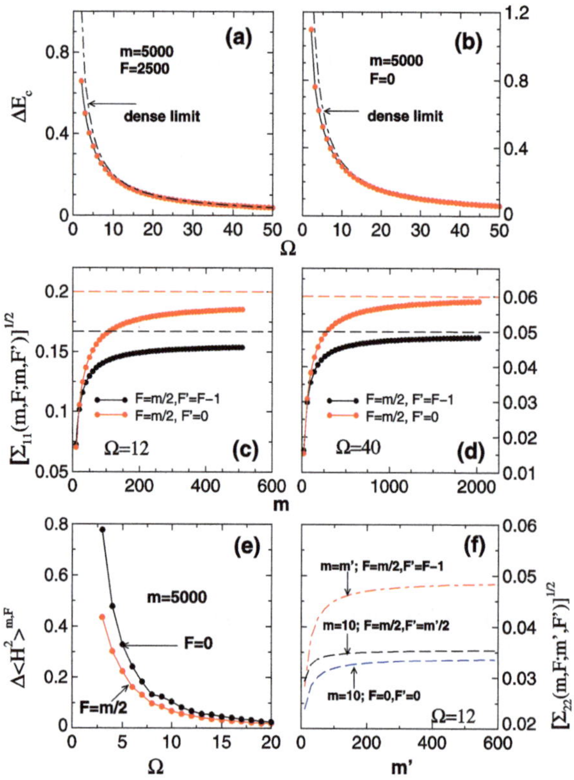

Fig. 12.5 Self and cross correlations in energy centroids $\hat{\Sigma}_{11}$ and spectral variances $\hat{\Sigma}_{22}$ for various BEGOE(2)-F systems. (**a**) Self-correlations in eigenvalue centroids giving width ΔE_c of the fluctuations in eigenvalue centroids scaled to the spectrum width, as a function of Ω for 5000 bosons with maximum F-spin ($F = 2500$). Dense limit (*dot-dashed*) *curve* corresponds to the results given by (12.8). (**b**) same as (**a**) but for $F = 0$. (**c**) Cross-correlations in eigenvalue centroids for BEGOE(2)-F systems with $\Omega = 12$. Shown are results for $\hat{\Sigma}_{11}^{1/2}$ vs m with $m = m'$ but different F-spins ($F \neq F'$). The *dashed lines* in (**c**) are the dense limit results. (**d**) Same as (**c**) but for $\Omega = 40$. (**e**) Self-correlations in spectral variances giving width $\Delta\langle H^2\rangle^{m,F}$ of the spectral variances as a function of Ω for 5000 bosons with $F = 0$ and 2500. (**f**) Three examples for cross-correlation in spectral variances with same or different particle numbers and same or different spins. All the results are obtained using 500 member ensembles. Figure is constructed from the results in [8]. Note that $\hat{\Sigma}$ is written as Σ in the figures

$m/\Omega \to \infty$ and F fixed, it can be shown that [with $F^2 = F(F+1)$],

$$
\begin{aligned}
\overline{\sigma_H^2(m,F)}/\overline{\sigma_H^2(m,F_{max})} &= \left[\frac{m/(m+2) + F^2/F_{max}^2}{m/(m+2)+1}\right]^2, \\
\overline{\langle H\rangle^{m,F}\langle H\rangle^{m',F'}} &= \frac{\lambda^2}{16\Omega^2}\Big[\big(m^2-4F^2\big)\big\{\big(m'\big)^2-4\big(F'\big)^2\big\} \\
&\quad + \big(3m^2+4F^2\big)\big\{3\big(m'\big)^2+4\big(F'\big)^2\big\}\Big].
\end{aligned}
\tag{12.7}
$$

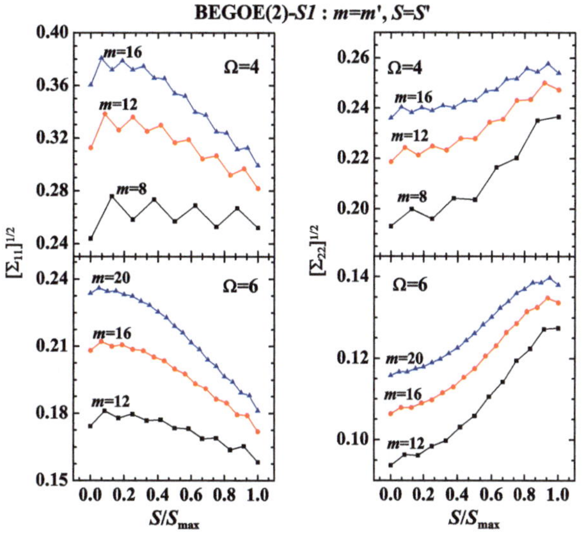

Fig. 12.6 Normalized self-correlations in energy centroids—$[\hat{\Sigma}_{11}]^{1/2}$ and in spectral variances—$[\hat{\Sigma}_{22}]^{1/2}$ as a function of spin S for a 200 member BEGOE(2)-$S1$ ensemble for various values of (m, S) with $\Omega = 4$ and 6. Note that $\hat{\Sigma}$ is written as Σ in the figures

Then, $[\hat{\Sigma}_{11}]^{1/2}$ with $m = m'$ and $F = F'$ giving ΔE_c is

$$[\hat{\Sigma}_{11}]^{1/2} = \Delta E_c = \frac{\sqrt{2(5m^4 + 8m^2F^2 + 16(F^2)^2)}}{\Omega(m^2 + 4F^2)}. \tag{12.8}$$

Equation (12.8) gives $[\Sigma_{11}]^{1/2}$ to be $\sqrt{10}/\Omega$ and $2/\Omega$ for $F = 0$ and $F = F_{max}$ and these dense limit results are well verified by the results in Fig. 12.5. Similarly, Eq. (12.7) will give $[\Sigma_{11}]^{1/2}$ to be $\sqrt{6}/\Omega$ for $(m = m' : F = F_{max}, F' = 0)$ and $2/\Omega$ for $(m = m' : F = F_{max}, F' = F_{max} - 1)$. The upper and lower dashed lines in Fig. 12.5c for $\Omega = 12$ and Fig. 12.5d or $\Omega = 40$ correspond to these two dense limit results respectively. It is seen that the dense limit results are close to exact results for $\Omega = 40$ but there are deviations for smaller Ω values. In general, for sufficiently large value of Ω and $m \gtrsim 5\Omega$, the dense limit result describes quite well the exact results. Finally, unlike for the covariances in eigenvalue centroids, at present complete analytical formulation for the covariances in spectral variances is not available.

For BEGOE(2)-$S1$ described in Sect. 10.2, calculating energy centroids and spectral variances for each member of the ensemble using the formulation given in Sect. 10.2.3, normalized covariances $\hat{\Sigma}_{pp}(m, S : m', S')$, $p = 1, 2$ have been studied in several examples and some results are shown in Fig. 12.6. In the numerical calculations used is $\lambda = \lambda_0 = \lambda_1 = \lambda_2$ so that $\hat{\Sigma}$ is independent of λ. In the figure, only for self correlations ($m = m'$ and $S = S'$) results are presented. It is seen that the centroid fluctuations ($\hat{\Sigma}_{11}$) are large for $S = 0$ with $m \gg \Omega$ and decreases with increase in S value. However, for small m, the variation with spin S is weak. For fixed value of m, there is decrease of $[\hat{\Sigma}_{11}]^{1/2}$ with increasing Ω value. On the other hand, $[\hat{\Sigma}_{22}]^{1/2}$ values are always smaller than $[\hat{\Sigma}_{11}]^{1/2}$ values just as for BEGOE(2)-F. For example, it is seen from Fig. 12.6 that for $\Omega = 6$, the width of the fluctuations in the variances are 10–13 % while the centroid fluctuations are ~18–24 %.

Before concluding, let us add that the EGOE(2)-J ensemble, relevant for nuclei and atoms, is introduced in Chap. 13 and there the results for cross correlations in this ensemble, as given in [2, 9], will be discussed.

References

1. P. Papenbrock, H.A. Weidenmüller, Origin of chaos in the spherical nuclear shell model: role of symmetries. Nucl. Phys. A **757**, 422–438 (2005)
2. P. Papenbrock, H.A. Weidenmüller, Two-body random ensemble in nuclei. Phys. Rev. C **73**, 014311 (2006)
3. L. Benet, T. Rupp, H.A. Weidenmüller, Spectral properties of the k-body embedded Gaussian ensembles of random matrices. Ann. Phys. **292**, 67–94 (2001)
4. M. Vyas, V.K.B. Kota, Spectral properties of embedded Gaussian unitary ensemble of random matrices with Wigner's $SU(4)$ symmetry. Ann. Phys. (N.Y.) **325**, 2451–2485 (2010)
5. M. Vyas, V.K.B. Kota, Embedded Gaussian unitary ensembles with $U(\Omega) \otimes SU(r)$ embedding generated by random two-body interactions with $SU(r)$ symmetry. J. Math. Phys. **53**, 123303 (2012)
6. V.K.B. Kota, N.D. Chavda, R. Sahu, One plus two-body random matrix ensemble with spin: analysis using spectral variances. Phys. Lett. A **359**, 381–389 (2006)
7. V.K.B. Kota, Two-body ensembles with group symmetries for chaos and regular structures. Int. J. Mod. Phys. E **15**, 1869–1883 (2006)
8. M. Vyas, N.D. Chavda, V.K.B. Kota, V. Potbhare, One plus two-body random matrix ensembles for boson systems with F-spin: analysis using spectral variances. J. Phys. A, Math. Theor. **45**, 265203 (2012)
9. T. Papenbrock, H.A. Weidenmüller, Random matrices and chaos in nuclear spectra. Rev. Mod. Phys. **79**, 997–1013 (2007)

Chapter 13
Further Extended Embedded Ensembles

In this chapter we will describe briefly several other embedded ensembles that are introduced in literature. Very few analytical results are available for these ensemble. However, there are some numerical studies because of their physical relevance. We will begin with the EGOE$(1+2)$-J ensemble that is of great interest in nuclear and atomic physics.

13.1 EGOE$(1+2)$-$(j_1, j_2, \ldots, j_r : J)$ Ensemble

In the nuclear shell model (NSM), structure of low-lying states of atomic nuclei is well described by the simple picture that protons and neutrons (i.e. nucleons) move in a mean-field single particle potential that can be approximated by a three dimensional harmonic oscillator [1]. With strong spin-orbit force, the sp states are labeled by the radial (n_r), orbital angular momentum (ℓ), total angular momentum (j) and its z-projection (m_j) quantum numbers; major oscillator shell number $\mathscr{N} = 2n_r + \ell$. Thus, nucleons (proton and neutrons) not only carry spin $(\mathbf{s})$ degree of freedom considered in Chap. 6 but also orbital (ℓ) angular momentum giving $\vec{j} = \vec{\ell} + \vec{\mathbf{s}}$. The spin-orbit force gives the magic numbers 2, 8, 20, 50, 82 and 126 (also semi-magic numbers 28 and 40) for protons and neutrons. Magic numbers define the completely filled j orbits (core) and the remaining valence nucleons (say m_p number of valence protons and m_n number of valence neutrons) move in a few j orbits. Structure of the low-lying states of a nucleus is determined by the mean-field sp energies of the valence j orbits $j_1, j_2, \ldots, j_r$ and an effective two-body interaction in the $(j_1, j_2, \ldots, j_r)^{m_p, m_n}$ space. Thus, here H is one plus two-body in nature and more importantly, H preserves total angular momentum J. There are many nuclei where $m_p = 0$ or $m_n = 0$ and then we have identical fermions in j orbits. For example, for ^{20}O the spectroscopic space is $({}^1d_{5/2}, {}^2s_{1/2}, {}^1d_{3/2})^{m_n=4,J}$; the superscript in the sp orbits is $n_r + 1$. Similarly for ^{44}Ca it is just $({}^1f_{7/2})^{m_n=4,J}$. On the other hand the space for ^{24}Mg is $({}^1d_{5/2}, {}^2s_{1/2}, {}^1d_{3/2})^{m_n=4, m_n=4, J}$ and similarly for ^{51}Mn it is $({}^1f_{7/2}, {}^2p_{3/2}, {}^2p_{1/2}, {}^1f_{5/2})^{m_p=5, m_n=6, J}$. A picture similar to NSM applies to atoms

V.K.B. Kota, *Embedded Random Matrix Ensembles in Quantum Physics*, Lecture Notes in Physics 884, DOI 10.1007/978-3-319-04567-2_13,

except that in atoms one has only electrons (one type of fermions). From all these, it should be clear that it is important to study random matrix ensembles generated by fermions in j orbits interacting with a two-body interaction that preserves the many fermion J value. For the situation with identical fermions, these ensembles are here-after called EGOE(1 + 2)-($j_1, j_2, \dots, j_r : J$) and this is appropriate for atomic systems and for certain nuclei like O, Ca and Ni isotopes.

13.1.1 Definition and Construction

Let us begin with EGOE(2)-($j : J$), i.e. EE for m identical fermions in a single j orbit [it will have $(2j+1)$ degenerate sp states] with H being a two-body operator preserving angular momentum J-symmetry. It should be added that EGOE(1 + 2)-($j : J$) will not exist as there is only one j orbit. Denoting the angular momentum of two fermions by J_2 and its J_Z eigenvalue by M_2 (for m fermions they are denoted by J and M respectively), the H operator is given by,

$$\hat{H}(2) = \sum_{J_2=\text{even}, M_2} V_{J_2} A\big(j^2; J_2 M_2\big)\big[A\big(j^2; J_2 M_2\big)\big]^{\dagger}, \tag{13.1}$$

where $V_{J_2} = \langle (j^2) J_2 M_2 | H | (j^2) J_2 M_2 \rangle$ are two-body matrix elements (TBME) independent of M_2 and $J_2 = 0, 2, 4, \dots, (2j-1)$. The operator $A(j^2; J_2 M_2)$ creates a normalized two particle state. The EGOE(2)-($j : J$) ensemble is now generated by assuming V_{J_2}'s to be independent Gaussian random variables $G(0,1)$. One simple way to construct the EGOE(2)-($j : J$) ensemble in m-particle spaces with a fixed-J value is as follows. Consider the $(2j+1)$ single particle states $|jm_j\rangle$, $m_j = -j, -j+1, \dots, j$. Now distributing m fermions in the m_j states in all possible ways will give the configurations $[m_\nu] = [n_{\nu_1}, n_{\nu_2}, \dots, n_{\nu_m}]$ where $(\nu_1, \nu_2, \dots, \nu_m)$ are the filled orbits so that $n_{\nu_i} = 1$. We can select configurations such that $M = \sum_{i=1}^{m} n_{\nu_i} m_{\nu_i} = 0$ for even m and $M = 1/2$ for odd m. Note that M are eigenvalues of the J_Z operator and the configurations $[m_\nu]$ are m-particle eigenstates of J_Z. The number of $[m_\nu]$'s for $M = 0$, with m even, is $D(m, M = 0) = \sum_{J=0}^{J_{max}} d(m, J)$ and similarly for odd m, $D(m, M = 1/2) = \sum_{J=1/2}^{J_{max}} d(m, J)$. Here $d(m, J)$ is the dimension of the (m, J) space without counting the $(2J+1)$ factor. Converting V_{J_2} into the $|jm'\rangle|jm''\rangle$ basis will give,

$$\begin{aligned} V_{m_1,m_2,m_3,m_4} &= \langle jm_3 jm_4 | \hat{H}(2) | jm_1 jm_2 \rangle \\ &= 2 \sum_{J_2=\text{even}: M_2} \langle jm_1 jm_2 | J_2 M_2 \rangle \langle jm_3 jm_4 | J_2 M_2 \rangle V_{J_2} \end{aligned} \tag{13.2}$$

where $M_2 = m_1 + m_2 = m_3 + m_4$. The TBME V_{m_1,m_2,m_3,m_4} together with the formalism used for EGOE(2) for spinless fermion systems (see Chap. 5) will give easily the H matrix in the $[m_\nu]$ basis. Starting with the J^2 operator and writing its one and two particle matrix elements in the $|jm'\rangle|jm''\rangle$ basis, it is possible to construct the

J^2 matrix in the $[m_\nu]$ basis. Diagonalizing this matrix will give (with $M_0 = 0$ for even m and 1/2 for odd m) the C-coefficients in the expansion,

$$\left|(j)^m \alpha J M_0\right\rangle = \sum_{[m_\nu]} C_{[m_\nu]}^{\alpha J M_0} \left|[m_\nu]\right\rangle \tag{13.3}$$

and we can identify the J-value of the eigenfunctions by using the J^2 eigenvalues $J(J+1)$. In Eq. (13.3), α are additional labels (they will be $d(m, J)$ in number) required for completely specifying the m fermion states with fixed J value. With this, matrix elements of the H matrix in the $|(j)^m \alpha J M_0\rangle$ basis are

$$\left\langle (j)^m \beta J M_0 \middle| H \middle| (j)^m \alpha J M_0 \right\rangle = \sum_{[m_\nu]_i, [m_\nu]_f} C_{[m_\nu]_i}^{\alpha J M_0} C_{[m_\nu]_f}^{\beta J M_0} \left\langle [m_\nu]_f \middle| V \middle| [m_\nu]_i \right\rangle. \tag{13.4}$$

The above procedure (called M-scheme in nuclear structure physics) can be implemented on a computer easily. An alternative is to construct the matrices directly in a good J-basis (called J-scheme) and this is used in the early years of developing nuclear shell model codes [2].

Now let us turn to EGOE(1 + 2)-($j_1, j_2, \ldots, j_r : J$) ensemble for m identical fermions in ($j_1, j_2, \ldots, j_r$) orbits. A one plus two-body H preserving total m particle angular momentum J is given by

$$\begin{aligned}
\widehat{H}(1+2) &= \hat{h}(1) + \widehat{V}(2) \\
&= \sum_{j_i} \varepsilon_{j_i} \hat{n}_{j_i} + \sum_{j_1 \geq j_2;\, j_3 \geq j_4;\, J_{12}, M_{12}} V_{j_1, j_2, j_3, j_4}^{J_{12}} A(j_3, j_4; J_{12}, M_{12}) \\
&\quad \times \left[A(j_1, j_2; J_{12}, M_{12})\right]^\dagger.
\end{aligned} \tag{13.5}$$

Here ε_{j_i} are sp energies, $\hat{n}_{j_i}$ is the number operator for the orbit j_i and $V_{j_1, j_2, j_3, j_4}^{J_{12}} = \langle (j_1, j_2) J_{12}, M_{12} | \widehat{V}(2) | (j_3, j_4) J_{12}, M_{12} \rangle$ are TBME independent of M_{12}. Note that J_{12} is even if $j_1 = j_2$ or $j_3 = j_4$. The $\widehat{V}(2)$ matrix $V(2)$ in two particle spaces is a direct sum of (in general more than 2) matrices [for EGOE(2)-($j : J$) they are one dimensional] one for each J_{12} value, $V(2) = \sum_{J_{12}} V^{J_{12}}(2) \oplus$. Then, representing $\{V^{J_{12}}(2)\}$ matrices by independent GOEs and propagating this $\{V(2)\}$ ensemble [along with $h(1)$] to the m particle spaces with a given J value by the shell model geometry [i.e. by the algebra $U(N) \supset G \supset SO_J(3)$, $N = \sum_i (2j_i + 1)$ with a suitable subalgebra G in between] will give EGOE(1 + 2)-($j_1, j_2, \ldots, j_r : J$) in (m, J) spaces. Let us consider the example of $j = (7/2, 5/2, 3/2, 1/2)$, i.e. the nuclear $2p1f$ shell. Here $J_{12} = 0, 1, 2, 3, 4, 5$ and 6 and the corresponding matrix dimensions are 4, 3, 8, 5, 6, 2 and 2 respectively. This gives 94 independent matrix elements for the $\{V(2)\}$ ensemble and one choice is to chose them to be independent $G(0, 1)$ variables. This version of EGOE(2)-($j_1, j_2, \ldots, j_r : J$), without $h(1)$, is often called TBRE. A different choice, considered in some studies is to represent $V^{J_{12}}(2)$ matrix by GOE with matrix elements variance a function of J_{12}. Shell model geometry was discussed in detail in [2, 3]. Then the m particle H matrix elements are linear combination of two-particle matrix elements with the expansion coefficients being the so called fractional parentage coefficients. For the $(2p1f)^{m=8}$ example, the dimensions $d(m,J)$ for the EGOE(1 + 2)-$(7/2,5/2,3/2,1/2:J)$ ensemble for $J=0$–14 are 347, 880, 1390, 1627, 1755, 1617, 1426, 1095, 808, 514, 311, 151, 73, 22 and 6 respectively. Both M-scheme and J-scheme approaches can be used to construct EGOE(2)-$(j_1,j_2,\dots,j_r:J)$ using modern nuclear shell model codes NUSHELL [4] and ANTONIE/NATHAN [5]. Similar shell model codes for atoms are also available [6–8]. When there is no confusion we refer the ensemble EGOE(1 + 2)-$(j_1,j_2,\dots,j_r:J)$ as just EGOE(1 + 2)-J.

EGOE(1+2)-J have been analyzed numerically by many groups for nuclei using shell model codes [9–13]. These showed that many of the results known for spinless EGOE(1 + 2) apply to EGOE(1 + 2)-J. More importantly, nuclear shell model calculations with specific effective interactions for nuclei [9–11] and also realistic atomic structure calculations [6–8, 14, 15] showed that these systems can be well represented by EGOE(1 + 2)-J. It has been verified that the form of the fixed-J eigenvalue density is close to a Gaussian, strength sums are close to a ratio of Gaussians, transition strength densities are close to bivariate Gaussian and level fluctuations follow GOE for strong enough interaction. As the interaction strength increases from zero, just as it is seen for EGOE(1 + 2) and EGOE(1 + 2)-**s**, there are three chaos markers exhibited by EGOE(1 + 2)-J. These numerical results in fact formed the basis for statistical nuclear spectroscopy [16]. It should be added that the results for high-J states show deviations from these results as matrix dimensions are small in all the shell model examples considered. Therefore, there are questions about the extension of EGOE(1 + 2) and EGOE(1 + 2)-**s** results to EGOE(1 + 2)-$(j_1,j_2,\dots,j_r:J)$ with J large. This issue deserves further studies. As the shell model geometry is in general complex, EGOE(1+2)-J ensemble is mathematically a very difficult ensemble and therefore, very few analytical results are available for this ensemble. Now we will discuss the available analytical results and then turn to further extensions of EGOE(1 + 2)-J.

13.1.2 Expansions for Dimensions, Energy Centroids and Spectral Variances

Let us begin with fixed-(m,J) dimensions $d(m,J)$. Following the discussion in the previous subsection it is easy to see that fixed-(m,M) dimension $D(m,M)$ will be essentially a Gaussian giving,

$$\begin{aligned}D(m,M) = \binom{N}{m} D_{\mathscr{G}}(m,M)\Bigg\{1 + \left[\frac{k_4(m:J_z)}{24} He_4(\hat{M})\right] \\ + \left[\frac{k_6(m:J_z)}{720} He_6(\hat{M}) + \frac{1}{2}\left(\frac{k_4(m:J_z)}{24}\right)^2 He_8(\hat{M})\right]\Bigg\} \quad (13.6)\end{aligned}$$

where $D_{\mathscr{G}}(m, M)$ is standardized Gaussian, $\hat{M}$ is the standardized M and k_r are cumulants for the m-particle J_Z spectrum. As J_Z is a simple one-body operator, it is easy to derive the following formulas,

$$\sigma^2_{J_z}(m) = \langle J_z^2\rangle^m = \frac{m(N-m)}{N-1}\langle J_z^2\rangle^1;$$

$$\begin{aligned} N &= \sum_{j_i}(2j_i+1), \qquad \langle J_z^2\rangle^1 = \frac{1}{3N}\sum_{j_i} j_i(j_i+1)(2j_i+1), \\ \langle J_z^4\rangle^m &= \frac{m(N-m)}{(N-1)(N-2)(N-3)}\big\{\big[N(N+1)-6m(N-m)\big]\langle J_z^4\rangle^1 \\ &\quad + 3N(m-1)(N-m-1)\big(\langle J_z^2\rangle^1\big)^2\big\}; \\ \langle J_z^4\rangle^1 &= \frac{1}{5N}\sum_{j_i}(2j_i+1)\bigg[\big\{j_i(j_i+1)\big\}^2 - \frac{j_i(j_i+1)}{3}\bigg]. \end{aligned} \tag{13.7}$$

Similarly, formula for $\langle J_z^6\rangle^1$ and hence for $\langle J_Z^6\rangle^m$ can be written down [16]. These will allow us to calculate easily $k_4(m : J_z)$ and $k_6(m : J_z)$. Equation (13.6) works extremely well in practice even for small number of j orbits and m is not too large.

Definition of normalized fixed-(E, M) density of states, which is a bivariate density, is

$$\rho^{H,m}(E,M) = \big\langle\delta(H-E)\delta(J_z-M)\big\rangle^m = \big\{d(m)\big\}^{-1}\big\langle\!\langle\delta(H-E)\delta(J_z-M)\big\rangle\!\rangle^m. \tag{13.8}$$

The operators H and J_z whose eigenvalues are E and M respectively, commute and therefore the bivariate moments of $\rho^{H,m}(E, M)$ are just $M_{rs}(m) = \langle H^r J_z^s\rangle^m$; note that we are considering only those effective Hamiltonians (true for nuclei and atoms) that are all J scalars. Many times we will drop m in $M_{rs}(m)$ and similarly in the corresponding bivariate cumulants $k_{rs}(m)$. It is plausible to use bivariate ED expansion for $\rho(E, M)$ [17]. Changing $\rho(E, M)$ to $\eta(\widehat{E}, \widehat{M})$ where the standardized variables $\widehat{E} = (E - E_c(m))/\sigma(m)$ and $\widehat{M} = M/\sigma_{J_z}(m)$ with $E_c(m) = \langle H\rangle^m$ and $\sigma^2(m) = \sigma^2_H(m) = \langle H^2\rangle^m - [E_c(m)]^2$, the ED expansion for $\eta(\widehat{E}, \widehat{M})$ is [17],

$$\begin{aligned} \eta(\widehat{E},\widehat{M}) = \eta_{\mathscr{G}}(\widehat{E})\eta_{\mathscr{G}}(\widehat{M})\bigg[1 &+ \bigg\{\frac{k_{30}(m)}{3!}He_3(\widehat{E}) + \frac{k_{12}(m)}{2!}He_1(\widehat{E})He_2(\widehat{M})\bigg\} \\ &+ \bigg\{\frac{k_{40}(m)}{4!}He_4(\widehat{E}) + \frac{k_{22}(m)}{2!2!}He_2(\widehat{E})He_2(\widehat{M}) + \frac{k_{04}(m)}{4!}He_4(\widehat{M}) \\ &+ \frac{[k_{30}(m)]^2}{2!3!3!}He_6(\widehat{E}) + \frac{[k_{12}(m)]^2}{(2!)^3}He_2(\widehat{E})He_4(\widehat{M}) \\ &+ \frac{k_{30}(m)k_{12}(m)}{3!2!}He_4(\widehat{E})He_2(\widehat{M})\bigg\} + O\big(1/m^{3/2}\big)\bigg]. \end{aligned} \tag{13.9}$$

Note that $He_{rs}(\widehat{E}, \widehat{M}) = He_r(\widehat{E})He_s(\widehat{M})$ as $\zeta = 0$. Fixed-J averages of a J invariant operator $\mathcal{O}$ follow from fixed-M averages using,

$$
\begin{aligned}
\langle \mathcal{O} \rangle^{m,J} &= \frac{\langle\langle \mathcal{O} \rangle\rangle^{m,M=J} - \langle\langle \mathcal{O} \rangle\rangle^{m,M=J+1}}{D(m, M=J) - D(m, M=J+1)} \\
&\simeq \left[-\frac{\partial D(m,M)}{\partial M}\bigg|_{M=J+1/2} \right]^{-1} \left[-\frac{\partial \langle\langle \mathcal{O} \rangle\rangle^{M}}{\partial M}\bigg|_{M=J+1/2} \right].
\end{aligned} \tag{13.10}
$$

Equations (13.9) and (13.10) will give expansions for fixed-(m, J) dimensions, energy centroids and spectral variances [18, 19]. Fixed-J dimension is given by,

$$
\begin{aligned}
d(m,J) &= \left[\binom{N}{m} (2J+1) \Big/ \sqrt{8\pi} \sigma_{J_z}^3(m) \right] \exp - \frac{(J+1/2)^2}{2\sigma_{J_z}^2(m)} \\
&\quad \times \left[1 + \frac{K_{04}(m)}{24} \left\{ \left[\frac{J(J+1)}{\sigma_{J_z}^2(m)} \right]^2 - 10 \frac{J(J+1)}{\sigma_{J_z}^2(m)} + 15 \right\} \right].
\end{aligned} \tag{13.11}
$$

Similarly, up to $[J(J+1)]^2$ correction, $E_c(m, J)$ is

$$
\begin{aligned}
\langle \widehat{H} \rangle^{m,J} &= \frac{E_c(m,J) - E_c(m)}{\sigma(m)} \\
&= \left[\frac{k_{12}(m)}{2} \left(-3 + \frac{1}{4\sigma_{J_z}^2(m)} \right) + \frac{k_{14}(m)}{8} \left(5 - \frac{5}{6\sigma_{J_z}^2(m)} + \frac{1}{48\sigma_{J_z}^4(m)} \right) \right. \\
&\quad \left. + \frac{k_{04}(m)k_{12}(m)}{4} \left(-5 + \frac{5}{4\sigma_{J_z}^2(m)} - \frac{1}{24\sigma_{J_z}^4(m)} \right) \right] \\
&\quad + \frac{J(J+1)}{\sigma_{J_z}^2(m)} \left\{ \frac{k_{12}(m)}{2} + \frac{k_{14}(m)}{12} \left(-5 + \frac{1}{4\sigma_{J_z}^2(m)} \right) \right. \\
&\quad \left. + \frac{k_{04}(m)k_{12}(m)}{4} \left(5 - \frac{1}{3\sigma_{J_z}^2(m)} \right) \right\} \\
&\quad + \frac{[J(J+1)]^2}{\sigma_{J_z}^4(m)} \left\{ \frac{k_{14}(m)}{24} - \frac{k_{04}(m)k_{12}(m)}{6} \right\} \\
&\simeq \left[-\frac{3k_{12}(m)}{2} \right] + \frac{k_{12}(m)}{2} \frac{J(J+1)}{\sigma_{J_z}^2(m)} \\
&\quad + \left\{ \frac{k_{14}(m)}{24} - \frac{k_{04}(m)k_{12}(m)}{6} \right\} \frac{[J(J+1)]^2}{\sigma_{J_z}^4(m)}.
\end{aligned} \tag{13.12}
$$

The last step here follows from the assumption that $\sigma_{J_z}^2(m) \gg 1$. Proceeding further, we will obtain $\sigma^2(m, J)$ with $[J(J+1)]^2$ correction,

$$
\begin{aligned}
&\sigma^2(m, J)/\sigma^2(m) \\
&\quad = \langle \widehat{H}^2 \rangle^{m,J} - \left(\langle \widehat{H} \rangle^{m,J}\right)^2 \\
&\quad = \left[1 - \frac{3k_{22}(m)}{2} + \frac{3[k_{12}(m)]^2}{2} + \frac{5k_{24}(m)}{8} - \frac{5k_{14}(m)k_{12}(m)}{2}\right. \\
&\qquad \left. - \frac{5k_{22}(m)k_{04}(m)}{4} + \frac{15k_{04}(m)[k_{12}(m)]^2}{4}\right] + \left[\frac{J(J+1)}{\sigma^2_{J_z}(m)} + \frac{1}{4\sigma^2_{J_z}(m)}\right] \\
&\qquad \times \left\{\frac{k_{22}(m)}{2} - \left[k_{12}(m)\right]^2 - \frac{5k_{24}(m)}{12} + \frac{5k_{14}(m)k_{12}(m)}{2} - 5k_{04}(m)\left[k_{12}(m)\right]^2\right. \\
&\qquad \left. + \frac{5k_{22}(m)k_{04}(m)}{4}\right\} + \left[\frac{J(J+1)}{\sigma^2_{J_z}(m)} + \frac{1}{4\sigma^2_{J_z}(m)}\right]^2 \left\{\frac{k_{24}(m)}{24} - \frac{k_{14}(m)k_{12}(m)}{3}\right. \\
&\qquad \left. - \frac{k_{22}(m)k_{04}(m)}{6} + \frac{5k_{04}(m)[k_{12}(m)]^2}{6}\right\} \\
&\quad \simeq \left[1 - \frac{3k_{22}(m)}{2}\right] + \left[\frac{k_{22}(m)}{2}\right]\frac{J(J+1)}{\sigma^2_{J_z}(m)} \\
&\qquad + \left\{\frac{k_{24}(m)}{24} - \frac{k_{22}(m)k_{04}(m)}{6}\right\}\left[\frac{J(J+1)}{\sigma^2_{J_z}(m)}\right]^2.
\end{aligned}
\tag{13.13}
$$

In the last step here, assuming that $\sigma^2_{J_z}(m) \gg 1$, neglected is the $1/4\sigma^2_{J_z}(m)$ terms and so also the terms with squares and products of cumulants that are expected to be small. Using trace propagation formalism, it is possible to derive the cumulants needed for $[J(J+1)]^2$ corrections for EGOE(2)-($j : J$) and up to $J(J+1)$ corrections for EGOE(2)-($j_1, j_2, \ldots, j_r : J$) ensemble [18, 19]. In order to gain some insight, let us consider EGOE(2)-($j : J$) energy centroids and spectral variances.

In the dilute limit with $m \to \infty$, $N \to \infty$ and $m/N \to 0$, the centroids $E_c(m, J)$ take a simple form. Firstly, the constant term in the expansion for $E_c(m, J)$ is

$$
E_c(m) - 3\sigma(m)\frac{k_{12}(m)}{2} \simeq \frac{m^2}{N^2}\sum_{J_2}(2J_2 + 1)V_{J_2}. \tag{13.14}
$$

Similarly, the $J(J+1)$ term is

$$
\begin{aligned}
&\sigma(m)\frac{k_{12}(m)}{2\sigma^2_{J_z}(m)} \simeq \frac{3}{2[j(j+1)]^2N^2}\sum_{J_2}(2J_2 + 1)V_{J_2}\left(J^2\right)^{\nu=2}_{J_2}; \\
&\left(J^2\right)^{\nu=2}_{J_2} = J_2(J_2 + 1) - (2j - 1)(j + 1) \simeq -2Y_j \begin{Bmatrix} j & j & J_2 \\ j & j & 1 \end{Bmatrix}, \\
&\qquad Y_j = j(j + 1)(2j + 1).
\end{aligned}
\tag{13.15}
$$

More remarkable is that the $[J(J+1)]^2$ term $\frac{\sigma(m)k_{14}(m)}{24\sigma^4_{J_z}(m)} - \frac{\sigma(m)k_{04}(m)k_{12}(m)}{6\sigma^4_{J_z}(m)}$ also takes a simple form. The expression for the first term is,

$$\begin{aligned}
\sigma(m)\frac{k_{14}(m)}{24\sigma^4_{J_z}(m)} &= \sum_{J_2}(2J_2+1)V^{\nu=2}_{J_2}S_{J_2}, \\
S_{J_2} &\simeq \frac{9}{40m^2(N-m)^2N^2[j(j+1)]^4}\{3\big[(J^2)^{\nu=2}_{J_2}\big]^2(N-2m)^2 \\
&\quad - 4(J^2)^{\nu=2}_{J_2}j(j+1)\big[2N^2-2Nm+2m^2\big]\}.
\end{aligned} \tag{13.16}$$

Note that $V^{\nu=2}_{J_2} = V_{J_2} - \overline{V}$ where $\overline{V}$ is the average of V_{J_2}'s. More importantly, the expression for the second term $\frac{\sigma(m)k_{04}(m)k_{12}(m)}{6\sigma^4_{J_z}(m)}$, in the dilute limit, reduces exactly to the second piece in the expression for S_{J_2} in Eq. (13.16). Therefore, in the dilute limit, the term multiplying $[J(J+1)]^2$ in the $E_c(m,J)$ expansion is,

$$\begin{aligned}
\frac{\sigma(m)}{\sigma^4_{J_z}(m)}\left\{\frac{k_{14}(m)}{24} - \frac{k_{04}(m)k_{12}(m)}{6}\right\} &= \sum_{J_2}(2J_2+1)V^{\nu=2}_{J_2}R_{J_2}, \\
R_{J_2} &\sim \frac{9(N-2m)^2}{40m^2(N-m)^2N^2[j(j+1)]^4}\{3\big[J_2(J_2+1)-2j(j+1)\big]^2\}.
\end{aligned} \tag{13.17}$$

The final formulas given by Eqs. (13.14), (13.15) and (13.17) [for the constant, $J(J+1)$ and $[J(J+1)]^2$ terms] are very close to those given by Mulhall et al. [20, 21]. They have used Fermi occupancies and cranking approximation for J projection.

13.1.3 Probability for Spin 0 Ground States and Distribution of Spectral Widths in $(j)^m$ Space

Equations (13.14), (13.15) and (13.17) can be used to write $E_c(m,J)$ in the form,

$$\begin{aligned}
&E_c(m,J) = c + aJ(J+1) + b\big[J(J+1)\big]^2; \\
&a = \sum_{J_2} a_{J_2}V_{J_2}, \qquad b = \sum_{J_2} b_{J_2}V_{J_2}.
\end{aligned} \tag{13.18}$$

Assuming that V_{J_2} are independent $G(0,\sigma_{J_2})$ variables, the probability distribution for (a,b) is

$$\begin{aligned}
&P(a,b) = \frac{1}{2\pi\sqrt{AB-D^2}}\exp -\frac{Ba^2-2Dab+Ab^2}{2(AB-D^2)}; \\
&A = \sum_{J_2} a^2_{J_2}\sigma^2_{J_2}, \qquad B = \sum_{J_2} b^2_{J_2}\sigma^2_{J_2}, \qquad D = \sum_{J_2} a_{J_2}b_{J_2}\sigma^2_{J_2}.
\end{aligned} \tag{13.19}$$

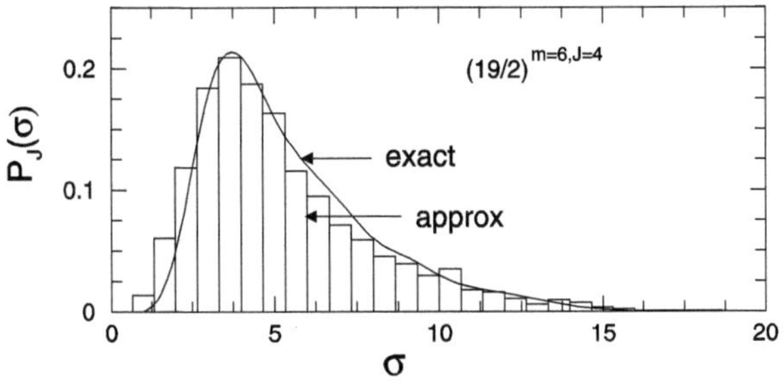

Fig. 13.1 Probability distribution for widths σ for a EGOE(2)-($j : J$) ensemble with six fermions in $j = 19/2$ orbit. In the figure, exact results (denoted by 'exact') are from [22] and they are obtained using Eq. (13.26). Results from the moment expansion to order $[J(J+1)]^2$, given by Eqs. (13.12) and (13.13), are denoted by 'approx'. Figure is taken from [18] with permission from World Scientific

Then the probability $f(0)$ for $E_c(m, J)$ with $J = 0$ to be lowest in energy is given by [20], with $J_{max}^2 = J_{max}(J_{max} + 1)$,

$$f(0) = \int_{a \geq 0, b \geq -a/J_{max}^2} P(a, b) da db$$
$$= \frac{1}{4} + \frac{1}{2\pi} \arctan\left[\frac{D + A/J_{max}^2}{AB - D^2}\right]. \tag{13.20}$$

Going to fixed-J variances, in the dilute limit, simplifying $k_{22}(m)$ and $\sigma^2(m)$ will give, by including the constant and $J(J+1)$ term,

$$\sigma^2(m, J) = \frac{m^2}{N^2} \sum_{J_2} (2J_2 + 1)\left(V_{J_2}^{\nu=2}\right)^2$$
$$+ \frac{3J(J+1)}{2N^2[j(j+1)]^2} \sum_{J_2} (2J_2 + 1)\left(V_{J_2}^{\nu=2}\right)^2 \left(J^2\right)_{J_2}^{\nu=2}. \tag{13.21}$$

The $[J(J+1)]^2$ correction term is given in [18] and it is not presented here as the formula for $k_{24}(m)$ is quite cumbersome. With the cumulants $k_{rs}(m)$ calculated using the formulas given in [18, 19] and Eqs. (13.12) and (13.13) for $E_c(m, J)$ and $\sigma^2(m, J)$ respectively, $\langle H^2 \rangle^{mJ}$ for each member of the ensemble can be obtained. Using these, probability distribution for widths $P_J(\sigma)$ vs σ curves for various J values are constructed and the results are shown in Fig. 13.1 for $J = 4$ for a $(19/2)^{m=6}$ system with 2500 members. In this example, $\sigma_{J_z}(m) = 12.124$ and $k_{04}(m) = -0.229$. Similarly, $\overline{k_{12}(m)}, \overline{k_{14}(m)} \sim 0$ as expected. However, $\overline{k_{22}(m)} = -0.053$ and $\overline{k_{24}(m)} = -0.114$. As $k_{22}(m)$ and $k_{24}(m)$ are large, the expansion to order $[J(J+1)]^2$ given by Eqs. (13.12) and (13.13) are needed. It is found that the expansions are good for $J < 30$ (note that $J_{\max} = 42$ for the example considered). It is seen from Fig. 13.1 that the calculated histogram is in good agreement with the exact curve given in [22]. Though not shown in the Fig. 13.1, it is seen that for $J=0$ the widths given by the exact results are larger than the numbers given by the expansion to $[J(J+1)]^2$ order and this could be because $J=0$ is an extreme J value.

Exact result for $P_J(\sigma)$ was derived by Papenbrock and Weidenmüller [22] using a different formulation and it is as follows. In terms of the TBME V_{J_2}, for EGOE(2)-J, the H matrix in (m, J) space can be written as,

$$H(m,J)=\sum_{J_2} V_{J_2} C_{J_2}(m,J). \tag{13.22}$$

The $C_{J_2}(m,J)$ are geometric factors transporting the TBME information from two particle space to m-particle spaces with a fixed J value. From now on we will use $v_\alpha = V_{2\alpha-2}$; $\alpha = 1, 2, \ldots, (j+1/2)$. Using Eq. (13.22), formula for the spectral widths is

$$\begin{aligned} \sigma^2(m,J) &= \big[d(m,J)\big]^{-1} \langle\langle H^2(m,J)\rangle\rangle \\ &= \big[d(m,J)\big]^{-1} \sum_{\alpha,\beta} v_\alpha v_\beta \langle\langle C_\alpha(m,J) C_\beta(m,J)\rangle\rangle. \end{aligned} \tag{13.23}$$

It is possible to diagonalize the overlap matrix $S_{\alpha,\beta} = [d(m,J)]^{-1} \langle\langle C_\alpha(m,J) \times C_\beta(m,J)\rangle\rangle$ using a unitary matrix $U(m,J)$ giving eigenvalues $s_\alpha^2(m,J)$. We can define $B_\alpha(m,J) = \sum_\beta U_{\beta\alpha}(m,J) C_\beta(m,J)$ and they have the important property

$$\big[d(m,J)\big]^{-1} \langle\langle B_\alpha(m,J) B_\beta(m,J)\rangle\rangle = \delta_{\alpha\beta} s_\alpha^2(m,J).$$

The $B_\alpha(m,J)$ act as unitary tensors in (m,J) space. Therefore, expanding $H(m,J)$ in terms of B's will give

$$H(m,J)=\sum_\alpha w_\alpha B_\alpha(m,J) \tag{13.24}$$

with the property that w_α are independent $G(0,1)$ variables given that V_{J_2} are independent $G(0,1)$ variables. Therefore we have,

$$\sigma^2(m,J)=\sum_\alpha w_\alpha^2 s_\alpha^2(m,J). \tag{13.25}$$

Thus, all the width information is in the eigenvalues $s_\alpha \geq 0$ and it is found in numerical calculations that one particular s_α is always large and the rest are very small. Equation (13.25) gives the exact formula for $P_J(\sigma)$ in an integral form involving s_α and the result is [22],

$$P_J(\sigma) = \frac{\sigma}{\pi} \int_{-\infty}^{\infty} dt \exp\big(it\sigma^2\big) \prod_\alpha \frac{\exp -\{(i/2)\arctan[2ts_\alpha^2(J)]\}}{(1+4t^2 s_\alpha^4(J))^{1/4}}. \tag{13.26}$$

Note that for brevity, here we have dropped m. Applications of Eqs. (13.20) and (13.26) will be discussed in Chap. 14.

13.1.4 Extensions of EGOE(1 + 2)-($j_1, j_2, \ldots, j_r : J$)

There are several ways EGOE(1 + 2)-($j_1, j_2, \ldots, j_r : J$) ensemble can be modified. One extension, appropriate for nuclei, is to consider proton and neutron j orbits separately. Then, $\hat{h}(1) \to \hat{h}(1:p) + \hat{h}(1:n)$ and $\hat{V}(2) \to \hat{V}(2:pp) + \hat{V}(2:nn) + \hat{V}(2:pn)$ in Eq. (13.5). Now, given number of protons to be m_p and neutrons to be m_n, one can construct H matrix in (m_p, m_n, J) spaces with J being the total angular momentum for the system. Choosing $V^{J_{12}}(pp)$, $V^{J_{12}}(nn)$ and $V^{J_{12}}(pn)$ matrices in two particle spaces to be independent GOEs, we have EGOE(1 + 2) in (m_p, m_n, J) spaces. This (dropping the labels of proton and neutron j orbits) ensemble is referred as EGOE(1 + 2)-(m_p, m_n, J).

For nuclei with protons and neutrons occupying the same j orbits, it is necessary to consider isospin (T) quantum number and H that preserves both J and T quantum numbers. Then, $\hat{h}(1)$ is independent of proton and neutron degrees of freedom and $\hat{V}(2)$ will have isospin $T_{12} = 0$ and $T_{12} = 1$ parts. Representing the $V^{J_{12},T_{12}}(2)$ matrices by independent GOEs, we have EGOE(1 + 2)-JT in (m, J, T) spaces. Note that $\hat{h}(1)$ is defined by the sp energies ε_j. Nuclear shell model codes [4, 5] allow one to construct EGOE(1 + 2)-JT ensemble numerically and analyze. For example for nuclear ($2s1d$) shell the sp orbits will have $j = 5/2$, $3/2$ and $1/2$. Then the total number of independent two-body matrix elements $\langle (j_1 j_2)J_{12}, T_{12} | V(2) | (j_3 j_4) J_{12}, T_{12} \rangle$ will be 63 in number. Choosing these 63 matrix elements to be independent $G(0, v^2)$ random variables and then using the shell model codes for each realization of $V(2)$ matrix in two particle spaces, one can construct the H matrices in (m, J, T) spaces. This then generates numerically EGOE(2)-(m, J, T) in a given (m, J, T) space if we choose the sp energies to be degenerate; note that here m is explicitly specified. An important application of this construction is discussed in Sect. 13.1.5.

Another extension, appropriate when L–S coupling is good, is to consider two-particle matrices in LSJ spaces to be independent GOEs and then generate H matrix ensembles in (m, L, S, J) spaces. This gives EGOE(1 + 2)-(m, L, S, J) and this ensemble is useful not only for nuclei but also for atoms. Another obvious extension is to EGOE(1 + 2)-(m, L, S, J, T) and this is useful for $N = Z$ nuclei. Another important extension is to consider 3-, 4- and general k-body Hamiltonians [Appendix G gives a short discussion on some properties of EGOE(3)] and construct EGOE ensembles with good J, JT, LST and so on and these k-body ($k > 2$) ensembles have so far received limited attention [9, 23]. All these extended ensembles are used in the analysis of regular structures generated by random interactions as discussed in the next chapter.

13.1.5 Cross Correlations in EGOE(2)-(m, J, T)

Constructing EGOE(2)-(m, J, T) for m nucleons occupying a given set of shell model j orbits, it is easy to see that this ensemble generates cross correlations just as the situation with various EE discussed in Chap. 12. This is because of the H matrix elements in (m, J, T) spaces are linear combination of the two-particle matrix elements and these defining matrix elements remain same (independent of m, J and T values) as long as the j orbits are not changed. Given the density of energy levels [then $(2J+1)(2T+1)$ degeneracy factor is not counted] $\rho(E:m,J,T)$, cross correlations are defined by the two-point function

$$S\bigl(E,W:m,J,T:m',J',T'\bigr) = \overline{\rho(E:m,J,T)\rho\bigl(W:m',J',T'\bigr)} - \overline{\rho(E:m,J,T)}\,\overline{\rho\bigl(W:m',J',T'\bigr)}. \quad (13.27)$$

As an example, considered in [12, 13] is the $(2s1d)$ shell mentioned in Sect. 13.1.4 and constructed a 400 member EGOE(2)-(m,J,T) for $(m=8, J=0, T=0)$, $(m=8, J=2, T=0)$ and $(m=6, J=0, T=0)$ with H matrix dimensions 325, 1206 and 71 respectively. Here, $m=8$ corresponds to ^{24}Mg and $m=6$ corresponds to ^{22}Ne. Using these, $\overline{\rho(E:8,0,0)\rho(W:8,2,0)}$, $\{\overline{\rho(E:8,0,0)}\}\{\overline{\rho(W:8,2,0)}\}$ and $S(E,W:8,0,0:8,2,0)$ are calculated and the results are shown in Fig. 13.2a. This gives cross correlations between spectra with different J values with fixed (m,T) values. It is seen from the figure that the cross correlations are ∼13 %. Similarly, shown in Fig. 13.2b are the results for $\overline{\rho(E:8,0,0)\rho(W:6,0,0)}$, $\{\overline{\rho(E:8,0,0)}\}\{\overline{\rho(W:6,0,0)}\}$ and $S(E,W:8,0,0:6,0,0)$. This gives cross correlations in spectra with different particle numbers with fixed JT values. Here, it is seen from the figure that the cross correlations are 6 %. As we are using EGOE(2)-JT, the final results will not depend on the variance v^2 of the two-particle matrix elements [they are $G(0,v^2)$ variables]. Let us mention that a structure quite similar to the one in Figs. 13.2a and b was also seen in a EGOE(2)-**s** calculation [24] and there the correlations are ∼1 %. Thus, consistent with the result seen in Table 12.1, cross correlations are enhanced considerably in EGOE(2)-JT as the JT symmetry reduces the number of independent two-particle matrix elements much more than the spin (**s**) symmetry. Another important observation, as seen from the Figs. 13.2a and b, is that the cross correlation has a minimum in the center of the two spectra. This indicates that significant correlations exist only in the tails of the spectra. As most of the low-lying levels will be in the tails of the spectra, this result is important. It would be of interest if the existence of these correlations can be verified in experimental data. Keeping this in mind, Papenbrock and Weidenmüller carried out a numerical analysis as follows. Firstly, the ensemble average is replaced by average over an ensemble of nuclei in the same shell. They have considered the nuclei $^{20-24}$Ne, $^{22-24}$Na, $^{24-26}$Mg, ^{26}Al, ^{30}Si, ^{34}P, ^{32,34}S and ^{36}Ar. Then, starting with a $(2s1d)$ shell two-particle interaction, studied for even (odd) mass nuclei are correlations of the $J=0$ and $J=2$ ($J=1/2$ and $J=5/2$) states. Labeling the nearest-neighbor spacings of levels with a given J consecutively, correlations between nearest-neighbor level spacings of the lowest few states with different spins were evaluated as in Eq. (13.27) with ensemble average being replaced by the running average over the set of nuclei listed above. Final results presented in Fig. 13.2c show that the correlations are ∼10 % for $(2s1d)$ shell nuclei. Experimental verification of this result requires new experiments measuring complete spectral sequences with at least ∼10 levels for several nuclei.

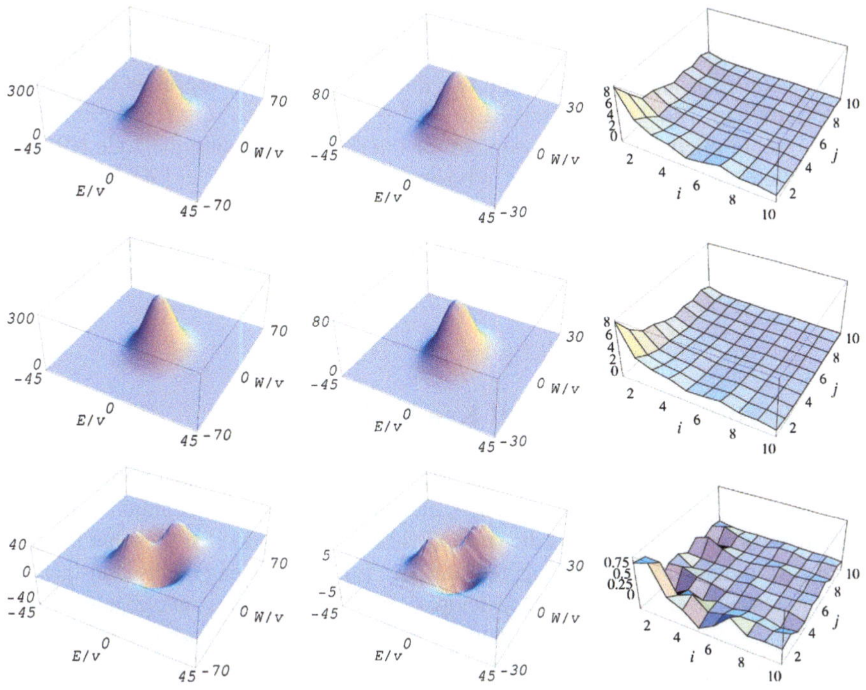

(a) **(b)** **(c)**

Fig. 13.2 (**a**) Correlations between levels with $J = 0$, $T = 0$ and $J = 2$, $T = 0$ for ^{24}Mg. *Top figure* is for the average of the product of the two level densities, *middle figure* is for the product of the averages of the level densities and the *bottom figure* gives their difference, i.e. the correlator S defined by Eq. (13.27). (**b**) same as (**a**) but for $J = 0$, $T = 0$ of ^{24}Mg and $J = 0$, $T = 0$ of ^{22}Na. (**c**) Same as (**a**) and (**b**) but for the correlations between pairs of nearest-neighbor spacings of low-lying levels with different J values and averaged over a number of $(2s1d)$ shell nuclei. The indices i and j label the spacings consecutively starting from the lowest level. See text for further details. Figure is taken from [13] with permission from American Physical Society

13.2 BEGOE(1 + 2)-($\ell_1, \ell_2, \ldots, \ell_r : L$) Ensembles

Interacting boson models (IBMs) are quite successful in nuclear and molecular physics. In one version of these models, bosons are assumed to carry angular momentum $\ell = 0^+$ (s), 2^+ (d), 4^+ (g), 1^- (p), 3^- (f) and so on. The Hamiltonians here preserve total angular momentum L of the bosons and they are usually considered to be $(1+2)$-body. Then the H operator can be written in the same form as in Eq. (13.5) by replacing j by ℓ and the fermion operators by boson operators appropriately. Again, choosing the boson TBME to be independent $G(0, v^2)$ variables and then constructing the H matrices in m boson spaces with fixed-L, we have BEGOE(1 + 2)-($\ell_1, \ell_2, \ldots, \ell_r : L$) ensemble. It is simplest to construct the ensemble for spIBM as here one can write down simple formulas for the matrix elements in (m, L) spaces using $|m_s, m_p, L\rangle$ basis with $m = m_s + m_p$ [25]. Appendix H gives details of BEGOE(1 + 2)-($s, p : L$). On the other hand, construction of these in (m, L) spaces is possible for sdIBM using Scholten code [26], for sdgIBM using Devi and Kota code [27] and for $sdpf$IBM using Kusnezov code [28]. It is also possible to consider boson systems with two kinds of bosons and represent them by F spin (see Chap. 10). Then one has BEGOE(1 + 2)-($\ell_1, \ell_2, \ldots, \ell_r : L, F$) ensemble. It is possible to construct this ensemble for $(sd)^m$ systems using the code NPBOS [29]. Other possible extensions are to bosons carrying isospin $T = 1$ degree of freedom and also spin-isospin degrees of freedom $(ST) = (10) + (01)$. There is interest in the sd boson version of these ensembles [30]. Many of these bosonic ensembles are used in the analysis of regular structures generated by random interactions, the topic of next chapter.

13.3 Partitioned EGOE and K + EGOE

In indefinitely large spectroscopic spaces a EGOE(1 + 2) generated with the same variance for all the two particle matrix elements that are zero centered Gaussian variables, is not appropriate as the space divides into distant subspaces that interact weakly (the interaction within a given subspace is usually strong). This is in-fact the situation whenever there is shell structure (examples are atoms, nuclei and atomic clusters). In these situations it is more appropriate to consider partitioned embedded ensembles (p-EE). For example, in many nuclear structure studies such as level densities, Gamow-Teller strength distributions, giant dipole resonance strengths and widths etc., it is important to include multi-$\hbar\omega$ excitations in the shell model spaces. In such nuclear structure calculations one should remember that the shell model stability ensures that the mixing between the distant multi-$\hbar\omega$ configurations is weak. Therefore, the H matrix in two particle spaces (we assume H to be two-body) will be a block structured matrix with each diagonal block denoting the configurations that are far apart (for light nuclei, they correspond to $0\hbar\omega$, $2\hbar\omega$, $4\hbar\omega$, ... excitations for states with parity same as the ground state) and the off-diagonal blocks giving the mixing between these distant configurations. Then we have p-EGOE(2) [or p-EGOE(1 + 2)] defined by a partitioned GOE in the two particle spaces [31–33]. Here, the variances of the TBME in each diagonal as well as off-diagonal blocks are different and also the centroid of the diagonal blocks are different (one-point function for the partitioned GOE was solved in [34]).

Let us consider an example with 3 shells Λ_{-1}, Λ_0 and Λ_{+1} with number of single particle states N_{-1}, N_0 and N_{+1} and parities $-$, $+$ and $-$ respectively. In general they denote, as shown in Fig. 13.3a, the closed, valance and open shells respectively. For simplicity, J, T and other quantum numbers are ignored. Spacing between the shells is $\Delta \sim \hbar\omega$ and it is much larger than the splitting between the sp states in a given shell. Now the H matrix in two-particle spaces will be a 6×6 block matrix as shown in Fig. 13.3b with the upper 4×4 matrix for +ve parity and the lower 2×2 matrix for $-$ve parity with no connection between the two (assuming parity is conserved). Note that the first block is H in the $0\hbar\omega$ space and the next three blocks correspond to $2\hbar\omega$. Thus, the off-diagonal block matrices A, B and C generate mixing between 0 and $2\hbar\omega$ +ve parity configurations. Therefore, the variance of

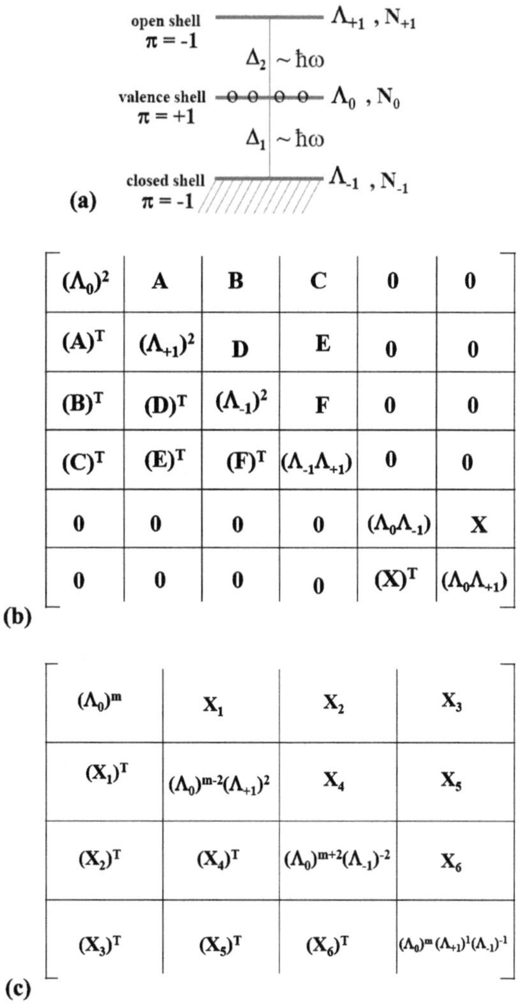

Fig. 13.3 (a) Single particle spectrum with closed, valance and open shells. As an example, 4 particles occupying the valance shell is shown. (b) Block structure of the H matrix in two particle spaces. (c) Block structure of the H matrix up to $(0+2) - \hbar\omega$ excitation in m particle spaces. See text for further details

(a) Single particle spectrum:
- open shell, $\pi = -1$: Λ_{+1}, N_{+1}
- $\Delta_2 \sim \hbar\omega$
- valence shell, $\pi = +1$: Λ_0, N_0
- $\Delta_1 \sim \hbar\omega$
- closed shell, $\pi = -1$: Λ_{-1}, N_{-1}

(b)

$(\Lambda_0)^2$	A	B	C	0	0
$(A)^T$	$(\Lambda_{+1})^2$	D	E	0	0
$(B)^T$	$(D)^T$	$(\Lambda_{-1})^2$	F	0	0
$(C)^T$	$(E)^T$	$(F)^T$	$(\Lambda_{-1}\Lambda_{+1})$	0	0
0	0	0	0	$(\Lambda_0\Lambda_{-1})$	X
0	0	0	0	$(X)^T$	$(\Lambda_0\Lambda_{+1})$

(c)

$(\Lambda_0)^m$	X_1	X_2	X_3
$(X_1)^T$	$(\Lambda_0)^{m-2}(\Lambda_{+1})^2$	X_4	X_5
$(X_2)^T$	$(X_4)^T$	$(\Lambda_0)^{m+2}(\Lambda_{-1})^{-2}$	X_6
$(X_3)^T$	$(X_5)^T$	$(X_6)^T$	$(\Lambda_0)^m (\Lambda_{+1})^1 (\Lambda_{-1})^{-1}$

the matrix elements in the diagonal blocks should be much larger than those of A, B and C matrix elements. Also the centroid of the diagonal matrix elements of the first block should differ by $\sim 2\Delta$ from those of the next three blocks in the figure. The matrices D, E and F will mix the three different $2\hbar\omega$ configurations and the matrix elements variance of these matrices can be comparable. For negative parity states, the structure, as shown in the figure, is much simpler. Choosing appropriate values for the centroids of the diagonal matrix elements of the diagonal blocks (for others the Gaussian variables are zero centered) and similarly the variances for the diagonal as well as the off-diagonal blocks, the H matrix ensemble (it will be a partitioned GOE) in two-particle spaces can be defined. Using this, in m (many) particle spaces, it is straightforward to construct p-EGOE(2) H matrices on a machine by applying the formulation given in Chap. 4 [this is similar to the EGOE$(1+2)$-π

construction discussed in Chap. 8]. In general one is interested first in the m particle H matrix in $(0+2)-\hbar\omega$ space. Then the p-EGOE(2) will be a 4×4 block structured matrix as shown in Fig. 13.3c. As there are too many parameters in two-particle spaces, the p-EGOE(2) ensemble is quite hard to be solved analytically. As the matrix dimensions will be very large (it is easy to write the dimensions of the matrices in Figs. 13.3b and c), it is also difficult to handle p-EGOE(2) numerically. Therefore, a highly simplified version of p-EGOE(2) was solved [32] and verified numerically in [33] showing that the eigenvalue density will be in general multi-modal. However, practically usable multi-modal forms are not yet available. This is a very important open problem. It is useful to add that a particular form of p-EGOE(1 + 2), called the layer model in [35], was solved for a system with $\sim$50 sp states in [36] to address the problem of localized to delocalized phase transition in Fock-space; see also [37–39].

Another ensemble that is useful is $\mathscr{K} + \alpha$ EE where for example $\mathscr{K}$ is a fixed Hamiltonian such as the pairing (P) or pairing plus quadrupole–quadrupole $(P + QQ)$ interaction which generate regular features seen in the low-lying states in atomic nuclei. The EE for example may be EGOE(2), EGOE(1 + 2) or EGOE(1 + 2)-J. Preliminary numerical studies of some $\mathscr{K} + \alpha$ EE are given in [40, 41].

References

1. I. Talmi, *Simple Models of Complex Nuclei: The Shell Model and Interacting Boson Model* (Harwood Academic Publishers, Chur, 1993)
2. J.B. French, E.C. Halbert, J.B. McGrory, S.S.M. Wong, Complex spectroscopy, in *Advances in Nuclear Physics*, vol. 3, ed. by M. Baranger, E. Vogt (Plenum, New York, 1969), pp. 193–257
3. P.J. Brussaard, P.W.M. Glaudemans, *Shell Model Applications in Nuclear Spectroscopy* (North-Holland, Amsterdam, 1977)
4. B.A. Brown, W.D.M. Rae, Nushell@MSU, MSU-NSCL Report (2007)
5. E. Caurier, F. Nowacki, Present status of shell model techniques. Acta Phys. Pol. B **30**, 705–714 (1999)
6. V.V. Flambaum, A.A. Gribakina, G.F. Gribakin, M.G. Kozlov, Structure of compound states in the chaotic spectrum of the Ce atom: localization properties, matrix elements, and enhancement of weak perturbations. Phys. Rev. A **50**, 267–296 (1994)
7. V.V. Flambaum, A.A. Gribakina, G.F. Gribakin, I.V. Ponomarev, Quantum chaos in many-body systems: what can we learn from the Ce atom. Physica D **131**, 205–220 (1999)
8. D. Angom, V.K.B. Kota, Signatures of two-body random matrix ensembles in Sm I. Phys. Rev. A **67**, 052508 (2003)
9. T.A. Brody, J. Flores, J.B. French, P.A. Mello, A. Pandey, S.S.M. Wong, Random matrix physics: spectrum and strength fluctuations. Rev. Mod. Phys. **53**, 385–479 (1981)
10. J.M.G. Gómez, K. Kar, V.K.B. Kota, R.A. Molina, A. Relaño, J. Retamosa, Many-body quantum chaos: recent developments and applications to nuclei. Phys. Rep. **499**, 103–226 (2011)
11. V. Zelevinsky, B.A. Brown, N. Frazier, M. Horoi, The nuclear shell model as a testing ground for many-body quantum chaos. Phys. Rep. **276**, 85–176 (1996)
12. T. Papenbrock, H.A. Weidenmüller, Random matrices and chaos in nuclear spectra. Rev. Mod. Phys. **79**, 997–1013 (2007)
13. P. Papenbrock, H.A. Weidenmüller, Two-body random ensemble in nuclei. Phys. Rev. C **73**, 014311 (2006)

14. S. Sahoo, G.F. Gribakin, V. Dzuba, Recombination of low energy electrons with U^{28+}, arXiv:physics/0401157v1 [physics.atom-ph]
15. D. Angom, V.K.B. Kota, Chaos and localization in the wavefunctions of complex atoms NdI, PmI and SmI. Phys. Rev. A **71**, 042504 (2005)
16. V.K.B. Kota, R.U. Haq, *Spectral Distributions in Nuclei and Statistical Spectroscopy* (World Scientific, Singapore, 2010)
17. V.K.B. Kota, Bivariate distributions in statistical spectroscopy studies: I. Fixed-J level densities, fixed-J averages and spin cut-off factors. Z. Phys. A **315**, 91–98 (1984)
18. V.K.B. Kota, M. Vyas, K.B.K. Mayya, Spectral distribution analysis of random interactions with J-symmetry and its extensions. Int. J. Mod. Phys. E **17**(Supp), 318–333 (2008)
19. M. Vyas, Some studies on two-body random matrix ensembles, Ph.D. Thesis, M.S. University of Baroda, India (2012)
20. D. Mulhall, A. Volya, V. Zelevinsky, Geometric chaoticity leads to ordered spectra for randomly interacting fermions. Phys. Rev. Lett. **85**, 4016–4019 (2000)
21. D. Mulhall, Quantum chaos and nuclear spectra, Ph.D. Thesis, Michigan State University, East Lansing, USA (2002)
22. T. Papenbrock, H.A. Weidenmüller, Distribution of spectral widths and preponderance of spin-0 ground states in nuclei. Phys. Rev. Lett. **93**, 132503 (2004)
23. A. Volya, Emergence of symmetry from random n-body interactions. Phys. Rev. Lett. **100**, 162501 (2008)
24. M. Vyas, Random interaction matrix ensembles in mesoscopic physics, in *Proceedings of the National Seminar on New Frontiers in Nuclear, Hadron and Mesoscopic Physics*, ed. by V.K.B. Kota, A. Pratap (Allied Publishers, New Delhi, 2010), pp. 23–37
25. D. Kusnezov, Two-body random ensembles: from nuclear spectra to random polynomials. Phys. Rev. Lett. **85**, 3773–3776 (2000)
26. O. Scholten, *The Program Package PHINT, KVI Report* (University of Groningen, 1990). https://www.kvi.nl/scholten/
27. Y.D. Devi, V.K.B. Kota, Fortran programmes for spectroscopic calculations in (sdg)—boson space: the package SDGIBM1, Physical Research Laboratory (Ahmedabad, India), Technical Report PRL-TN-90-68 (1990)
28. D.F. Kusnezov, Nuclear collective quadrupole-octupole excitations in the $U(16)$ $spdf$ interacting boson model, Ph.D. Thesis, Yale University, USA (1988)
29. T. Otsuka, N. Yoshida, Users's manual of program NPBOS, Japan Atomic Energy Research Institute, Report JAERI-M/85-094 (1985)
30. V.K.B. Kota, Two-body ensembles with group symmetries for chaos and regular structures. Int. J. Mod. Phys. E **15**, 1869–1883 (2006)
31. J.B. French, V.K.B. Kota, Nuclear level densities and partition functions with interactions. Phys. Rev. Lett. **51**, 2183–2186 (1983)
32. J.B. French, in *Mathematical and Computational Methods in Nuclear Physics*, ed. by J.S. Dehesa, J.M.G. Gomez, A. Polls (Springer, Berlin, 1984), pp. 100–121
33. V.K.B. Kota, D. Majumdar, R. Haq, R.J. Leclair, Shell model tests of the bimodal partial state densities in a 2×2 partitioned embedded random matrix ensemble. Can. J. Phys. **77**, 893–901 (1999)
34. Z. Pluhař, H.A. Weidenmüller, Approximation for shell-model level densities. Phys. Rev. C **38**, 1046–1057 (1988)
35. B. Georgeot, D.L. Shepelyansky, Breit-Wigner width and inverse participation ratio in finite interacting Fermi systems. Phys. Rev. Lett. **79**, 4365–4368 (1997)
36. X. Leyronas, J. Tworzydlo, C.W.J. Beenakker, Non-Cayley-tree model for quasiparticle decay in a quantum dot. Phys. Rev. Lett. **82**, 4894–4897 (1999)
37. B.L. Altshuler, Y. Gefen, A. Kamenev, L.S. Levitov, Quasiparticle lifetime in a finite system: a nonperturbative approach. Phys. Rev. Lett. **78**, 2803–2806 (1997)
38. C. Mejía-Monasterio, J. Richert, T. Rupp, H.A. Weidenmüller, Properties of low-lying states in a diffusive quantum dot and Fock-space localization. Phys. Rev. Lett. **81**, 5189–5192 (1998)

39. V.K.B. Kota, Embedded random matrix ensembles for complexity and chaos in finite interacting particle systems. Phys. Rep. **347**, 223–288 (2001)
40. A. Cortes, R.U. Haq, A.P. Zuker, Transition between random and collective behaviour in spectra generated by two-body forces. Phys. Lett. B **115**, 1–6 (1982)
41. V. Velázquez, A.P. Zuker, Spectroscopy with random and displaced random ensembles. Phys. Rev. Lett. **88**, 072502 (2002)

Chapter 14
Regular Structures with Random Interactions: A New Paradigm

14.1 Introduction

Embedded random matrix ensembles opened up a new paradigm of regular structures with random interactions in isolated finite quantum systems. For the first time in 1998, Johnson, Bertsch and Dean, using the nuclear shell model, noticed that random two-body interactions lead to ground states, for even-even nuclei, having spin 0^+ with very high probability [1]. Similarly, Bijker and Frank [2] using interacting boson model of atomic nuclei found that random interactions generate vibrational and rotational structures with high probability. Examples are shown in Figs. 14.1 and 14.2 and Table 14.1. Later studies in nuclear structure with random interactions revealed statistical predominance of odd-even staggering in binding energies, the seniority pairing gap and 0^+, 2^+, 4^+, ... yrast sequence. Also seen are regularities in parity distributions in ground states of even-even, odd-A and odd-odd nuclei, in energy centroids, spectral variances and in many other quantities. On the other hand it is also found that random interactions generate, for systems with even number of fermions, spin zero ground states preferentially (see the discussion in Sect. 7.1.1 and Fig. 7.2) giving rise to delay in Stoner instability in itinerant systems and odd-even staggering in ground state energies in nm scale metallic grains (see the discussion in Sect. 7.1.2 and Figs. 7.3 and 7.4). The result that regular features can arise due to random interactions (with rotational and other symmetries) is opposed to the conventional ideas of using regular (or coherent) interactions like pairing in understanding the structure of nuclear and other systems. As Zelevinsky and Volya state [3], this is not limited to nuclear physics. Atomic clusters, particles in traps, quantum dots, disordered systems such as quantum spin glasses, are just a few examples where the same questions are to be answered—to what extent a realistic interaction can be random but still give the ground state and the levels near the yrast line to be realistic? References [3, 4] give early reviews on the topic of regular structures from random interactions.

V.K.B. Kota, *Embedded Random Matrix Ensembles in Quantum Physics*, Lecture Notes in Physics 884, DOI 10.1007/978-3-319-04567-2_14,

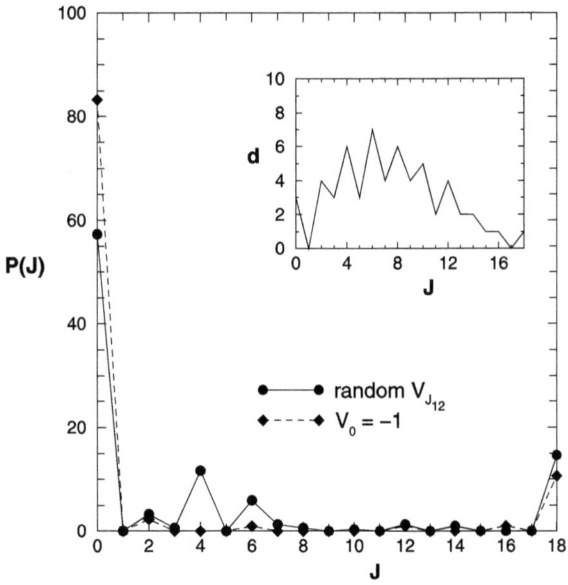

Fig. 14.1 Probability for spin-0 ground states for $(\frac{11}{2})^{m=6}$ system. Calculations use 1000 samples of random interactions preserving J, i.e. EGOE(2)-($j = \frac{11}{2} : J$) is used with 1000 members. Results obtained by putting $V_0 = -1$ (rest of the $V_{J_{12}}$ being Gaussian random variables) are also shown in the figure (results obtained by putting $V_0 = 0$ are almost same as those given by random interactions). In addition, the matrix dimensions $d(m = 6, J)$ are also shown in the insect figure. Figure is constructed using the results in [5]

Large number of numerical calculations for many particle systems are carried out using nuclear shell model, fermions in a single-j shell or two-j shells, bosons in a single ℓ orbit and interacting boson models (IBMs) for nuclei and molecules (spIBM, sdIBM, sdgIBM etc.) using ensembles of random interactions. For the preponderance of $J^\pi = 0^+$ states in even-even nuclei: (i) Zelevinsky and Volya [3] proposed the idea of 'geometric chaos' as the source of regularities seen in nuclear shell model results; (ii) Zhao et al. [4] proposed a empirical rule for describing the results for $(j)^m$ fermion and $(\ell)^m$ boson systems with extensions to more complicated systems; (iii) Papenbrock and Weidenmüller [6] proposed an explanation in terms of fixed-J spectral variances. On the other hand Kusnezov [7] showed that an approach based on random polynomials will apply if the H matrix is tridiagonal with analytical forms for the diagonal and the off-diagonal matrix elements known. This method is used to describe, completely analytically, the results for spIBM. Similarly, Bijker and Frank [8] employed mean-field methods for near quantitative understanding of the results for spIBM and sdIBM. The mean-field approach has been generalized to IBMs for two-level systems (with degeneracies n_1 and n_2 respectively) with $SO(n_1) \oplus SO(n_2)$ symmetry by Kota [9]. Similarly, regularities in energy centroids and spectral variances defined over symmetry subspaces generated by nuclear shell model and IBMs have been studied in a number of examples using trace propagation formulas. These group theoretical examples opened a new window to the study of regularities of many-body systems in the presence of random forces. We will now describe in this chapter these and other results on regularities generated by random interactions.

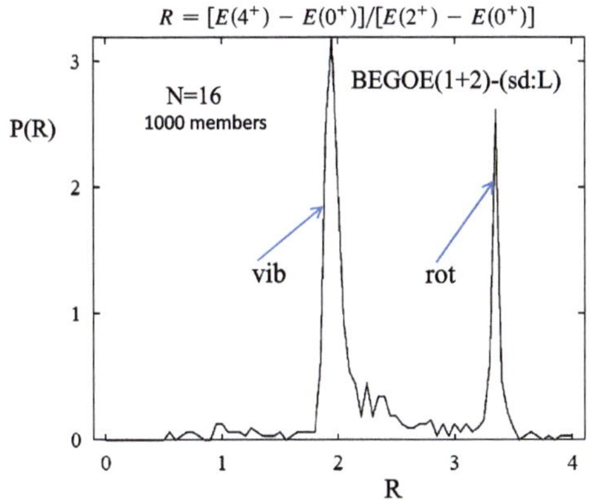

Fig. 14.2 Probability distribution $P(R)$ for the ratio $[E(4^+) - E(2^+)]/[E(2^+) - E(0^+)]$ with $\int P(R)dR = 1$ for a 1000 member BEGOE(1 + 2)-$(sd : L)$ for $m = 16$ bosons, i.e. for a 16 boson system in sdIBM with random one plus two-body interactions. Results in the figure clearly show that most members are either vibrational $[P(R) \sim 2]$ or rotational $[P(R) \sim 3.33]$. In the calculations, $P(R)$ is used for those members that gave $L = 0^+$ ground states. Figure is taken from [2] with permission from American Physical Society (Color figure online)

Table 14.1 Probabilities for ground states with $J^\pi = 0^+$ generated by random interactions in some $(2s1d)$ and $(2p1f)$ shell nuclei. Results are shown for EGOE(2)-J and its modifications as described in Sect. 14.2. Last column gives the percentage of 0^+ states in the model space and this is denoted by $d(0)/d$ in the table. All numbers in the table are in (%). Table is taken from [4]. See Sect. 14.2 for more details

Nuclei	EGOE(2)-J	RQE	RQE-NP	RQE-SPE	$d(0)/d$
^{20}O	50	68	50	49	11.1
^{22}O	71	72	68	77	9.8
^{24}O	55	66	51	78	11.1
^{44}Ca	41	70	46	70	5
^{46}Ca	56	76	59	74	3.5
^{48}Ca	58	72	53	71	2.9

14.2 Basic Shell Model and IBM Results for Regular Structures

Johnson et al. [1] considered examples of even-even nuclei in $(2s1d)$-shell with nucleons in $^1d_{5/2}$, $^2s_{1/2}$ and $^1d_{3/2}$ orbits with 63 independent TBME. Generating 1000 random interactions in this space, they have constructed Hamiltonian matrices for all allowed J values in many nucleon spaces using nuclear shell model codes. Using these, they have calculated the probability for the ground state to be $J^\pi = 0^+$. They have used EGOE(2)-J with degenerate sp energies for the three sp orbits. In addition, used are also three modified versions of EGOE(2)-J. One of them is with variance (v^2) of the two-particle matrix elements to be dependent on the two particle J_{12} and T_{12} values with $v^2(V^{J_{12},T_{12}}) = 1/[(2J_{12}+1)(2T_{12}+1)]$ as given by particle-hole symmetry. This is called random quasi-particle ensemble (RQE). The other two are RQE without monopole pairing part (called RQE-NP) and RQE with non degenerate sp energies (call RQE-SPE). All the calculations are also repeated for some examples in $(2p1f)$-shell with nucleons in ${}^1f_{7/2}$, ${}^2p_{3/2}$, ${}^2p_{1/2}$ and ${}^1f_{5/2}$ orbits with 195 independent TBME. Results of this study are shown in Table 14.1. It is seen that 0^+ gs appears with probability ∼40–70 % although the fraction of 0^+ states in the total space is less than 12 % in all the examples. Several other shell model examples have been given by Zelevinsky et al. with similar results [3]. Extensive analysis using fermions in one and two j-orbits and similarly bosons in a single ℓ-orbit, sp orbits (i.e. spIBM), sd orbits and sdg orbits have been carried out by Zhao et al. [4] and Zelevinsky et al. [3]. In order to understand preponderance of spin-0 ground states with EGOE(2)-J, Zhao et al. [4] gave a simple procedure. Say there are $\mathscr{K}$ number of TBME. Then, put one of the TBME to -1, the rest to zero and calculate the spectrum. Repeat this procedure $\mathscr{K}$ times putting one of the TBME to -1 each time. Say K_J is number of times the ground state is found to have spin J. Then, the probability to find spin-0 ground states is $K_0/\mathscr{K}$. This prescription seem to work quite well as verified in many examples [4]. On the other hand, Zelevinsky et al. [3] invoked the idea of more attractive "geometric chaos". We will discuss this in Sect. 14.4 ahead. Papenbrock and Weidenmüller [6] used spectral radius R_J , the distance between lowest and highest state of levels with a given J, and its relation to spectral width σ_J. Numerical results showed that

$$R_J \sim r_J \sigma_J \tag{14.1}$$

with r_J approximately a constant independent of J. It is easy to see that for Gaussian density of eigenvalues, r_J will depend logarithmically on the matrix dimension. Distribution of σ_J discussed in Sect. 13.1.3 clearly show that spin-0 width will be relatively large compared to other J-widths (also with small fluctuations) and this gives preponderance of spin-0 ground states in shell model. In a recent investigation, Johnson [10] also emphasized the importance of spectral widths in understanding the preponderance of spin-0 ground states.

Kirson et al. [11] made an analysis of isospin structure of the ground states with random interactions. They have carried out nuclear shell model studies using EGOE(1 + 2)-JT in $(2s1d)$ space with 6 and 8 nucleons and varying $|N - Z|$. Figure 14.3 shows the main result of this work. It is clearly seen that random interactions distinguish between the ground state structure of even-even and odd-odd nuclei with the later having $J = 1$ ground states more predominantly while it is $J = 0$ for even-even nuclei. In addition, random interactions generate predominantly $T = T_{min}$ ground states and also natural isospin ordering.

14.3 Regularities in Ground State Structure in Two-Level Boson Systems: Mean-Field Theory

Large class of IBMs (see Figs. 14.4 and 14.5) admit two-level structure with degeneracy n_1 and n_2 respectively for the levels #1 and #2. One of the general group

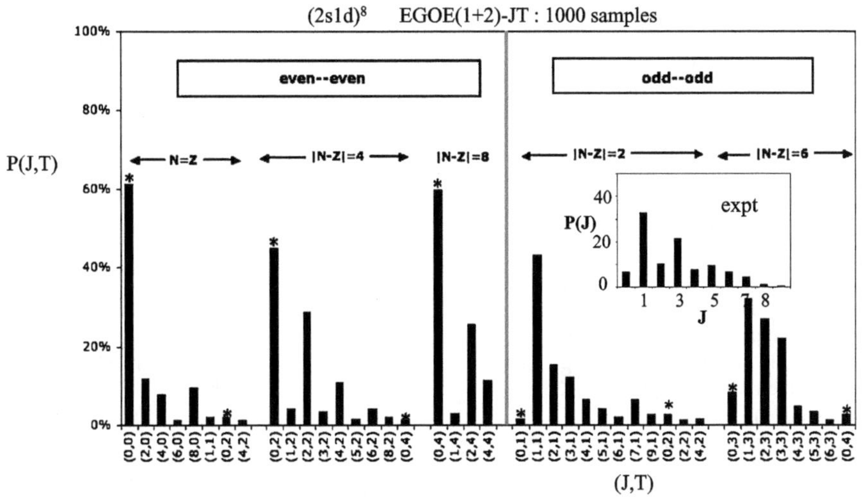

Fig. 14.3 Probability for ground states with (J, T) quantum numbers for $m = 6$ nucleons in $(2s1d)$ shell for a 9000 member EGOE(2)-JT. Note that in $(2s1d)$ shell there are 63 independent two-particle matrix elements. Results are shown for both even-even and odd-odd nuclei with $m = 6$. The $J = 0$ ground states are marked with a *asterisks*. Note that for odd-odd nuclei $J = 1$ ground states are more probable while for even-even nuclei (as expected) $J = 0$ ground states. Insect figure shows experimental data for the distribution of ground states with spin J for 276 odd-odd nuclei having positive parity ground states. Both the main figure and the insect figure are taken from [11] with permission from American Physical Society

structures generated by two-level models is $U(n) \supset G \supset SO(n_1) \oplus SO(n_2) \supset K$, $n_1 + n_2 = n$. Then, it is of interest to address the question of with what probability a given $SO(n_1) \oplus SO(n_2)$ irrep $[\omega_1] \oplus [\omega_2]$ will be the ground state in even-even nuclei with the Hamiltonians preserving $SO(n_1) \oplus SO(n_2)$ symmetry. There are two group-subgroup chains with this general structure. The group chains and the corresponding quantum numbers (irrep labels) for a m boson system for $n_1 \geq 3$ and $n_2 \geq 3$ situation are,

$$
\begin{aligned}
&\text{(A):} \quad \left| \begin{matrix} U(n) \supset U(n_1) \oplus U(n_2) \supset SO(n_1) \oplus SO(n_2) \supset K \\ \{m\} \quad\quad \{m_1\} \quad\quad \{m_2\} \quad\quad [\omega_1] \quad\quad [\omega_2] \quad\quad \alpha \end{matrix} \right\rangle \\
&\quad m_1 = 0, 1, 2, \ldots, m; m_2 = m - m_1 \\
&\quad \omega_1 = m_1, m_1 - 2, \ldots, 0 \text{ or } 1, \omega_2 = m_2, m_2 - 2, \ldots, 0 \text{ or } 1 \\
&\text{(B):} \quad \left| \begin{matrix} U(n) \supset SO(n) \supset SO(n_1) \oplus SO(n_2) \supset K \\ \{m\} \quad\quad [\omega] \quad\quad [\omega_1] \quad\quad [\omega_2] \quad\quad \alpha \end{matrix} \right\rangle \\
&\quad \omega = m, m - 2, \ldots, 0 \text{ or } 1, \omega_1 + \omega_2 = \omega, \omega - 2, \ldots, 0 \text{ or } 1.
\end{aligned}
\tag{14.2}
$$

Note that m_1 and m_2 denote number of bosons in levels #1 and # 2. Also, the algebra K in (14.2) is irrelevant for the discussion in this section. A general two-body Hamiltonian that mixes the states of these two chains but preserves the $[\omega_1]$ and $[\omega_2]$

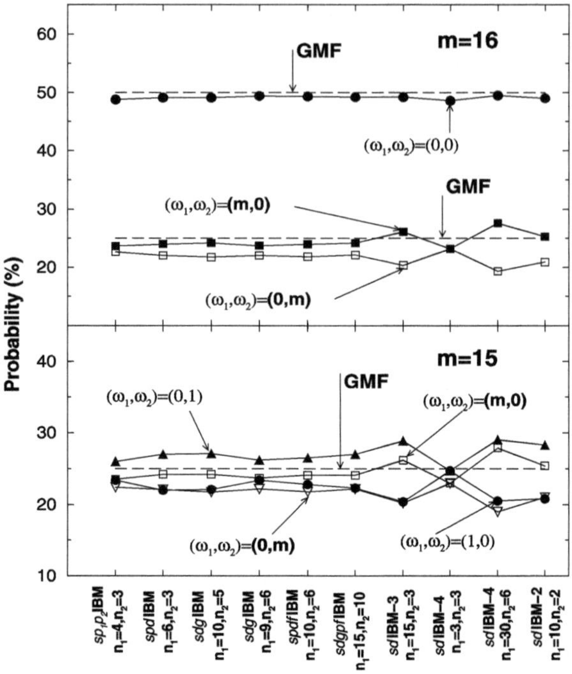

Fig. 14.4 Probabilities (in percentage) for $(\omega_1, \omega_2) = (0, 0)$, $(1, 0)$, (01), $(m, 0)$ and $(0, m)$ to be ground state irreps for various interacting boson models with $n_1 \geq 3$ and $n_2 \geq 2$. Note that for $pn - sd$IBM in the figure, $n_2 = 2$ and $(0, m) = (0, m) \oplus (0, -m)$. Calculations use 1000 members and the most general one plus two-body Hamiltonian interpolating the two symmetry limits. Figure is taken from [12] with permission from American Institute of Physics

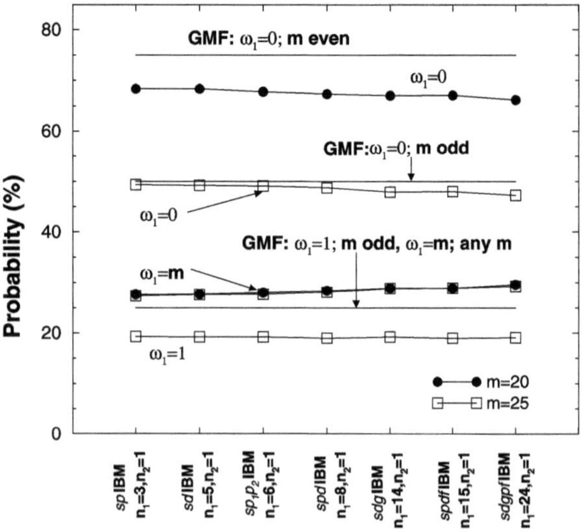

Fig. 14.5 Probabilities (in percentage) for $\omega_1 = 0$, 1 (only for odd m) and m to be ground state irreps for various interacting boson models with $n_1 \geq 3$ and $n_2 = 1$. Calculations use 1000 members and the most general one plus two-body Hamiltonian interpolating the two symmetry limits. Figure is taken from [12] with permission from American Institute of Physics

quantum numbers of $SO(n_1)$ and $SO(n_2)$ respectively [hereafter called $(\omega_1\omega_2)$] is,

$$\begin{aligned}
H^{AB} &= \frac{1}{m}\left[\alpha_1\mathscr{C}_1(U(n_1)) + \alpha_2\mathscr{C}_1(U(n_2))\right] \\
&\quad + \frac{1}{m(m-1)}\left[\alpha_3\mathscr{C}_2(U(n_1)) + \alpha_4\mathscr{C}_2(U(n_2)) + \alpha_5\mathscr{C}_1(U(n_1))\mathscr{C}_1(U(n_2))\right. \\
&\quad \left. + \alpha_6\mathscr{C}_2(SO(n)) + \alpha_7\mathscr{C}_2(SO(n_1)) + \alpha_8\mathscr{C}_2(SO(n_2))\right]. \qquad (14.3)
\end{aligned}$$

Note that $\mathscr{C}_2(G)$ is the quadratic Casimir operator of G and $\mathscr{C}_1(U(r))$ is the number operator for the level r. In the basis (A), the α_6 term in Eq. (14.3) is the mixing part and all others are diagonal. Note that, in the situation $n_1 \geq 3$ and $n_2 = 1$, the $SO(n_2)$ algebra will not exist and hence $\mathscr{C}_2(SO(n_2))$ will not appear in Eq. (14.3). Then $(\omega_1\omega_2) \to (\omega_1)$ with $\omega_1 = 0, 1, 2, \ldots, \omega$ for chain (B). Similarly, for $n_1 \geq 3$ and $n_2 = 2$ one has $\omega_2 = \pm m_2, \pm(m_2 - 2), \ldots, \pm 1$ or 0 for chain (A) and $\omega_1 + |\omega_2| = \omega, \omega - 2, \ldots, 0$ or 1 for (B); thus, here $(\omega_1, \omega_2) \to (\omega_1, \pm\omega_2)$. Reference [13] gives more details. Starting with the basis defined by (A) and using the transformation brackets between (A) and (B) given in analytical form in [13], the matrix of H^{AB} can be constructed easily for a given m and $(\omega_1\omega_2)$. Calculations have been carried out for boson numbers $m = 10$–25 for *sd*IBM [14], *sp*IBM [15], *spd*IBM [16], *sdg*IBM [17], *sdpf*IBM [18], *sdgpf*IBM [19], *sd*IBM-2 [14], *sd*IBM-3 [20, 21] and *sd*IBM-4 [22] by choosing the parameters in Eq. (14.3) to be independent Gaussian variables with zero mean and variance unity. Some results obtained using 1000 samples of random interactions are given in Figs. 14.4 and 14.5.

A mean-field (MF) theory was developed in [8] for explaining the *sp*IBM and *sd*IBM results. Its generalization explains the results for all IBMs [9]. In this generalized mean-field theory (GMF), intrinsic bosons y and z that correspond to the two levels are defined as $y_0^\dagger = \frac{1}{\sqrt{p}}\sum_{i=1}^{p} b_{\ell_i,0}^\dagger$ with $\sum_{i=1}^{p}(2\ell_i + 1) = n_1$ and $z_0^\dagger = \frac{1}{\sqrt{q}}\sum_{j=1}^{q} b_{\ell_j',0}^\dagger$ with $\sum_{j=1}^{q}(2\ell_j' + 1) = n_2$. The angular momenta ℓ are real or fictitious. Then, the coherent state (CS) or the intrinsic state is

$$|m\alpha\rangle = \frac{1}{\sqrt{m!}}\left(\cos\alpha\, y_0^\dagger + \sin\alpha\, z_0^\dagger\right)^m |0\rangle \qquad (14.4)$$

where α is a parameter with $-\pi/2 < \alpha \leq \pi/2$. Now, let us consider the simpler one parameter Hamiltonian [with only the α_2 and α_6 terms in Eq. (14.3)],

$$\begin{aligned}
H &= \frac{1}{m}\cos\chi\,\hat{n}_2 + \frac{1}{m(m-1)}\sin\chi\, S_+ S_-, \\
S_+ &= S_+(1) - S_+(2) = \sum_{i=1}^{p} b_{\ell_i}^\dagger \cdot b_{\ell_i}^\dagger - \sum_{j=1}^{q} b_{\ell_j'}^\dagger \cdot b_{\ell_j'}^\dagger, \; S_- = (S_+)^\dagger .
\end{aligned} \qquad (14.5)$$

The range of χ is $-\pi/2 < \chi \leq 3\pi/2$. Now, the CS expectation value of H or the energy functional $E(\alpha)$ is,

$$E(\alpha) = \cos\chi \sin^2\alpha + \frac{1}{4}\sin\chi \cos^2 2\alpha. \tag{14.6}$$

The minimum of E divides (α, χ) into three ranges and they are: (i) $\alpha = 0$ for $-\pi/2 < \chi \leq \pi/4$; (ii) $\cos 2\alpha = \cot\chi$ for $\pi/4 \leq \chi \leq 3\pi/4$; (iii) $\alpha = \pi/2$ for $3\pi/4 < \chi \leq 3\pi/2$. Note that $\alpha = 0$ gives y-boson condensate with energy $E(\alpha = 0) \propto -\sin\chi\,\omega_1(\omega_1 + n_1 - 2)$. Then for even m, the ground state irreps are $(\omega_1\omega_2) = (00)$ with 25 % and $(\omega_1\omega_2) = (m0)$ with 12.5 % probability. Similarly $\alpha = \pi/2$ gives z-boson condensate with energy $E(\alpha = \pi/2) \propto -\sin\chi\,\omega_2(\omega_2 + n_2 - 2)$ and then the ground state irreps are $(\omega_1\omega_2) = (00)$ with 25 % and $(\omega_1\omega_2) = (0m)$ with 12.5 % probability. In the situation $\cos 2\alpha = \cot\chi$, cranking has to be done with respect to both $SO(n_1)$ and $SO(n_2)$. Evaluating moment of inertia, by an extension of the ordinary $SO(3)$ cranking, gives [9]

$$\begin{aligned} E &\propto \left[\frac{\omega_1(\omega_1 + n_1 - 2)}{A_+}\right] + \left[\frac{\omega_2(\omega_2 + n_2 - 2)}{A_-}\right]; \\ A_\pm &= \mp\frac{\sin\chi \pm \cos\chi}{\cos\chi \sin\chi}. \end{aligned} \tag{14.7}$$

This gives, $(m0)$ and $(0m)$ irreps to be ground states each with 12.5 % probability. Combining all the results will give for even m systems, $(\omega_1\omega_2) = (00)$, $(m0)$ and $(0m)$ irreps to be ground states with 50 %, 25 % and 25 % probability. For odd m, the y and z boson condensates give (10) and (01) irreps in place of (00) irrep. Therefore, for odd N systems, $(\omega_1\omega_2) = (10)$, (01), $(m0)$ and $(0m)$ irreps will be ground states with 25 % probability each. These GMF results for even and odd m are well verified in many examples for different IBMs as shown in Fig. 14.4. All these results are valid only for $n_1 \geq 3$ and $n_2 \geq 3$. However, these will also give the results for the situation with $n_1 \geq 3$ but $n_2 = 1$ with the following changes. With $n_2 = 1$, the irrep $[\omega_2]$ will not exist and then the irreps $(01) \to \omega_1 = 0$ and $(0m) \to \omega_1 = 0$. Therefore, for $n_1 \geq 3$ and $n_2 = 1$ the results are: (i) for even m, ground states will be $\omega_1 = 0$ and m with probability 75 % and 25 % respectively; (ii) for odd m, ground states will be $\omega_1 = 0$, 1 and m with probabilities 50 %, 25 % and 25 % respectively. These GMF results are well conformed in many numerical examples as shown in Fig. 14.5.

14.4 Regularities in Energy Centroids Defined over Group Irreps

Energy centroids defined over group irreps form simplest quantities for studying regularities generated by random interactions as it is possible to write simple formulas (exact in many situations and approximate in some) for these. Examples are already given in the previous chapters. Energy centroids were first discussed by Mulhall et al. [23]. Later, Zhao et al. [24] analyzed fixed-L energy centroids in sdgIBM and also fixed-J energy centroids in shell model spaces using random interactions. Essentially here one is using EGOE(1 + 2)-J or EGOE(1 + 2)-JT in shell model and EGOE(1 + 2)-L in IBM's. They found that L_{min} (or J_{min}) and L_{max} (or J_{max}) will be lowest with largest probabilities and centroids with other L (or J) values are lowest with very small probability. Following these, in a number of examples, energy centroids with fixed spin or isospin, with fixed irreps of various group symmetries of both shell model and interacting boson models, in the presence of random two-body and three-body interactions have been studied by recognizing that simple propagation formulas can be written for energy centroids in many situations [25–29]. The examples studied are:

1. $U_{sd}(6) \supset SU_{sd}(3)$ energy centroids $\overline{E_{m,(\lambda\mu)}}$ in sdIBM with two- and three-body interactions; $(\lambda\mu)$ are $SU(3)$ irreps,
2. $\overline{E_{(m_1\omega_1,m_2\omega_2,\ldots)}}$ of $U(\mathscr{N}) \supset \sum_i [U(\mathscr{N}_i) \supset SO(\mathscr{N}_i)]\oplus$ in IBM's [$m = \sum_i m_i$ and ω_i are the irreps of $SO(\mathscr{N}_i)$] with the specific example of sdgIBM,
3. $\overline{E_{m,(\lambda\mu),T}}$ of $U(3\mathscr{N}) \supset U(\mathscr{N}) \otimes [SU_T(3) \supset O_T(3)]$ in IBM-T [i.e. IBM with bosons carrying isospin $T = 1$ degree of freedom and this is also called IBM-3] with the specific example of sdIBM-T with both two- and three-body interactions,
4. $\overline{E_{m,\{f\},[\sigma]}}$ of $U(6\mathscr{N}) \supset U(\mathscr{N}) \otimes [SU_{ST}(6) \supset O_{ST}(6)]$ in IBM-ST [i.e. IBM with the bosons carrying spin-isospin degrees of freedom $(ST) = (10) \oplus (01)$ and this is also called IBM-4] with the specific example of sdIBM-ST,
5. $\overline{E_{n_{sd}(\lambda_{sd}\mu_{sd}):n_{pf}(\lambda_{pf}\mu_{pf})}}$ of $U_{sdpf}(16) \supset [U_{sd}(6) \supset SU_{sd}(3)] \oplus [U_{pf}(10) \supset SU_{pf}(3)]$ in $sdpf$IBM,
6. $\overline{E_{m,\omega}}$ in $U(\mathscr{N}) \supset SO(\mathscr{N})$ of IBMs; $[\omega]$ are irreps of $SO(\mathscr{N})$,
7. $\overline{E_{\{f\}(ST)}}$ of $U(24) \supset U(6) \otimes [SU_{\{f\}}(4) \supset SU_S(2) \otimes SU_T(2)]$ in shell model for $(2s1d)$ shell nuclei,
8. $\overline{E_{m,T}}$ of $U(2\mathscr{N}) \supset U(\mathscr{N}) \otimes SU_T(2)$ in shell model spaces with two- and three-body interactions,
9. $\overline{E_{m,\{f\},(\lambda\mu)}}$ of $U(24) \supset [U(6) \supset SU(3)] \otimes SU_{\{f\}}(4)$ for $(2s1d)$ shell nuclei,
10. $\overline{E_{m,J}}$ for $(j)^m$ system of fermions and $\overline{E_{m,L}}$ for $(\ell)^m$ system of boson with two- and three-body interactions (here approximate formulas given in Sect. 13.1.2 are used).

In all these examples it is seen that, with random interactions, the energy centroids over highest and lowest irreps are lowest in energy with large ($\gtrsim$90 %) probability. For illustration, let us consider the example of $\overline{E_{m,\omega}}$ where $[\omega]$ are irreps of $SO(\mathscr{N})$ in $U(\mathscr{N}) \supset SO(\mathscr{N})$ of IBM's; $\mathscr{N} = 6$ for sdIBM, 15 for sdgIBM and 16 for $sdpf$IBM. With $\omega = m, m-2, \ldots, 0$ or 1 and the matrix elements of boson pairing operator H_P being $\frac{1}{4}(m-\omega)(m+\omega+\mathscr{N}-2)$, we have, $\overline{E_{m,\omega}} = E_0(m) + [(m-\omega)(m+\omega+\mathscr{N}-2)/2\mathscr{N}][\overline{E_{2,0}} - \overline{E_{2,2}}]$. Then clearly energy centroids with highest and lowest ω will be lowest in energy with 50 % probability each. In another example, consider fixed isospin centroids $\overline{E_{(m,T)}}$ generated by a 3-body Hamiltonian. It is easily seen that, $\overline{E_{(m,T)}} = \{\frac{m^3-6m^2+8}{12} - \frac{2T(T+1)}{3} + \frac{mT(T+1)}{3}\}\overline{E_{3,\frac{3}{2}}} + \{\frac{m^3-4m}{12} + \frac{2T(T+1)}{3} - \frac{mT(T+1)}{3}\}\overline{E_{3,\frac{1}{2}}}$ and this implies

$$\overline{E_{(m,T_{max})}} - \overline{E_{(m,T)}} = \{\overline{E_{3,\frac{3}{2}}} - \overline{E_{3,\frac{1}{2}}}\}\left(\frac{m-2}{3}\right)\big\{T_{max}(T_{max}+1) - T(T+1)\big\}. \tag{14.8}$$

Equation (14.8) shows that for m nucleons, even with random 3-body Hamiltonians, just as with 2-body Hamiltonians, $T = T_{max}$ and $T = 0$ energy centroids will be lowest in energy each with 50 % probability. Now we will discuss some select non trivial examples.

14.4.1 *sdg*IBM Energy Centroids

Spectrum generating algebra for sdgIBM is $U(15)$ and one of the decompositions of the m boson space is according to $(m_s, m_d, v_d, m_g, v_g)$ where m_s, m_d and m_g are s, d and g boson numbers with the total boson number $m = m_s + m_d + m_g$. Similarly v_d and v_g are the d and g boson seniority quantum numbers, $v_d = m_d, m_d - 2, \ldots, 0$ or 1 and $v_g = m_g, m_g - 2, \ldots, 0$ or 1. Then, it is possible to consider regularities in fixed-$(m_s, m_d, v_d, m_g, v_g)$ energy centroids using the propagation formula [26],

$$\begin{aligned}
\overline{E_{(m_s,m_d,v_d,m_g,v_g)}} = {} & \sum_i m_i\varepsilon_i + \sum_{i>j}\overline{V_{ij}}m_im_j + \sum_i \frac{m_i(m_i-1)}{2}\langle V\rangle^{m_i'=2,\omega_i=2} \\
& + \sum_i \frac{\langle V\rangle^{m_i'=2,\omega_i=0} - \langle V\rangle^{m_i'=2,\omega_i=2}}{2\mathcal{N}_i}(m_i - v_i) \\
& \times (m_i + v_i + \mathcal{N}_i - 2); \\
\overline{V_{ij}} = {} & \big\{\big[\mathcal{N}_i(\mathcal{N}_j + \delta_{ij})\big]/(1+\delta_{ij})\big\}^{-1}\sum_L V^L_{\ell_i\ell_j\ell_i\ell_j}(2L+1), \\
\langle V\rangle^{m_i'=2,\omega_i=0} = {} & \big\langle(\ell_i\ell_i)L_i = 0\big|V\big|(\ell_i\ell_i)L_i = 0\big\rangle, \\
\langle V\rangle^{m_i'=2,\omega_i=2} = {} & \left[\frac{\mathcal{N}_i(\mathcal{N}_i+1)}{2} - 1\right]^{-1}\left[\frac{\mathcal{N}_i(\mathcal{N}_i+1)}{2}\overline{V_{ii}} - \langle V\rangle^{m_i=2,\omega_i=0}\right].
\end{aligned} \tag{14.9}$$

Here, $i = s$, d and g and $\mathcal{N}_s = 1$, $\mathcal{N}_d = 5$ and $\mathcal{N}_g = 9$. Also, for two particles, m_i is denoted by m_i' and v_s, v_d and v_g are denoted by ω_s, ω_d and ω_g respectively. Similarly, ε_i are the 3 sp energies $(\varepsilon_s, \varepsilon_d, \varepsilon_g)$ and $V^L(\ell_1, \ell_2, \ell_1, \ell_2)$ the 16 diagonal two-particle matrix elements. For the s orbit, $m_s = 2$ and $\omega_s = 2$ and there will be no two-boson state with $\omega_s = 0$. Therefore, the fourth term in the centroid formula is only for $i = d$ and g. Note that for m_s bosons, trivially $v_s = m_s$ and hence it is not specified. The 19 parameters (3 sp energies and 16 TBME) in sdgIBM are chosen to be Gaussian random variables with zero center and unit variance. To maintain proper scaling, the sp energies are divided by m and the two

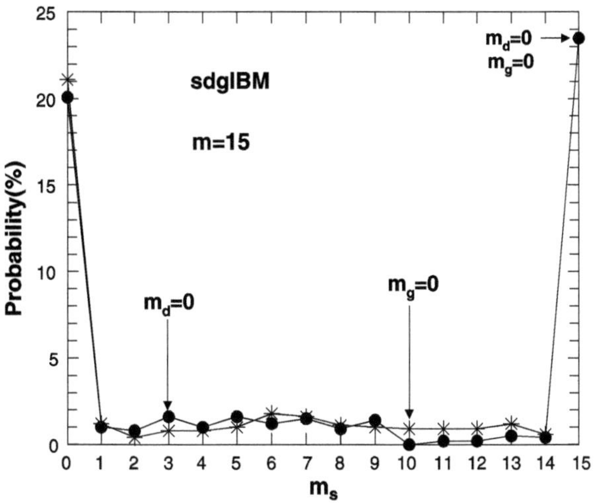

Fig. 14.6 Probabilities for sdgIBM fixed-$(m_s, m_d, v_d, m_g, v_g)$ energy centroids to be lowest in energy vs m_s for a system of 15 bosons ($m = 15$). For each m_s, the probability shown is the sum of the probabilities for the irreps with the seniority quantum number lowest ($v_\ell = v_\ell^{min}$) and highest ($v_\ell = m_\ell$). *Filled circles* and *stars* are for configurations with $m_d = 0$ and $m_g = 0$ respectively; they are joined by *lines* to guide the eye. Figure is taken from [26] with permission from American Physical Society

particle matrix elements by $m(m-1)$. Numerical results obtained for a 1000 member ensemble with $m = 15$ are given in Fig. 14.6. Let us denote $v_d = 0$ or 1 (for m_d is even or odd respectively) by v_d^{min} and similarly v_g^{min} is defined. As seen from the figure, configurations $(m_s, m_d = v_d = m - m_s, m_g = v_g = 0)$, $(m_s, m_d = m - m_s, v_d = v_d^{min}, m_g = v_g = 0)$, $(m_s, m_d = v_d = 0, m_g = v_g = m - m_s)$ and $(m_s, m_d = v_d = 0, m_g = m - m_s, v_g = v_g^{min})$ exhaust about 91 % probability. In this, the configurations with $m_s = m_d = 0$ carry $\sim$20 %, $m_s = m_g = 0$ carry $\sim$21 % and $m_s = m$ carry $\sim$24 % probability. Thus the configurations with $m_s = 0, m$ are most probable but others give non negligible probability for being the lowest.

14.4.2 sdIBM-T Energy Centroids with 3-Body Forces

In the second example we will consider $\overline{E_{m,(\lambda\mu),T}}$ of $U_{sd}(18) \supset U(6) \otimes [SU_T(3) \supset SO_T(3)]$ in sdIBM-3 with three body interactions. Firstly the $U_{sd}(18)$ irrep is the totally symmetric irrep $\{m\}$ and the $SU_T(3)$ irreps [same as those of $U(6)$] are $(\lambda, \mu) = ((f_1 - f_2), (f_2 - f_3))$ where $f_1 + f_2 + f_3 = m$ and $f_1 \geq f_2 \geq f_3 \geq 0$. The $(\lambda\mu) \to T$ reductions follow from Elliott's rules [30] given by Eq. (10.32). Counting of number of irreps $(\lambda\mu)T$ for $m \leq 3$ shows that besides the operators 1, $\hat{n}$, $\hat{C}_2$, $\hat{C}_3$ and $\hat{T}^2$, we need one extra $SO_T(3)$ scalar in $SU_T(3)$. The well known $SU(3) \supset SO(3)$ integrity basis operator $\hat{X}_3$ defined by Eq. (10.36), which is three-body, is useful here. Fixed-$(\lambda\mu)T$ averages of the $\hat{X}_3$ operator are given by [31]

$$X_3\big((\lambda\mu)T\big) = \langle \hat{X}_3 \rangle^{(\lambda\mu)T} = (-1)^{1-\delta_{\mu,0}} \frac{[3-4T(T+1)]\sqrt{T(T+1)}}{\sqrt{(2T-1)(2T+3)}} \times \sqrt{C_2(\lambda\mu)} \frac{\sum_k \langle (\lambda\mu)kT(11)2 \mid\mid (\lambda\mu)kT \rangle_{\rho=1}}{\sum_{k'} 1}. \tag{14.10}$$

In Eq. (14.10), $\langle --- \mid\mid --- \rangle$ are $SU_T(3) \supset SO_T(3)$ reduced Wigner coefficients and these coefficients can be calculated, as stated in Chap. 10, using the programs in [32]. Using $\hat{X}_3$ averages for $m \leq 3$, propagation formula for $\overline{E_{m,(\lambda\mu),T}}$ has been derived in [29] and the result is [with $T^2 = T(T+1)$ and $X_3((\lambda\mu)T) = \langle \hat{X}_3 \rangle^{(\lambda\mu)T}$],

$$\begin{aligned}
\overline{E_{m,(\lambda\mu),T}} &= \left[\frac{7m}{9} - \frac{7m^2}{18} - \frac{7C_2(\lambda\mu)}{90} - \frac{7T^2}{30} + \frac{7m^3}{162} + \frac{7mC_2(\lambda\mu)}{270} + \frac{7mT^2}{90} \right. \\
&\quad \left. - \frac{C_3(\lambda\mu)}{90} + \frac{X_3((\lambda\mu)T)}{45}\right] \overline{E_{3,(30),3}} \\
&\quad + \left[\frac{m}{3} - \frac{m^2}{6} - \frac{C_2(\lambda\mu)}{5} + \frac{7T^2}{30} + \frac{m^3}{54} + \frac{mC_2(\lambda\mu)}{15} - \frac{7mT^2}{90} + \frac{C_3(\lambda\mu)}{15} \right. \\
&\quad \left. - \frac{X_3((\lambda\mu)T)}{45}\right] \overline{E_{3,(30),1}} \\
&\quad + \left[-\frac{m}{3} + \frac{C_2(\lambda\mu)}{6} - \frac{T^2}{12} + \frac{m^3}{27} - \frac{mT^2}{18} - \frac{C_3(\lambda\mu)}{6} + \frac{X_3((\lambda\mu)T)}{18}\right] \overline{E_{3,(11),1}} \\
&\quad + \left[-\frac{5m}{9} + \frac{C_2(\lambda\mu)}{6} + \frac{T^2}{12} + \frac{5m^3}{81} - \frac{2mC_2(\lambda\mu)}{27} + \frac{mT^2}{18} + \frac{C_3(\lambda\mu)}{18} \right. \\
&\quad \left. - \frac{X_3((\lambda\mu)T)}{18}\right] \overline{E_{3,(11),2}} \\
&\quad + \left[\frac{m}{9} + \frac{m^2}{18} - \frac{C_2(\lambda\mu)}{18} + \frac{m^3}{162} - \frac{mC_2(\lambda\mu)}{54} + \frac{C_3(\lambda\mu)}{18}\right] \overline{E_{3,(00),0}}.
\end{aligned} \tag{14.11}$$

Using Eq. (14.11) calculations have been carried out in [29] for boson numbers $m = 10-20$ bosons using a 1000 member random 3-body ensemble obtained by treating $\overline{E_{3,(\lambda\mu),T}}$ as Gaussian random variables with zero center and unit variance and the results are shown in Fig. 14.7. Energy centroids of highest [according to $C_2(\lambda\mu)$ value] $(\lambda\mu)$ with lowest and highest T values and the lowest $(\lambda\mu)$ carry ~88 % probability for being lowest in energy. The only other irrep that carries significant probability (~9 %) is $(0, \frac{m}{2})T = 0$ for m even and $(1, \frac{m-1}{2})T = 1$ for m odd. Thus, random 3-body interactions generate regularities in energy centroids.

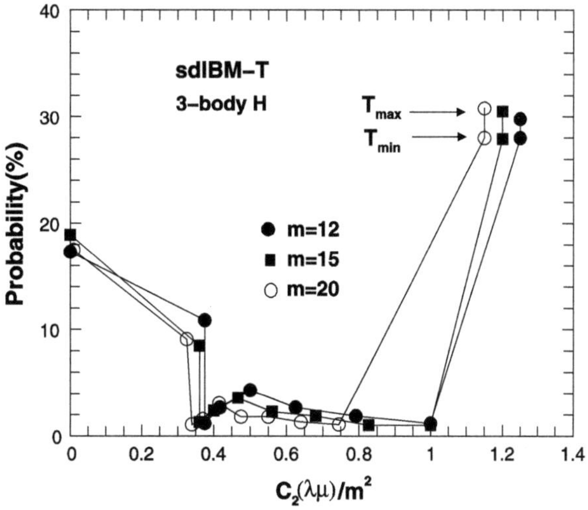

Fig. 14.7 Probabilities for the sdIBM-T energy centroids $\overline{E_{m,(\lambda\mu),T}}$ to be lowest in energy vs $C_2(\lambda\mu)/m^2$ for boson systems with $m = 12$, 15 and 20. For the highest $(\lambda\mu)$, the probabilities for both highest and lowest T are shown and for the lowest $(\lambda\mu)$ only one T value is possible. For the irreps not shown in the figure, the probability is less than 0.1 %. All the points for a given m are joined by *lines* to guide the eye. Figure is taken from [29] with permission from World Scientific

14.4.3 SU(4)-ST Energy Centroids

For $(2s1d)$ shell nuclei, $U(24)$ is the spectrum generating algebra and the spin-isospin (ST) supermultiplet $SU(4)$ algebra appears in the subalgebra $U(24) \supset U(6) \otimes \{SU(4) \supset SU_S(2) \otimes SU_T(2)\}$; note that $U(6)$ generates the orbital part. For a given number of nucleons m, the allowed $U(4)$ irreps are $\{f\} = \{f_1, f_2, f_3, f_4\}$ with $f_1 \geq f_2 \geq f_3 \geq f_4 \geq 0$, $f_1 \leq 6$ and $f_1 + f_2 + f_3 + f_4 = m$ and the $U(6)$ irreps, by direct product nature, are $\{\tilde{f}\}$, the transpose of $\{f\}$. It is important to note that the equivalent $SU(4)$ irreps are $\{f_1 - f_4, f_2 - f_4, f_3 - f_4\}$. With these, from now on we will use $U(4)$ and the irreps $\{f\}$. It is well known that a totally symmetric $U(4)$ irrep $\{\lambda\} \to (ST) = (\frac{\lambda}{2}, \frac{\lambda}{2}), (\frac{\lambda}{2} - 1, \frac{\lambda}{2} - 1), \ldots, (00)$ or $(\frac{1}{2}, \frac{1}{2})$. Using this result and expanding a given $U(4)$ irrep into totally symmetric $U(4)$ irreps will give easily $\{f\} \to (ST)$ reductions. Just as the fixed-T energy centroids propagate, the fixed $\{f\}(ST)$ energy centroids $\overline{E_{\{f\}(ST)}}$ for a one plus two-body Hamiltonian propagate as the available scalars of maximum body rank 2 are 1, $\hat{n}$, $\hat{n}^2$, $C_2(U(4))$, S^2 and T^2 and the centroids for $m \leq 2$ are also six in number. The propagation equation, with $C_2(\{f\}) = \sum_i f_i^2 + 3f_1 + f_2 - f_3 - 3f_4$ where $C_2(\{f\})$ gives the eigenvalues of the quadratic Casimir invariant of $U(4)$, is [33]

$$
\begin{aligned}
\overline{E_{\{f\}(ST)}} = {} & \big(1 - 3m + m^2\big)\langle H\rangle^{\{0\}(00)} + \big(2m - m^2\big)\langle H\rangle^{\{1\}(\frac{1}{2}\frac{1}{2})} \\
& + \left[-\frac{9}{8}m + \frac{1}{4}m^2 + \frac{1}{8}C_2\big(\{f\}\big) + \frac{1}{4}S(S+1) + \frac{1}{4}T(T+1)\right]\langle H\rangle^{\{2\}(11)} \\
& + \left[-\frac{1}{8}m + \frac{1}{8}C_2\big(\{f\}\big) - \frac{1}{4}S(S+1) - \frac{1}{4}T(T+1)\right]\langle H\rangle^{\{2\}(00)}
\end{aligned}
$$

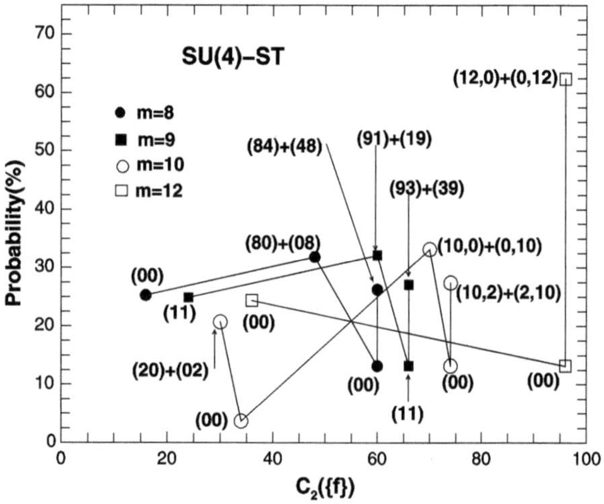

Fig. 14.8 Probabilities for the $(2s\,1d)$ shell energy centroids $\overline{E_{\{f\}(ST)}}$ to be lowest in energy vs $C_2(\{f\})$. Results are shown for nucleon numbers $m = 8$, 9, 10 and 12. The $U(4)$ irreps $\{f\}$ for the results in the figure are given in the text. The corresponding (ST) values are shown in the figure. All the *points* for a given m are joined by *lines* to guide the eye. Figure is taken from [28] with permission from American Physical Society

$$+\left[\frac{3}{8}m + \frac{1}{8}m^2 - \frac{1}{8}C_2(\{f\}) + \frac{1}{4}S(S+1) - \frac{1}{4}T(T+1)\right]\langle H\rangle^{\{1^2\}(10)}$$
$$+\left[\frac{3}{8}m + \frac{1}{8}m^2 - \frac{1}{8}C_2(\{f\}) - \frac{1}{4}S(S+1) + \frac{1}{4}T(T+1)\right]\langle H\rangle^{\{1^2\}(01)}. \tag{14.12}$$

Considering the basic energy centroids $\langle H\rangle^{\{f\}(ST)}$ with $m \leq 2$ as independent zero centered (with unit variance) Gaussian random variables, instead of using ε_i and $V_{ijkl}^{J,t=0.1}$ as random variables, the $\{f\}(ST)$ structure of the ground states has been studied in [28]. Figure 14.8 shows results obtained using 1000 samples for $m = 8$–12. The probabilities split into three $U(4)$ irreps (other irreps carry less than 1 % probability and they are not shown in the figure) for $n = 8$, 9 and 10 and the corresponding (ST) values are as shown in the figure. Energy centroids with the lowest and highest $U(4)$ irreps carry ~25 % and ~40 % respectively. The lowest irreps are $\{2^4\}$, $\{32^3\}$ and $\{3^22^2\}$ for $n = 8$, 9 and 10 respectively and the highest irreps are $\{6, n-6\}$. The third irreps $\{4^2\}$, $\{54\}$ and $\{5^2\}$, with probability ~32 %, for $n = 8$, 9 and 10 respectively are those that carry $S = n/2$ or $T = n/2$; for $n = 10$ the irrep $[3^31](00)$ carries 3.7 % probability. For the mid-shell example with $n = 12$, the probabilities split into the lowest $\{3^4\}$ and highest $\{6^2\}$ irreps with ~25 % and ~75 % respectively. The lowest irrep supports only $(ST) = (00)$ and the probability for the highest irrep splits into ~13 % and ~62 % for $(ST) = (00)$ and $(12,0) + (0,12)$. Figure also shows that the probability for the energy centroid with lowest $U(4)$ irrep to be lowest is only ~25 % and it should be noted that the corresponding $SU(4)$ irreps are $\{0\}(00)$, $\{1\}(\frac{1}{2}\frac{1}{2})$ and $\{1^2\}(10) + (01)$ respectively for $n = 4k$, $4k + 1$ and $4k + 2$ with k being a positive integer. This result is in agreement with EGOE(1 + 2)-JT calculations carried out using nuclear shell model codes in [34].

14.4.4 $(j)^m$ and $(\ell)^m$ Systems with 2- and 3-Body Interactions: Geometric Chaos

In the final example we will consider spin J centroids $\overline{E_{(m,J)}}$ generated by random 2-body and 3-body Hamiltonians for m identical fermions in a single j shell, i.e. EGOE(2)-$(j:J)$ and EGOE(3)-$(j:J)$ energy centroids. The $\overline{E_{(m,J)}}$'s correspond to averages of H over the space defined by the irreps m and J of $U(2j+1)$ and $SO(3)$ respectively in $U(2j+1) \supset SO(3)$. As discussed in Sect. 13.1.2, $\overline{E_{(m,J)}}$ can be expanded in powers of $J(J+1)$ and to a good approximation one can truncate the expansion to $[J(J+1)]^2$ term. Then, for a 2-body H, the probability P_0 for $\overline{E_{(m,J=0)}}$ to be lowest in energy is given by Eq. (13.20). Application of this shows that P_0 is close to 50 %. This result is in direct correlation with the numerically observed result (see Table 14.1) that the probability for $J=0$ ground states is ~50 % with random 2-body interactions. To the extent that only lower order moments of the eigenvalue density $\rho(E)$ determine the ground states, one can argue that the regularities of the energy centroids and spectral variances (see next section) result in regularities in $J=0$ ground states. The energy centroids and spectral variances average out many J-couplings in m particle spaces—a geometric effect—giving propagation equations (exact or approximate). Thus, it is possible to argue that the preponderance of $J=0$ ground states (similarly other regularities) generated by random interactions is a geometric effect and in [23] this is termed *geometric chaos*. It should be added that a precise definition of *geometric chaos* is still lacking. It is good to recall here that for EE, in addition to a classical ensemble in the defining space (2-particle space for two-body interactions), there is information propagation from the defining space to m particle spaces. This geometric aspect is absent in classical ensembles.

Turning to 3-body H, Eq. (13.12) gives the formula for $\overline{E_{(m,J)}}$ to order $J(J+1)$ to be,

$$\overline{E_{(m,J)}} = \left[\langle H(3)\rangle^m - \frac{3}{2}\frac{\langle J_z^2 \widetilde{H(3)}\rangle^m}{\langle J_z^2\rangle^m}\right] + \frac{1}{2}\frac{\langle J_z^2 \widetilde{H(3)}\rangle^m}{[\langle J_z^2\rangle^m]^2} J(J+1) \tag{14.13}$$

and this is good for $m \gg 3$, $j \gg m$ and j large. In Eq. (14.13), $\widetilde{H}$ is H with the average part $\langle H\rangle^m$ removed. Denoting three particle antisymmetric states by $|(j)^3;\alpha J_3\rangle$ with α being the extra label required to completely specify the states, diagonal 3-particle matrix elements of $H(3)$ are $G_{\alpha J_3} = \langle (j)^3;\alpha J_3 \mid H(3) \mid (j)^3;\alpha J_3\rangle$. It is easy to see that,

$$\begin{aligned} \langle H(3)\rangle^m &= \binom{m}{3}\langle H(3)\rangle^3 = \binom{m}{3}\binom{2j+1}{3}^{-1} \sum_{\alpha,J_3} G_{\alpha J_3}(2J_3+1), \\ \langle J_z^2\rangle^m &= \frac{1}{3}\langle J^2\rangle^m = \frac{1}{6}m(2j+1-m)(j+1) \simeq \frac{m}{3}j(j+1). \end{aligned} \tag{14.14}$$

Tensorial decomposition of J^2 and H operators with respect to $U(2j+1)$ will give $J^2 = (J^2)^{\nu=0} + (J^2)^{\nu=2}$ and $H(3) = H^{\nu=0}(3) + H^{\nu=2}(3) + H^{\nu=3}(3)$.

Single most important property of this decomposition is that it is orthogonal with respect to m particle averages. Then, $\widetilde{H} = H - H^{\nu=0}$ gives $\langle J^2 \widetilde{H(3)} \rangle^m = \langle (J^2)^{\nu=2} H^{\nu=2}(3) \rangle^m = \langle (J^2)^{\nu=2} H(3) \rangle^m$. Also, Eq. (4.18) gives $H^{\nu=2}(3) = (\hat{n} - 2) F^{\nu=2}(2)$ where F is a two-body operator with rank $\nu = 2$ and $\hat{n}$ is number operator. These will give $\langle (J^2)^{\nu=2} H(3) \rangle^m = \frac{m(m-1)(m-2)(2j+1-m)(2j-m)}{6(2j-2)(2j-3)} \times \langle (J^2)^{\nu=2} H(3) \rangle^3$. Now the final formula for $\overline{E_{(m,J)}}$ is [28],

$$\overline{E_{(m,J)}} \simeq E_0 + \frac{3m}{2} \left[\frac{\sum_{\alpha, J_3} \{J_3(J_3+1) - 3j(j+1)\} G_{\alpha J_3} (2J_3+1)}{[j(j+1)(2j+1)]^2 (2j+1)} \right] J(J+1). \tag{14.15}$$

Thus, the $J(J+1)$ term will have linear m dependence. An interesting observation in many numerical calculations (not only with single j but also multi-j and JT centroids) is $\langle \overline{E_{(m,J)}} \rangle_{min} \sim E_0 + CJ(J+1)$ where $\langle \overline{E_{(m,J)}} \rangle_{min}$ is the average of $\overline{E_{(m,J)}}$ over the members of the EGOE(3)-($j : J$) ensemble for which $\overline{E_{(m,J)}}$ with $J \sim J_{min}$ is lowest in energy. The coefficient C follows from Eq. (14.15) and it is given by

$$C = \sqrt{\frac{2}{\pi}} \sqrt{\sum_{\alpha, J_3} \left[\{J_3(J_3+1) - 3j(j+1)\}(2J_3+1)\right]^2} \times \frac{3m}{2[j(j+1)(2j+1)]^2 (2j+1)}. \tag{14.16}$$

For two-body interactions, i.e. for EGOE(2)-($j : J$), C will be independent of m and this follows from Eq. (13.15). Thus 3-body H's give m dependence to C that is absent for a two-body H. Finally, Eq. (14.15) extends easily to $(\ell)^m$ boson systems giving [28],

$$\overline{E_{(m,L)}} \simeq E_0 + 6m \left[\frac{\sum_{\alpha, L_3} \{L_3(L_3+1) - 3\ell(\ell+2)\} G_{\alpha L_3} (2L_3+1)}{[\ell^2(2\ell+1)(2\ell+2)(2\ell+3)(2\ell+4)(2\ell+5)]} \right] L(L+1) \tag{14.17}$$

where $G_{\alpha L_3}$ are three-body matrix elements for bosons with spin L_3. This gives $C \sim 0.033m$ for d boson systems and compares well with the numerical calculations with a 1000 member BEGOE(3)-($\ell = 2 : L$) that gave $0.035m$ as reported in [28].

Going beyond 2- and 3-body ensembles, Volya [35] has analyzed EGOE(k)-($j : J$) ensembles for $(j)^m$ systems, with $k < m$, and argued using the numerical results that symmetries emerge out of random interactions. This and the related argument [36] that symmetries are responsible for chaos or random matrix behavior in nuclear shell model certainly deserve much further study.

14.5 Regularities in Spectral Variances over Group Irreps with Random Interactions

Going beyond energy centroids, many different types of correlations involving spectral variances can be studied in order to understand the origin of regular structures generated by random interactions. Some studies of spectral variances defined over good symmetry subspaces, i.e. fixed-m variances for EGOE(k)/EGUE(k), fixed-(m, S) variances for EGOE(2)-**s** and fixed-(m, J) variances for EGOE(2)-J are already discussed in Chaps. 4, 6 and 9–13. In addition, variances $\sigma^2(\Gamma)$ over subspaces (Γ) defined over broken symmetries of nuclear shell model and the interacting boson models (similarly for other finite quantum systems) yield valuable information. Here Γ are the irreps of G in $U(N) \supset G \supset G_f$ with G_f being a symmetry of H such as $SO_J(3)$ and G is a broken symmetry such as the configuration symmetry in nuclear shell model. Important point here being that the variances $\sigma^2(\Gamma) = \langle (H - \langle H \rangle^\Gamma)^2 \rangle^\Gamma$ determine much of the statistical behavior of strength functions or partial densities $\langle \delta(H - E) \rangle^\Gamma$. Here, we will present results for $\sigma^2(\Gamma)$ for some EGOEs. Firstly it is important to note that $\sigma^2(\Gamma)$, just as energy centroids, propagate in a simple manner in many situations. For example, for shell model spherical configurations $(\mathbf{m}) = (m_1, m_2, \ldots)$ where m_i is number of particles in the shell model j_i orbit and similarly for interacting boson models (with or without internal degrees of freedom), the configuration variances $\sigma^2(\mathbf{m})$ are given by,

$$\sigma^2(\mathbf{m}) = \sum_{i \geq j, k \geq \ell} \frac{m_i(m_j - \delta_{ij})(N_k \mp m_k)(N_\ell \mp m_\ell \mp \delta_{k\ell})}{N_i(N_j \mp \delta_{ij})(N_k \mp \delta_{ki} \mp \delta_{kj})(N_\ell \mp \delta_{\ell i} \mp \delta_{\ell j} \mp \delta_{\ell k})} \times \sum_\Gamma \big(\tilde{V}^\Gamma_{ijkl}\big)^2 [\Gamma]. \tag{14.18}$$

In Eq. (14.18), N_i is the degeneracy of the i-th orbit, Γ is two-particle J or JT in shell model and L (or LT or LST) in IBMs. Similarly, $[\Gamma]$ is the dimension of the Γ space and for example, $[J] = (2J + 1)$ and $[JT] = (2J + 1)(2T + 1)$. With $\overline{V}_{ij}$ the average two-particle matrix element for particles in the orbits i and j, we have $\tilde{V}^\Gamma_{ijij} = V^\Gamma_{ijij} - \overline{V}_{ij}$ and for the rest $\tilde{V}^\Gamma_{ijkl} = V^\Gamma_{ijkl}$. Finally in $\mp$ in Eq. (14.18), the upper sign is for fermions (shell model) and the lower sign is for bosons (IBMs).

Some of the other situations where it is possible to write propagation equations are [37, 38]:

1. $U_{sd}(6) \supset SU_{sd}(3)$ variances $\sigma^2(m, (\lambda\mu))$ in sdIBM,
2. $\sigma^2(m, [\omega])$ of $U(\mathcal{N}) \supset SO(\mathcal{N})$ in sdIBM, sdgIBM, sdIBM-T, $sdpf$IBM etc. [$[\omega]$ are irreps of $SO(\mathcal{N})$],
3. $\sigma^2(\mathbf{m}, \boldsymbol{\omega}) = \sigma^2(m_1, \omega_1; m_2, \omega_2; \ldots)$ of $U(\mathcal{N}) \supset \sum_i [U(\mathcal{N}_i) \supset SO(\mathcal{N}_i)]\oplus$ in IBMs [ω_i are the irreps of $SO(\mathcal{N}_i)$] and similarly in shell model with $SO(\mathcal{N}_i)$ replaced by $Sp(\mathcal{N}_i)$. The $\sigma^2(\mathbf{m})$ in Eq. (14.18) corresponds to $U(\mathcal{N}) \supset \sum_i U(\mathcal{N}_i)\oplus$,

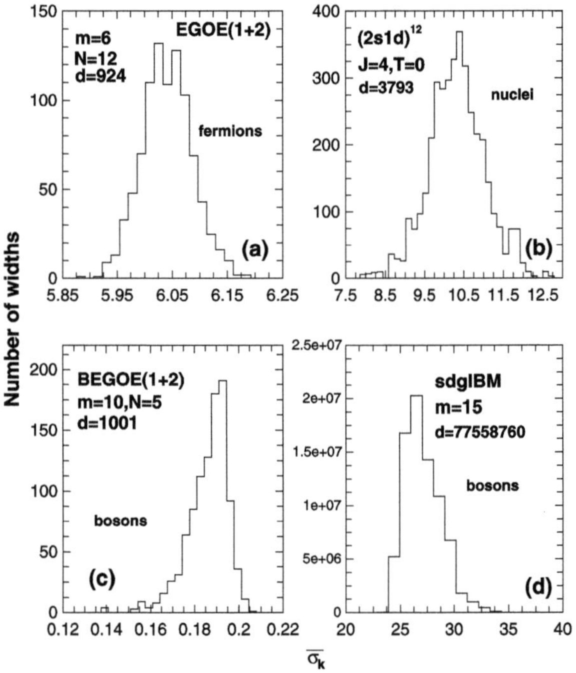

Fig. 14.9 Distribution of ensemble averaged (except in (**b**)) widths over the irreps k in 4 examples. In the histograms, at the center of each bin, the number of widths in the corresponding bin gives the height of the bin. Adding the number of widths will give the matrix dimension d. Values of d are given in the figures. Results are shown for: (**a**) EGOE(1 + 2) for spinless fermions; (**b**) a $(2s1d)$ nuclear shell model example; (**c**) BEGOE(1 + 2) for spinless bosons; (**d**) BEGOE(1 + 2)-L for sdgIBM. In (**b**), the widths are in units of MeV

4. $\sigma^2(m, (\lambda\mu), T)$ of $U(3\mathscr{N}) \supset U(\mathscr{N}) \otimes [SU_T(3) \supset O_T(3)]$ in IBM-T (or IBM-3); here one has to use the $\hat{X}_3$ and $\hat{X}_4$ integrity basis operators of $SU_T(3) \supset SO_T(3)$,
5. $\sigma^2(m, \{f\})$ of $U(6\mathscr{N}) \supset U(\mathscr{N}) \otimes SU_{ST}(6)$ in IBM-ST (or IBM-4),
6. $\sigma^2(m, \{f\}ST)$ of $U(24) \supset U(6) \otimes [SU_{ST}(4) \supset SU_S(2) \otimes SU_T(2)]$ for $(2s1d)$ shell nuclei and similarly for $(2p1f)$ shell nuclei [also just $\sigma^2(m, \{f\})$],
7. $\sigma^2(m, T)$ of $U(2\mathscr{N}) \supset U(\mathscr{N}) \otimes SU_T(2)$ in shell model spaces,
8. $\sigma^2(\mathbf{m}, T)$ and $\sigma^2(\mathbf{m}, \mathbf{T})$ in shell model; $\mathbf{T} = (T_1, T_2, \ldots)$ with T_i being the isospin of m_i nucleons in a j_i orbit,
9. $\sigma^2(m, J)$ for $(j)^m$ system of fermions and $\sigma^2(m, L)$ for $(\ell)^m$ system of bosons [here, expansions in powers of $J(J+1)$ are possible as discussed in Chap. 13]. Similarly, though much more complicated, also for multi-j shell fermion and multi-ℓ shell boson systems [39].

Using the propagation equations one can calculate for each member of EGOEs, $\sigma^2(\Gamma)$ without H matrix diagonalization and therefore it is easy to obtain $\overline{\sigma(\Gamma)} = \sqrt{\overline{\sigma^2(\Gamma)}}$ where the bar denotes average over the appropriate EGOE ensemble. Figure 14.9 gives $\overline{\sigma_k}$, $k = \Gamma$ in 4 examples: (a) EGOE(1+2) for spinless fermions with $\{H\} = h(1) + \lambda\{V(2)\}$ and 6 fermions in 12 sp states. Here $\{V(2)\}$ is GOE in two particle spaces with unit variance for the matrix elements. For the single particle energies defining $h(1)$ and other details, see Chap. 5. In the calculations $\lambda = 0.3$ and number of members is 50. The irreps k are the $h(1)$ basis states. (b) Shell model with k being shell model basis states for the $(2s1d)^{m=12,J=4,T=0}$ system with H defined by ^{17}O single particle energies and a two-body interaction called KLS. See [38] for details and note that the shell model results can be viewed as the results for a typical member of EGOE(1 + 2)-JT. (c) BEGOE(1 + 2) for bosons with $\{H\} = h(1) + \lambda\{V(2)\}$ and 10 bosons in 5 sp states. Here $\{V(2)\}$ is GOE in two particle spaces with unit variance for the matrix elements. For the single particle energies defining $h(1)$ and other details, see Chap. 9. In the calculations $\lambda = 0.1$ and number of members is 20. The irreps k are $h(1)$ basis states. (d) BEGOE(1 + 2)-$(sdg : L)$ constructed for sdgIBM [17]. Here, in $\{H\} = h(1) + \lambda\{V(2)\}$, $V(2)$ preserves L. There are 32 two-body matrix elements defining $V(2)$ in sdgIBM and they are chosen to be independent Gaussian variables with zero center and unit variance. The k's in this example are the configurations defined by (m_s, m_d, m_g). Therefore, s, d and g boson single particle energies will not contribute to the k-variances; see Eq. (14.18). In calculating the number of widths, the dimensions $\binom{n_d+4}{4}\binom{n_g+8}{8}$ of the configurations (m_s, m_d, m_g) is taken into account. Calculations are for 15 bosons and number of members in the ensemble is 500. It is clearly seen from Fig. 14.9 that in all the examples the ensemble averaged fixed irrep widths, i.e. $\overline{\sigma_k}$, are nearly constant with respect to k and the fluctuation (∼5–10 %) in the widths $\overline{\sigma_k}$ is Gaussian distributed for fermion systems while it is asymmetric for bosons. Constancy of variances appear to be a generic property of EE (see also Chap. 12).

14.6 Results from EGOE(1 + 2)-s, EGOE(1 + 2)-π, BEGOE(1 + 2)-F and BEGOE(1 + 2)-$S1$ Ensembles

14.6.1 EGOE(1 + 2)-s *Results*

In Chap. 7 we have already shown that random interactions with EGOE(1 + 2)-s generate two important ordered structures: (i) spin $S = 0$ ground states; (ii) odd-even staggering in ground state energies. These features are also seen in EGOE(1 + 2)-J and EGOE(1 + 2)-JT [3]. Going beyond these, regularities with random interactions have been studies via pairing operator expectation values. In the eigenfunctions defined by the EGOE(1 + 2)-s Hamiltonian

$$H = h(1) + \lambda\big[\big\{V^{s=0}(2)\big\} + \big\{V^{s=1}(2)\big\}\big], \qquad (14.19)$$

expectation values of the pairing Hamiltonian H_p given by Eq. (6.36) are calculated in a number of examples in [40]. In Fig. 14.10 results are shown for a 50 member EGOE(1 + 2)-s ensemble with 8 fermions ($m = 8$) in 8 orbits ($\Omega = 8$) and for $S = 0$ and 1. The exact results are compared with the EGOE formula given by Eq. (4.82) both with and without Edgeworth corrections. In this example $\lambda_c = 0.05$ and for this λ value we have (with $K = H_p$) for the K densities: $\varepsilon_K, |\gamma_1(K)| \sim 0$, $\sigma_K \sim 1.06$, $\gamma_2(K) \sim -0.33$ and $\langle K\rangle^{m,S} \sim 2.22$ for $S = 0$. Similarly, $\gamma_2(K) \sim -0.37$ and $\langle K\rangle^{m,S} \sim 2.00$ for $S = 1$. For $\lambda = 0.3 \gg \lambda_F$, we have

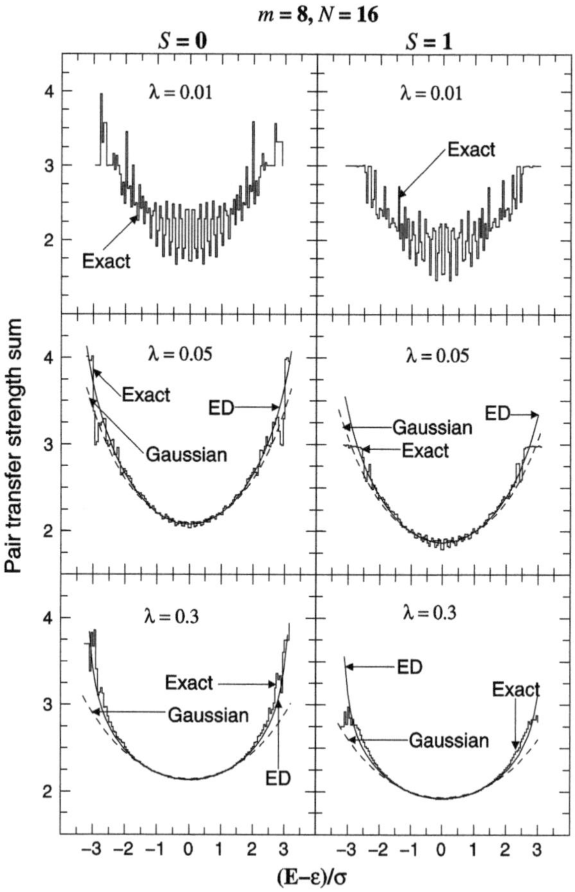

Fig. 14.10 Pairing expectation value or pair transfer strength sum $\langle PP^{\dagger}\rangle^{E} = \langle H_P\rangle^{E}$ vs $\hat{E} = (E-\varepsilon)/\sigma$ for a 500 member EGOE(1 + 2)-s ensemble with $\Omega = m = 8$ (number of sp states $N = 16$) and total spins $S = 0$ and 1; ε and σ are centroid and width of the eigenvalues E. Results (called 'exact' in the figure) are shown for various values of the strength λ of the two-body part of H; H is defined by Eq. (14.19). Results are compared with the EGOE formula given by Eq. (4.82) with Gaussian forms and also with Edgeworth corrected Gaussians (called ED in the figure)

$\gamma_2(K) \sim -0.44$ for $S = 0$ and -0.47 for $S = 1$. As seen from the figure, pair expectation values follow, in the chaotic domain ($\lambda \geq \lambda_c$) the simple EGOE law with little fluctuations. More importantly, at low energies the pair expectation value is large (still much smaller than the that for the pure pairing Hamiltonian) and then deceases as we go to the center (after that it will again increase as the space is finite). Also the expectation value in ground state domain for $S = 0$ is always larger than for $S = 1$. Thus, random interactions, even in the chaotic domain, exhibit stronger pairing correlations in the ground state region and they decrease as we go up in the energy. To probe pairing generated by random interactions further, one can use fixed seniority (v) partial densities $I^{m,v,S}$. Then, $f(v) = I^{m,v,S}(E)/I^{m,S}(E)$ gives the fraction of the intensity of the states with a given v in the eigenstate with energy E. For the random Hamiltonian given by Eq. (14.19), for $\lambda = 0.3$ in Fig. 14.10, $f(v)$ for $v = 0, 2, 4$ and 6 are 7 %, 33 %, 42 % and 18 % for $\hat{E} = -3$ and 12 %, 44 %, 37 % and 7 % for $\hat{E} = -3.1$. Thus in the ground state domain, although the pair expectation values are enhanced, the wavefunctions have relatively small strength for $v = 0$ states, i.e. they are not close to pure H_p eigenstates. This result is consistent with the EGOE(1 + 2)-J and EGOE(1 + 2)-JT results obtained using nuclear shell model codes [3, 34]. Let us add that in [41], the splitting of sp energies was shown to play important role in EGOE(1 + 2)-J generating pair structure in low-lying states.

14.6.2 EGOE(1 + 2)-π *Results*

Experimental data on the parity of the ground states show that all known even-even nuclei have +ve parity ground states without any exception. In the compilation used in [4], there are 346 odd-A nuclei with A > 120 where parity of ground states is known. The shell model space for these involve sp states with both parities. In the data it is seen that there are 182 nuclei with +ve parity ground states clearly identified and 164 nuclei with −ve parity. Similarly, there are 146 odd-odd nuclei (with A > 120) with 68 of them having +ve parity ground states and the remaining having −ve parity. Thus, data shows preponderance of +ve parity ground states in even-even nuclei even when sp states of both parity are present in the shell model space appropriate for these and similarly, there is near equilibration of both parities for odd-A and odd-odd nuclei. Using shell model spaces $(f_{5/2}p_{1/2}g_{9/2})^{m_p,m_n}$, $(h_{11/2}s_{1/2}d_{3/2})^{m_p,m_n}$, $(f_{5/2}p_{1/2}g_{9/2})^{m_p}(g_{7/2}d_{5/2})^{m_n}$ and for many different values of proton (m_p) and neutron (m_n) numbers, EGOE(1 + 2)-J calculations have been performed in [34] and it is found that they generate ground states with parities having pattern almost close to that found in experimental data. Then, an important question is how to understand the shell model results using much simpler EE that include parity degree of freedom.

Towards this end, Papenbrock and Weidenmüller [42] used EGOE(1 + 2)-π ensemble introduced in Chap. 8 with $\tau \to \infty$, $\alpha = \tau$ and studied the probability (R_+) for +ve parity ground states over the ensemble for several (N_+, N_-, m) systems. Their numerical calculations showed considerable variation (18–84 %) in R_+. In addition, they gave a plausible proof that in the dilute limit $[m \ll (N_+, N_-)]$, R_+ will approach 50 %. Combining these, they argued that the observed preponderance of +ve parity ground states could be a finite size (finite N_+, N_-, m) effect. However, for the general EGOE(1 + 2)-π considered in Chap. 8, it was shown in [43] that R_+ can reach 100 % by varying the α and τ parameters and we will turn to these results briefly.

For EGOE(1 + 2)-π with $\tau \sim 0$, clearly one will get $R_+ = 100$ % for even m (with $m \ll N_+, N_-$). Going beyond this, calculations have been carried out for a 200 member ensemble for $(N_+, N_-, m) = (6, 6, 6)$ and a 100 member ensembles for $(8, 8, 5)$, $(6, 6, 6)$, $(6, 10, 4)$ and $(6, 10, 5)$ systems using $\alpha = \tau$ and 1.5τ. The results are shown in Fig. 14.11. For $\alpha = \tau$, the results are as follows. For $\tau \lesssim 0.04$, we have $R_+ \sim 100$ % and then R_+ starts decreasing with some fluctuations between $\tau = 0.1$ and 0.2; τ is restricted to the realistic range of $\tau \leq 1$. It is seen that $R_+ \gtrsim 50$ % for $\tau \leq 0.3$ independent of (N_+, N_-, m) and then it decreases much faster reaching ∼30 % for $\tau = 0.5$ for $(N_+, N_-, m) = (6, 6, 6)$. For $m < (N_+, N_-)$, the decrease in R_+ is slower. If we increase α, we can easily infer that the width of the lowest +ve

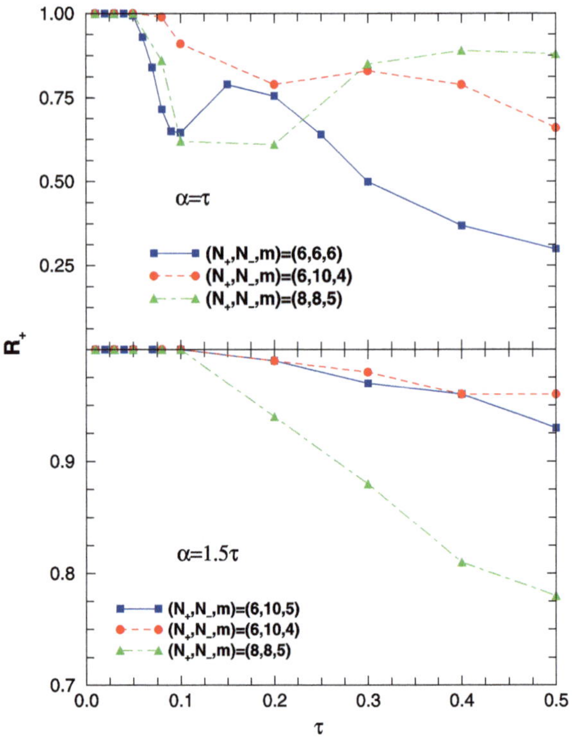

Fig. 14.11 Probability (R_+) for +ve parity ground states for various (τ, α) values and for various (N_+, N_-, m) systems in EGOE(1 + 2)-π. Figure is taken from [43] with permission from American Physical Society

parity (m_1, m_2) unitary configuration becomes much larger compared to the lowest −ve parity unitary configuration. Therefore with increasing α, R_+ is expected to increase and this is clearly seen in Fig. 14.11. Thus $\alpha \gtrsim \tau$ is required for R_+ to be large. A quantitative description of R_+ requires the construction of +ve and −ve parity state densities accurately in the tail region and this calls for more detailed analytical study of EGOE(1 + 2)-π.

14.6.3 Results from BEGOE(1 + 2)-*F and* BEGOE(1 + 2)-*S*1

Turning to BEGOE, in Chap. 10 some of the ordered structures generated by random interactions in BEGOE(1 + 2)-F are presented and they are: (i) $F = F_{max}$ ground states; (ii) natural F-spin ordering. In addition, just as in EGOE(1 + 2)-**s**, BEGOE(1 + 2)-F also generates ground states with relatively large value for the expectation value of the pairing Hamiltonian H_p; H_p is defined by Eq. (F.2). Expectation values $\langle H_p \rangle^{m,F,E}$ of the pairing Hamiltonian in the eigenstates generated by BEGOE(1 + 2)-F carry signatures of pairing. It is useful to note that, given the eigenvalues E_p of the pairing operator and eigenvalues E of the Hamiltonian operator, pairing expectation values are nothing but the centroids of the conditional density $\rho^\Gamma(E_p|E) = \rho^\Gamma(E_p, E)/\rho^\Gamma(E)$ defined over

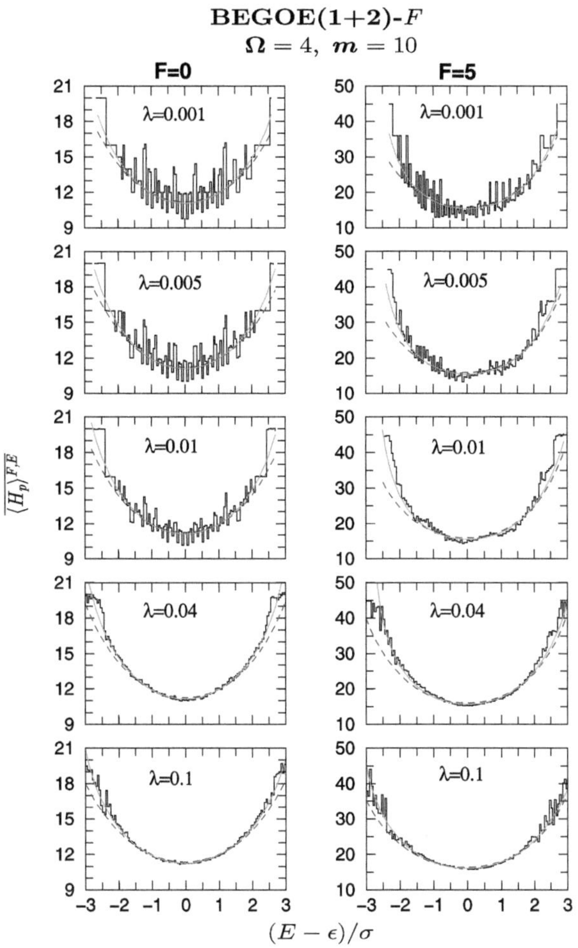

Fig. 14.12 Pairing expectation value $\langle H_P \rangle^{m,F,E}$ vs $\hat{E} = (E - \varepsilon)/\sigma$ for a 100 member BEGOE(1 + 2)-F ensemble with $\Omega = 4$ and $m = 10$ (number of sp states $N = 8$); ε and σ are centroid and width of the eigenvalues E. Ensemble results (histograms in the figure) are shown for various values of the strength λ of the two-body part of H; H is defined by Eq. (10.3) with $\lambda_0 = \lambda_1 = \lambda$. The sp energies are $\varepsilon_i = i + 1/i$ as used in Chap. 10. Results for $F = 0$ and 5 are compared with Eq. (4.82) with Gaussian forms (*red dashed curves*) and also with Edgeworth corrected Gaussians (*green continuous curves*)

fixed-$\Gamma = F$ spaces. Numerical results for a 100 member BEGOE(1 + 2)-F are shown in Fig. 14.12. Results are similar to those in Fig. 14.10. Firstly pairing expectation values are largest near the ground state. Secondly the EGOE formula, ratio of Gaussians as given by Eq. (4.82), is seen to apply to BEGOE(1 + 2)-F. Application of Eq. (F.9) shows clearly (see also Table F.1) that the maximum value of the H_p eigenvalues increases with F-spin for a fixed-m. The values are 28, 32, 34, 42, 48 and 60 for $F = 0$–5 respectively, for $\Omega = 4$ and $m = 10$. Numerical results in Fig. 14.12 also show that for states near the lowest eigenvalue (near the ground state) increases with F-spin. Thus random interactions preserve this regular property of the pairing Hamiltonian in addition to generating $F = F_{max}$ ground states as discussed in Sect. 10.1.4.

There are some preliminary investigations of regular structures generated by BEGOE(1 + 2)-$S1$ in [44]. As an example, shown in Fig. 14.13 are results for expectation values of the two pairing Hamiltonians $H_{\mathscr{P}}$ and H_P (see Sects. F.2.1

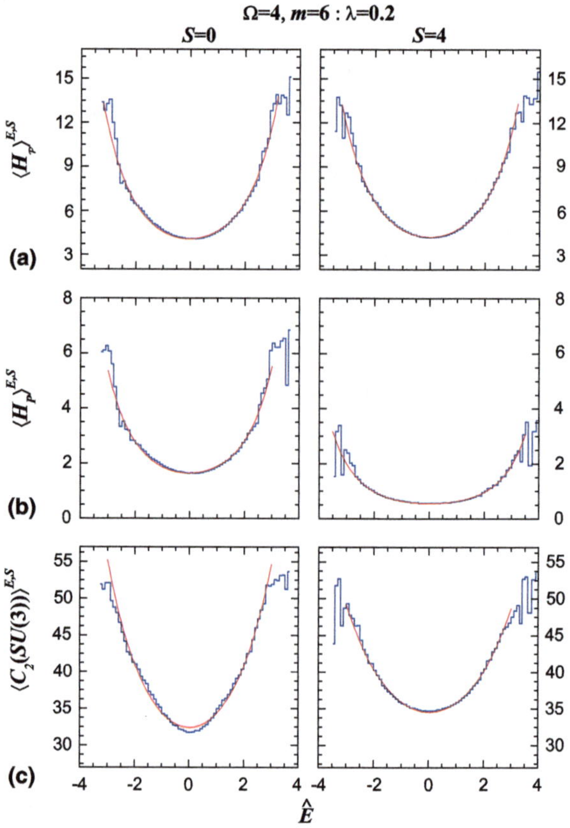

Fig. 14.13 Expectation values of the two pairing Hamiltonians $H_{\mathscr{P}}$ and H_P and $\hat{C}_2(SU(3))$ vs $\hat{E} = (E - \varepsilon)/\sigma$ for a 250 member BEGOE(1 + 2)-$S1$ systems with H defined by Eq. (10.17) and ($\Omega = 4, m = 6$). Results are shown for spins $S = 0$ and $S = 4$. (**a**) expectation values of $H_{\mathscr{P}}$, (**b**) expectation values of H_P and (**c**) expectation value of $\hat{C}_2(SU(3))$. Ensemble averaged results are shown by histograms while (*red*) *continuous curves* are from theory (ratio of Edgeworth corrected Gaussians) given by Eq. (4.82)

and F.2.2) and also $\hat{C}_2(SU(3))$ in the eigenstates of the BEGOE(1 + 2) Hamiltonian defined by Eq. (10.17). We have chosen the parameters in the region of chaos, i.e. $\lambda_0 = \lambda_1 = \lambda_2 = \lambda = 0.2$ so that fluctuations in the expectation values will be minimal. It is seen that the expectation values are largest near the ground states and then decrease as we move towards the center of the spectrum. The calculated results are in good agreement with the prediction that for boson systems also expectation values will be ratio of Gaussians. Results in the figure show that with repulsive pairing, ground states will be dominated by low seniority structure (small value for ω or $\omega_1 + \omega_2 + \omega_3$). In addition, results in Fig. 14.13c show that random interactions give ground states with large value for the expectation value of $\hat{C}_2(SU(3))$. Moreover, for the irrep $(m, 0) = (6, 0)$, we have easily $\langle \hat{C}_2(SU(3)) \rangle^{m=6,(6,0),S} = 54$ and from the figure one can then infer that ground states will be dominated by the $SU(3)$ irrep $(\lambda\mu) = (m, 0) = (6, 0)$. This result is of importance for IBM-3 model of atomic nuclei [21].

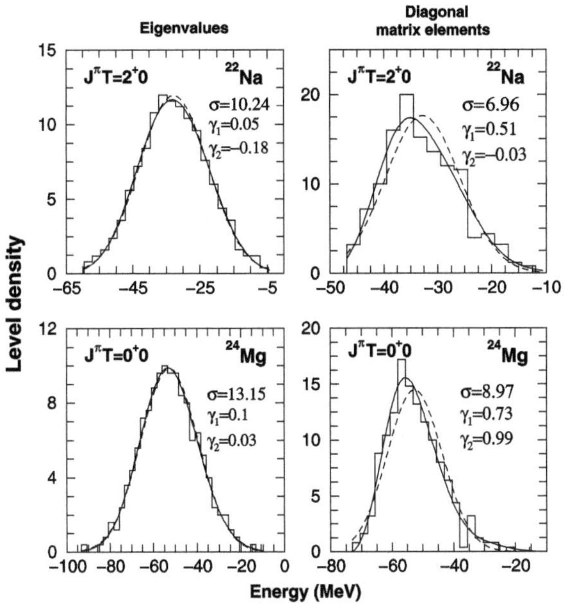

Fig. 14.14 Plot showing density of eigenvalues and density of diagonal matrix elements for the Hamiltonian matrices of ^{22}Na and ^{24}Mg nuclei obtained using nuclear shell model in $(2s\,1d)$ space. Values of the widths σ, skewness γ_1 and excess γ_2 are given in the figures. The units for σ are MeV. The centroid $E_c = -32.77$ MeV for ^{22}Na and -52.59 MeV for ^{24}Mg. Histograms are the exact results with bin size 2.5 MeV for all the examples. The *dashed curves* are the Gaussians with centroid E_c given above and width σ whose value is given in the figure. Similarly *continuous curves* are Edgeworth corrected Gaussians. Figure is taken from [45] with permission from Springer

14.7 Correlations Between Diagonal *H* Matrix Elements and Eigenvalues

Large number of numerical calculations have shown [46, 47] that the joint probability distribution $\rho(E, e_k)$ of the diagonal matrix elements e_k and eigenvalues E of a typical nuclear shell model H matrix is close to a bivariate Gaussian and this has its origin in a similar result valid more generally for EGOE$(1+2)$-J (or JT). Therefore the marginal densities $\rho(E)$ and $\rho(e_k)$ will be close to Gaussians with same centroids but different widths and the widths of the conditional densities $\rho(E|e_k)$ will be independent of e_k. These results have been used to derive a formula for the number of principal components and information entropy in wavefunctions as given in Chap. 5. The close to Gaussian form of $\rho(E)$ and $\rho(e_k)$ imply that the eigenvalues E and the diagonal elements of the H matrix (or equivalently the basis state energies) will be correlated. Flambaum et al. examined, for CeI, eigenvalue spectrum vs the spectrum generated by e_k [48] and they found a close correlation between the two spectra. As an example, density of eigenvalues and density of diagonal matrix elements for the Hamiltonian matrices of ^{22}Na and ^{24}Mg nuclei are shown in Fig. 14.14. These distributions are compared with the Gaussian form ($\rho_{\mathscr{G}}$) and the Edgeworth (ED) corrected Gaussian form (ρ_{ED}). It is clearly seen that the eigenvalue distributions for the two nuclear examples are quite close to $\rho_{\mathscr{G}}$ while the densities of the diagonal matrix elements are, with some deviations, close to ρ_{ED}. These results reconfirm [45] that in the nuclear examples, the eigenvalues and the diagonal matrix elements of the H matrix are highly correlated and their distributions are close to Gaussian forms. In atomic examples much larger differences are found to exist [45] and this could be because atomic examples are much further from EGOE$(1+2)$-J. It should be added that, more recently Zhao et al. have argued [49, 50], using many EGOE$(1+2)$-J and EGOE$(1+2)$-JT numerical examples, that high correlation between eigenvalues and diagonal matrix elements is a much more a general phenomena. Using this, an extrapolation scheme was proposed by Zhao et al. [51, 52] for determining the energies of ground state and other low-lying states within nuclear shell model without diagonalizing huge matrices.

14.8 Collectivity and Random Interactions

Following the initial result of Bijker and Frank [2] that sdIBM with random interactions generate vibrational and rotational spectra, as shown in Fig. 14.2, there are many investigations within nuclear shell model (i.e. using fermion systems) to understand the origin of collective motion in atomic nuclei. To this end, studied using random interactions in some shell model spaces are: (i) predominance of prolate nuclear deformation [53]; (ii) origin of quadrupole collectivity in nuclei [54]; (iii) generation of pairing seniority structure and quadrupole vibrations and rotations [55]; (iv) generation of vibrational and rotational structure within the FDSM model which is a truncated version of the shell model [56]. Although numerical results do indicate that random interactions generate collectivities, there is no good analytical understanding yet. In an another interesting application, EGOE$(1+2)$-J and EGOE$(1+2)$-JT are used by some groups to identify important parts of the two-body interaction in the configuration-interaction shell model [53, 57]. Finally, it is also found in numerical calculations that random interactions in sdIBM of atomic nuclei generate strong correlations between energy levels generating many different regular structures such as preponderance of ground states with $L=0^+$, an-harmonic vibrations, d-boson condensation, rotational motion and so on [58, 59].

In conclusion, results of various studies on regular structures generated by random interactions, discussed in some detail in this chapter, confirm the statement of Zelevinsky and Volya [3]: *Standard textbook ideas of the factors that form the low-lying structure of a closed self-sustaining mesoscopic systems are insufficient. The quantum numbers of the ground states and some regularities of spectra emerge not necessarily due to the corresponding coherent parts of the inter-particle interaction.*

References

1. C.W. Johnson, G.F. Bertsch, D.J. Dean, Orderly spectra from random interactions. Phys. Rev. Lett. **80**, 2749–2753 (1998)

2. R. Bijker, A. Frank, Band structure from random interactions. Phys. Rev. Lett. **84**, 420–422 (2000)
3. V. Zelevinsky, A. Volya, Nuclear structure, random interactions and mesoscopic physics. Phys. Rep. **391**, 311–352 (2004)
4. Y.M. Zhao, A. Arima, N. Yoshinag, Regularities of many-body systems interacting by a two-body random ensemble. Phys. Rep. **400**, 1–66 (2004)
5. D. Mulhall, Quantum chaos and nuclear spectra, Ph.D. Thesis, Michigan State University, East Lansing, USA (2002)
6. T. Papenbrock, H.A. Weidenmüller, Distribution of spectral widths and preponderance of spin-0 ground states in nuclei. Phys. Rev. Lett. **93**, 132503 (2004)
7. D. Kusnezov, Two-body random ensembles: from nuclear spectra to random polynomials. Phys. Rev. Lett. **85**, 3773–3776 (2000)
8. R. Bijker, A. Frank, Mean-field analysis of interacting boson models with random interactions. Phys. Rev. C **64**, 061303(R) (2001)
9. V.K.B. Kota, Random interactions in nuclei and extension of 0^+ dominance in ground states to irreps of group symmetries. High Energy Phys. Nucl. Phys. (China) **28**, 1307–1312 (2004)
10. C.W. Johnson, Random matrices, symmetries and many-body states (2011). arXiv:1103.4161
11. M.W. Kirson, J.A. Mizrahi, Random interactions with isospin. Phys. Rev. C **76**, 064305 (2007)
12. V.K.B. Kota, Interacting boson model applications to exotic nuclear structure, in *Recent Trends in Nuclear Physics—2012*, AIP Conf. Proc., vol. 1524 (2013), pp. 52–57
13. V.K.B. Kota, Transformation brackets between $U(N) \supset SO(N) \supset SO(N_a) \oplus SO(N_b)$ and $U(N) \supset U(N_a) \oplus U(N_b) \supset SO(N_a) \oplus SO(N_b)$. J. Math. Phys. **38**, 6639–6647 (1997)
14. F. Iachello, A. Arima, *The Interacting Boson Model* (Cambridge University Press, Cambridge, 1987)
15. F. Iachello, R.D. Levine, *Algebraic Theory of Molecules* (Oxford University Press, New York, 1995)
16. D.J. Rowe, F. Iachello, Group theoretical models of giant resonance splittings in deformed nuclei. Phys. Lett. B **130**, 231–234 (1983)
17. Y.D. Devi, V.K.B. Kota, sdg interacting boson model: hexadecupole degree of freedom in nuclear structure. Pramana—J. Phys. **39**, 413–491 (1992)
18. D.F. Kusnezov, Nuclear collective quadrupole-octupole excitations in the $U(16)$ $spdf$ interacting boson model, Ph.D. Thesis, Yale University, USA (1988)
19. G.L. Long, W.L. Zhang, H.Y. Li, E.G. Zhao, Sci. China, Ser. A, Math. Phys. Astron. **41**, 1296–1301 (1998)
20. V.K.B. Kota, Spectra and E2 transition strengths for $N = Z$ even-even nuclei in IBM-3 dynamical symmetry limits with good s and d boson isospins. Ann. Phys. (N.Y.) **265**, 101–133 (1998)
21. J.E. García-Ramos, P. Van Isacker, The interacting boson model with $SU(3)$ charge symmetry and its applications to even-even $N \approx Z$ nuclei. Ann. Phys. (N.Y.) **274**, 45–75 (1999)
22. V.K.B. Kota, $O(36)$ symmetry limit of IBM-4 with good s, d and sd boson spin-isospin Wigner's $SU(4) \sim O(6)$ symmetries for $N \approx Z$ odd-odd nuclei. Ann. Phys. (N.Y.) **280**, 1–34 (2000)
23. D. Mulhall, A. Volya, V. Zelevinsky, Geometric chaoticity leads to ordered spectra for randomly interacting fermions. Phys. Rev. Lett. **85**, 4016–4019 (2000)
24. Y.M. Zhao, A. Arima, K. Ogawa, Energy centroids of spin I states by random two-body interactions. Phys. Rev. C **71**, 017304 (2005)
25. V.K.B. Kota, K. Kar, Group symmetries in two-body random matrix ensembles generating order out of complexity. Phys. Rev. E **65**, 026130 (2002)
26. V.K.B. Kota, Regularities with random interactions in energy centroids defined by group symmetries. Phys. Rev. C **71**, 041304(R) (2005)
27. V.K.B. Kota, Group theoretical and statistical properties of interacting boson models of atomic nuclei: recent developments, in *Focus on Boson Research*, ed. by A.V. Ling (Nova Science Publishers Inc., New York, 2006), pp. 57–105

28. Y.M. Zhao, A. Arima, N. Yoshida, K. Ogawa, N. Yoshinaga, V.K.B. Kota, Robustness of regularities for energy centroids in the presence of random interactions. Phys. Rev. C **72**, 064314 (2005)
29. V.K.B. Kota, Two-body ensembles with group symmetries for chaos and regular structures. Int. J. Mod. Phys. E **15**, 1869–1883 (2006)
30. J.P. Elliott, Collective motion in the nuclear shell model. I. Classification schemes for states of mixed configurations. Proc. R. Soc. Lond. Ser. A **245**, 128–145 (1958)
31. J.P. Draayer, G. Rosensteel, $U(3) \rightarrow R(3)$ integrity-basis spectroscopy. Nucl. Phys. A **439**, 61–85 (1985)
32. Y. Akiyama, J.P. Draayer, A users guide to Fortran programs for Wigner and Racah coefficients of SU_3. Comput. Phys. Commun. **5**, 405–415 (1973)
33. R.U. Haq, J.C. Parikh, Space symmetry in light nuclei: (II). $SU(4)$ isospin-spin averages. Nucl. Phys. A **220**, 349–366 (1974)
34. Y.M. Zhao, A. Arima, N. Shimizu, K. Ogawa, N. Yoshinaga, O. Scholten, Patterns of the ground states in the presence of random interactions: nucleon systems. Phys. Rev. C **70**, 054322 (2004)
35. A. Volya, Emergence of symmetry from random n-body interactions. Phys. Rev. Lett. **100**, 162501 (2008)
36. P. Papenbrock, H.A. Weidenmüller, Origin of chaos in the spherical nuclear shell model: role of symmetries. Nucl. Phys. A **757**, 422–438 (2005)
37. V.K.B. Kota, R.U. Haq, *Spectral Distributions in Nuclei and Statistical Spectroscopy* (World Scientific, Singapore, 2010)
38. V.K.B. Kota, M. Vyas, K.B.K. Mayya, Spectral distribution analysis of random interactions with J-symmetry and its extensions. Int. J. Mod. Phys. E **17**(Supp), 318–333 (2008)
39. S.S.M. Wong, *Nuclear Statistical Spectroscopy* (Oxford University Press, New York, 1986)
40. M. Vyas, V.K.B. Kota, N.D. Chavda, One-plus two-body random matrix ensembles with spin: results for pairing correlations. Phys. Lett. A **373**, 1434–1443 (2009)
41. Y. Lei, Z.Y. Xu, Y.M. Zhao, S. Pittel, A. Arima, Emergence of generalized seniority in low-lying states with random interactions. Phys. Rev. C **83**, 024303 (2011)
42. T. Papenbrock, H.A. Weidenmüller, Abundance of ground states with positive parity. Phys. Rev. C **78**, 054305 (2008)
43. M. Vyas, V.K.B. Kota, P.C. Srivastava, One plus two-body random matrix ensembles with parity: density of states and parity ratios. Phys. Rev. C **83**, 064301 (2011)
44. H. Deota, N.D. Chavda, V.K.B. Kota, V. Potbhare, M. Vyas, Random matrix ensemble with random two-body interactions in the presence of a mean-field for spin one boson systems. Phys. Rev. E **88**, 022130 (2013)
45. M. Vyas, V.K.B. Kota, Random matrix structure of nuclear shell model Hamiltonian matrices and comparison with an atomic example. Eur. Phys. J. A **45**, 111–120 (2010)
46. J.B. French, V.K.B. Kota, Nuclear level densities and partition functions with interactions. Phys. Rev. Lett. **51**, 2183–2186 (1983)
47. V.K.B. Kota, K. Kar, Spectral distributions in nuclei: general principles and applications. Pramana—J. Phys. **32**, 647–692 (1989)
48. V.V. Flambaum, A.A. Gribakina, G.F. Gribakin, M.G. Kozlov, Structure of compound states in the chaotic spectrum of the Ce atom: localization properties, matrix elements, and enhancement of weak perturbations. Phys. Rev. A **50**, 267–296 (1994)
49. J.J. Shen, A. Arima, Y.M. Zhao, N. Yoshinaga, Strong correlation between eigenvalues and diagonal matrix elements. Phys. Rev. C **78**, 044305 (2008)
50. N. Yoshinaga, A. Arima, J.J. Shen, Y.M. Zhao, Correlation between eigenvalues and sorted diagonal elements of a large dimensional matrix. Phys. Rev. C **79**, 017301 (2009)
51. J.J. Shen, Y.M. Zhao, A. Arima, Lowest eigenvalue of the nuclear shell model matrix. Phys. Rev. C **82**, 014309 (2010)
52. J.J. Shen, Y.M. Zhao, A. Arima, N. Yoshinaga, New extrapolation method for low-lying states of nuclei in the sd and the pf shells. Phys. Rev. C **83**, 044322 (2011)

53. M. Horoi, V. Zelevinsky, Random interactions explore the nuclear landscape: predominance of prolate nuclear deformation. Phys. Rev. C **81**, 034306 (2010)
54. V. Abramkina, A. Volya, Quadrupole collectivity in the two-body random interaction. Phys. Rev. C **84**, 024322 (2011)
55. C.W. Johnson, H.A. Nam, New puzzle for many-body systems with random two-body interactions. Phys. Rev. C **75**, 047305 (2007)
56. Y.M. Zhao, J.L. Ping, A. Arima, Collectivity of low-lying states under random two-body interactions. Phys. Rev. C **76**, 054318 (2007)
57. C.W. Johnson, P.G. Krastev, Sensitivity of random two-body interactions. Phys. Rev. C **81**, 054303 (2010)
58. J. Barea, R. Bijker, A. Frank, Eigenvalue correlations and the distribution of ground state angular momentum for random many-body quantum systems. Phys. Rev. C **79**, 054302 (2009)
59. Y. Lei, Y.M. Zhao, N. Yoshida, A. Arima, Correlations of excited states for *sd* bosons in the presence of random interactions. Phys. Rev. C **83**, 044302 (2011)

Chapter 15
Time Dynamics and Entropy Production to Thermalization in EGOE

In this final chapter, we will consider time evolution of isolated finite many-particle systems with random two-body interactions in presence of a mean-field. As pointed out first by Flambaum [1], results here will be useful in the study of the stability of a quantum computer against quantum chaos. Similarly, as Lea Santos and others have pointed out [2–4], they are important in the study of issues related to thermalization in isolated finite quantum systems. It is also possible to address fidelity and Loschmidt echoes in many-particle quantum systems [5]. We will discuss the available results briefly in the next three sections.

15.1 Time Dynamics in BW and Gaussian Regions in EGOE(1 + 2) and BEGOE(1 + 2)

Let us consider a system of m spinless particles (fermions or bosons) in N sp states with the Hamiltonian consisting of a mean-field [generated by a one-body part $h(1)$] and a random two-body interaction $V(2)$ with strength λ,

$$H = h(1) + \lambda V(2). \tag{15.1}$$

Note that $V(2)$ is represented by EGOE(2) or BEGOE(2) with GOE(1) representation in two-particle space. Say the system is prepared in a state $|k\rangle$ and this is assumed to be an eigenstate of $h(1)$. Then at time $t = 0$,

$$\Psi(t = 0) = |k\rangle = \sum_E C_k^E |E\rangle. \tag{15.2}$$

After time 't', the state changes to $\psi(t)$,

$$\Psi(t) = \big|k(t)\big\rangle = \exp -iHt|k\rangle. \tag{15.3}$$

V.K.B. Kota, *Embedded Random Matrix Ensembles in Quantum Physics*,
Lecture Notes in Physics 884, DOI 10.1007/978-3-319-04567-2_15,

Here we are putting $\hbar = 1$ so that t is in E^{-1} units. Applying Eq. (15.2) will give,

$$\begin{aligned} \Psi(t) = \big|k(t)\big\rangle &= \sum_{E} C_k^E \exp -iEt|E\rangle \\ &= \sum_{E,f} C_k^E C_f^E \exp -iEt|f\rangle \end{aligned} \tag{15.4}$$

where $|f\rangle$ are the complete set of eigenstates of $h(1)$ with $|k\rangle$ being one of them. Thus, the probability that the state $|k\rangle$ changes to the state $|f\rangle$ is $W_{k\to f}(t)$ where

$$\begin{aligned} W_{k\to f}(t) &= \big|\langle f|\exp -iHt|k\rangle\big|^2 = \big|A_{k\to f}(t)\big|^2; \\ A_{k\to f}(t) &= \sum_{E} C_k^E C_f^E \exp -iEt. \end{aligned} \tag{15.5}$$

Now, the survival or return probability is

$$W_{k\to k}(t) = \big|A_{k\to k}(t)\big|^2 = \left|\sum_{E} \big[C_k^E\big]^2 \exp -iEt\right|^2. \tag{15.6}$$

The $A_{k\to k}$ can be written as an integral using the discrete form of the strength function $F_k(E)$,

$$A_{k\to k}(t) = \int F_k(E) \exp -iEt\, dE. \tag{15.7}$$

With H represented by EGOE(1 + 2) [or BEGOE(1 + 2)] and the interaction strength $\lambda > \lambda_c$, level and strength fluctuations follow GOE and hence in this region one can replace to a good approximation $F_k(E)$ by its smoothed form. Thus, the first important results is that in most situations the smoothed form of strength functions determine time evolution in EE. As established in Chaps. 5 and 9, $F_k(E)$ changes from BW to Gaussian form as λ increases from λ_c. With this, we will consider four situations: (i) small 't' limit where we can apply perturbation theory; (ii) BW limit of EGOE(1+2) and BEGOE(1+2); (iii) Gaussian region of EGOE(1+2) and BEGOE(1+2); (iv) region intermediate to BW and Gaussian forms for $F_k(E)$. Several of the results for (i)–(iii) were given first by Flambaum and Izrailev [1, 6].

15.1.1 Small 't' Limit: Perturbation Theory

For small 't', we can write $\exp -iEt \simeq [\exp -ih(1)t][\exp -iV(2)t]$. Then,

$$\begin{aligned} A_{k\to k}(t) &= \langle k|\exp -iHt|k\rangle = [\exp -iE_k t]\langle k|\exp -iV(2)t|k\rangle \\ &= [\exp -iE_k t]\langle k|1 - iV(2)t - \big[V(2)\big]^2 t^2/2 + \cdots |k\rangle \\ &\simeq [\exp -iE_k t]\big[1 - \sigma_k^2 t^2/2\big] \simeq \exp\big[-iE_k t - \big(\sigma_k^2 t^2/2\big)\big]. \end{aligned} \tag{15.8}$$

Here, we have used the results $E_k = \langle k|H|k\rangle \simeq \langle k|h(1)|k\rangle$ and $\sigma_k^2 = \langle k|H^2|k\rangle - E_k^2 \simeq \langle k|[V(2)]^2|k\rangle$; see [7] for details. Thus, in the small 't' region we have for the return probability,

$$W_{k\to k}(t) = \exp -\sigma_k^2 t^2. \tag{15.9}$$

Some numerical examples testing Eq. (15.9) are shown in Figs. 15.1a for fermions and 15.1c for boson systems

15.1.2 Breit-Wigner Region

In the BW region with λ, the strength of the two-body interaction, not far from λ_c, the strength function will be of BW form with level and strength fluctuations following GOE. In this situation, replacing $F_k(E)$ by BW form (with spreading width Γ) in Eq. (15.7) we obtain,

$$A_{k\to k}(t) = \int_{-\infty}^{+\infty} \frac{\Gamma}{2\pi[(E-E_k)^2 + \frac{\Gamma^2}{4}]} \exp -iEt\, dE. \tag{15.10}$$

This is nothing but the Fourier transform of the BW function and the result for this is well known [8, 9]. Applying this gives,

$$A_{k\to k}(t) = \exp -\left[iE_k t + \frac{\Gamma}{2} t \right]. \tag{15.11}$$

Therefore, for BW the return probability will follow exponential law,

$$W_{k\to k}(t) \overset{BW\ region}{\longrightarrow} \exp -\Gamma t. \tag{15.12}$$

Note that, when t is in $[\sigma_H]^{-1}$ units, the spreading width Γ will be in σ_H units. Some numerical examples testing Eq. (15.12) are shown in Figs. 15.1b for fermions and d for boson systems. In all the calculations (presented in Figs. 15.1–15.3), the $|k\rangle$ states are the mean-field states obtained by the distributing m particles in the given N sp states. Similarly, the basis state energies E_k are the diagonal matrix elements of H in the m-particle basis states giving $E_k = \langle k|h(1) + \lambda V(2)|k\rangle$. Note that the centroids of the E_k energies are same as that of the eigenvalue (E) spectra but their widths are different. In the calculations E and E_k are zero centered (for each member) and scaled by the spectrum width. In all the calculations, the sp energies are taken as independent Gaussian random variables. In order to calculate ensemble averaged $W_{k\to f}$, for each member at a given time t, $|A_{k\to f}(t)|^2$ are summed over the basis states $|k\rangle$ and $|f\rangle$ in the energy windows $E_k \pm \delta$ and $E_f \pm \Delta$. Then, ensemble averaged $W_{k\to f}(t)$ for fixed k is obtained by binning. In Fig. 15.1, results are shown for $W_{k\to k}(t)$ for $E_k = 0$ with $\delta = \Delta = 0.005$.

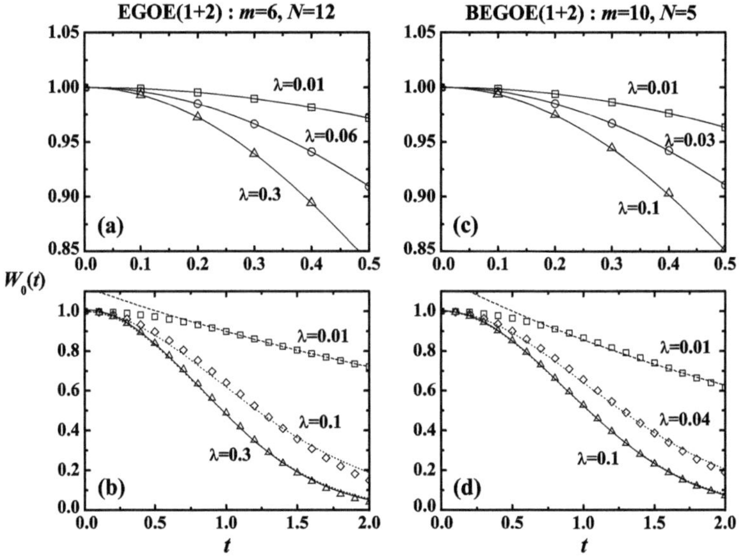

Fig. 15.1 Results for different values of λ for the return probability $W_0(t)$ vs t, i.e. $W_{k\to k}(t)$ vs t with $E_k = 0$. (**a**) For a EGOE(1 + 2) system with t up to 0.5 and the ensemble averaged results (*open symbols*) are compared with Eq. (15.9) (*continuous curves*). (**b**) For the same EGOE(1 + 2) system as in (**a**) but for t up to 2 and the ensemble averaged results (*open symbols*) are compared with theoretical results. Here, the *dashed* and *continuous curves* are obtained by using Eqs. (15.12) and (15.15) respectively. The *dotted curves* are due to Eq. (15.18). Values of the parameters in Eq. (15.18) for good fits are as follows: $\alpha = 3.5$ and $\beta = 0.6$ for $\lambda = 0.1$ and similarly, $\alpha = 17$ and $\beta = 1.36$ for $\lambda = 0.3$. (**c**) For a BEGOE(1 + 2) system and other details are same as in (**a**) except the λ values are different. (**d**) For the same BEGOE(1 + 2) system as in (**c**) and other details are as in (**b**). Here, the values of the parameters in Eq. (15.18) for good fits are as follows: $\alpha = 4$ and $\beta = 0.59$ for $\lambda = 0.04$ and similarly, $\alpha = 17$ and $\beta = 1.2$ for $\lambda = 0.1$. In all the calculation 50 member ensembles are used

15.1.3 Gaussian Region

In the Gaussian region with λ, the strength of the two-body interaction, much greater than λ_F, the strength function will be of Gaussian form. In this situation, replacing $F_k(E)$ by Gaussian form (with width σ_k) in Eq. (15.7), we obtain

$$A_{k\to k}(t) = \int_{-\infty}^{+\infty} dE \exp -iEt \frac{1}{\sqrt{2\pi}\sigma_k} \exp -\frac{(E-E_k)^2}{2\sigma_k^2}. \tag{15.13}$$

Carrying out the integral by treating 'iEt' as if it is real will give correctly the final result,

$$A_{k\to k}(t) = \exp -\left[iE_k t + \frac{\sigma_k^2 t^2}{2} \right]. \tag{15.14}$$

Therefore, for Gaussian strength functions the return probability will follow Gaussian law,

$$W_{k\to k}(t) \xrightarrow{\textit{Gaussian region}} \exp -\sigma_k^2 t^2. \tag{15.15}$$

Note that, when t is in σ_H^{-1} units, the spectral width σ_k will be in σ_H units. Thus the decay law in the BW and Gaussian regions are different in EE with $\ln W$ being linear in t for BW and quadratic for Gaussian. Some numerical examples testing Eq. (15.15) are shown in Figs. 15.1b and d for fermion and boson systems.

15.1.4 Region Intermediate to BW and Gaussian Forms for $F_k(E)$

In the BW to Gaussian transition region, as discussed in Chap. 5, it is possible to represent $F_k(E)$ by the student-t distribution defined in terms of the shape parameter α and scale parameter β as given by Eq. (5.27). With the transformations $\alpha = (\nu + 1)/2$ and $(E - E_k) = \sqrt{\frac{\beta(\nu+1)}{2\nu}}x$, the $F_k(E)$ in the transition region transforms to $F_k(x:\nu)$ where,

$$F_k(x:\nu) = \frac{\Gamma(\frac{\nu+1}{2})}{\sqrt{\pi}\sqrt{\nu}\Gamma(\frac{\nu}{2})} \frac{dx}{(\frac{x^2}{\nu}+1)^{\frac{\nu+1}{2}}}. \tag{15.16}$$

Substituting this in Eq. (15.7) gives,

$$A_{k\to k}(t) = \exp -iE_k t \int_{-\infty}^{+\infty} dx \left[\exp -i\left[\sqrt{\frac{\beta(\nu+1)}{2\nu}}t\right]x\right] F_k(x:\nu). \tag{15.17}$$

The integral in Eq. (15.17) was a subject of many investigations in statistics literature and an easily usable form was given very recently in [10]. Then the final result is,

$$\begin{aligned}
W_{k\to k}(t) &\xrightarrow{\textit{transition region}} \left|A_{k\to k}(t;\nu,\beta)\right|^2; \\
A_{k\to k}(t:\nu,\beta) &= [\exp -iE_k t]\frac{2^\nu(\sqrt{\nu})^\nu}{\Gamma(\nu)} \int_0^\infty dx\left[x\left(x+\left|t'\right|\right)\right]^{(\nu-1)/2} \\
&\quad \times \exp -\sqrt{\nu}\left(2x+\left|t'\right|\right), \\
t' &= \sqrt{\frac{\beta(\nu+1)}{2\nu}}t.
\end{aligned} \tag{15.18}$$

Note that, for $\nu = 1$ we have BW form for $F_k(E)$ with $\beta = \Gamma^2/4$ and for $\nu \to \infty$ we have Gaussian form with $\sigma_k^2 = \beta/2$. Now the results in [10] and Eq. (15.18) clearly show that we will correctly recover the results given by Eqs. (15.12) and (15.15) for BW and Gaussian limits respectively. Some numerical examples testing Eq. (15.18) are shown in Figs. 15.1b and d for fermion and boson systems.

15.2 Entropy Production with Time in EGOE(1 + 2) and BEGOE(1 + 2): Cascade Model and Statistical Relaxation

Complexity generated with time can be studied by examining the time evolution of entropy. For simplicity of notation, from now on we will denote $W_{k\to f}(t)$ by $W_f(t)$ so that $W_0(t) = W_{k\to k}(t)$. Also assume that there are total $d+1$ states so that $f = 1, 2, \ldots, d$ with $|f = 0\rangle = |k\rangle$, the state in which the system is prepared at time $t = 0$. Now, entropy after time t is,

$$S_{(k)}(t) = -\sum_{f=0}^{d} W_f(t) \ln W_f(t). \tag{15.19}$$

From now on we will drop the subscript (k) in $S_{(k)}(t)$. Using Eq. (15.5), it is easy to prove the following important equality,

$$\sum_{f=0}^{d} W_f(t) = 1. \tag{15.20}$$

Before going into details of $S(t)$, let us examine $W_f(t)$. Using Eq. (15.5) we have,

$$\begin{aligned} W_f(t) &= \sum_E |C_0^E|^2 |C_f^E|^2 + 2\sum_{E>E'} C_0^E C_f^E C_0^{E'} C_f^{E'} \cos(E-E')t \\ &= W_f^{avg}(t) + W_f^{flu}(t). \end{aligned} \tag{15.21}$$

Note that the first term (W^{avg}) is independent of t and the second term (W^{flu}) is a fluctuating term. In three situations it is possible to simplify Eq. (15.21). First one is for $f = 0$ and we have already derived formulas for $W_0(t)$ fully taking into account both W^{avg} and W^{flu}. Next, in the small t limit we have simply

$$W_f(t) \overset{small\ t}{\longrightarrow} \left|\langle f|\exp -iHt|0\rangle\right|^2 \simeq \left|\langle f|H|0\rangle\right|^2 t^2 = H_{0f}^2 t^2. \tag{15.22}$$

Thirdly, for t large it is plausible to argue that the second term approaches zero and then $W_f(t) \approx W_f^{avg}(t)$, a constant [the first term in Eq. (15.21)]. More specifically, in the long time limit we have,

$$\begin{aligned} W_f(t) &= \sum_E |C_0^E|^2 |C_f^E|^2 \\ &\simeq \int_{-\infty}^{\infty} \frac{dE}{\rho(E)} F_0(E) F_f(E). \end{aligned} \tag{15.23}$$

In the situation that the strength functions are of BW form, one can replace $\rho(E)$ in Eq. (15.23) by its average value $\overline{\rho(E)}$ and move it outside the integral. Then, one is

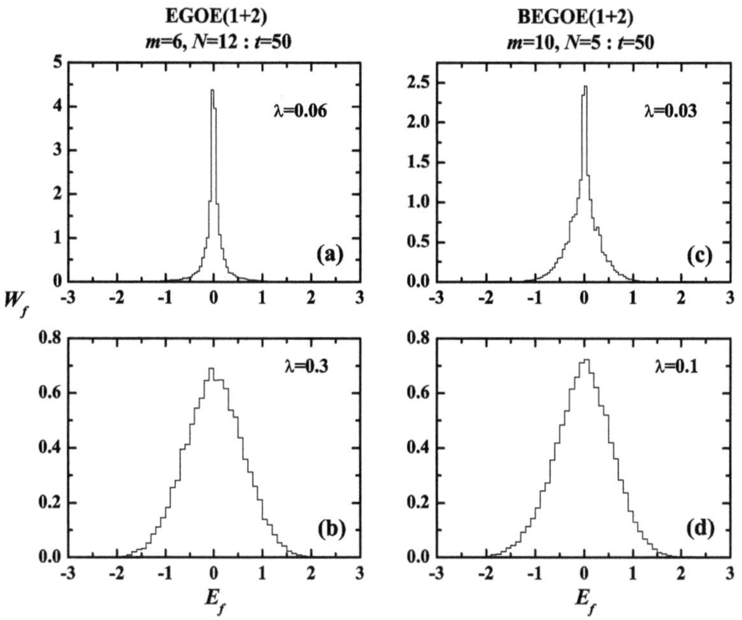

Fig. 15.2 $W_f(t)$ vs E_f for $t = 50$ value. (**a**) For a EGOE(1 + 2) system with $\lambda = 0.06$ (here BW form for strength functions applies). (**b**) For the same EGOE(1 + 2) system as in (**a**) but for $\lambda = 0.3$ (here Gaussian form for strength functions applies). (**c**) For a BEGOE(1 + 2) system with $\lambda = 0.03$ (here BW form for strength functions applies). (**d**) For the same BEGOE(1 + 2) system as in (**c**) but for $\lambda = 0.1$ (here Gaussian form for strength functions applies). In all the calculations, 25 member ensembles are used and histograms are obtained for $E_k = 0$. In the plots $\int W_f dE_f = 1$. Note that the bin sizes [see the discussion below Eq. (15.12)] used in constructing the histograms are $\Delta = 0.05$ in the BW examples and $\Delta = 0.1$ in the Gaussian examples; in all the examples $\delta = 0.005$

left with an integral that is a convolution of two BW functions giving

$$W_f(t) \simeq \frac{1}{2\pi \overline{\rho(E)}} \frac{\Gamma_t}{(E_0 - E_f)^2 + \dfrac{\Gamma_t^2}{4}}; \quad \Gamma_t = \Gamma_0 + \Gamma_f. \tag{15.24}$$

Similarly, in the situation that the strength functions are Gaussians, it is possible to evaluate the integral in Eq. (15.23) as $\rho(E)$ is a Gaussian (assumed to be zero centered with unit width) for EE. This gives,

$$W_f(t) = \frac{1}{\sqrt{\sigma_0^2 + \sigma_f^2 - \sigma_0^2\sigma_f^2}} \exp - \frac{1}{2(\sigma_0^2 + \sigma_f^2 - \sigma_0^2\sigma_f^2)} \left[(E_0 - E_f)^2 - \sigma_f^2 E_0^2 - \sigma_0^2 E_f^2\right]. \tag{15.25}$$

Figures 15.2a and b show the results for W_f vs E_f for a large value of t for EGOE(1 + 2) in BW and Gaussian regions. Similarly, Figs. 15.2c and d show the results for BEGOE(1 + 2). It is seen that the results in Fig. 15.2 are consistent with Eqs. (15.24) and (15.25). In conclusion, for large t, W_f is expected to be independent of t giving the result that entropy saturates after a large t value.

In the situation $f \neq 0$ and t is neither small or large, good knowledge of both terms in Eq. (15.21) is needed. For EE, no exact or good approximate formulas for $S(t)$ are available at present because of the complexity associated with the second term in Eq. (15.21). It is useful to note that $\sigma_k \sim \sigma_V$ and in σ_H units, it is nothing but the correlation coefficient ζ introduced in the context of NPC in EGOE(1 + 2). Also $|\langle f|\exp -iHt|0\rangle|^2$ is like strengths. Therefore, ideas based on transition strength density theory (see Chap. 5) may be useful in simplifying Eq. (15.21) in the Gaussian domain and ζ is likely to play an important role.

15.2.1 Cascade Model and Statistical Relaxation

With $S(t)$ approaching a constant as $t \to \infty$, it is important to understand the approach of $S(t)$ to saturation. As a good theory for $S(t)$ is not yet available for EE, Flambaum and Izrailev [6] introduced a Cascade model. In this model, $S(t)$ increases linearly with t. Before discussing this result, let us first consider small 't' limit result for $S(t)$. Using Eqs. (15.22) and (15.9) we have,

$$\begin{aligned} S(t) &= -W_0(t)\ln W_0(t) - \sum_f W_f(t)\ln W_f(t) \\ &\overset{small\ t}{\longrightarrow} \sigma_0^2 t^2 - t^2 \sum_{f=1}^{d} H_{0f}^2 \ln\{H_{0f}^2 t^2\}. \end{aligned} \tag{15.26}$$

Thus in the small t limit, entropy $S(t)$ will be quadratic in t.

In the cascade model of Flambaum and Izrailev [6], firstly the basis states are divided into sub-classes. The first class contains those N_1 states that are directly coupled to the initial state $|0\rangle$; i.e states for which $H_{0f} \neq 0$. The second class then contains N_2 sates that are coupled to the initial state by second order of the perturbation, i.e states for which $H_{0\alpha}H_{\alpha f} \neq 0$ with α any basis state. Continuing this cascading, let us say there are n classes. It is further assumed that all the W_f that belong to a given class will have small fluctuations and therefore, they can be replaced by their class average $\overline{W_r}$. Then $C_r = N_r\overline{W_r}$ and $\sum_0^\infty C_r = 1$. As N_n is expected to grow with n, one may put $N_n \approx M^n$ with M some constant. As N_n grows with n, this justifies neglecting the return probability to the previous classes and then (assuming BW form for strength functions),

$$\frac{dC_r}{dt} = -\Gamma\{C_r - C_{r-1}\}. \tag{15.27}$$

The first term here is the probability for the system to be in class r and the second term is the flux from the previous class. Solution of Eq. (15.27) is

$$C_r = \frac{(\Gamma t)^r}{r!} \exp - \Gamma t. \tag{15.28}$$

Assuming infinite number of classes, we have the identity $\sum_{r=0}^{\infty} C_r = 1$. Now, the entropy after time t is

$$\begin{aligned} S_0(t) &\approx - \sum_{r=0}^{\infty} C_r \ln\left(\frac{C_r}{N_r}\right) \\ &= - \sum_{r=0}^{\infty} \frac{(\Gamma t)^r}{r!} [\exp - \Gamma t] \left\{ -r \ln M - \Gamma t + \ln \frac{(\Gamma t)^r}{r!} \right\} \\ &= t\big[\Gamma(1 + \ln M)\big] - \left[(\exp - \Gamma t) \sum_{r=0}^{\infty} \frac{(\Gamma t)^r}{r!} \ln \frac{(\Gamma t)^r}{r!} \right]. \end{aligned} \tag{15.29}$$

In the last equality, the second term on the right-hand side is much smaller than the first term giving

$$S_0(t) \approx t \Gamma \ln M. \tag{15.30}$$

For Gaussian strength functions it is plausible to use Eq. (15.30) by replacing Γ by σ_0. Most important observation that follows from Eq. (15.30) is that the entropy after a small time will increase linearly with t. Thus it is expected, in the BW and Gaussian domains of EE, that with increasing time, the entropy will have initial quadratic growth as given by perturbation theory, then the linear behavior as given by the cascade model and finally saturation (saturation value will be the GOE value $\ln 0.48 d_{eff}$ where d_{eff} is an 'effective' dimension). Numerical results for EGOE(1 + 2) and BEGOE(1 + 2) are shown in Figs. 15.3a–d and they exhibit the expected behavior. For more quantitative description, it is possible to use Eqs. (15.19) and (15.20) with the assumption that in the sum only f's in an energy shell will contribute and within the shell, the variation of $W_f(t)$ is small. Then, with N_s the number of f's inside the energy shell, we have

$$S_0(t) = -W_0(t) \ln W_0(t) - \big[1 - W_0(t)\big] \ln\left(\frac{1 - W_0(t)}{N_s}\right). \tag{15.31}$$

The N_s in Eq. (15.31) can be determined numerically from $N_s \sim \langle \exp S \rangle$ where the average can be taken over a long time interval. This and Eq. (15.31) give a good description of the numerical results in Fig. 15.3. Equation (15.31) is expected to be good when the number of classes n is small. This appears to be true in practice as shown in some examples in [2]. Note that the condition $dW_n/dt = 0$ gives $n = \Gamma t$ (therefore for $t \ll 1/\Gamma$, there will be flow only into the first class). Another important observation is that in the BW region, $S(t)$ exhibits oscillations after the

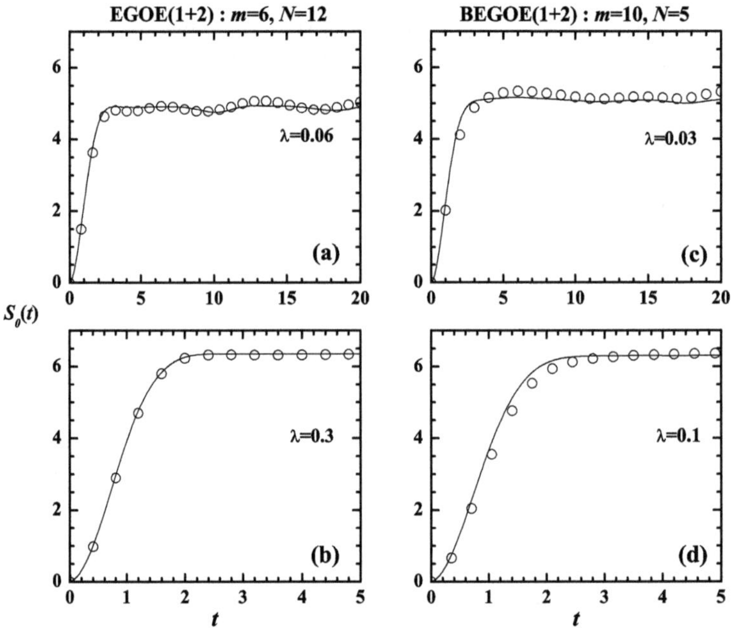

Fig. 15.3 Entropy $S_0(t)$ vs t for k states with $E_k = 0$ for EGOE(1 + 2) and BEGOE(1 + 2) examples. Ensemble averaged numerical results (*open circles*) are compared with the results from Eq. (15.31). See Fig. 15.2 for further details

linear increase. Flambaum and Izrailev [6] gave an explanation for this result but there is no formal derivation yet for this.

It is useful to add that interacting spin 1/2 fermions or hard-core bosons on 1D lattice and bosons on a 1D ring are also shown [2, 4] to exhibit statistical relaxation [$S_0(t)$ vs t behavior] similar to the results for spinless EE shown in Fig. 15.3. Going beyond these systems, it will be useful to study in future the role of spin in statistical relaxation by analyzing spin ensembles described in Chaps. 6 and 10.

15.3 Ergodicity Principle for Expectation Values of Observables: EGOE(1 + 2) Results

In recent years, study of the equilibration and thermalization mechanisms in isolated finite quantum systems has attracted great interest partly because the non-equilibrium dynamics, after an external perturbation is applied, has become experimentally accessible for ultra-cold quantum gases and electrons in mesoscopic systems such as quantum dots [11–13]. Advances in technology makes it possible to induce sharp changes in the parameters controlling the system and then observe the subsequent time evolution, which is essentially unitary because on short and intermediate time scales the perturbed system is almost isolated from the environment.

Thus, one can experimentally study if an isolated system, after a sharp perturbation, thermalizes or retains memory of the initial conditions. Here, it seems that the so-called eigenstate thermalization hypothesis (ETH) [14, 15] plays a fundamental role. The ETH states that thermalization occurs at the level of individual eigenstates whenever they satisfy Berry's conjecture [16] on chaotic eigenfunctions, i.e., whenever they behave as (quasi) random superpositions of the basis states. For this and other reasons, the role played by quantum chaos and chaotic wavefunctions in thermalization has been investigated using some 1D and 2D fermionic and bosonic systems by Rigol and others [2, 17–21].

In another line of approach, the EGOE(1 + 2), BEGOE(1 + 2) and their spin versions, as described in Chaps. 5, 6, 9 and 10, have been used, together with some related models, such as nuclear shell model, to perform different studies on thermalization of isolated fermionic and bosonic systems. As already discussed in the previous chapters and in the reviews [22, 23], the thermalization criteria used in these studies were based on the equivalence between different definitions of entropy, different definitions of temperature and representability of occupancies by Fermi-Dirac distribution (Bose-Einstein distribution for bosons). There are also some calculations of expectation values using the canonical distribution [24]. However, to get a deeper understanding of the role of quantum chaos, it is important that the ergodicity principle [25, 26], which is the cornerstone for thermalization, and clearly more precise and general than the aforementioned criteria, is tested. Here below, we will present the results from a EGOE(1 + 2) study of ergodicity principle using expectation values of operators.

15.3.1 Long-Time Average and Micro-canonical Average of Expectation Values

Let us begin with EGOE(1 + 2) for m spinless fermions in N sp states as in Sect. 15.1 with H defined by Eq. (15.1). Say EGOE(1 + 2) generates eigenvalues E_μ, $\mu = 1, 2, \ldots, d$ where the dimension $d = \binom{N}{m}$. Further let us introduce the following notations,

$$\begin{aligned} \Delta E_{\mu,\nu} &= E_\nu - E_\mu, \\ D_\mu(\mathcal{O}) &= \langle E_\mu|\mathcal{O}|E_\mu\rangle, \\ R_{\mu,\nu}(\mathcal{O}) &= \langle E_\mu|\mathcal{O}|E_\nu\rangle, \quad \mu \neq \nu. \end{aligned} \tag{15.32}$$

Now, consider a quantum system modeled by EGOE(1 + 2) and prepared at time $t = 0$ in the state $|\Psi(0)\rangle$. Say that this state is localized in a 'narrow' energy window around energy E,

$$\Psi(0) = \sum_\mu{}' C_\mu|E_\mu\rangle. \tag{15.33}$$

In order to make this precise, we need models for C_μ and we will return to this later. The system is said to thermalizes if the long time average of 'any reasonable observable' $\mathcal{O}$

$$\langle \mathcal{O}\rangle^t_{\Delta t} = \frac{1}{2\Delta t}\int_{t-\Delta t}^{t+\Delta t} \big\langle \Psi(t)\big|\mathcal{O}\big|\Psi(t)\big\rangle, \tag{15.34}$$

converges to a constant value predictable by an appropriate statistical ensemble, like for example micro-canonical ensemble. Using the time evolution of the eigenstates, as discussed in the previous two sections, we can introduce the density operators $\rho(\Psi,t)$, $\rho_D(\Psi)$ and $\rho_{ND}(\Psi,t)$,

$$\begin{aligned} \rho(\Psi,t) &= \big|\Psi(t)\big\rangle\big\langle\Psi(t)\big| \\ &= \sum_\mu |C_\mu|^2 |E_\mu\rangle\langle E_\mu| + \sum_{\mu\neq\nu}^{1,2,\ldots,d} C^*_\mu C_\nu \exp\left\{\frac{i}{\hbar}\Delta E_{\mu,\nu} t\right\}|E_\mu\rangle\langle E_\nu| \\ &= \rho_D(\Psi) + \rho_{ND}(\Psi,t) \end{aligned} \tag{15.35}$$

and then, the expectation value of an operator $\mathcal{O}$ is

$$\big\langle\Psi(t)\big|\mathcal{O}\big|\Psi(t)\big\rangle = \big\langle\!\big\langle\mathcal{O}\rho(\Psi,t)\big\rangle\!\big\rangle = \big\langle\!\big\langle\mathcal{O}\rho_D(\Psi)\big\rangle\!\big\rangle + \big\langle\!\big\langle\mathcal{O}\rho_{ND}(\Psi,t)\big\rangle\!\big\rangle. \tag{15.36}$$

If we consider long time average of the expectation values, the second term in Eq. (15.36) will tend to zero, i.e. the non-diagonal term vanishes. This is seen as follows,

$$\begin{aligned} \frac{1}{2\Delta t}\int_{t-\Delta t}^{t+\Delta t} \big\langle\!\big\langle\mathcal{O}\rho_{ND}(\Psi,t)\big\rangle\!\big\rangle &= \sum_{\mu\neq\nu}^{1,2,\ldots,d} C^*_\mu C_\nu R_{\mu,\nu}(\mathcal{O})\frac{1}{2\Delta t} \\ &\quad\times \int_{t-\Delta t}^{t+\Delta t} \exp\left\{\frac{i}{\hbar}\Delta E_{\mu,\nu} t'\right\} dt'; \\ \frac{1}{2\Delta t}\int_{t-\Delta t}^{t+\Delta t} \exp\left\{\frac{i}{\hbar}\Delta E_{\mu,\nu} t'\right\} dt' &= \exp\left\{\frac{i}{\hbar}\Delta E_{\mu,\nu} t\right\}\frac{\sin[\Delta E_{\mu,\nu}\Delta t/\hbar]}{[\Delta E_{\mu,\nu}\Delta t/\hbar]}, \\ \frac{\sin[\Delta E_{\mu,\nu}\Delta t/\hbar]}{[\Delta E_{\mu,\nu}\Delta t/\hbar]} &\overset{\Delta t\gg 1}{\sim} 0. \end{aligned} \tag{15.37}$$

Therefore the long time average $(t-av)$ gives the diagonal approximation,

$$\langle\mathcal{O}\rangle_{t-av} \overset{\Delta t\gg 1}{\sim} \big\langle\!\big\langle\mathcal{O}\rho_D(\Psi)\big\rangle\!\big\rangle = \sum_\mu |C_\mu|^2 D_\mu(\mathcal{O}) = \sum_\mu |C_\mu|^2 \langle E_\mu|\mathcal{O}|E_\mu\rangle. \tag{15.38}$$

It is easy to see that we can write the last form in terms of the 'strength function' $F_{\Psi(0)}(E)$ and the expectation value density $\rho_{\mathcal{O}}(E_\mu) = \langle E_\mu|\mathcal{O}|E_\mu\rangle\rho(E_\mu)$.

For isolated systems, micro-canonical ensemble is expected to be more appropriate. Then, the ensemble averaged expectation value is obtained by averaging the expectation value $\langle E_\mu|\mathcal{O}|E_\mu\rangle$ in a energy window $E_0 \pm \Delta E$. The ΔE is sufficiently small compared to the spectrum span but large enough to contain many eigenstates [note that level (and strength) fluctuations over the window $E_0 \pm \Delta E$ average out]. With say d' levels in the energy shell $W = \{|E_\mu\rangle; E_\mu \in [E_0 - \Delta E, E_0 + \Delta E]\}$, this gives ρ_{stat} where

$$\rho_{stat} = \frac{1}{d'} \sum_\mu{}' |E_\mu\rangle\langle E_\mu|. \tag{15.39}$$

The symbol Σ' means the sum is restricted to eigenstates belonging to W. Then the corresponding micro-canonical average is,

$$\langle\mathcal{O}\rangle_{stat} = \frac{1}{d'} \sum_{E_\mu=E_0-\Delta E}^{E_0+\Delta E} \langle E_\mu|\mathcal{O}|E_\mu\rangle. \tag{15.40}$$

For thermalization, we need

$$\langle\mathcal{O}\rangle_{t-av} \approx \langle\mathcal{O}\rangle_{stat}. \tag{15.41}$$

To find the region of thermalization in EGOE(1+2) (essentially checking the goodness of the λ_t marker determined in Chaps. 5, 6 and 9), the following measure $\Delta_\mathcal{O}$ has been considered in [3],

$$\Delta_\mathcal{O} = \left| \frac{\langle\mathcal{O}\rangle_{t-av} - \langle\mathcal{O}\rangle_{stat}}{\langle\mathcal{O}\rangle_{stat}} \right| \tag{15.42}$$

and then

$$\Delta_\mathcal{O} \approx 0 \tag{15.43}$$

corresponds to thermalization.

15.3.2 Thermalization from Expectation Values

In order to verify $\Delta_o \to 0$ in some limit, four types of operators are considered in [3]:

- diagonal one-body operators $\mathcal{O}_d(1) = \sum_k \theta_k a_k^\dagger a_k$,
- general one-body operators $\mathcal{O}(1) = \sum_{k,l} \theta_{kl} a_k^\dagger a_l$,
- general two-body operators $\mathcal{O}(2) = \sum_{k<l,p<q} \theta_{klpq} a_k^\dagger a_l^\dagger a_q a_p$,
- strength function operators $\mathcal{O}_{sf} = \mathcal{O}^T(1)\mathcal{O}(1)$,

where the parameters θ_k, θ_{kl} and θ_{klpq} are taken as random variables. To see how the initial conditions affect thermalization process, the system has been allowed to evolve from three different types of initial states, defined as:

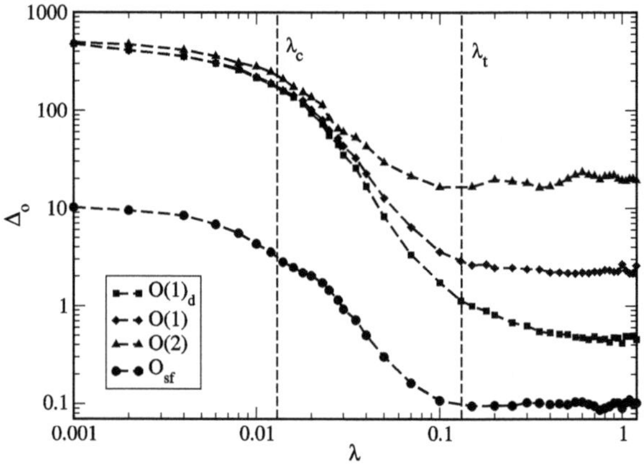

Fig. 15.4 Variation with the interaction strength λ of the ensemble averaged $\Delta_{o_d(1)}$ (*squares*), $\Delta_{o(1)}$ (*diamonds*), $\Delta_{o(2)}$ (*triangles*), and $\Delta_{o_{sf}}$ (*dots*), given in percent, for a 60 member EGOE(1 + 2) with $(m, N) = (6, 16)$ initially prepared in a state $\psi^{(1)}(0) \in W$. Figure is taken from [3] with permission from IOP publishing

- $|\Psi^{(1)}(0)\rangle \propto P_W|k_0\rangle$, where P_W is the projector onto W, and $|k_0\rangle$ is the mean-field state with energy E_0.
- $|\Psi^{(2)}(0)\rangle \propto \sum_\mu' C_\mu|E_\mu\rangle$, with the coefficients C_μ being $G(0, 1)$ variables.
- $|\Psi^{(3)}(0)\rangle \propto \sum_\mu' C_\mu|E_\mu\rangle$, where $C_\mu = \exp\{-\alpha(\frac{E_\mu - E_0}{\Delta E})^2\}\ G(0, 1)$.

The states $\Psi^{(2)}(0)$ and $\Psi^{(3)}(0)$ are random superpositions of the eigenstates belonging to W, but due to the Gaussian factor the distribution of the C_μ coefficients is wider for $\Psi^{(3)}(0)$. As we shall see below, distribution of the state amplitudes inside W is one of the factors that affects the thermalization process. In the calculations, W is chosen such that $E_0 = 0$ and $\Delta E = 0.1$ (all energies being zero centered and normalized to unit spectral width).

In the EGOE(1 + 2) calculations carried out, the sp energies ϵ_k are chosen to be independent Gaussian variables with $\overline{\epsilon_k} = k$ and $\overline{(\epsilon_k - k)^2} = 1/2$. Figure 15.4 shows the results for $\Delta_{\mathcal{O}}$ for the four operators as a function of the interaction strength λ for a 60 member EGOE(1 + 2) with $(m, N) = (6, 16)$ (matrix dimension being 8008). In the figure, the two vertical lines give the positions of λ_c and λ_t. In all the cases $\Delta_{\mathcal{O}}$ becomes smaller as the interaction strength increases up to $\lambda \approx \lambda_t$. It is very important to realize that the transition from Poisson to GOE spectral fluctuations, which is considered the most relevant signature of quantum chaos, occurs at $\lambda \approx \lambda_c$ and does not modify this trend. On the contrary, for $\lambda > \lambda_t$ the relative errors either remain essentially constant or the decreasing rate is much smaller. Recall that λ_t defines a region where the three entropies S^{ther}, S^{info} and S^{sp} take essentially the same values, and signals the point at which the wave functions start becoming very much delocalized in the mean-field basis. Beyond λ_t, $\Delta_{\mathcal{O}}$ becomes clearly smaller than one percent only for two operators, namely $\mathcal{O}_d(1)$ and $\mathcal{O}_{sf}$. Their errors are $\approx$0.5 % and $\approx$0.1 %, respectively. Thus, as long as the system is prepared in an initial state $\Psi^{(1)}(0) \in W$ and $\lambda > \lambda_t$, Eq. (15.41) approximately holds for the observables $\mathcal{O}_d(1)$ and $\mathcal{O}_{sf}$. Thus the system thermalizes relative to these two observables. This is not the case of the observables $\mathcal{O}(1)$ and $\mathcal{O}(2)$. It is worth noting that the main difference between $\mathcal{O}(1)$, $\mathcal{O}(2)$ in one hand, and $\mathcal{O}_d(1)$, $\mathcal{O}_{sf}$ in the other, is that the latter have meaningful smoothed form for large λ, as discussed in Chap. 5.

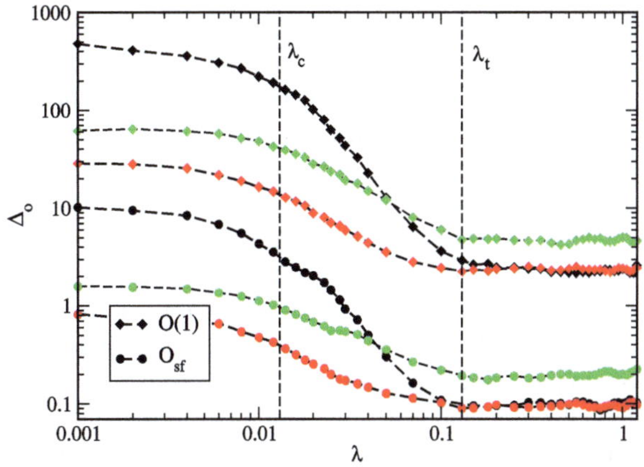

Fig. 15.5 Values of ensemble averaged $\Delta_{o(1)}$ (*diamonds*) and $\Delta_{o_{sf}}$ (*dots*), expressed in percent, as function of λ for a 60 member EGOE(1 + 2) with $(m, N) = (6, 16)$, and initial conditions given by $\Psi^{(1)}(0)$ (*black*), $\Psi^{(2)}(0)$ (*red*), and $\Psi^{(3)}(0)$ (*green*). In all cases the initial state belongs to the energy shell W. Figure is taken from [3] with permission from IOP publishing

The results corresponding to other choices of $\Psi(0) \in W$ are shown in Fig. 15.5. For simplicity, results are shown only for $\mathscr{O}_{sf}$ and $\mathscr{O}(1)$. Curves in black, red and green correspond to $\Psi^{(1)}(0)$, $\Psi^{(2)}(0)$ and $\Psi^{(3)}(0)$, respectively. In all cases the initial state belongs to the energy shell W. We see that the choice of the initial conditions does not affect the main trend: $\Delta_{o(1)}$, and $\Delta_{o_{sf}}$ diminish progressively as the strength λ is increased, and for $\lambda > \lambda_t$ their values remain essentially constant. However, the precise values are quite different. When $\lambda > \lambda_t$, the initial states $\Psi^{(1)}(0)$ and $\Psi^{(2)}(0)$ give rise to very similar results while the error corresponding to $\Psi^{(3)}(0)$ is clearly larger. Moreover, as in Fig. 15.4, expectation values of the operator $\mathscr{O}(1)$ do not thermalize independent of the initial condition while $\mathscr{O}_{sf}$ thermalizes.

In order to obtain some analytical insight into the behavior of $\Delta_{\mathscr{O}}$, it is plausible to consider C_μ to be Gaussian random variables lying on a unit sphere in W. Then the fluctuation properties of C_μ follow P-T law and $\overline{|C_\mu|^2} = 1/d'$. Also, it is possible to assume that C_μ and $A_\mu = \langle E_\mu|\mathscr{O}|E_\mu\rangle$ are independent. Then clearly,

$$
\begin{aligned}
|\Delta_{\mathscr{O}}|^2 &= \frac{\overline{\{\sum_{\mu=1}^{d'}(|C_\mu|^2 - \frac{1}{d'})A_\mu\}^2}}{\overline{[\{\frac{1}{d'}\sum_{\mu=1}^{d'} A_\mu\}]^2}} \\
&= \frac{2}{d'}\,\frac{\overline{\frac{1}{d'}\sum_{\mu=1}^{d'}(A_\mu - \overline{A_\mu})^2}}{\overline{[\{\frac{1}{d'}\sum_{\mu=1}^{d'} A_\mu\}]^2}}.
\end{aligned}
\tag{15.44}
$$

For $\mathscr{O} = \mathscr{O}_{sf} = \mathscr{O}^T(1)\mathscr{O}(1)$, the factor multiplying $\frac{2}{d'}$ in Eq. (15.44) is nothing but the inverse of the NPC in transition strengths $(\xi^{(s)}_{2:\mathscr{O}(1)})$ generated by the one-body operator $\mathscr{O}(1)$. This result is discussed in Sect. 5.5 with the final result following from Eq. (5.48),

$$
|\Delta_{\mathscr{O}}| = \sqrt{\frac{4}{3d'}}\left[\xi^{(s)}_{2:\mathscr{O}(1)}(E_0)\right]^{-1/2}. \tag{15.45}
$$

This establishes a connection between the thermalization of the system, relative to an observable $\mathscr{O}_{sf} = \mathscr{O}^T(1)\mathscr{O}(1)$, and the value of the NPC for the transition strengths generated by $\mathscr{O}(1)$ acting on the eigenstate with energy E_0. An important outcome is that for chaotic systems the NPC is expected to be large and hence these systems will thermalize, while for regular systems NPC has to be small and thus thermalization will be hindered. More details of EGOE(1 + 2) study of thermalization using expectation values is given in [3].

In summary, it is found that the λ_t marker indeed marks the region of thermalization and thermalization occurs only for certain types of operators such as occupancies (and their linear combinations) and strength function operators. It is also seen in the numerical calculations (by varying W value) that spectrum edges hinder thermalization and also large Hilbert space enhances thermalization process.

Before concluding this chapter, it should be added that besides the first studies on statistical relaxation and thermalization using EGOE(1 + 2) [also BEGOE(1 + 2)], there is also an attempt by Seligman et al. [5] to study Loschmidt echoes in EGOE(1 + 2). Their study showed that the fidelity amplitude displays 'freeze'. This freeze, typically present for most realizations (most members) of EGOE(1 + 2), is found to vanish on average. More detailed studies of this may give new information on the ergodic properties of EGOE(1 + 2).

References

1. V.V. Flambaum, Time dynamics in chaotic many-body systems: can chaos destroy a quantum computer? Aust. J. Phys. **53**, 489–497 (2000)
2. L.F. Santos, F. Borgonovi, F.M. Izrailev, Onset of chaos and relaxation in isolated systems of interacting spins: energy shell approach. Phys. Rev. E **85**, 036209 (2012)
3. V.K.B. Kota, A. Relaño, J. Retamosa, M. Vyas, Thermalization in the two-body random ensemble, J. Stat. Mech. P10028 (2011)
4. G.P. Berman, F. Borgonovi, F.M. Izrailev, A. Smerzi, Irregular dynamics in a one-dimensional Bose system. Phys. Rev. Lett. **92**, 030404 (2004)
5. I. Pižorn, T. Prosen, T.H. Seligman, Loschmidt echoes in two-body random matrix ensembles. Phys. Rev. B **76**, 035122 (2007)
6. V.V. Flambaum, F.M. Izrailev, Entropy production and wave packet dynamics in the Fock space of closed chaotic many-body systems. Phys. Rev. E **64**, 036220 (2001)
7. V.K.B. Kota, R. Sahu, Structure of wavefunctions in (1 + 2)-body random matrix ensembles. Phys. Rev. E **64**, 016219 (2001)
8. M. Abramowtiz, I.A. Stegun (eds.), *Handbook of Mathematical Functions*, NBS Applied Mathematics Series, vol. 55 (U.S. Govt. Printing Office, Washington, D.C., 1972)
9. A. Bohr, B.R. Mottelson, *Nuclear Structure*, Single-Particle Motion, vol. I (Benjamin, New York, 1969)
10. I. Dreier, S. Kotz, A note on the characteristic function of the t-distribution. Stat. Probab. Lett. **57**, 221–224 (2002)
11. R. Jördens, N. Strohmaier, K. Günter, H. Moritz, T. Esslinger, A Mott insulator of fermionic atoms in an optical lattice. Nature **455**, 204–207 (2008)
12. U. Schneider et al., Metallic and insulating phases of repulsively interacting fermions in a 3D optical lattice. Science **322**, 1520–1525 (2008)
13. L. Perfetti et al., Time evolution of the electronic structure of 1T-TaS2 through the insulator-metal transition. Phys. Rev. Lett. **97**, 067402 (2006)

14. M. Srednicki, Chaos and quantum thermalization. Phys. Rev. E **50**, 888–901 (1994)
15. J.M. Deutsch, Quantum statistical mechanics in a closed system. Phys. Rev. A **43**, 2046–2049 (1991)
16. M.V. Berry, Regular and irregular semiclassical wavefunctions. J. Phys. A **10**, 2083–2091 (1977)
17. M. Rigol, V. Dunjko, M. Olshanii, Thermalization and its mechanism for generic isolated quantum systems. Nature (London) **452**, 854–858 (2008)
18. M. Rigol, Breakdown of thermalization in finite one-dimensional systems. Phys. Rev. Lett. **103**, 100403 (2009)
19. L.F. Santos, M. Rigol, Onset of quantum chaos in one dimensional bosonic and fermionic systems and its relation to thermalization. Phys. Rev. E **81**, 036206 (2010)
20. L.F. Santos, M. Rigol, Localization and the effects of symmetries in the thermalization properties of one-dimensional quantum systems. Phys. Rev. E **82**, 031130 (2010)
21. L.F. Santos, F. Borgonovi, F.M. Izrailev, Chaos and statistical relaxation in quantum systems of interacting particles. Phys. Rev. Lett. **108**, 094102 (2012)
22. V.K.B. Kota, Embedded random matrix ensembles for complexity and chaos in finite interacting particle systems. Phys. Rep. **347**, 223–288 (2001)
23. J.M.G. Gómez, K. Kar, V.K.B. Kota, R.A. Molina, A. Relaño, J. Retamosa, Many-body quantum chaos: recent developments and applications to nuclei. Phys. Rep. **499**, 103–226 (2011)
24. V.V. Flambaum, A.A. Gribakina, G.F. Gribakin, I.V. Ponomarev, Quantum chaos in many-body systems: what can we learn from the Ce atom. Physica D **131**, 205–220 (1999)
25. P. Reimann, Foundation of statistical mechanics under experimentally realistic conditions. Phys. Rev. Lett. **101**, 190403 (2008)
26. S. Goldstein, J.L. Lebowitz, C. Mastrodonato, R. Tumulka, N. Zanghi, Approach to thermal equilibrium of macroscopic quantum systems. Phys. Rev. E **81**, 011109 (2010)

Chapter 16
Brief Summary and Outlook

Embedded random matrix ensembles, introduced in 1970 and being explored in considerable detail since 1994, have started occupying an important position in quantum physics in the study of isolated finite quantum many-particle systems. Borrowing from H.A. Weidenmüller [1], let us state that: *although used with increasing frequency in many branches of Physics, random matrix ensembles sometimes are too unspecific to account for important features of the physical system at hand. One refinement which retains the basic stochastic approach but allows for such features consists in the use of embedded ensembles*. Significantly, the study of embedded random matrix ensembles is still developing. Partly this is due to the fact that mathematical tractability of these ensembles is still a problem and computational analysis for more than say 10 particles (fermions or bosons) is challenging even with most modern computers. Nevertheless, these random matrix ensembles are still being investigated as they are proved to be rich in their content and wide in their scope. We have tried to give in this book an up to date discussion of the properties and applications of a variety of embedded ensembles defined by various embedding algebras both for fermion and boson systems. Figure 16.1 shows a summary of various embedded ensembles based on Lie symmetries that are considered in the literature and described in this book.

After a general introduction in Chap. 1 emphasizing the importance of random matrix theory in quantum physics in general and embedded random matrix ensembles in particular, some important results for the classical random matrix ensembles GOE and GUE and their interpolations are described in Chaps. 2 and 3. These are used to define various statistical quantities, notations and so on that are needed for easy reading of the remaining chapters to follow. For spinless fermion systems, properties of EGOE(k), $k \geq 2$ are described in Chap. 4 and the results due to the presence of a mean-field are described in Chap. 5. Similarly Chap. 6 gives results for EE for fermions with spin degree of freedom, Chap. 8 for fermions with parity degree of freedom, Chap. 9 for spinless bosons, Chap. 10 for bosons with spin (spin-1/2 and spin-one) and Chap. 13 is on other EE. All the analytical results in these chapters are derived using: (i) perturbation theory; (ii) trace propagation methods; (iii) binary correlation approximation. Going beyond these, analytical formula-

V.K.B. Kota, *Embedded Random Matrix Ensembles in Quantum Physics*, Lecture Notes in Physics 884, DOI 10.1007/978-3-319-04567-2_16,

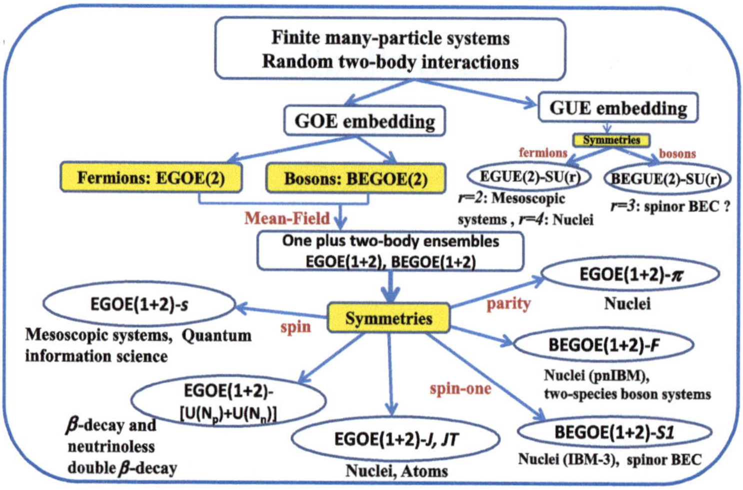

Fig. 16.1 Embedded random matrix ensembles with symmetries. EGOE and EGUE correspond to fermionic systems and BEGOE and BEGUE correspond to bosonic systems. Physical systems where a given ensemble applies are mentioned below the ensemble. See Chaps. 4–14 for details of various ensembles. Also note that for $r = 1$ in EGUE(2)-$SU(r)$ [BEGUE(2)-$SU(r)$], the ensemble is for spinless fermions [bosons]. Similarly, N_p and N_n correspond to number of sp states for protons and neutrons respectively

tion in terms of the Wigner-Racah algebra of the embedding algebra for EGUE(2)-$SU(r)$ and BEGUE(2)-$SU(r)$ ensembles with examples from $r = 1$–4 is described in Chap. 11. Applications to nuclei and mesoscopic systems is given in Chap. 7 and the new signature of cross-correlations in EE has been discussed in Chap. 12. Similarly Chap. 14 is on the new paradigm of regular structures from random interactions, a topic being investigated in considerable detail in nuclear structure physics. Finally, Chap. 15 is on time evolution and thermalization in EE.

One of the most significant result derived, via extensive numerical calculations supplemented by the results from perturbation theory and some asymptotic formulas, is that random interactions in the presence of a mean-field generate three chaos markers. This, establishes that EE provide a framework for defining and understanding, for what one may call quantum many-body chaos. The three chaos markers are established for systems of spinless fermions (Chap. 5), fermions with spin (Chap. 6) and spinless bosons (Chap. 9). There is also good evidence for the existence of these markers for bosons with spin (see Chap. 10) although some more investigations for these ensembles are needed. Figure 16.2 gives a description of the chaos markers.

Although impressive progress has been made in the investigation of a variety of EE and in applying some them, there are yet some outstanding problems waiting to be solved. Some of these are: (i) developing binary correlation method (or some other methods) to derive asymptotic results for EGOE$(1+2)$-**s** and also for bosonic ensembles; (ii) analysis of embedded symplectic ensembles (EGSE) [so far no EGSE has bee analyzed]; (iii) deriving formulas for the U-coefficients needed for analyzing higher moments of EGUEs and this calls for new advances in the general

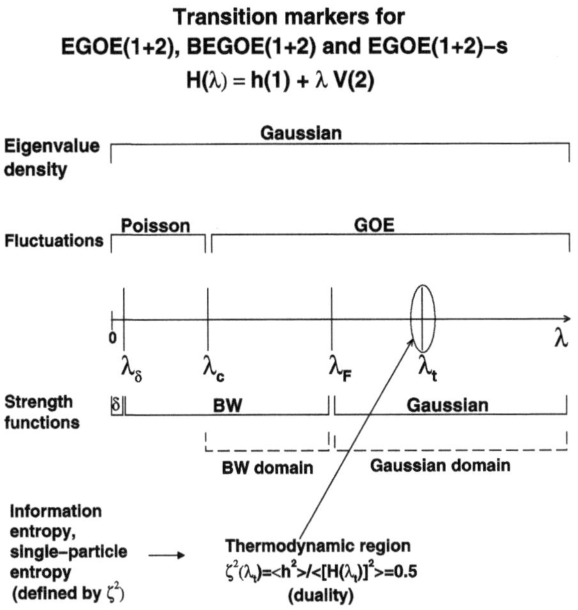

Fig. 16.2 Transition (chaos) markers generated by EE. Results in the figure are verified by extensive numerical calculations for EGOE$(1+2)$, BEGOE$(1+2)$ and EGOE$(1+2)$-s ensembles. Note that for EGOE$(1+2)$ and BEGOE$(1+2)$, the $V(2)$ is GOE(1) in two-particle spaces. Similarly, for EGOE$(1+2)$-s used is $V(2) = V^{s=0}(2) + V^{s=1}(2)$ with independent GOE(1) representations for $V^{s=0}(2)$ and $V^{s=1}(2)$ in two-particle spaces. Also, λ is in the units of the average spacing Δ of the sp levels defining $h(1)$. As discussed in Chaps. 5, 6 and 9, strength functions take δ-function form (denoted by δ in the figure) for $\lambda < \lambda_\delta$ with $\lambda_\delta \ll \lambda_c$ and they start taking BW form as λ crosses λ_δ. The BW domain is defined by $\lambda_c < \lambda < \lambda_F$ and here the strength functions take BW form and the fluctuations follow GOE. Similarly, in the Gaussian domain, defined by $\lambda > \lambda_F$, the strength functions take Gaussian form and the fluctuations follow GOE. Also in this region, the information entropy and single-particle entropy are defined by the ζ^2 parameter given in Eq. (5.21). Note that λ_t defines thermodynamic (also duality) region

$SU(N)$ Wigner-Racah algebra; (iv) more higher level "computational experiments" than those that are carried out till now. However, the most important unsolved problem in the subject of EE is the derivation of the two-point correlation function. This is not known even for the simplest spinless EGOE(2) and BEGOE(2). So far this has resisted all attempts. It should be stressed that there is considerable numerical evidence suggesting that level and strength fluctuations in EGOE (BEGOE) in the dilute limit (in the dense limit) follow GOE. To strengthen this result, recently level fluctuations in EE have been investigated using a new measure that is independent of the unfolding function and unfolding procedure [2]. As mentioned in Chap. 4 and also in many examples discussed in the later chapters, EGOEs are found to follow GOE, for level fluctuations, only when spectral unfolding is used but not for ensemble unfolding. Also for EE, determining the unfolding function is fraught with uncertainties in particular near the spectrum ends. In 2007, Oganesyan and Huse [3] suggested the use of statistics of quotients of successive spacings and this is

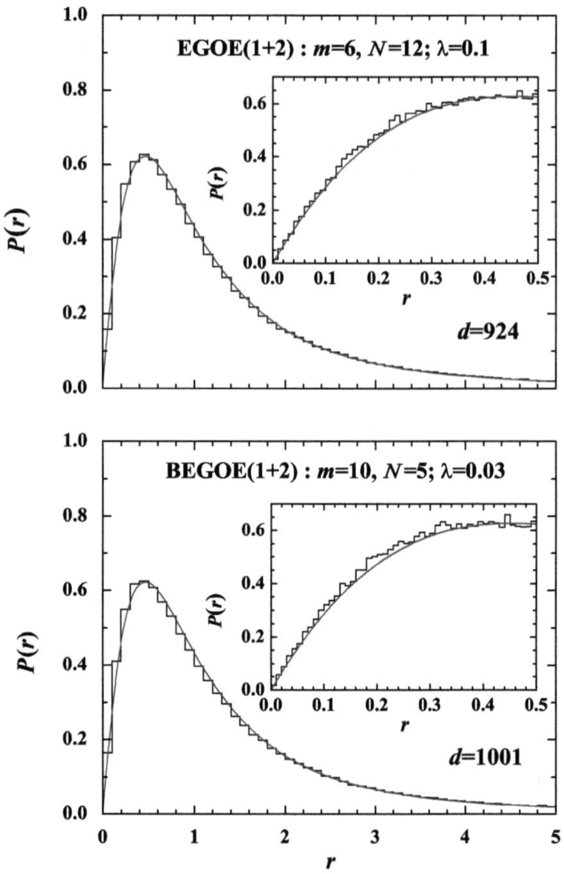

Fig. 16.3 Histogram, in the *upper panel*, represents $P(r)$ the probability distribution of the ratio of consecutive level spacings for a 500 member EGOE$(1+2)$ ensemble. The *red smoothed curve* is due to the surmise $P_W(r)$ given by Eq. (16.2). In the *inset* figure in the *upper panel* shown are results for $P(r)$ for $r \leq 0.5$. Similarly, *lower panel* shows the same results but for a BEGOE$(1+2)$ example. Figure is constructed from the results in [2] (Color figure online)

independent of unfolding. They have applied this new statistic to investigate numerically many-body localization [4–6] and this measure was also used to quantify the distance from integrability on finite size lattices [7, 8]. Given an ordered set of eigenvalues (energy levels) E_n, where $n = 1, 2, \ldots, d$, the nearest-neighbor spacing is given by $s_n = E_{n+1} - E_n$ and then, the ratio of two consecutive level spacings is $r_n = s_{n+1}/s_n$. The probability distribution for consecutive level spacings is denoted by $P(r)dr$ and it is easy to see that for Poisson systems $P(r)$ is [denoted by $P_P(r)$],

$$P_P(r) = \frac{1}{(1+r)^2}. \tag{16.1}$$

Similarly, for GOE, derived using 3×3 real symmetric matrices, the $P(r)$ is given by a Wigner-like surmise [9],

$$P_W(r) = \frac{27}{8} \frac{r + r^2}{(1 + r + r^2)^{5/2}}. \tag{16.2}$$

Some examples for $P(r)$ for EGOE(1 + 2) and BEGOE(1 + 2) are shown in Fig. 16.3. It is seen that the agreement with the GOE form given by Eq. (16.2) is excellent. Further examples for EGOE(1 + 2)-**s**, BEGOE(1 + 2)-F and BEGOE(1 + 2)-$S1$ are given in [2]. All these confirm that the probability distribution $P(r)$ of ratio of consecutive level spacings for embedded random matrix ensembles follow GOE for strong enough two-body interaction and this also establishes that level fluctuations in many-particle systems with strong enough interaction are universal. Let us add that $P(r)$ gives a better handle on the statistics at the edges of the spectrum where unfolding presents the biggest problem (Refs. [2] and [10] give more details).

Finally, future in the subject of embedded random matrix ensembles in quantum physics is exciting with enormous scope for developing new mathematical methods for their analysis and with the possibility of their applications in, besides atomic, nuclear and mesoscopic physics, newer areas of research such as quantum information science, Bose gasses, statistical mechanics of isolated finite quantum systems and quantum many-body chaos. In conclusion, with the fact that RMT already became a part of quantum mechanics course in many places, we believe that in the near future, with all the developments described in this monograph, embedded ensembles will become a part of quantum many-body physics course.

References

1. H.A. Weidenmüller, Private communication
2. N.D. Chavda, V.K.B. Kota, Probability distribution of the ratio of consecutive level spacings in interacting particle systems. Phys. Lett. A **377**, 3009–3015 (2013)
3. V. Oganesyan, D.A. Huse, Localization of interacting fermions at high temperature. Phys. Rev. B **75**, 155111 (2007)
4. V. Oganesyan, A. Pal, D.A. Huse, Energy transport in disordered classical spin chains. Phys. Rev. B **80**, 115104 (2009)
5. A. Pal, D.A. Huse, Many-body localization transition. Phys. Rev. B **82**, 174411 (2010)
6. S. Iyer, V. Oganesyan, G. Refael, D.A. Huse, Many-body localization in a quasiperiodic system. Phys. Rev. B **87**, 134202 (2013)
7. C. Kollath, G. Roux, G. Biroli, A.M. Läuchli, Statistical properties of the spectrum of the extended Bose-Hubbard model. J. Stat. Mech. P08011 (2010)
8. M. Collura, H. Aufderheide, G. Roux, D. Karevski, Entangling many-body bound states with propagative modes in Bose-Hubbard systems. Phys. Rev. A **86**, 013615 (2012)
9. Y.Y. Atas, E. Bogomolny, O. Giraud, G. Roux, Distribution of the ratio of consecutive level spacings in random matrix ensembles. Phys. Rev. Lett. **110**, 084101 (2013)
10. Y.Y. Atas, E. Bogomolny, O. Giraud, P. Vivo, E. Vivo, Joint probability densities of level spacing ratios in random matrices, arXiv:1305.7156 [quant-ph]

Appendix A
Time Reversal in Quantum Mechanics

A.1 General Structure of Time Reversal Operator

General reference here is the book by R.G. Sachs [1]. The transformation $t \to t' = -t$ of the time variable is expected to transform state vectors $\psi(t)$ and observables Q of any system in the Schrödinger picture as

$$T\psi(t) = \psi'(t'), \qquad TQT^{-1} = Q' \tag{A.1}$$

and in the Heisenberg picture as

$$T|\psi\rangle = |\psi'\rangle, \qquad TQ(t)T^{-1} = Q'(t'). \tag{A.2}$$

Then, for example, $TxT^{-1} = x'$, $TpT^{-1} = p'$ and $T\sigma T^{-1} = \sigma'$. We require that T must be a kinematically admissible transformation and therefore, it should be consistent with the commutation relations. In other words, under time reversal $[x_i, p_j] = i\hbar\delta_{ij}$ and $[J_i, J_j] = i\varepsilon_{ijk}J_k$ must retain their structure. We also impose that the classical conditions hold good, giving

$$TxT^{-1} = x, \qquad TpT^{-1} = -p, \qquad T\sigma T^{-1} = -\sigma. \tag{A.3}$$

Then, $T[x_i, p_j]T^{-1} = [x'_i, p'_j] = [x_i, -p_j] = -[x_i, p_j]$ and applying the relation $[x_i, p_j] = i\hbar\delta_{ij}$ gives the important result,

$$TiT^{-1} = -i. \tag{A.4}$$

Therefore T *must include* the operator K which takes any complex number z into its complex conjugate,

$$KzK^{-1} = z^* \quad \Rightarrow \quad K^{-1} = K. \tag{A.5}$$

As commutation relations are invariant under any linear transformation, T can be written as

$$T = UK, \tag{A.6}$$

V.K.B. Kota, *Embedded Random Matrix Ensembles in Quantum Physics*,
Lecture Notes in Physics 884, DOI 10.1007/978-3-319-04567-2,

where U is a linear transformation. If $|\psi\rangle$ is a state vector, then $T|\psi\rangle = |\psi'\rangle$ also should be a state vector belonging to the same unitary space and therefore U is unitary.

Explicit form of U depends on the complete set of kinematic variables defining the Hilbert space and examples for this are given ahead. As T should be a kinematically admissible transformation, we have the condition that in the absence of interactions (zero forces) dynamic equations must be left invariant under T-transformation. Then,

$$i\hbar\frac{\partial}{\partial t}|\psi\rangle = H_0|\psi\rangle \quad \Rightarrow \quad i\hbar\frac{\partial}{\partial t'}|\psi'\rangle = H_0|\psi'\rangle, \quad t' = -t. \tag{A.7}$$

Here H_0 is the Hamiltonian operator that includes only the kinetic energy terms of the full Hamiltonian. Now, applying $T = UK$ on both sides of Eq. (A.7) will give,

$$\begin{aligned}
&Ti\hbar T^{-1}T\frac{\partial}{\partial t}T^{-1}T|\psi\rangle = TH_0T^{-1}T|\psi\rangle \\
\Rightarrow \quad &UKi\hbar K^{-1}U^{-1}UK\frac{\partial}{\partial t}K^{-1}U^{-1}|\psi'\rangle = (TH_0T^{-1})|\psi'\rangle \\
\Rightarrow \quad &-i\hbar\frac{\partial}{\partial t}|\psi'\rangle = i\hbar\frac{\partial}{\partial t'}|\psi'\rangle = TH_0T^{-1}|\psi'\rangle.
\end{aligned} \tag{A.8}$$

Thus, $T = UK$ if $TH_0T^{-1} = H_0$. As H_0 contains only terms that are quadratic in momentum, the condition $TH_0T^{-1} = H_0$ is satisfied due to Eq. (A.3) and therefore $T = UK$.

Most important outcome of Eq. (A.8) is that T is anti-unitary as $T \neq U$ but $T = UK$. The T operator in fact belongs to a class of operators that Wigner calls "involutional". These operators, when repeated restore the original state. For such operators F, one has $F^2 = nI$ where $n = e^{i\theta}$ and I is the identity operator. Then, by F^2 operation we will get back the same state to within a phase. For T operator, Wigner showed that $T^2 = \eta I$, $\eta = \pm 1$. To prove this, consider T^2 with U being unitary in $T = UK$,

$$\begin{aligned}
T^2 &= UKUK \\
&= UKUK^{-1} \\
&= UU^*, \\
U^{-1} = U^\dagger &= \widetilde{U}^* \quad \Rightarrow \quad U^* = \widetilde{U}^{-1} \quad \Rightarrow \quad T^2 = U\widetilde{U}^{-1} = \eta I, \\
& \qquad\qquad\quad\ \Rightarrow \quad U = \eta\widetilde{U} \quad \text{or} \quad \widetilde{U} = \eta U.
\end{aligned} \tag{A.9}$$

Then,

$$U = \eta\widetilde{U} = \eta^2 U \quad \Rightarrow \quad \eta = \pm 1 \quad \Rightarrow \quad T^2 = \pm I. \tag{A.10}$$

For Eq. (A.10) is to be valid, η should be real and this is proved as follows,

$$\begin{aligned} T^2 &= \eta I \quad \Rightarrow \quad T\eta I T^{-1} = T T^2 T^{-1}, \quad \text{then} \\ \eta^* I = T^2 &= \eta I \quad \Rightarrow \quad \eta = \eta^* \\ & \qquad\quad \Rightarrow \quad \eta \text{ is real.} \end{aligned} \tag{A.11}$$

Therefore, there are even systems ($T^2 = 1$) and odd systems ($T^2 = -1$). The sign of T^2 is determines the properties of U. A fundamental property of anti-unitary operator is $\langle T\psi|T\phi\rangle = \langle\psi|\phi\rangle^*$ and this can be seen as follows,

$$\begin{aligned} \langle T\psi|T\phi\rangle &= \langle UK\psi|UK\phi\rangle \\ &= \langle U\psi^*|U\phi^*\rangle = \langle\psi^*|U^\dagger U\phi^*\rangle = \langle\psi^*|\phi^*\rangle \\ &= \langle\psi|\phi\rangle^* = \langle\phi|\psi\rangle. \end{aligned} \tag{A.12}$$

This result is true for any form of U. Equation (A.12) shows that under T, the probability $|\langle\psi|\phi\rangle|^2$ is preserved. We will now derive one more property that is true for systems with $T^2 = -1$. Consider the states $|\psi\rangle$ and $|T\psi\rangle$. Then,

$$\langle T\psi|\psi\rangle = \langle T\psi|T^2\psi\rangle = -\langle T\psi|\psi\rangle \quad \Rightarrow \quad \langle T\psi|\psi\rangle = 0. \tag{A.13}$$

Here, in the first step used is Eq. (A.12) and in the next step $T^2 = -1$ is used. Thus $|\psi\rangle$ and $|T\psi\rangle$ must be different and orthogonal (note that nothing new happens for $T^2 = 1$). This is the basis for Kramer degeneracy. To prove this, consider T invariant H, i.e. $THT^{-1} = H$. Then, applying T on both sides of $H|\psi\rangle = E|\psi\rangle$ we get

$$THT^{-1}T|\psi\rangle = ET|\psi\rangle \overset{THT^{-1}=H}{\longrightarrow} H\big(T|\psi\rangle\big) = E\big(T|\psi\rangle\big), \tag{A.14}$$

where we have used the result that E is real. Equation (A.14) shows that $|\psi\rangle$ and $T|\psi\rangle$ will have the same energy E if $THT^{-1} = H$ and they are orthogonal for $T^2 = -1$. Therefore, for such systems there will be at least two fold degeneracy—called Kramer degeneracy.

A.2 Structure of U in $T = UK$

A.2.1 Spinless Particle

For a spinless particle, the only variables are $\vec{x}$ and $\vec{p}$ and then $TxT^{-1} = x$ gives,

$$UKx(UK)^{-1} = UKxK^{-1}U = UxU^{-1} \quad \Rightarrow \quad UxU^{-1} = x. \tag{A.15}$$

Thus, in the position representation the components of the position vectors are real. In the same position representation, non-vanishing matrix elements of the momentum operator are purely imaginary. Then, $TpT^{-1} = -p$ gives $UpU^{-1} = p$. As

(x, p) are the only observables, $U = e^{i\lambda} I$ is the only possibility and we can choose $\lambda = 0$. Thus,

$$T = K, \qquad U = I. \tag{A.16}$$

in the position representation.

A.2.2 Spin $\frac{1}{2}$ Particle

For a spin $\frac{1}{2}$ particle, a basic property is

$$T \vec{\sigma} T^{-1} = -\vec{\sigma}. \tag{A.17}$$

In the σ_z diagonal representation the (x, y, z) components of $\vec{\sigma}$ are Pauli matrices,

$$\begin{aligned} &\sigma_x = \begin{pmatrix} 0 & 1 \\ 1 & 0 \end{pmatrix}, \qquad \sigma_y = \begin{pmatrix} 0 & -i \\ i & 0 \end{pmatrix}, \qquad \sigma_z = \begin{pmatrix} 1 & 0 \\ 0 & -1 \end{pmatrix} \\ &\Rightarrow \quad K\sigma_x K^{-1} = \sigma_x, \qquad K\sigma_y K^{-1} = -\sigma_y, \qquad K\sigma_z K^{-1} = \sigma_z. \end{aligned} \tag{A.18}$$

Using $T\sigma_i T^{-1} = -\sigma_i$, Eq. (A.18) and the relation $T = UK$, it is easy to derive

$$\begin{aligned} &U\sigma_x U^{-1} = -\sigma_x, \qquad U\sigma_y U^{-1} = \sigma_y, \qquad U\sigma_z U^{-1} = -\sigma_z \\ &\Rightarrow \quad U\sigma_x + \sigma_x U = 0, \qquad U\sigma_y - \sigma_y U = 0, \qquad U\sigma_z + \sigma_z U = 0 \\ &\Rightarrow \quad U = C_0 \sigma_y. \end{aligned} \tag{A.19}$$

As U is unitary, $|U|^2 = 1$. Therefore $|C_0| = 1$ and choosing $C_0 = 1$ gives

$$T = \sigma_y K \tag{A.20}$$

for a spin-($\frac{1}{2}$) particle in position representation $|x, m_s\rangle$; m_s is the eigenvalue of σ_z. Note that $T^2 = \sigma_y K \sigma_y K = \sigma_y K \sigma_y K^{-1} = -\sigma_y^2 = -1$ as required. Similarly, $T^2 = K^2 = 1$ for spinless particle and therefore spinless particle is even and spin-$\frac{1}{2}$ particle is odd.

A.2.3 Many Particle Systems

For a many particle (say N particles) system in the basis

$$\left| x^1, x^2, \ldots, x^N; m_s^1, m_s^2, \ldots, m_s^N \right\rangle,$$

it is easy to see that $U = \sigma_y^1 \sigma_y^2 \cdots \sigma_y^N$. Then, using $K\sigma_y K = -\sigma_y$ and $\sigma_y^2 = 1$,

$$\begin{aligned} T^2 &= \sigma_y^1 \sigma_y^2 \cdots \sigma_y^N K \sigma_y^1 \sigma_y^2 \cdots \sigma_y^N K \\ &= (-1)^N I \\ &= +I \text{ for } N\text{-even} \\ &= -I \text{ for } N\text{-odd}. \end{aligned} \tag{A.21}$$

An alternative to Eq. (A.20) is the choice [with $C_0 = i$ in Eq. (A.19)],

$$T = i\sigma_y K = \left(\exp i \frac{\pi}{2} \sigma_y \right) K. \tag{A.22}$$

Then, for A number of particles

$$\begin{aligned} T &= i\sigma_y^1 \, i\sigma_y^2 \cdots i\sigma_y^N K \\ &= (\exp i\pi S_y) K. \end{aligned} \tag{A.23}$$

Note that S_y is the y component of the total spin operator. Equation (A.23) gives correctly $T^2 = (\exp i 2\pi S_y) = +1$ for A even and -1 for A odd. The second form in (A.23) is used in Sect. 2.1.

A.3 Applications

Some applications of the structure of T operator are: (i) deriving phase relations between Wigner (or Clebsch-Gordan) coefficients for $m \to -m$; (ii) classifying types of operator that are rotational scalars, T invariant and parity invariant; (iii) determining different types of random matrices; (iv) nature of electric dipole moments under T. Here, we will consider (i) and (ii). In Chap. 2, the application (iii) is discussed in some detail and we refer the readers to Ref. [1] for (iv).

A.3.1 A Phase Relation Between Wigner Coefficients

Let us consider the action of T operator on angular momentum eigenstates $|jm\rangle$. Firstly, J_z, $J_\pm$ are real in the $|jm\rangle$ basis and therefore $TJ_iT^{-1} = -J_i$, $i = x, y, z$. Then $UKJ_zK^{-1}U^{-1} = -J_z = UJ_zU^{-1}$. Also $TJ_+T^{-1} = T(J_x + iJ_y)T^{-1} = (-J_x) - i(-J_y) = -J_-$. Therefore,

$$\begin{aligned} &UJ_z + J_zU = 0, \qquad UJ_+U^{-1} = -J_-, \qquad UJ_-U^{-1} = -J_+, \\ &UJ^2 - J^2U = 0. \end{aligned} \tag{A.24}$$

Using these we can find $T|jm\rangle$. Applying Eq. (A.24) we have,

$$\begin{aligned} &\langle m'|UJ_z + J_zU|m\rangle = 0 \\ &\Rightarrow \quad (m+m')\langle m'|U|m\rangle = 0. \end{aligned} \tag{A.25}$$

Then $m' = -m$ or $\langle m'|U|m\rangle = 0$ for $m' \neq -m$. Similarly $\langle m'|UJ_+ + J_-U|m\rangle = 0$ gives $\langle m'|U|m+1\rangle\sqrt{(j-m)(j+m+1)} + \langle m'+1|U|m\rangle\sqrt{(j-m')(j+m'+1)} = 0$. Now putting $m' = -m-1$ gives,

$$\frac{\langle -(m+1)|U|(m+1)\rangle}{\langle -m|U|m\rangle} = -1. \tag{A.26}$$

By iteration we have,

$$\frac{\langle m'|U|-m'\rangle}{\langle m|U|-m\rangle} = (-1)^{m'-m} = i^{2(m'-m)}. \tag{A.27}$$

We can choose $\langle m|U|-m\rangle = (i)^{2m}$ and therefore

$$T|jm\rangle = (-1)^m|j-m\rangle. \tag{A.28}$$

Equation (A.28) is standard for integer j values, i.e. $j = \ell$. In co-ordinate representation $|lm\rangle = Y^l_m(\theta,\phi)$ and $T = K$. Then,

$$Y^{l*}_m(\theta,\phi) = T|lm\rangle = (i)^{2m}|l-m\rangle \quad \Rightarrow \quad Y^{l*}_m(\theta,\phi) = (-1)^m Y^l_{-m}(\theta,\phi). \tag{A.29}$$

This choice must be chosen for consistency with T. For half-odd-integer j, it is conventional to use $(-1)^{j-m}$ [note that for a half-integer j, $(-1)^m = i$ or $-i$]. This corresponds to the choice $T|j,+j\rangle = |j,-j\rangle$ and then $(J_+)^{j-m}T = (-1)^{j-m}T(J_-)^{j-m}$ gives

$$T|jm\rangle = (-1)^{j-m}|j-m\rangle. \tag{A.30}$$

Then, we have the important relation

$$\langle j_1m_1j_2m_2|j_3m_3\rangle = (-1)^{j_1+j_2-j_3}\langle j_1-m_1j_2-m_2|j_3-m_3\rangle. \tag{A.31}$$

A.3.2 Restrictions on Hamiltonians

With only two-body interactions, for two particles the variables are

$$\left(\vec{X}^{\alpha\beta}, \vec{p}^{\alpha\beta}, \vec{\sigma}^{\alpha}, \vec{\sigma}^{\beta}\right)$$

with $\vec{X}^{\alpha\beta} = \vec{X}^{\alpha} - \vec{X}^{\beta}$. In keeping with non-relativistic approximation, interaction dependence on velocity (or p) will be restricted to terms that are at most quadratic.

Therefore, translational invariance and Galilean invariance (invariance under the transformation to a reference frame moving at a constant velocity with respect to the original frame) will restrict the interactions to the form,

$$H_{int} = \sum_{\alpha<\beta} V_{\alpha\beta}\left(\vec{X}^{\alpha\beta}, \vec{p}^{\alpha\beta}, \vec{\sigma}^{\alpha}, \vec{\sigma}^{\beta}\right). \tag{A.32}$$

With $X^{\alpha\beta} = |\vec{X}^{\alpha\beta}|$, $V_{\alpha\beta}$ might be expected to be a function of $X_{\alpha\beta}$ giving shape of the potential and $S_{\alpha\beta}$ that is a rotational scalar and P (Parity) and T invariant,

$$V_{\alpha\beta} = V\left(X^{\alpha\beta}\right) S_{\alpha\beta}. \tag{A.33}$$

Then the only possibilities for $S^i_{\alpha\beta}$ are: (i) $S_{\alpha\beta} = 1$; (ii) $\vec{\sigma}^{\alpha} \cdot \vec{\sigma}^{\beta}$; (iii, iv) $(\vec{X}^{\alpha\beta} \times \vec{p}^{\alpha\beta}) \cdot (\vec{\sigma}^{\alpha} \pm \vec{\sigma}^{\beta})$; (v) $(\vec{\sigma}^{\alpha} \cdot \vec{X}^{\alpha\beta})(\vec{\sigma}^{\beta} \cdot \vec{X}^{\alpha\beta})$; (vi) $(\vec{\sigma}^{\alpha} \cdot \vec{p}^{\alpha\beta})(\vec{\sigma}^{\beta} \cdot \vec{p}^{\alpha\beta})$. These 6 possibilities are denoted by $S^i_{\alpha\beta}$, $i = 1, 2, \ldots, 6$. For example $\vec{X}^{\alpha\beta} \cdot (\vec{\sigma}^{\alpha} \pm \vec{\sigma}^{\beta})$, $\overrightarrow{p^{\alpha\beta}} \cdot (\vec{\sigma}^{\alpha} \pm \vec{\sigma}^{\beta})$ and $(\vec{X}^{\alpha\beta} \times \vec{p}^{\alpha\beta}) \cdot (\vec{\sigma}^{\alpha} \times \vec{\sigma}^{\beta})$ are not possible. For identical particles, $(\vec{X}^{\alpha\beta} \times \vec{p}^{\alpha\beta}) \cdot (\vec{\sigma}^{\alpha} - \vec{\sigma}^{\beta}) = 0$ as this is anti-symmetric in α and β variables.

References

1. R.G. Sachs, *The Physics of Time Reversal* (University of Chicago Press, Chicago, 1987)

Appendix B
Univariate and Bivariate Moments and Cumulants

Say the operator H is defined in a model space of m particles (fermions or bosons) and the eigenvalues of H are E_r with $E_r < E_{r+1}$ and $r = 1, 2, \dots, l$. The eigenvectors of H are denoted by $\psi_{r,i}$ where $i = 1, 2, \dots, g_r$ distinguishes between degenerate states and the model space dimensionality is $d = \sum_{r=1}^{l} g_r$. Then the eigenvalue density $I(x)$ (or $\rho(x)$ with unit normalization) that corresponds to a univariate distribution and the corresponding distribution function $F(x)$ are,

$$\begin{aligned}
I^H(x) \quad &\Leftrightarrow \quad I(x) = d\rho(x) = \sum_{r=1}^{l} g_r \delta(x - E_r) = \sum_{r,i} \langle ri | \delta(H - x) | ri \rangle \\
&= \langle\langle \delta(H - x) \rangle\rangle^m = d \langle \delta(H - x) \rangle^m, \\
\int_{-\infty}^{+\infty} I(x)dx &= d; \qquad \int_{-\infty}^{+\infty} \rho(x)dx = 1, \\
F(x) &= \int_{-\infty}^{x} \rho(z)dz; \qquad F(-\infty) = 0, \qquad F(+\infty) = 1.
\end{aligned} \tag{B.1}$$

We have written things for finite d; no difficulties will arise with $d \to \infty$. Observe that in I, ρ, and F, we have quickly dropped the label which describes the variable, as it does in I^H. The moments of ρ are M_p and they are the expansion coefficients in the Taylor expansion of the Fourier transform of ρ, the characteristic function ϕ. Then,

$$\begin{aligned}
M_p &= \int \rho(x) x^p dx = d^{-1} \sum_r g_r (E_r)^p = \langle H^p \rangle^m, \\
\phi(t) &= \int \rho(x) \exp(itx) dx = \sum_p \frac{(it)^p}{p!} M_p = \langle \exp(itH) \rangle^m, \\
M_0 &= 1, \qquad M_1 = \xi, \qquad M_2 - (M_1)^2 = \sigma^2, \qquad \phi(0) = 1.
\end{aligned} \tag{B.2}$$

V.K.B. Kota, *Embedded Random Matrix Ensembles in Quantum Physics*,
Lecture Notes in Physics 884, DOI 10.1007/978-3-319-04567-2,

Gaussian density $\rho_{\mathscr{G}}(x)$ and the standard Gaussian $\eta_{\mathscr{G}}(\widehat{x})$ that involves standardized variable $\widehat{x}$ are,

$$\begin{aligned}
\rho_{\mathscr{G}}(x) &= \frac{1}{\sqrt{2\pi}\sigma}\exp\left\{-\frac{(x-\xi)^2}{2\sigma^2}\right\};\\
\eta_{\mathscr{G}}(\widehat{x}) &= \frac{1}{\sqrt{2\pi}}\exp\left\{-\frac{\widehat{x}^2}{2}\right\},\\
\widehat{x} &= (x-M_1)/\left(M_2-M_1^2\right)^{1/2} = (x-\xi)/\sigma,\\
\rho_{\mathscr{G}}(x)dx &= \eta_{\mathscr{G}}(\widehat{x})d\widehat{x}.
\end{aligned} \tag{B.3}$$

For $\rho_{\mathscr{G}}(x)$, the characteristic function is $\phi_{\mathscr{G}}(t;\xi,\sigma^2) = \exp(it\xi - \frac{\sigma^2 t^2}{2})$. In Eq. (B.3), ξ is the centroid which fixes the location of the distribution and the width σ fixes the scale. Moments obtained by shifting the centroid to zero are called central moments $\mathscr{M}_p$. Just as the moments enter into the Taylor expansion of $\phi(t)$, the cumulants K_p enter into the expansion of its logarithm:

$$\ln\phi(t) = \sum_{p=1}^{\infty}\frac{(it)^p}{p!}K_p; \qquad \phi(t) = \exp\left(\sum\frac{K_p(it)^p}{p!}\right). \tag{B.4}$$

Note that $K_1 = \xi$, $K_2 = \sigma^2$ (as long as these moments exist). For a Gaussian we have $K_r = 0$ for $r > 2$. In practice reduced moments μ_r and reduced cumulants k_p are more useful as they are scale free. Note that $\mu_r = \mathscr{M}_r/\sigma^r$ and $k_p = K_p/\sigma^p$. The shape parameter $k_3 = \gamma_1$ is called skewness and $k_4 = \gamma_2$ the excess. Broadly speaking $k_3 > 0$ defines a distribution which extends more in the $(x > \xi)$ domain than in $(x < \xi)$, and $k_4 > 0$ a distribution more sharply peaked than the Gaussian. One can write k_p in terms of μ_r and similarly μ_r in terms of k_p. For example,

$$\begin{aligned}
&k_2 = \mu_2, \qquad k_3 = \mu_3, \qquad k_4 = \mu_4 - 3,\\
&k_5 = \mu_5 - 10\mu_3, \qquad k_6 = \mu_6 - 15\mu_4 - 10(\mu_3)^2 + 30,\\
&k_7 = \mu_7 - 21\mu_5 - 35\mu_4\mu_3 + 210\mu_3,\\
&k_8 = \mu_8 - 28\mu_6 - 56\mu_5\mu_3 - 35(\mu_4)^2 + 420\mu_4 + 560(\mu_3)^2 - 630.
\end{aligned} \tag{B.5}$$

Given the standardized variable $\widehat{x}$ and the corresponding Gaussian density, it was argued by Edgeworth that $\eta(\widehat{x})$,

$$\eta(\widehat{x}) = \exp\left\{\sum_{\nu\geq 3}(-1)^{\nu}\frac{k_\nu}{\nu!}\frac{\partial^\nu}{\partial\widehat{x}^\nu}\right\}\eta_{\mathscr{G}}(\widehat{x}); \qquad \eta_{\mathscr{G}}(\widehat{x}) = \frac{1}{\sqrt{2\pi}}\exp-\frac{\widehat{x}^2}{2}, \tag{B.6}$$

is a "true and unique law that represents the frequency curve of a magnitude that depends on a number of independent elements" [1]. If $k_\nu \propto \Upsilon^{-(\nu-2)/2}$, we can get, from Eq. (B.6), an expansion in powers of $1/\sqrt{\Upsilon}$ and it is called Edgeworth (ED) expansion. The ED expansion to order $1/\Upsilon$ is, with $k'_\nu = k_\nu/\nu!$,

$$\eta_{ED}(\widehat{x}) = \eta_{\mathscr{G}}(\widehat{x})\left\{1 + \left[k'_3 He_3(\widehat{x})\right] + \left[k'_4 He_4(\widehat{x}) + \frac{(k'_3)^2}{2!} He_6(\widehat{x})\right]\right\}. \tag{B.7}$$

Similarly, He are the Hermite polynomials: $He_3(x) = x^3 - 3x$, $He_4(x) = x^4 - 6x^2 + 3$ and $He_6(x) = x^6 - 15x^4 + 45x^2 - 15$. Now we will consider bivariate distributions.

Bivariate distributions $\rho(x, y)$, i.e. distributions in two variables, are encountered for example when we deal with two-point function in eigenvalues (see Chaps. 2, 5, 9, 11, 12) and transition strength densities (see Chaps. 4, 5, 7). Just as with the univariate distributions, it is possible to introduce bivariate moments M_{rs}, central moments $\mathscr{M}_{rs}$, reduced moments μ_{rs}, cumulants K_{rs} and reduced cumulants k_{rs}. Integral of $\rho(x, y)$ over y gives the marginal density $\rho_1(x)$ and similarly integral over x gives the marginal $\rho_2(y)$. The centroids and variances of these are called marginal centroids and variances. Let us say that they are $(\varepsilon_1, \sigma_1^2)$ for $\rho_1(x)$ and similarly $(\varepsilon_2, \sigma_2^2)$ for $\rho_2(y)$. Then M_{rs} and $\mathscr{M}_{rs}$ are,

$$\begin{aligned} M_{rs} &= \int_{-\infty}^{\infty}\int_{-\infty}^{\infty} x^r y^s \rho(x, y)\, dx\, dy, \\ \mathscr{M}_{rs} &= \int_{-\infty}^{\infty}\int_{-\infty}^{\infty} (x - \varepsilon_1)^r (y - \varepsilon_2)^s \rho(x, y)\, dx\, dy. \end{aligned} \tag{B.8}$$

Bivariate Gaussian in terms of the standardized variables $\widehat{x} = (x - \varepsilon_1)/\sigma_1$ and $\widehat{y} = (y - \varepsilon_2)/\sigma_2$ is given by,

$$\eta_{\mathscr{G}}(\widehat{x}, \widehat{y}) = \frac{1}{2\pi\sqrt{(1 - \zeta^2)}} \exp\left\{-\frac{\widehat{x}^2 - 2\zeta\widehat{x}\widehat{y} + \widehat{y}^2}{2(1 - \zeta^2)}\right\}. \tag{B.9}$$

Here ζ is the correlation coefficient. Thus, a bivariate Gaussian is defined by the five variables $(\varepsilon_1, \varepsilon_2, \sigma_1, \sigma_2, \zeta)$. For completeness, we give here the bivariate cumulants K_{rs} for $r + s \leq 6$ in terms of the central moments $\mathscr{M}_{r'+s'}$ [2],

$$K_{30} = \mathscr{M}_{30},$$
$$K_{21} = \mathscr{M}_{21},$$

$$
\begin{aligned}
K_{40} &= \mathcal{M}_{40} - 3\mathcal{M}_{20}^2, \\
K_{31} &= \mathcal{M}_{31} - 3\mathcal{M}_{20}\mathcal{M}_{11}, \\
K_{22} &= \mathcal{M}_{22} - \mathcal{M}_{20}\mathcal{M}02 - 2\mathcal{M}_{11}^2, \\
K_{50} &= \mathcal{M}_{50} - 10\mathcal{M}_{30}\mathcal{M}_{20}, \\
K_{41} &= \mathcal{M}_{41} - 4\mathcal{M}_{30}\mathcal{M}_{11} - 6\mathcal{M}_{21}\mathcal{M}_{20}, \\
K_{32} &= \mathcal{M}_{32} - \mathcal{M}_{30}\mathcal{M}_{02} - 6\mathcal{M}_{21}\mathcal{M}_{11} - 3\mathcal{M}_{20}\mathcal{M}_{12}, \\
K_{60} &= \mathcal{M}_{60} - 15\mathcal{M}_{40}\mathcal{M}_{20} - 10\mathcal{M}_{30}^2 + 30\mathcal{M}_{20}^3, \\
K_{51} &= \mathcal{M}_{51} - 5\mathcal{M}_{40}\mathcal{M}_{11} - 10\mathcal{M}_{31}\mathcal{M}_{20} - 10\mathcal{M}_{30}\mathcal{M}_{21} + 30\mathcal{M}_{20}^2\mathcal{M}_{11}, \\
K_{42} &= \mathcal{M}_{42} - \mathcal{M}_{40}\mathcal{M}_{02} - 8\mathcal{M}_{31}\mathcal{M}_{11} - 4\mathcal{M}_{30}\mathcal{M}_{12} - 6\mathcal{M}_{22}\mathcal{M}_{20} \\
&\quad - 6\mathcal{M}_{21}^2 + 6\mathcal{M}_{20}^2\mathcal{M}_{02} + 24\mathcal{M}_{20}\mathcal{M}_{11}^2, \\
K_{33} &= \mathcal{M}_{33} - 3\mathcal{M}_{31}\mathcal{M}_{02} - \mathcal{M}_{30}\mathcal{M}_{03} - 9\mathcal{M}_{22}\mathcal{M}_{11} - 9\mathcal{M}_{21}\mathcal{M}_{12} \\
&\quad - 3\mathcal{M}_{20}\mathcal{M}_{13} + 18\mathcal{M}_{20}\mathcal{M}_{11}\mathcal{M}_{02} + 12\mathcal{M}_{11}^3.
\end{aligned}
\tag{B.10}
$$

Note that $K_{rs} \to K_{sr}$ with $\mathcal{M}_{r's'} \to \mathcal{M}_{s'r'}$. Similarly $K_{20} = \sigma_{20}^2 = \mathcal{M}_{20} = \sigma_1^2$ and $K_{02} = \sigma_{02}^2 = \mathcal{M}_{02} = \sigma_2^2$. The reduced cumulants $k_{rs} = K_{rs}/[\{K_{20}\}^{r/2}\{K_{02}\}^{s/2}]$. Finally, the correlation coefficient $\zeta = k_{11} = \mathcal{M}_{11}/[\mathcal{M}_{20}\mathcal{M}_{02}]^{1/2}$.

Let us add that the probability densities $\rho_{12}(x|y)$ and $\rho_{21}(y|x)$, that are referred as the conditional densities, are given by

$$
\rho_{12}(x|y) = \rho(x,y)/\rho_2(y), \qquad \rho_{21}(y|x) = \rho(x,y)/\rho_1(x). \tag{B.11}
$$

Note that $\rho_1(x)$ and $\rho_2(y)$ are the marginal densities. The conditional moments $M_p(y)$ of ρ_{12} (similarly $M_p(x)$ of ρ_{21}) are,

$$
\begin{aligned}
M_p(y) &= \int_{-\infty}^{\infty} x^p \rho_{12}(x|y)\, dx \\
&= \int_{-\infty}^{\infty} x^p \rho(x,y)\, dx \Big/ \int_{-\infty}^{\infty} \rho(x,y)\, dx.
\end{aligned}
\tag{B.12}
$$

An important result is that for the bivariate Gaussian defined by Eq. (B.9), the corresponding conditional densities will be Gaussians,

$$
\eta_{12:\mathcal{G}}(\widehat{x}|\widehat{y}) = \frac{1}{\sqrt{2\pi(1-\zeta^2)}} \exp\left\{ -\frac{(\widehat{x} - \zeta\widehat{y})^2}{2(1-\zeta^2)} \right\}. \tag{B.13}
$$

Therefore, for a bivariate Gaussian, the conditional centroids are linear in the fixed variable with a slope given by the correlation coefficient. Similarly the conditional width is a constant defined by the correlation coefficient and independent of the fixed variable.

Given the bivariate Gaussian $\eta_{\mathscr{G}}(\widehat{x},\widehat{y})$, the bivariate Edgeworth expansion for any bivariate distribution $\eta(\widehat{x},\widehat{y})$ follows from

$$\eta(\widehat{x},\widehat{y}) = \exp\left\{\sum_{r+s\geq 3}(-1)^{r+s}\frac{k_{rs}}{r!s!}\frac{\partial^r}{\partial\widehat{x}^r}\frac{\partial^s}{\partial\widehat{y}^s}\right\}\eta_{\mathscr{G}}(\widehat{x},\widehat{y}). \tag{B.14}$$

Assuming that the bivariate reduced cumulants k_{rs} behave as $k_{r+s}\propto \Upsilon^{-(r+s-2)/2}$ where Υ is a system parameter, and collecting in the expansion of Eq. (B.14) all the terms that behave as $\Upsilon^{-P/2}$, $P=1,2,\ldots$, we obtain the bivariate ED expansion. To order $1/P$ we have [2, 3],

$$\begin{aligned}
\eta_{biv-ED}(\widehat{x},\widehat{y}) = &\Bigg\{1+\left(\frac{k_{30}}{6}He_{30}(\widehat{x},\widehat{y})+\frac{k_{21}}{2}He_{21}(\widehat{x},\widehat{y})\right.\\
&\left.+\frac{k_{12}}{2}He_{12}(\widehat{x},\widehat{y})+\frac{k_{03}}{6}He_{03}(\widehat{x},\widehat{y})\right)\\
&+\Bigg(\Bigg\{\frac{k_{40}}{24}He_{40}(\widehat{x},\widehat{y})+\frac{k_{31}}{6}He_{31}(\widehat{x},\widehat{y})\\
&+\frac{k_{22}}{4}He_{22}(\widehat{x},\widehat{y})+\frac{k_{13}}{6}He_{13}(\widehat{x},\widehat{y})+\frac{k_{04}}{24}He_{04}(\widehat{x},\widehat{y})\Bigg\}\\
&+\Bigg\{\frac{k_{30}^2}{72}He_{60}(\widehat{x},\widehat{y})+\frac{k_{30}k_{21}}{12}He_{51}(\widehat{x},\widehat{y})\\
&+\left[\frac{k_{21}^2}{8}+\frac{k_{30}k_{12}}{12}\right]He_{42}(\widehat{x},\widehat{y})\\
&+\left[\frac{k_{30}k_{03}}{36}+\frac{k_{12}k_{21}}{4}\right]He_{33}(\widehat{x},\widehat{y})\\
&+\left[\frac{k_{12}^2}{8}+\frac{k_{21}k_{03}}{12}\right]He_{24}(\widehat{x},\widehat{y})+\frac{k_{12}k_{03}}{12}He_{15}(\widehat{x},\widehat{y})\\
&+\frac{k_{03}^2}{72}He_{06}(\widehat{x},\widehat{y})\Bigg\}\Bigg)\Bigg\}\eta_{\mathscr{G}}(\widehat{x},\widehat{y}).
\end{aligned} \tag{B.15}$$

The bivariate Hermite polynomials $He_{m_1m_2}(\widehat{x},\widehat{y})$ in Eq. (B.15) satisfy the recursion relation,

$$\begin{aligned}
(1-\zeta^2)He_{m_1+1,m_2}(\widehat{x},\widehat{y}) = &(\widehat{x}-\zeta\widehat{y})He_{m_1,m_2}(\widehat{x},\widehat{y})\\
&-m_1He_{m_1-1,m_2}(\widehat{x},\widehat{y})+m_2\zeta He_{m_1,m_2-1}(\widehat{x},\widehat{y}).
\end{aligned} \tag{B.16}$$

The polynomials $He_{m_1 m_2}$ with $m_1 + m_2 \leq 2$ are

$$\begin{aligned} He_{00}(\widehat{x},\widehat{y}) &= 1, \\ He_{10}(\widehat{x},\widehat{y}) &= (\widehat{x} - \zeta\widehat{y})/\left(1-\zeta^2\right), \\ He_{20}(\widehat{x},\widehat{y}) &= \frac{(\widehat{x} - \zeta\widehat{y})^2}{(1-\zeta^2)^2} - \frac{1}{(1-\zeta^2)}, \\ He_{11}(\widehat{x},\widehat{y}) &= \frac{(\widehat{x} - \zeta\widehat{y})(\widehat{y} - \zeta\widehat{x})}{(1-\zeta^2)^2} + \frac{\zeta}{1-\zeta^2}. \end{aligned} \tag{B.17}$$

Note that $He_{m_1 m_2}(\widehat{x},\widehat{y}) = He_{m_2 m_1}(\widehat{y},\widehat{x})$. Finally, we refer the readers to [2] for additional details on the properties of univariate and bivariate distributions.

References

1. A.L. Bowley, *F.Y. Edgeworth's Contributions to Mathematical Statistics* (Augustus M. Kelley, Clifton, 1972)
2. A. Stuart, J.K. Ord, *Kendall's Advanced Theory of Statistics: Distribution Theory* (Oxford University Press, New York, 1987)
3. V.K.B. Kota, Bivariate distributions in statistical spectroscopy studies: I. Fixed-J level densities, fixed-J averages and spin cut-off factors. Z. Phys. A **315**, 91–98 (1984)

Appendix C
Dyson-Mehta $\overline{\Delta_3(\overline{n})}$ Statistic as an Integral Involving $\Sigma^2(r)$

Detailed derivation of the relation between $\overline{\Delta_3}(\overline{n})$ and $\Sigma^2(r)$ is as follows. The result is due to Pandey [1]. Firstly, starting from the definition of $\overline{\Delta_3}(\overline{n})$ given by Eq. (2.71),

$$\frac{d\overline{\Delta_3}(2L)}{dA} = 0, \qquad \frac{d\overline{\Delta_3}(2L)}{dB} = 0 \quad \Rightarrow$$
$$\int_{x-L}^{x+L} \big(dF(y) - Ay - B\big) y\,dy = 0, \qquad \int_{x-L}^{x+L} \big(dF(y) - Ay - B\big)\,dy = 0. \tag{C.1}$$

Say $F = (1/2L)\int_{x-L}^{x+L} dF(y) y\,dy$ and $G = (1/2L)\int_{x-L}^{x+L} dF(y)\,dy$. Then Eq. (C.1) gives,

$$F - A\frac{y^3}{6L}\bigg|_{x-L}^{x+L} - B\frac{y^2}{4L}\bigg|_{x-L}^{x+L} = F - \frac{A}{3}\big(3x^2 + L^2\big) - Bx = 0,$$
$$G - \frac{A}{2L}\frac{y^2}{2}\bigg|_{x-L}^{x+L} - \frac{By}{2L}\bigg|_{x-L}^{x+L} = G - Ax - B = 0 \tag{C.2}$$
$$\Rightarrow \quad (F - Gx) - A\frac{L^2}{3} = 0.$$

Now Eq. (C.2) gives,

$$A = \frac{3}{L^2}(F - Gx), \qquad Ax + B = G. \tag{C.3}$$

Ergodicity and stationary properties of the Gaussian ensembles [2] imply that $\overline{\Delta_3}(\overline{n})$ should be independent of x in Eq. (2.71). Therefore, putting $x = 0$ in Eqs. (C.2) and (C.3) will give the following relationships between the parameters (A, B) and

V.K.B. Kota, *Embedded Random Matrix Ensembles in Quantum Physics*, Lecture Notes in Physics 884, DOI 10.1007/978-3-319-04567-2,

the integrals (F, G),

$$A = \frac{3}{L^2} F, \qquad F = \frac{1}{2L} \int_{-L}^{L} d\, F(y) y\, dy, \qquad B = G = \frac{1}{2L} \int_{-L}^{L} d\, F(y)\, dy. \tag{C.4}$$

Substituting Eq. (C.4) in Eq. (2.71) leads to,

$$\begin{aligned}
\overline{\Delta_3}(\overline{n}) &= \frac{1}{2L} \int_{-L}^{L} \left[d^2 F^2(y) + A^2 y^2 + B^2 - 2dF(y)Ay - 2dF(y)B + 2ABy \right] dy \\
&= \frac{1}{2L} \int_{-L}^{L} d^2 F^2(y)\, dy - \frac{2dA}{2L} \int_{-L}^{L} yF(y)\, dy - \frac{2dB}{2L} \int_{-L}^{L} F(y)\, dy \\
&\quad + \frac{A^2}{2L} \frac{y^3}{3} \bigg|_{-L}^{L} + \frac{B^2 y}{2L} \bigg|_{-L}^{L} + \frac{2AB}{2L} \frac{y^2}{2} \bigg|_{-L}^{L} \\
&= \frac{1}{2L} \int_{-L}^{L} d^2\, F^2(y)\, dy - \frac{2dA}{2L} \int_{-L}^{L} yF(y)\, dy - \frac{2dB}{2L} \int_{-L}^{L} F(y)\, dy \\
&\quad + \left[\frac{A^2 L^2}{3} + B^2 \right] \\
&= \frac{1}{2L} \int_{-L}^{L} d^2 F^2(y)\, dy - \frac{3d}{L^3} F \int_{-L}^{L} yF(y)\, dy - \frac{d}{L} G \int_{-L}^{L} F(y)\, dy \\
&\quad + \left[\frac{3F^2}{L^2} + G^2 \right] \\
&= \frac{d^2}{2L} \int_{-L}^{L} F^2(y)\, dy - \frac{3d^2}{2L^4} \int_{-L}^{L} yF(y) dy \int_{-L}^{L} zF(z)\, dz \\
&\quad - \frac{d^2}{2L^2} \int_{-L}^{L} F(y)\, dy \int_{-L}^{L} F(z)\, dz + \frac{3d^2}{4L^4} \left[\int_{-L}^{L} yF(y)\, dy \right]^2 \\
&\quad + \frac{d^2}{4L^2} \left[\int_{-L}^{L} F(y)\, dy \right]^2 \\
&= \frac{d^2}{2L} \int_{-L}^{L} F^2(y)\, dy - \frac{3d^2}{4L^4} \int_{-L}^{L} \int_{-L}^{L} yzF(y)F(z)\, dydz \\
&\quad - \frac{d^2}{4L^2} \int_{-L}^{L} \int_{-L}^{L} F(y)F(z)\, dydz.
\end{aligned} \tag{C.5}$$

Simplifying Eq. (C.5) using the identity

$$\int_{-L}^{L} \int_{-L}^{L} \left(F(y) - F(z) \right)^2 dydz = 4L \int_{-L}^{L} F^2(y) dy - 2 \int_{-L}^{L} \int_{-L}^{L} F(y)F(z)\, dydz, \tag{C.6}$$

yields,

$$\overline{\Delta_3}(\overline{n}) = \frac{d^2}{8L^2}\left[\int_{-L}^{L}\int_{-L}^{L}\left[1+\frac{3yz}{L^2}\right]\big(F(y)-F(z)\big)^2\,dydz\right]. \tag{C.7}$$

Note that $L = \overline{n}/2$ in Eq. (C.7). Now, let us put $d^2(F(y)-F(z))^2 = \Sigma^2(|y-z|) + r^2 = f(y,z) = f(r)$ where $r = |y-z|$ and $X = y+z$. The range of X is from $\overline{n}-r$ to $-(\overline{n}-r)$ and that of r is 0 to $\overline{n}$. With these we have,

$$\begin{aligned}
\overline{\Delta_3}(\overline{n}) &= \frac{1}{2\overline{n}^2}\int_{-L}^{L}\int_{-L}^{L}\left(1+\frac{12yz}{\overline{n}^2}\right)f(y,z)\,dydz \\
&= \frac{1}{2\overline{n}^2}\int_0^{\overline{n}} dr \int_{-(\overline{n}-r)}^{\overline{n}-r}\left[1+\frac{3}{\overline{n}^2}\left(X^2-r^2\right)\right]f(r)\,dXdr \\
&= \frac{1}{2\overline{n}^2}\int_0^{\overline{n}}\left[X+\frac{X^3}{\overline{n}^2}-\frac{3}{\overline{n}^2}Xr^2\right]_{-(\overline{n}-r)}^{\overline{n}-r} f(r)\,dr \\
&= \frac{1}{\overline{n}^2}\int_0^{\overline{n}}(\overline{n}-r)\left\{1+\frac{(\overline{n}-r)^2}{\overline{n}^2}-\frac{3r^2}{\overline{n}^2}\right\}f(r)\,dr \\
&= \frac{2}{\overline{n}^4}\int_0^{\overline{n}}\left(\overline{n}^3-2\overline{n}^2r+r^3\right)\left[\Sigma^2(r)+r^2\right]dr
\end{aligned}$$

$$\Rightarrow \quad \overline{\Delta_3}(\overline{n}) = \frac{2}{\overline{n}^4}\int_0^{\overline{n}}\left(\overline{n}^3-2\overline{n}^2r+r^3\right)\Sigma^2(r)\,dr. \tag{C.8}$$

Note that Σ^2 in Eq. (C.8) is the ensemble averaged Σ^2. Last line of Eq. (C.8) gives the final result.

References

1. A. Pandey, Fluctuation, stationarity and ergodic properties of random-matrix ensembles, Ph.D. Thesis, University of Rochester (1978)
2. A. Pandey, Statistical properties of many-particle spectra III. Ergodic behavior in random matrix ensembles. Ann. Phys. (N.Y.) **119**, 170–191 (1979)

Appendix D
Breit-Wigner Form for Strength Functions or Partial Densities

Let us say that for a m particle system we have a d dimensional Hilbert space. Then the system is defined by d number of basis states and say the eigenvalues (E) of a Hamiltonian H acting in this space are E_α with $\alpha = 1, 2, \ldots, d$. Now, consider a basis state $|k\rangle$. With $|k\rangle = \sum_E C_k^E |E\rangle$, the strength function $F_k(E) = \sum_{E'} |C_k^E|^2 \delta(E - E')$. To derive the form for $|C_k^E|^2$, we denote the first basis state as $|k\rangle$, with energy $\overline{E_k} = \int_{-\infty}^{+\infty} F_k(E) E dE$, and this state couples to the remaining $d - 1$ states $|\nu\rangle$. Diagonalization of the $(d-1) \times (d-1)$ sub-matrix gives the eigenvalues ε_ν and eigenvectors $|B_\nu\rangle$, $\nu = 1, 2, \ldots, d-1$. The H matrix in the $|k\rangle \oplus |B_\nu\rangle$ basis takes the form,

$$H = \begin{pmatrix} \overline{E_k} & V_{k1} & V_{k2} & \ldots & \ldots \\ V_{k1} & \varepsilon_1 & 0 & 0 & \ldots \\ V_{k2} & 0 & \varepsilon_2 & 0 & \ldots \\ \ldots & \ldots & \ldots & \ldots & \ldots \end{pmatrix}. \tag{D.1}$$

The eigenvalue equation for this H can be easily solved for 2×2 and 3×3 matrices and extend them to any $d \times d$ matrix giving,

$$E_\alpha - \overline{E_k} = \sum_\nu \frac{|V_{k\nu}|^2}{(E_\alpha - \varepsilon_\nu)}. \tag{D.2}$$

Similar procedure for the eigenvectors yields,

$$|C_k^\alpha|^2 = \frac{1}{1 + \sum_\nu \frac{|V_{k\nu}|^2}{(E_\alpha - \varepsilon_\nu)^2}}. \tag{D.3}$$

To proceed further, the following assumptions are made (this is also referred as the standard model [1]):

(i) The background spectrum, defined by the $d - 1 \times d - 1$ sub-matrix, is dense and rigid, and therefore it can be replaced by an uniform spectrum with mean spacing $\overline{D}$ i.e. $\varepsilon_\nu = \overline{n}\overline{D}, \overline{n} = -\infty$ to $+\infty$.

V.K.B. Kota, *Embedded Random Matrix Ensembles in Quantum Physics*, Lecture Notes in Physics 884, DOI 10.1007/978-3-319-04567-2,

(ii) The coupling matrix elements $V_{k\nu}^2$ are uncorrelated with the energies ε_ν and weakly fluctuate around the mean value. Then $(V_{k\nu})^2 = \overline{(V_{k\nu})^2} = V^2$.

(iii) The mixing is 'sufficiently' strong so that $V_{k\nu}^2 \gg \overline{D}^2$.

Constraints (i) and (ii) simplify the eigenvalue equation (D.2) and give the following result,

$$E_\alpha - \overline{E_k} = \frac{\pi V^2}{\overline{D}} \cot \frac{\pi E_\alpha}{\overline{D}}. \tag{D.4}$$

Here we made use of the formula $\cot z = \sum_{k=-\infty}^{+\infty} 1/(z - k\pi)$. Substituting this in Eq. (D.3) and using the identity $\operatorname{cosec}^2 z = \sum_{k=-\infty}^{\infty} 1/(z - k\pi)^2$, we obtain

$$\left|C_k^\alpha\right|^2 = \left[1 + \frac{\pi^2 V^2}{\overline{D}^2}\left\{1 + \cot^2 \frac{\pi E}{\overline{D}}\right\}\right]^{-1}. \tag{D.5}$$

Further simplifications using assumption (iii) give,

$$\left|C_k^\alpha\right|^2 = \frac{\overline{D}}{2\pi} \frac{\Gamma}{(E_\alpha - \overline{E_k})^2 + \Gamma^2/4}; \quad \Gamma = \frac{2\pi V^2}{\overline{D}}. \tag{D.6}$$

As $F_k(E) = (\overline{D})^{-1}|C_k^\alpha|^2$, Eq. (D.6) leads to BW form for $F_k(E)$,

$$F_{k:BW}(E) = \frac{1}{2\pi} \frac{\Gamma}{(E - \overline{E_k})^2 + \frac{\Gamma^2}{4}}. \tag{D.7}$$

Here, Γ is the spreading width. Note that the assumptions of the standard model are no longer valid in the "strong coupling" limit, i.e. when the spreading width grows larger than the energy interval ΔE where the level density can be considered as approximately constant.

Some properties of the Breit-Wigner (BW) or Cauchy distribution given by Eq. (D.7) are: (i) $F_{k:BW}(E)$ is normalized to unity; (ii) the moments for BW are undefined; (iii) the median/mode is $\overline{E_k}$; (iv) the distance between the lower and upper quartiles is the spreading width Γ; (v) with $\lambda = \Gamma/2$, $\mu = \overline{E_k}$, the characteristic function $\phi(t : \lambda, \mu) = \exp(i\mu t - \lambda|t|)$; (v) under convolution of two Cauchy's, the spreading widths add, i.e. $\mu = \mu_1 + \mu_2$ and $\lambda = \lambda_1 + \lambda_2$ as $\phi = \phi_1\phi_2$ [2].

A non-canonical correction to BW has been suggested recently and this is called modified BW (MBW). The MBW form, being positive definite always and used recently to describe shell model configuration partial densities, is defined by [3]

$$F_{k:MBW}(E) = \frac{\mathcal{N}}{W^3} \frac{(E - E_{min})^2(E_{max} - E)^2}{(E - E_0)^2 + W^2}. \tag{D.8}$$

Here $\mathcal{N}$ is the normalization factor and E_{min} and E_{max} are the end points of the eigenvalues, $E_{min} \le E \le E_{max}$. The four parameters $(E_0, E_{min}, E_{max}, W)$ can be determined from the centroid, variance, γ_1 and γ_2 of F_k.

References

1. A. Bohr, B.R. Mottelson, *Nuclear Structure*, Single-Particle Motion, vol. I (Benjamin, New York, 1969)
2. H. Cramer, *Mathematical Methods of Statistics* (Princeton University Press, Princeton, 1974)
3. E. Terán, C.W. Johnson, Simple models for shell-model configuration densities. Phys. Rev. C **74**, 067302 (2006)

Appendix E
Random Matrix Theory for Open Quantum Systems: 2 x 2 Matrix Results and Modified P-T Distribution

RMT is usually applied to isolated (closed) finite quantum systems where the coupling to the environment can be neglected. However, many systems of current interest such as quantum dots, compound nuclear resonances, micro lasers cavities, microwave billiards and so on coupling of the quantum system to the environment must be explicitly taken into account. Properties of the open and marginally stable quantum many-body systems can be studied in a general fashion using the effective Hamiltonian [1],

$$H_{eff} = H_0 - \frac{i}{2} V V^{\dagger}. \tag{E.1}$$

Here H_0 gives the discrete spectrum and $VV^{\dagger}$ represents the coupling to the continuum. With N discrete states coupled to M open channels ($N \gg M$), H_0 is a $N \times N$ matrix and V is a $N \times M$ matrix. We restrict the discussion here to time-reversal and rotationally invariant systems. Therefore, H_0 is real symmetric matrix and similarly, the matrix elements of V are real. Thus, in the random matrix approach, H_{eff} will be a random matrix ensemble with H_0 represented by a GOE and V matrix elements are chosen to be independent Gaussian variables with zero center and variance say $1/\eta$,

$$\{H_{eff}\} = \{H_0\} - \frac{i}{2} \{V V^{\dagger}\} \tag{E.2}$$

where $\{--\}$ represents ensemble. Due to the V part, the eigenvalues of H_{eff} will be complex and we can write them as $E - \frac{i}{2}\Gamma$. For example for resonance states, E represents the resonance positions and Γ their width. Using the ensemble defined by Eq. (E.2) one can study for example the statistics of neutron resonance spacings in the region where the resonance widths are not very small compared to the average resonance spacing and similarly the modifications to the resonance width distribution (i.e. modification to P-T law). First we consider NNSD for the resonance spacings.

Let us consider the simplest situation of $N = 2$ and $M = 1$, i.e. two bound states coupled to a single open channel. Then the H_{eff} (hereafter we call it simply H)

V.K.B. Kota, *Embedded Random Matrix Ensembles in Quantum Physics*, Lecture Notes in Physics 884, DOI 10.1007/978-3-319-04567-2,

matrix structure is,

$$H = \begin{pmatrix} a & b \\ b & c \end{pmatrix} - i \begin{pmatrix} x_1^2 & x_1 x_2 \\ x_1 x_2 & x_2^2 \end{pmatrix} \tag{E.3}$$

where a, b, c, x_1 and x_2 are independent $G(0, 2v^2)$, $G(0, v^2)$, $G(0, 2v^2)$, $G(0, \sigma^2)$ and $G(0, \sigma^2)$ variables. In the H_0 diagonal basis, with E_1^0 and E_2^0 being its eigenvalues, the structure of V will remain unaltered, i.e. $(x_1, x_2) \to (X_1, X_2)$ with X_1 and X_2 being independent $G(0, \sigma^2)$ variables. Then,

$$H = \begin{pmatrix} E_1^0 & 0 \\ 0 & E_2^0 \end{pmatrix} - i \begin{pmatrix} X_1^2 & X_1 X_2 \\ X_1 X_2 & X_2^2 \end{pmatrix}. \tag{E.4}$$

Let us define $c_1 = X_1^2$ and $c_2 = X_2^2$. Then it is easy to see that the joint probability distribution $P(E_1^0, E_2^0, c_1, c_2)$ is,

$$\begin{aligned} &P\left(E_1^0, E_2^0, c_1, c_2\right) dE_1^0 \, dE_2^0 \, dc_1 \, dc_2 \\ &\quad \propto \frac{|E_1^0 - E_2^0|}{\sqrt{c_1 c_2}} \exp -\left\{ \frac{(E_1^0)^2 + (E_2^0)^2}{4v^2} + \frac{c_1 + c_2}{2\sigma^2} \right\} dE_1^0 \, dE_2^0 \, dc_1 \, dc_2. \end{aligned} \tag{E.5}$$

Denoting the two eigenvalues of H as $\mathscr{E}_1 = E_1^R + iE_1^I$ and $\mathscr{E}_2 = E_2^R + iE_2^I$, in order to derive the joint distribution $P(E_1^R, E_2^R, E_1^I, E_2^I) \, dE_1^R \, dE_2^R \, dE_1^I \, dE_2^I$ we need the Jacobian determinant

$$\left| \frac{\partial(E_1^0, E_2^0, c_1, c_2)}{\partial(E_1^R, E_2^R, E_1^I, E_2^I)} \right|.$$

Diagonalizing the matrix given by Eq. (E.4), formulas for $\mathscr{E}_1$ and $\mathscr{E}_2$ in terms of (E_1^0, E_2^0, c_1, c_2) are easily obtained. Now equating the real and imaginary parts of the sum of the eigenvalues and similarly the sum of the squares of the eigenvalues will give,

$$\begin{aligned} X &= E_1^0 + E_2^0 = E_1^R + E_2^R, \\ Y^2 &= \left(E_1^0 - E_2^0\right)^2 = \left(E_1^R - E_2^R\right)^2 + 4E_1^I E_2^I, \\ c_1 &= -\frac{E_1^I + E_2^I}{2} - \frac{(E_1^R - E_2^R)(E_1^I - E_2^I)}{2Y}, \\ c_2 &= -\frac{E_1^I + E_2^I}{2} + \frac{(E_1^R - E_2^R)(E_1^I - E_2^I)}{2Y}. \end{aligned} \tag{E.6}$$

These relations will lead to the result,

$$\left| \frac{\partial(X, Y, c_1, c_2)}{\partial(E_1^R, E_2^R, E_1^I, E_2^I)} \right| = -\frac{2}{Y^2} \left[\left(E_1^R - E_2^R\right)^2 + \left(E_1^I - E_2^I\right)^2 \right]. \tag{E.7}$$

Combing Eqs. (E.5) and (E.7) will give $P(E_1, E_2, \Gamma_1, \Gamma_2)$. Defining the eigenvalues of H to be $\mathscr{E}_i = E_i - \frac{i}{2}\Gamma_i$, $i = 1, 2$ we have $E_i^R = E_i$ and $E_i^I = -\Gamma_i/2$. Similarly, putting $A = 1/4v^2$ and $\sigma^2 = 1/2\eta$, the final result is,

$$P(E_1, E_2, \Gamma_1, \Gamma_2)\, dE_1 dE_2 d\Gamma_1 d\Gamma_2$$
$$\propto \frac{[(E_1 - E_2)^2 + \frac{1}{4}(\Gamma_1 - \Gamma_2)^2]}{[(E_1 - E_2)^2 + \frac{1}{4}(\Gamma_1 + \Gamma_2)^2]^{1/2}} \frac{1}{(\Gamma_1 \Gamma_2)^{1/2}}$$
$$\times \exp\left[-A\left(E_1^2 + E_2^2 + \frac{\Gamma_1 \Gamma_2}{2}\right) - \frac{\eta}{2}(\Gamma_1 + \Gamma_2)\right] dE_1 dE_2 d\Gamma_1 d\Gamma_2. \tag{E.8}$$

Now, changing the variables E_1 and E_2 to $Z = E_1 + E_2$ and $s = E_1 - E_2$ and integrating over Z gives,

$$P(s, \Gamma_1, \Gamma_2)\, ds\, d\Gamma_1\, d\Gamma_2 \propto \frac{s^2 + \frac{1}{4}(\Gamma_1 - \Gamma_2)^2}{[s^2 + \frac{1}{4}(\Gamma_1 + \Gamma_2)^2]^{1/2}} \frac{1}{(\Gamma_1 \Gamma_2)^{1/2}}$$
$$\times \exp\left[-\frac{A}{2}(s^2 + \Gamma_1 \Gamma_2) - \frac{\eta}{2}(\Gamma_1 + \Gamma_2)\right] ds\, d\Gamma_1 d\Gamma_2. \tag{E.9}$$

This result agrees with Eq. (4) of [2]. Now, integrating over Γ_1 and Γ_2, we will obtain the distribution of the spacings between the real parts of the eigenvalues,

$$P(\hat{S} : \Lambda)\, d\hat{S} = \mathcal{N} d\hat{S} \exp -\frac{\hat{S}^2}{2} \int_0^\infty \left\{ \frac{dx}{\sqrt{\hat{S}^2 + x^2/4}} \left[\exp\left(-\frac{x^2}{32} - \frac{\Lambda}{2}x\right)\right]\right.$$
$$\left. \times \left[(8\hat{S}^2 + x^2) I_0\left(\frac{x^2}{32}\right) + x^2 I_1\left(\frac{x^2}{32}\right)\right]\right\}. \tag{E.10}$$

In Eq. (E.10), $\hat{S} = \sqrt{A}s$, $\Lambda = \eta/\sqrt{A}$ and I_n are modified Bessel functions of first kind. The constant $\mathcal{N}$ follows from the normalization condition $\int_0^\infty P(\hat{S} : \Lambda) d\hat{S} = 1$. Note that $\sqrt{A} \propto 1/\Delta$ where Δ is the man level spacing of the closed system (defined by H_0). Similarly the transition parameter $1/\Lambda \propto \overline{\Gamma}/\Delta$ where $\overline{\Gamma}$ is the average width. In [2] it was shown that Eq. (E.10) applies to the general situation with any N and M by treating Λ as a effective parameter and similarly extensions to GUE and GSE H_{eff} are also given for the first time. Further, Eq. (E.10) shows that level repulsion is suppressed for open systems as there is finite probability for zero spacings. Thus the real parts of the eigenvalues will be attracted due to the coupling to the environment. The non zero probability for $s \sim 0$ is clearly seen in open chaotic 2D microwave cavity experiments [2].

Turning to the more general situation with any N and $M = 1$, let us denote the real part of the N eigenvalues $E - \frac{i}{2}\Gamma$ by $\overrightarrow{E} = (E_1, E_2, \ldots, E_N)$ and similarly the imaginary parts by $\overrightarrow{\Gamma} = (\Gamma_1, \Gamma_2, \ldots, \Gamma_N)$. The joint distributions $P(\overrightarrow{E}, \overrightarrow{\Gamma})$ in the real and imaginary parts of the eigenvalues for any N and $M = 1$ has been derived explicitly in [3] for all the three GOE, GUE and GSE ensembles while for GOE it was given much earlier in [4, 5]. However, it is easy write the final result for GOE just by inspection using the 2×2 matrix result given by Eq. (E.8),

$$
\begin{aligned}
&P(\vec{E},\vec{\Gamma})\,d\vec{E}\,d\vec{\Gamma} \\
&\quad \propto \prod_{m<n} \frac{(E_m - E_n)^2 + \frac{(\Gamma_m - \Gamma_n)^2}{4}}{[(E_m - E_n)^2 + \frac{(\Gamma_m + \Gamma_n)^2}{4}]^{1/2}} \prod_n \frac{1}{(\Gamma_n)^{1/2}} \exp - F(\vec{E},\vec{\Gamma})\,d\vec{E}\,d\vec{\Gamma}; \\
&F(\vec{E},\vec{\Gamma}) = \frac{N}{a^2}\sum_n E_n^2 + \frac{N}{2a^2}\sum_{m<n} \Gamma_m \Gamma_n + \frac{1}{2\gamma}\sum_n \Gamma_n .
\end{aligned}
\tag{E.11}
$$

With $\overline{D}$ the average spacing between the real parts of the eigenvalues and $\overline{\Gamma}$ the average width, the parameters a and γ in Eq. (E.11) are $a = N\overline{D}/2$ and $\gamma = \overline{\Gamma}$. Using Eq. (E.11) and integrating all the unwanted variables one can obtain the width distribution $P(\Gamma)\,d\Gamma$ and it will be an extension of P-T law taking into account the coupling to the environment. Using mean-field approximation and saddle point integration, it was shown recently in [6] that the modified P-T law can be presented as,

$$
P(\Gamma)\,d\Gamma \propto \frac{1}{\sqrt{(N\gamma - \Gamma)(\Gamma)}} \exp\bigl[-\bigl(N/2q^2\bigr)(N\gamma - \Gamma)\Gamma\bigr] \left(\frac{\sinh \kappa k_0}{\kappa k_0}\right)^{1/2} d\Gamma .
\tag{E.12}
$$

Here, $\kappa = \pi\Gamma/2\overline{D}$ and $k_0 = 1 - [\Gamma/(\gamma N)]$. The parameter q in Eq. (E.12) determines the standard deviation of Γ. The first factor here is essentially P-T and the second factor determines the deviations from P-T (the region of interest is with $\Gamma \ll N\gamma$). It is also possible, as shown in [6], to derive Eq. (E.12) using the approach adopted in Appendix D by suitably modifying Eqs. (D.1) and (D.2). Recently, new questions on the applicability of P-T law for slow neutron resonance widths have been raised [7, 8] and the approach outlined in this appendix, taking into account the coupling to the continuum, with Eq. (E.12) explains the source of the deviations from P-T [6, 9]; see also [10].

References

1. N. Auerbach, V. Zelevinsky, Super-radiant dynamics, doorways and resonances in nuclei and other open mesoscopic systems. Rep. Prog. Phys. **74**, 106301 (2011)
2. C. Poli, G.A. Luna-Acosta, H.-J. Stöckmann, Nearest level spacing statistics in open chaotic systems: generalization of the Wigner surmise. Phys. Rev. Lett. **108**, 174101 (2012)
3. H.-J. Stöckmann, P. Šeba, The joint energy distribution function for the Hamiltonian $H = H_0 - iWW^\dagger$ for the one-channel case. J. Phys. A, Math. Gen. **31**, 3439–3449 (1998)
4. N. Ullah, On a generalized distribution of the Poles of the unitary collision matrix. J. Math. Phys. **10**, 2099–2104 (1969)
5. V.V. Sokolov, V.G. Zelevinsky, Dynamics and statistics of unstable quantum states. Nucl. Phys. A **504**, 562–588 (1989)

6. G. Shchedrin, V. Zelevinsky, Resonance width distribution for open quantum systems. Phys. Rev. C **86**, 044602 (2012)
7. P.E. Koehler, F. Becvar, M. Krticka, J.A. Harvey, K.H. Guber, Anomalous fluctuations of s-wave reduced neutron widths of $^{192;194}$Pt resonances. Phys. Rev. Lett. **105**, 075502 (2010)
8. E.S. Reich, Nuclear theory nudged. Nature (London) **466**, 1034 (2010)
9. G.L. Celardo, N. Auerbach, F.M. Izrailev, V.G. Zelevinsky, Distribution of resonance widths and dynamics of continuum coupling. Phys. Rev. Lett. **106**, 042501 (2011)
10. H.A. Weidenmüller, Distribution of partial neutron widths for nuclei close to a maximum of the neutron strength function. Phys. Rev. Lett. **105**, 232501 (2010)

Appendix F
Pairing Symmetries in BEGOE(1 + 2)-F and BEGOE(1 + 2)-$S1$

Here, some results for the pairing algebra in (Ω, m, F) spaces of BEGOE(1 + 2)-F and for the pairing algebras in the (Ω, m, S) spaces of BEGOE(1 + 2)-$S1$ are presented. It is important to mention that two different pairing algebras are possible for BEGOE(1 + 2)-$S1$.

F.1 Pairing Algebra in BEGOE(1 + 2)-F Space

In BEGOE(2)-F, F-spin is a good symmetry and then following the $SO(5)$ pairing algebra for fermions [1], it is possible to consider pairs that are vectors in spin space. The pair creation operators $P_{i:\mu}$ for the level i and the generalized pair creation operators (over the Ω levels) P_μ, with $\mu = -1, 0, 1$, in F-spin coupled representation, are

$$P_\mu = \frac{1}{\sqrt{2}} \sum_i \left(b_i^\dagger b_i^\dagger\right)^1_\mu = \sum_i P_{i:\mu}, \qquad (P_\mu)^\dagger = \frac{1}{\sqrt{2}} \sum_i (-1)^{1-\mu} (\tilde{b}_i \tilde{b}_i)^1_{-\mu}. \tag{F.1}$$

Therefore, in the space defining BEGOE(1 + 2)-F, the pairing operator or the pairing Hamiltonian H_p and its two-particle matrix elements are,

$$H_p = \sum_\mu P_\mu (P_\mu)^\dagger, \qquad \left\langle (k\ell) f \middle| H_p \middle| (ij) f \right\rangle = \delta_{f,1} \delta_{i,j} \delta_{k,\ell}. \tag{F.2}$$

With this, we will proceed to identify and analyze the pairing algebra. In Sect. 10.1, the $U(2\Omega) \supset U(\Omega) \otimes SU(2)$ algebra was discussed in detail. Following this, it is easy to verify that the $\Omega(\Omega - 1)/2$ number of operators $C_{ij} = A^0_{ij} - A^0_{ji}$, $i > j$ generate a $SO(\Omega)$ subalgebra of the $U(\Omega)$ algebra. Therefore we have $U(2\Omega) \supset [U(\Omega) \supset SO(\Omega)] \otimes SU(2)$. The operators A_{ij} are defined by Eq. (10.8). The irreps of $SO(\Omega)$ algebra are uniquely labeled by the seniority quantum number v and a reduced spin $\tilde{f}$ similar to the reduced isospin introduced in the context of nuclear

V.K.B. Kota, *Embedded Random Matrix Ensembles in Quantum Physics*, Lecture Notes in Physics 884, DOI 10.1007/978-3-319-04567-2,

shell model and they in turn define the eigenvalues of H_p. To see this, first let us consider the quadratic Casimir operator of $SO(\Omega)$,

$$C_2\big[SO(\Omega)\big] = 2\sum_{i>j} C_{ij} \cdot C_{ji}. \tag{F.3}$$

Carrying out angular momentum algebra gives,

$$C_2\big[SO(\Omega)\big] = C_2\big[U(\Omega)\big] - 2H_p - \hat{n}. \tag{F.4}$$

The quadratic Casimir operator of the $U(\Omega)$ algebra is given by Eqs. (10.10) and its eigenvalues in (m, F) spaces by Eq. (10.11). Before giving the result for the eigenvalues of H_p, we need the irreps of $SO(\Omega)$ given the two-rowed $U(\Omega)$ irreps $\{m_1, m_2\}$; $m_1 + m_2 = m$, $m_1 - m_2 = 2F$. Firstly, it is clear that the $SO(\Omega)$ irreps should be of $[v_1, v_2]$ type and for later simplicity we use $v_1 + v_2 = v$ and $v_1 - v_2 = 2\tilde{f}$. The quantum number v is called seniority and $\tilde{f}$ is called reduced F-spin. The $SO(\Omega)$ irreps for a given $\{m_1, m_2\}$ can be obtained by first expanding the $U(\Omega)$ irrep $\{m_1, m_2\}$ in terms of totally symmetric irreps,

$$\{m_1, m_2\} = \{m_1\} \times \{m_2\} - \{m_1 + 1\} \times \{m_2 - 1\}. \tag{F.5}$$

Note that the irrep multiplication in (F.5) is a Kronecker multiplication [2, 3]. Second result to be used is the reduction of a totally symmetric $U(\Omega)$ irrep $\{m'\}$ to the $SO(\Omega)$ irreps and this is given by the well-known result

$$\big\{m'\big\} \to [v] = \big[m'\big] \oplus \big[m' - 2\big] \oplus \cdots \oplus [0] \text{ or } [1]. \tag{F.6}$$

Third result to be used is the reduction of the Kronecker product of two symmetric $SO(\Omega)$ irreps $[v_1]$ and $[v_2]$, $\Omega > 3$ into $SO(\Omega)$ irreps $[v_1, v_2]$ and this is given by (for $v_1 \geq v_2$) [2, 3],

$$[v_1] \times [v_2] = \sum_{k=0}^{v_2} \sum_{r=0}^{v_2-k} [v_1 - v_2 + k + 2r, k] \oplus . \tag{F.7}$$

Combining Eqs. (F.5), (F.6) and (F.7) will give the $\{m_1, m_2\} \to [v_1, v_2]$ reductions. It is easy to implement this procedure on a computer.

Given the space defined by $|\{m_1, m_2\}, [v_1, v_2], \alpha\rangle$, with α denoting extra labels needed for a complete specification of the state, the eigenvalues of $C_2[SO(\Omega)]$ are [2]

$$\big\langle C_2\big[SO(\Omega)\big]\big\rangle^{\{m_1,m_2\},[v_1,v_2]} = v_1(v_1 + \Omega - 2) + v_2(v_2 + \Omega - 4). \tag{F.8}$$

Equations (F.4), (F.8) and (10.11) will give the formula for the eigenvalues of the pairing Hamiltonian H_p. Changing $\{m_1, m_2\}$ to (m, F) and $[v_1, v_2]$ to $(v, \tilde{f})$, H_p eigenvalues are given by

$$\begin{aligned} E_p(m, F, v, \tilde{f}) = \langle H_p\rangle^{m,F,v,\tilde{f}} &= \frac{1}{4}(m - v)(2\Omega - 6 + m + v) \\ &\quad + \big[F(F + 1) - \tilde{f}(\tilde{f} + 1)\big]. \end{aligned} \tag{F.9}$$

This is same as the result that follows from Eq. (18) of [1] for fermions by using $\Omega \to -\Omega$ symmetry. From now on, we denote the $U(\Omega)$ irreps by (m, F) and $SO(\Omega)$ irreps by $(v, \tilde{f})$. In Table F.1, for $(\Omega, m) = (4, 10)$ and $(5, 8)$ systems, given are the $(m, F) \to (v, \tilde{f})$ reductions, the pairing eigenvalues given by (F.9) in the spaces defined by these irreps and also the dimensions of the $U(\Omega)$ and $SO(\Omega)$ irreps. The dimensions $d_b(\Omega, m, F)$ of the $U(\Omega)$ irreps (m, F) are given by (10.4). Similarly, the dimensions $\mathbf{d}(v_1, v_2) \Leftrightarrow \mathbf{d}(v, \tilde{f})$ of the $SO(\Omega)$ irreps $[v_1, v_2]$ follow from Eqs. (F.6) and (F.7) and then,

$$\begin{aligned} \mathbf{d}(v_1, v_2) &= \mathbf{d}(v_1)\mathbf{d}(v_2) - \sum_{k=0}^{v_2-1} \sum_{r=0}^{v_2-k} \mathbf{d}(v_1 - v_2 + k + 2r, k); \\ \mathbf{d}(v) &= \binom{\Omega + v - 1}{v} - \binom{\Omega + v - 3}{v - 2}. \end{aligned} \tag{F.10}$$

Note that in general the $SO(\Omega)$ irreps $(v, \tilde{f})$ can appear more than once in the reduction of $U(\Omega)$ irreps (m, F). For example, (2, 1) irrep of $SO(\Omega)$ appears twice in the reduction of the $U(\Omega)$ irrep (10, 1).

F.2 Pairing Algebras in BEGOE(1 + 2)-$S1$ Spaces

In the BEGOE(1 + 2)-$S1$ space it is possible to identify two different pairing algebras (each defining a particular type of pairing) and they follow from the results in [4, 5, 6]. One of them corresponds to the $SO(\Omega)$ algebra in $U(3\Omega) \supset [U(\Omega) \supset SO(\Omega)] \otimes [SU(3) \supset SO(3)]$ and we refer to this as $SO(\Omega) - SU(3)$ pairing. The other corresponds to the $SO(3\Omega)$ in $U(3\Omega) \supset SO(3\Omega) \supset SO(\Omega) \otimes SO(3)$. Note that, both the algebras have the $SO(3)$ subalgebra that generates the spin S quantum number. Here below, we will give some details of these pairing algebras. Inclusion of pairing terms in the BEGOE(1 + 2)-$S1$ Hamiltonian will alter the structure of the ground states.

F.2.1 $SO(\Omega)$–$SU(3)$ Pairing

Following the results given in [4, 5, 6], it is easy to identify the $\Omega(\Omega - 1)/2$ number of generators $U(i, j)$, $i < j$ of $SO(\Omega)$ in $U(3\Omega) \supset [U(\Omega) \supset SO(\Omega)] \otimes [SU(3) \supset SO(3)]$,

$$\begin{aligned} &U(i, j) = \sqrt{\alpha(i, j)}\big[g(i, j) + \alpha(i, j)g(j, i)\big], \quad i < j; \\ &\big|\alpha(i, j)\big|^2 = 1, \qquad \alpha(i, j) = \alpha(j, i), \qquad \alpha(i, j)\alpha(j, k) = -\alpha(i, k). \end{aligned} \tag{F.11}$$

Table F.1 Classification of states in the $U(2\Omega) \supset [U(\Omega) \supset SO(\Omega)] \otimes SU_F(2)$ limit for $(\Omega, m) = (4, 10)$ with $F \leq 2$ and $(\Omega, m) = (5, 8)$ with $F \leq 3$. Given are $U(\Omega)$ labels (m, F) and $SO(\Omega)$ labels $(v, \tilde{f})$ with the corresponding dimensions $d(\Omega, m, F)$ and $\mathbf{d}(v, \tilde{f})$ respectively and also the pairing eigenvalues $E_p(m, F, v, \tilde{f})$. Note that $\sum_{v,\tilde{f}} r\mathbf{d}(v, \tilde{f}) = d(\Omega, m, F)$; here r denotes multiplicity of the $SO(\Omega)$ irreps and in the table, they are shown only for the cases when $r > 1$

Ω	m	$(m, F)_{d(\Omega,m,F)}$	$(v, \tilde{f})^r_{\mathbf{d}(v,\tilde{f})}$	$E_p(m, F, v, \tilde{f})$	Ω	m	$(m, F)_{d(\Omega,m,F)}$	$(v, \tilde{f})^r_{\mathbf{d}(v,\tilde{f})}$	$E_p(m, F, v, \tilde{f})$
4	10	$(10, 0)_{196}$	$(2, 0)_{6}$	28	5	8	$(8, 0)_{490}$	$(0, 0)_{1}$	24
			$(4, 1)_{30}$	22				$(2, 1)_{14}$	19
			$(6, 2)_{70}$	12				$(4, 2)_{55}$	10
			$(6, 0)_{14}$	18				$(4, 0)_{35}$	16
			$(8, 1)_{54}$	8				$(6, 1)_{220}$	7
			$(10, 0)_{22}$	0				$(8, 0)_{165}$	0
		$(10, 1)_{540}$	$(2, 1)^2_{9}$	28			$(8, 1)_{1260}$	$(2, 1)_{14}$	21
			$(4, 2)^2_{25}$	20				$(4, 2)_{55}$	12
			$(6, 3)_{49}$	8				$(4, 1)^2_{81}$	16
			$(4, 1)_{30}$	24				$(6, 2)_{260}$	5
			$(6, 2)_{70}$	14				$(6, 1)_{220}$	9
			$(6, 1)^2_{42}$	18				$(8, 1)_{455}$	0
			$(8, 2)_{90}$	6				$(2, 0)_{10}$	23
			$(8, 1)_{54}$	10				$(6, 0)_{84}$	11
			$(10, 1)_{66}$	0			$(8, 2)_{1500}$	$(4, 2)^2_{55}$	16
			$(0, 0)_{1}$	32				$(6, 3)_{140}$	3
			$(4, 0)_{10}$	26				$(6, 2)_{260}$	9
			$(8, 0)_{18}$	12				$(8, 2)_{625}$	0
		$(10, 2)_{750}$	$(4, 2)^2_{25}$	24				$(2, 1)^2_{14}$	25
			$(6, 3)_{49}$	12				$(4, 1)_{81}$	20
			$(6, 2)^2_{70}$	18				$(6, 1)_{220}$	13
			$(8, 3)_{126}$	4				$(0, 0)_{1}$	30
			$(8, 2)_{90}$	10				$(4, 0)_{35}$	22
			$(10, 2)_{110}$	0			$(8, 3)_{1155}$	$(6, 3)_{140}$	9
			$(2, 1)_{9}$	32				$(8, 3)_{595}$	0
			$(4, 1)^2_{30}$	28				$(4, 2)_{55}$	22
			$(6, 1)_{42}$	22				$(6, 2)_{260}$	15
			$(8, 1)_{54}$	14				$(2, 1)_{14}$	31
			$(2, 0)_{6}$	34				$(4, 1)_{81}$	26
			$(6, 0)_{14}$	24				$(2, 0)_{10}$	33

Note that $g(i, j)$ are defined in Eq. (10.24). The quadratic Casimir invariant of $SO(\Omega)$ is,

$$\hat{C}_2\big(SO(\Omega)\big) = \sum_{i<j} U(i, j) \cdot U(j, i). \tag{F.12}$$

Applying Eq. (F.11) now gives,

$$\begin{aligned}
\hat{C}_2\big(SO(\Omega)\big) &= \sum_{i<j} \alpha(i, j)\big[g(i, j) \cdot g(i, j) + g(j, i) \cdot g(j, i) \\
&\quad + 2\alpha(i, j) g(i, j) \cdot g(j, i)\big] \\
&= \sum_{i \neq j} g(i, j) \cdot g(j, i) + \sum_{i \neq j} \alpha(i, j) g(i, j) \cdot g(i, j) \\
&= \hat{C}_2\big(U(\Omega)\big) - \sum_{i,j} \beta_i \beta_j g(i, j) \cdot g(i, j); \\
\beta_i \beta_j &= -\alpha(i, j), \quad \text{for } i \neq j,\ |\beta_i|^2 = 1.
\end{aligned} \tag{F.13}$$

Here we have introduced β_i's and the $\alpha(i, j)$ are defined in Eq. (F.11). Now defining the pairing operator $\mathscr{P}^k_q$, $k = 0, 2$ as

$$\mathscr{P}^k_q = \sum_i \beta_i \left(b^\dagger_{i;1} b^\dagger_{i;1}\right)^k_q; \quad k = 0, 2 \tag{F.14}$$

it is easy to see that,

$$\begin{aligned}
H_{\mathscr{P}} = \sum_{k=0,2;q} \mathscr{P}^k_q \big(\mathscr{P}^k_q\big)^\dagger &= \hat{C}_2\big(U(\Omega)\big) - \hat{C}_2\big(SO(\Omega)\big) - \hat{n} \\
&= \frac{2}{3}\hat{C}_2\big(SU(3)\big) - \hat{C}_2\big(SO(\Omega)\big) - (\Omega - 4)\hat{n} + \frac{\hat{n}^2}{3}.
\end{aligned} \tag{F.15}$$

In the final form above we have used Eqs. (10.27). Thus, the pairing Hamiltonian in the $U(3\Omega) \supset [U(\Omega) \supset SO(\Omega)] \otimes [SU(3) \supset SO(3)]$ algebra is a sum of $k = 0$ and 2 pairs and it is simply related to the $SO(\Omega)$ and $SU(3)$ algebras. The two-particle matrix elements of $H_{\mathscr{P}}$ are $V^{s=0}_{iijj} = 1$, $V^{s=2}_{iijj} = 1$ and all other matrix elements are zero.

With the $U(3\Omega) \supset SU(\Omega) \otimes SU(3)$ structure, the irreps of $U(\Omega)$ will be three rowed in Young tableaux notation. We can write the irreps as $\{f_1, f_2, f_3\}$ with $f_1 \geq f_2 \geq f_3 \geq 0$ and $f_1 + f_2 + f_3 = m$. As $U(\Omega)$ irreps are three rowed, the $SO(\Omega)$ irreps will be of maximum three rows and we can write them as $[v_1, v_2, v_3]$. For a given m, it is possible to enumerate the irreps $[v_1, v_2, v_3]$ of $SO(\Omega)$ given a $U(\Omega)$ irrep $\{f_1, f_2, f_3\}$ [or equivalently $SU(3)$ irrep $(\lambda = f_1 - f_2, \mu = f_2 - f_3)$]. See [6] and references therein for detailed results. Let us add that the eigenvalues of $\hat{C}_2(SO(\Omega))$ over a given $[v_1, v_2, v_3]$ space are given by [6],

$$\big\langle \hat{C}_2\big(SO(\Omega)\big)\big\rangle^{[v_1,v_2,v_3]} = v_1(v_1 + \Omega - 2) + v_2(v_2 + \Omega - 4) + v_3(v_3 + \Omega - 6). \tag{F.16}$$

Finally, in terms of the irrep labels $[v_1, v_2, v_3]$, m and the $SU(3)$ labels $(\lambda\mu)$, eigenvalues for $H_{\mathscr{P}}$ will follow from Eq. (F.15).

F.2.2 $SO(3\Omega)$ Pairing

Second pairing algebra in BEGOE(2)-$S1$ space follows from the recognition that $U(3\Omega)$ admits $SO(3\Omega)$ subalgebra and as we will see ahead, the pairing here is generated by $k = 0$ pairs $b_i^\dagger \cdot b_i^\dagger$ only. Following the results in [5], the generators of $SO(3\Omega)$ are easy to identify and they are,

$$
\begin{aligned}
& u_q^{k=1}(i,i); \quad i = 1, 2, \ldots, \Omega, \\
& V_q^k(i,j) = \sqrt{(-1)^k \alpha(i,j)}\big[u_q^k(i,j) + \alpha(i,j)(-1)^k u_q^k(j,i)\big], \quad i < j; \qquad \text{(F.17)} \\
& \big|\alpha(i,j)\big|^2 = 1, \qquad \alpha(i,j) = \alpha(j,i), \qquad \alpha(i,j)\alpha(j,k) = -\alpha(i,k).
\end{aligned}
$$

The operators u_q^k are defined by Eq. (10.19). Carrying out angular momentum algebra, the following relation between the quadratic Casimir invariants $\hat{C}_2(SO(3\Omega))$ and $\hat{C}_2(U(3\Omega))$ can be established using Eqs. (F.17) and (10.20),

$$
\begin{aligned}
\hat{C}_2\big(SO(3\Omega)\big) &= 2\sum_i u^1(i,i) \cdot u^1(i,i) + \sum_{i<j;k} V^k(i,j) \cdot V^k(i,j) \\
&= \hat{C}_2\big(U(3\Omega)\big) - \sum_{i,k} (-1)^k u^k(i,i) \cdot u^k(i,i) \\
&\quad + \sum_{i \neq j;k} (-1)^k \alpha(i,j) u^k(i,j) \cdot u^k(i,j). \qquad \text{(F.18)}
\end{aligned}
$$

Introducing the pairing operator P_+,

$$
P_+ = \sum_i \gamma_i P_+(i) = \frac{1}{2}\sum_i \gamma_i b_{i;1}^\dagger \cdot b_{i;1}^\dagger, \qquad P_- = (P_+)^\dagger \qquad \text{(F.19)}
$$

we can prove the following relationship between $\hat{C}_2(SO(3\Omega))$ and the pairing Hamiltonian $H_P = 4P_+P_-$,

$$
\begin{aligned}
4H_P = 4P_+P_- &= -\hat{n} + \hat{C}_2\big(U(3\Omega)\big) - \hat{C}_2\big(SO(3\Omega)\big) \\
&= \hat{n}(\hat{n} + 3\Omega - 2) - \hat{C}_2\big(SO(3\Omega)\big); \qquad \text{(F.20)} \\
\gamma_i\gamma_j &= -\alpha(i,j), \quad \text{for } i \neq j, \qquad |\gamma_i|^2 = 1.
\end{aligned}
$$

The $\gamma \leftrightarrow \alpha$ relation is needed for the correspondence between H_P and $\hat{C}_2(SO(3\Omega))$. Important point now being that the three operators P_+, P_- and $P_0 = (\Omega + \hat{n})/2$ will form a $SU(1,1)$ algebra complimentary to $SO(3\Omega)$. Thus the $SO(3\Omega)$ pairing is much simpler. With $U(3\Omega)$ irreps being $\{m\}$, the $SO(3\Omega)$ irreps are labeled by the seniority quantum number ω where,

$$\omega = m, m-2, \ldots, 0 \text{ or } 1 \tag{F.21}$$

and H_P eigenvalues are

$$\langle H_P \rangle^{m,\omega} = \frac{1}{4}(m-\omega)(m+\omega+3\Omega-2). \tag{F.22}$$

The two particle matrix elements of H_P are simply $V^{s=0}_{iijj} = 1$ and all other matrix elements are zero. With two different pairings in the BEGOE(1+2)-$S1$ space, analysis of spin one boson systems, for ground state spin structure and pair expectation values, with the following extended Hamiltonian H_{ext} will be interesting and useful,

$$\begin{aligned}\{\widehat{H}_{ext}\} = {} & \widehat{h}(1) + \lambda_0\{\widehat{V}^{s=0}(2)\} + \lambda_1\{\widehat{V}^{s=1}(2)\} + \lambda_2\{\widehat{V}^{s=2}(2)\} \\ & + \lambda_{p1} H_{\mathscr{P}} + \lambda_{p2} H_P + \lambda_{S_1} \hat{C}_2\big(SU(3)\big) + \lambda_{S_2} \hat{S}^2.\end{aligned} \tag{F.23}$$

Finally, just as the $SU(2)$ algebra for pairing in EGOE(1 + 2)-**s** (see Chap. 6), there will be simpler complimentary algebras [7] that correspond to the $SO(\Omega)$ pairing algebra of BEGOE(1 + 2)-F and the $SO(\Omega)$–$SU(3)$ pairing algebra of BEGOE(1+2)-$S1$ [as stated above, the algebra complementary to $SO(3\Omega)$ pairing in BEGOE(1 + 2)-$S1$ is $SU(1,1)$]. It will be interesting to explore these complimentary algebras in future.

References

1. B.H. Flowers, S. Szpikowski, A generalized quasi-spin formalism. Proc. Phys. Soc. **84**, 193–199 (1964)
2. V.K.B. Kota, Group theoretical and statistical properties of interacting boson models of atomic nuclei: recent developments, in *Focus on Boson Research*, ed. by A.V. Ling (Nova Science Publishers Inc., New York, 2006), pp. 57–105
3. B.G. Wybourne, *Symmetry Principles and Atomic Spectroscopy* (Wiley, New York, 1970)
4. V.K.B. Kota, Spectra and E2 transition strengths for $N = Z$ even-even nuclei in IBM-3 dynamical symmetry limits with good s and d boson isospins. Ann. Phys. (N.Y.) **265**, 101–133 (1998)
5. V.K.B. Kota, $O(36)$ symmetry limit of IBM-4 with good s, d and sd boson spin-isospin Wigner's $SU(4) \sim O(6)$ symmetries for $N \approx Z$ odd-odd nuclei. Ann. Phys. (N.Y.) **280**, 1–34 (2000)
6. V.K.B. Kota, J.A. Castilho Alcarás, Classification of states in $SO(8)$ proton-neutron pairing model. Nucl. Phys. A **764**, 181–204 (2006)
7. D.J. Rowe, M.J. Carvalho, J. Repka, Dual pairing of symmetry and dynamical groups in physics. Rev. Mod. Phys. **84**, 711–757 (2012)

Appendix G
Embedded GOE for Spinless Fermion Systems with Three-Body Interactions

For m spinless fermions occupying N sp states $|\nu_i\rangle$ and interacting via three-body forces, extending the formulation in Sect. 4.1, the three-body Hamiltonian operator is,

$$\widehat{H}(3) = \sum_{\nu_i<\nu_j<\nu_k;\,\nu_p<\nu_q<\nu_r} \langle \nu_p\nu_q\nu_r|\widehat{H}(3)|\nu_i\nu_j\nu_k\rangle a^\dagger_{\nu_p} a^\dagger_{\nu_q} a^\dagger_{\nu_r} a_{\nu_k} a_{\nu_j} a_{\nu_i}. \tag{G.1}$$

In Eq. (G.1), $\langle \nu_p\nu_q\nu_r|\widehat{H}(3)|\nu_i\nu_j\nu_k\rangle$ are antisymmetrized three particle matrix elements and symmetries of these matrix elements under the interchange of the indices i, j, k, p, q and r are easy to write down. Hamiltonian matrix $H(m)$ in the m-particle basis $|\nu_1\nu_2\cdots\nu_m\rangle$ contains four different types of non-zero matrix elements (all other matrix elements are zero due to three-body selection rules). Note that in the m particle basis states, the ν_i will be ordered and all ν_i will be different as we have spinless fermions. Explicit formulas for the four classes of non-zero matrix elements are,

$$\begin{aligned}
\langle \nu_1\nu_2\cdots\nu_m|\widehat{H}|\nu_1\nu_2\cdots\nu_m\rangle &= \sum_{\nu_i<\nu_j<\nu_k\le\nu_m} \langle \nu_i\nu_j\nu_k|\widehat{H}(3)|\nu_i\nu_j\nu_k\rangle,\\
\langle \nu_p\nu_2\nu_3\cdots\nu_m|\widehat{H}(3)|\nu_q\nu_2\cdots\nu_m\rangle &= \sum_{\nu_i<\nu_j;\,i\ge 2} \langle \nu_p\nu_i\nu_j|\widehat{H}(3)|\nu_q\nu_i\nu_j\rangle,\\
\langle \nu_p\nu_q\nu_3\cdots\nu_m|\widehat{H}(3)|\nu_r\nu_s\nu_3\cdots\nu_m\rangle &= \sum_{\nu_i;\,i\ge 3} \langle \nu_p\nu_q\nu_i|\widehat{H}(3)|\nu_r\nu_s\nu_i\rangle,\\
\langle \nu_p\nu_q\nu_r\nu_4\cdots\nu_m|\widehat{H}(3)|\nu_s\nu_t\nu_u\nu_4\cdots\nu_m\rangle &= \langle \nu_p\nu_q\nu_r|\widehat{H}(3)|\nu_s\nu_t\nu_u\rangle.
\end{aligned} \tag{G.2}$$

In Eq. (G.2), $p \neq q$ in the second equation, $p \neq q \neq r \neq s$ in the third equation and $p \neq q \neq r \neq s \neq t \neq u$ in the fourth equation. The EGOE(3) ensemble is defined by Eqs. (G.2) with GOE(1) representation for $H(3)$ in the three-particle spaces, i.e. $\langle \nu_p\nu_q\nu_r|\widehat{H}(3)|\nu_i\nu_j\nu_k\rangle$ with $\nu_i < \nu_j < \nu_k$ and $\nu_p < \nu_q < \nu_r$ being independent Gaussian variables with zero center and variance unity (variance is 2 for the diago-

V.K.B. Kota, *Embedded Random Matrix Ensembles in Quantum Physics*, Lecture Notes in Physics 884, DOI 10.1007/978-3-319-04567-2,

nal matrix elements). Using Eq. (G.2), numerical construction of EGOE(3) is easy. Adding one and two-body parts will give EGOE$(1+2+3)$.

Let us consider EGOE$(1+3)$, i.e. EGOE for spinless fermion systems generated by random three-body interactions in the presence of a mean-field one-body part $h(1)$. Then $H=h(1)+\lambda_3 V(3)$ where $V(3)$ is defined by Eq. (G.2) and say, $h(1)$ is defined a sp energies ε_i, $i=1,2,\ldots,N$ with average spacing $\Delta=1$. As λ_3 (in units of Δ) increases, the m fermion system exhibits Poisson to GOE transition in energy level fluctuations. Now we can apply the AJS criterion of Sect. 5.3.1. Then, the transition point $\lambda_{3:c}$ is determined by the spacing between states directly coupled by $V(3)$. Given a typical m particle state, the energy span (Δ_c) of the states directly connected by $V(3)$ can be estimated by putting 3 fermion at the bottom of the sp spectrum and similarly, at the top of the sp spectrum. This gives $\Delta_c \propto N$. Similarly the number (K) of states connected by $V(3)$, following Eq. (G.2), is given by $K=1+m(N-m)+[m(m-1)(N-m)(N-m-1)/4]+[m(m-1)(m-2)(N-m)(N-m-1)(N-m-2)/36] \propto m^3N^3$. Thus, in the dilute limit we have $\lambda_{3:c} \propto 1/(m^3N^2)$. Comparing with $\lambda_c \propto 1/(m^2N)$ given by Eq. (5.17), it is seen that EGOE$(1+3)$ generates, as the interaction strength λ increases, Poisson to GOE transition much faster (by a factor mN) than EGOE$(1+2)$.

Another interesting and possibly useful aspect is that it is possible to write down propagation equations for the energy centroids and spectral variances generated by $\widehat{H}(3)=V(3)$. These can be used to study centroid and variance fluctuations in EGOE(3). Denoting the three particle matrix elements in Eq. (G.1) by $V_{pqr,ijk}$, formula for energy centroids is,

$$E_c(m)=\big\langle\widehat{H}(3)\big\rangle^m=\binom{m}{3}V_c; \qquad V_c=\binom{N}{3}^{-1}\sum_{i<j<p} V_{ijp,ijp}. \tag{G.3}$$

For spectral variances, we need the decomposition of $\widehat{H}(3)$ with respect to $U(N)$ and it will have $\nu=0$, 1, 2 and 3 parts. Note that $E_c(m)$ in Eq. (G.3) is generated by the $\nu=0$ part and $H^{\nu=0}(3)=\binom{\hat{n}}{3}V_c$. Effective one-body matrix elements λ_{ij} and effective two-body matrix elements w_{ijrs} generated by $\widehat{H}(3)$ are defined by

$$\begin{aligned} \lambda_{ij} &= \binom{N-2}{2}^{-1}\sum_{q<r} V_{iqr,jqr}, \\ w_{ijrs} &= (N-4)^{-1}\sum_{q} V_{ijq,rsq}. \end{aligned} \tag{G.4}$$

Now, the $\nu=1$ part of $\widehat{H}(3)$ is

$$H^{\nu=1}(3)=\binom{m-1}{2}\tilde{\lambda}_{ij}a^\dagger a_j \tag{G.5}$$

where the traceless one-body matrix elements $\tilde{\lambda}_{ij}$ are defined by

$$\tilde{\lambda}_{ij} = \lambda_{ij} - \delta_{ij} \frac{1}{N} \sum_i \lambda_{ii} \,. \tag{G.6}$$

Similarly, the $\nu = 2$ part $H^{\nu=2}(3)$ is defined by the traceless two-body matrix elements $\tilde{w}_{ijrs}$ where,

$$\begin{aligned} \tilde{w}_{ijrs} = w_{ijrs} &- \frac{N-3}{N-4} (\tilde{\lambda}_{ir}\delta_{js} + \tilde{\lambda}_{js}\delta_{ir} - \tilde{\lambda}_{is}\delta_{jr} - \tilde{\lambda}_{jr}\delta_{is}) \\ &- \frac{N-2}{N-4} V_c (\delta_{ir}\delta_{js} - \delta_{is}\delta jr). \end{aligned} \tag{G.7}$$

Subtraction of $\nu = 0$, $\nu = 1$ and $\nu = 2$ parts from $\widehat{H}(3)$ gives $H^{\nu=3}(3)$. Let us add that Eqs. (G.3)–(G.7) are reported first in [1] and Ref. [2] gives the general formulation for the unitary decomposition; see also Eq. (4.18). Now, propagation equation for the spectral variances is [1, 2],

$$\begin{aligned} \langle [H(3)]^2 \rangle^m = \left[\binom{m}{3} \right]^2 (V_c)^2 + P(1, m) \left[\binom{m-1}{2} \right]^2 \sum_{i,j} [\tilde{\lambda}_{ij}]^2 \\ + P(2, m)(m-2)^2 \sum_{i<j, p<q} [\tilde{w}_{ijpq}]^2 + P(3, m) \langle [H^{\nu=3}(3)]^2 \rangle^3; \\ P(\nu, m) = \frac{\binom{m}{\nu}\binom{N-m}{\nu}}{\binom{N}{\nu}\binom{N-\nu}{\nu}}. \end{aligned} \tag{G.8}$$

The only unknown in Eq. (G.8) is $\langle [H^{\nu=3}(3)]^2 \rangle^3$ and this follows easily from by combining this equation with $\langle [H(3)]^2 \rangle^3$. Note that $\langle [H(3)]^2 \rangle^3$ is nothing but the sum of the squares of all the three particle matrix elements $V_{pqr,ijk}$, $i > j > k$ and $p > q > r$. It is possible to combine the results in Chap. 5 with Eqs. (G.3)–(G.8) and obtain $\langle \widehat{H}(1+2+3) \rangle^m$ and $\langle [\widehat{H}(1+2+3)]^2 \rangle^m$ for each member of the more general EGOE(1 + 2 + 3). This can be used to study ensemble averaged spectral variances as a function of (m, N) without H matrix construction and similarly, analyze the structure of fluctuations in energy centroids and spectral variances. Finally, just as EGOE(1 + 2 + 3), it is possible to define and analyze BEGOE(1 + 2 + 3) for boson systems with three-body forces.

References

1. K.D. Launey, T. Dytrych, J.P.D. Draayer, Similarity renormalization group and many-body effects in multiparticle systems. Phys. Rev. C **85**, 044003 (2012)
2. F.S. Chang, J.B. French, T.H. Thio, Distribution methods for nuclear energies, level densities and excitation strengths. Ann. Phys. (N.Y.) **66**, 137–188 (1971)

Appendix H
Bosonic Embedded GOE Ensemble for (s, p) Boson Systems

Given a system of interacting bosons carrying angular momentum $\ell^\pi = 0^+$ (s bosons) and $\ell^\pi = 1^-$ (p bosons) degrees of freedom with the Hamiltonian preserving total angular momentum L of the bosons, it is possible to define and construct easily BEGOE(1 + 2)-($s, p : L$). Firstly, let us say that the s and p boson creation and annihilation operators are $(s^\dagger, p^\dagger_\mu)$ and (s, p_μ) respectively with $\mu = 1, 0, -1$. Then, two particle states are $|m_s = 2, L = 0, M = 0\rangle = \frac{1}{\sqrt{2}}(s^\dagger s^\dagger)|0\rangle$, $|m_p = 2, L, M\rangle = \pm\frac{1}{\sqrt{2}}(p^\dagger p^\dagger)^L_M|0\rangle$ and $|m_s = 1, m_p = 1, L = 1, M\rangle = s^\dagger p^\dagger_M|0\rangle$. Note that m_s is number of s bosons, m_p is number of p bosons and then the total number of bosons $m = m_s + m_p$. For the $m_p = 2$ state, to be consistent with the results in [1], we choose $-$ve sign. It is useful to introduce $\tilde{p}^1_\mu = (-1)^{1+\mu} p_{-\mu}$ and then $[(p^\dagger p^\dagger)^k_r]^\dagger = (-1)^r (\tilde{p}\tilde{p})^k_{-r}$. Also, the p-boson number operator $\hat{n}_p = \sqrt{3}(p^\dagger \tilde{p})^0$ and the angular momentum operator $L^1_q = \sqrt{2}(p^\dagger \tilde{p})^1_q$. With all these, a general one plus two-body Hamiltonian preserving m and L can be written as,

$$\begin{aligned} H = {} & \varepsilon_s \hat{n}_s + \varepsilon_p \hat{n}_p + V^0_{ssss}\frac{1}{2}s^\dagger s^\dagger ss + V^1_{spsp}\sum_\mu s^\dagger p^\dagger_\mu \left(s^\dagger p^\dagger_\mu\right)^\dagger \\ & + \sum_{k=0,2} V^k_{pppp}\frac{1}{2}\left\{\sum_\mu (p^\dagger p^\dagger)^k_\mu \left[(p^\dagger p^\dagger)^k_\mu\right]^\dagger\right\} \\ & + V^0_{sspp}\left(-\frac{1}{2}\right)\left\{(s^\dagger s^\dagger)\left[(p^\dagger p^\dagger)^0\right]^\dagger + h.c.\right\}. \end{aligned} \tag{H.1}$$

A simple many particle basis that can be used for H matrix construction is defined by $|m, m_p, L, M\rangle$ where $m_s = m - m_p$ and $m_p = L, L+2, \dots, N$ or $N-1$. Parity of this state is $\pi = (-1)^L$. From now on, the M quantum number ($-L \le M \le L$) will be dropped as H matrix elements are independent of M. The H matrix dimension for a given (m, L) is

$$d(m, L) = \left[\frac{m - L}{2}\right] + 1 \tag{H.2}$$

V.K.B. Kota, *Embedded Random Matrix Ensembles in Quantum Physics*, Lecture Notes in Physics 884, DOI 10.1007/978-3-319-04567-2,

where $[X]$ denotes the integer part of X. It is seen from Eq. (H.1) that the H matrix in the $|m, m_p, L\rangle$ basis will be a tri-diagonal matrix when we arrange the basis states in order with respect to m_p value. Using the algebra described in [1, 2], the diagonal matrix elements of H are given by,

$$\begin{aligned}
&\langle m, m_p, L|H|m, m_p, L\rangle \\
&= \varepsilon_s(m - m_p) + \varepsilon_p m_p + V^0_{ssss}\binom{m - m_p}{2} + V^1_{spsp}(m - m_p)m_p \\
&\quad + V^0_{pppp}\frac{(n_p - L)(n_p + L + 1)}{6} + V^2_{pppp}\frac{2n_p(n_p - 2) + L(L + 1)}{6}.
\end{aligned} \tag{H.3}$$

Similarly, the off-diagonal matrix elements are given by,

$$\begin{aligned}
&\langle m, m_p - 2, L|H|m, m_p, L\rangle = \langle m, m_p, L|H|m, m_p - 2, L\rangle \\
&= \frac{1}{2\sqrt{3}} V^0_{sspp}\sqrt{(m - m_p + 1)(m - m_p + 2)(m_p - L)(m_p + L + 1)}.
\end{aligned} \tag{H.4}$$

Choosing the two-particle matrix elements V^0_{ssss}, V^1_{spsp}, V^0_{pppp}, V^2_{pppp} and V^0_{sspp} to be independent Gaussian variable with zero center and variance $2v^2$, $2v^2$, $2v^2$, $2v^2$ and v^2 respectively and constructing the H matrix using Eqs. (H.3) and (H.4) with $\varepsilon_s = \varepsilon_p = 0$ will give BEGOE(2)-$(s, p : L)$ ensemble. Adding the one-body part $\varepsilon_s(m - m_p) + \varepsilon_p m_p$ to the diagonal matrix elements with ε_s and ε_p to be fixed or random, will then gives BEGOE(1 + 2)-$(s, p : L)$. Finally, as the H matrix is tri-diagonal with simple formulas for both the diagonal and off-diagonal matrix elements, it is possible to use the results of [3] and study fluctuations in the ground state energies of BEGOE(1 + 2)-$(s, p : L)$. This was investigated in [4].

References

1. A. Frank, P. Van Isacker, *Algebraic Methods in Molecular and Nuclear Physics* (Wiley, New York, 1994)
2. F. Iachello, R.D. Levine, *Algebraic Theory of Molecules* (Oxford University Press, New York, 1995)
3. L.C.L. Hollenberg, N.S. Witte, Analytic solution for the ground-state energy of the extensive many-body problem. Phys. Rev. B **54**, 16309–16312 (1996)
4. D. Kusnezov, Two-body random ensembles: from nuclear spectra to random polynomials. Phys. Rev. Lett. **85**, 3773–3776 (2000)

MIX
Papier aus verantwortungsvollen Quellen
Paper from responsible sources
FSC® C105338
FSC
www.fsc.org

If you have any concerns about our products,
you can contact us on
ProductSafety@springernature.com

In case Publisher is established outside the EU,
the EU authorized representative is:
Springer Nature Customer Service Center GmbH
Europaplatz 3, 69115 Heidelberg, Germany

Printed by Libri Plureos GmbH
in Hamburg, Germany